# The Genus *Aspergillus*

From Taxonomy and Genetics to Industrial Application

**FEDERATION OF EUROPEAN MICROBIOLOGICAL SOCIETIES SYMPOSIUM SERIES**

*Recent FEMS Symposium volumes published by Plenum Press*

1991 • GENETICS AND PRODUCT FORMATION IN *STREPTOMYCES*
Edited by Simon Baumberg, Hans Krügel, and Dieter Noack
(FEMS Symposium No. 55)

1991 • THE BIOLOGY OF *ACINETOBACTER*: Taxonomy, Clinical Importance, Molecular Biology, Physiology, Industrial Relevance
Edited by K. J. Towner. E. Bergogne-Bérézin, and C. A. Fewson
(FEMS Symposium No. 57)

1991 • MOLECULAR PATHOGENESIS OF GASTROINTESTINAL INFECTIONS
Edited by T. Wadström, P. H. Mäkelä. A.-M. Svennerholm, and H. Wolf-Watz
(FEMS Symposium No. 58)

1992 • MOLECULAR RECOGNITION IN HOST-PARASITE INTERACTIONS
Edited by Timo K. Korhonen, Tapani Hovi, and P. Helena Mäkelä
(FEMS Symposium No. 61)

1992 • THE RELEASE OF GENETICALLY MODIFIED MICROORGANISMS– REGEM 2
Edited by Duncan E. S. Stewart-Tull and Max Sussman
(FEMS Symposium No. 63)

1993 • RAPID DIAGNOSIS OF MYCOPLASMAS
Edited by Itzhak Kahane and Amiram Adoni
(FEMS Symposium No. 62)

1993 • BACTERIAL GROWTH AND LYSIS: Metabolism and Structure of the Bacterial Sacculus
Edited by M. A. de Pedro, J.-V. Höltje, and W. Löffelhardt
(FEMS Symposium No. 65)

1994 • THE GENUS *ASPERGILLUS*: From Taxonomy and Genetics to Industrial Application
Edited by Keith A. Powell, Annabel Renwick, and John F. Peberdy
(FEMS Symposium No. 69)

# The Genus *Aspergillus*

## From Taxonomy and Genetics to Industrial Application

Edited by

*Keith A. Powell*
*Annabel Renwick*
Zeneca Agrochemicals
Bracknell, United Kingdom

and

*John F. Peberdy*
University of Nottingham
Nottingham, United Kingdom



PLENUM PRESS • NEW YORK AND LONDON

Library of Congress Cataloging-in-Publication Data

The Genus Aspergillus : from taxonomy and genetics to industrial application / edited by Keith A. Powell, Annabel Renwick, and John F. Peberdy.
p. cm. -- (FEMS symposium ; no. 69)
"Proceedings of a symposium held under the auspices of the Federation of European Microbiological Societies, April 5-8, 1993, in Canterbury, Kent, United Kingdom"--T.p. verso.
Includes bibliographical references and index.
ISBN 0-306-44701-0
1. Aspergillus--Congresses. 2. Aspergillus--Industrial applications--Congresses. I. Powell, Keith A. II. Renwick, Annabel. III. Peberdy, John F., 1937- . IV. Federation of European Microbiological Societies. V. Series.
QK625.M7G46 1994
589.2'3--dc20
94-15373
CIP

Proceedings of a symposium held under the auspices of the Federation of European Microbiological Societies, April 5–8, 1993, in Canterbury, Kent, United Kingdom

ISBN 0-306-44701-0


A Division of Plenum Publishing Corporation
233 Spring Street, New York, N.Y. 10013



Printed in the United States of America

## PREFACE

Every sixteen years or so it is appropriate to review progress in the understanding of a genus, its pathogenicity, practical utility, genetics and taxonomy, secondary metabolites and enzyme production. This book is the second attempt at such a work on the genus *Aspergillus*. It is a compilation of papers from a conference organised by the British Mycological Society and sponsored by the Federation of European Microbiological Societies. Additional sponsorship came from Pfizer, SmithKline Beecham and Zeneca Agrochemicals.

The purpose of the conference, held from 5th to 8th April 1993 was to enable a cross-discipline discussion of the properties of the genus. As can be seen from the chapters which follow the conference was broad and gave a wide coverage of topics. We are delighted to be able to produce this book relatively rapidly after the conference and hope that readers will find it both interesting and a useful source of reference as indeed was the first *Aspergillus* book produced by the British Mycological Society in 1977.

This text contains a comprehensive review of the fungal genus *Aspergillus* with contributions from a diverse group of experts including taxonomy, molecular genetics, medical pathology, industrial fermentation, food and agriculture; offering the reader a current overview of the subject.

The aspergilli have been used for many years by the fermentation industry for the production of citric and other organic acids, and have been used for centuries in the preparation of soy sauce. More recently the aspergilli have been exploited for the production of enzymes widely used in industry for the manufacture of a variety of materials.

A wide range of secondary metabolites are produced by these fungi including the potent carcinogens aflatoxins. Different aspects of this important topic are covered in several chapters. Cotty, Baymna, Egel and Elias have produced a comprehensive review of the occurrence of *Aspergillus* and aflatoxins in agriculture. Other chapters cover food spoilage in animal and human feed.

*Aspergillus nidulans* is one of the few filamentous fungi for which molecular genetics tools are being developed. The use of *A. nidulans* for heterologous gene expression is discussed which is of importance both when considering the industrial utility of these fungi and as a model system. The regulation of gene expression is illustrated by proline utilisation and *A. nidulans* is being used as a model for the genetic analysis of secondary metabolism in fungi as illustrated by Turner using penicillin biosynthesis. An area of current interest is the mechanism regulation of cell division. *A. nidulans* is one of the few fungal organisms being studied in detail. The role of two genes *bimB* and *bimD* are discussed especially with regard to their function in mitosis by May *et al.*

*Aspergillus* is one of the few fungal genera well characterised by traditional taxonomic methods based on morphological features. The advent of biochemical and molecular genetics techniques has offered new approaches to the detection and identification of different species of the aspergilli. Bainbridge reviews and compares the suitability of the different molecular approaches to understanding the taxonomy of this genus.

The aspergilli are major fungal pathogens of mammals. Campbell reviews the forms of aspergillosis which can occur in both mammals and birds. The nature of *Aspergillus* pathogenicity is discussed in two papers, Hearn and Latge *et al.* Hearn's paper covers the role of the cell wall in infection and host response, while Latge *et al.* investigate the role of molecules produced by *A. fumigatus* and how these affect host defense mechanisms. Both papers stress the importance of development of a diagnostic technique for the detection of early infection.

It is clear from the papers produced for this book that there will be much more to learn about *Aspergillus* in the future. The next sixteen years will no doubt warrant a future conference on the genus!

K.A.Powell, J.F.Peberdy, A.Renwick

September 1993

## REFERENCES

1. Genetics and Physiology of *Aspergillus* (Eds. J. E. Smith and J. A. Pateman). The British Mycological Society Symposium Series No.1. Academic Press, London, 1977.

## ACKNOWLEDGEMENTS

The BMS would like to thank the following organisations for their support:

FEMS
SmithKline Beecham
Pfizer
Zeneca Agrochemicals

## CONTENTS

# AGRICULTURE, AFLATOXINS AND *ASPERGILLUS*

P.J. Cotty, P. Bayman, D.S. Egel and K.S. Elias

Southern Regional Research Center
Agricultural Research Service
United Stated Department of Agriculture
P.O. Box 19687
New Orleans, Louisiana 70179

## INTRODUCTION

Human activities affect both the size and structure of fungal populations. Construction, war, recreation, and agriculture disrupt large expanses of vegetation and soil; disruption causes redistribution of fungal propagules and makes nutrients available to fungi. Many fungi, including the aspergilli, exploit these human engineered resources. This results in the association of large fungal populations with various human activities, especially agriculture. When crops are grown or animals raised, fungi are also grown. From a human perspective, most fungi associated with cultivation increase inadvertently. Human activity, however, partly dictates which and how many fungi occur and the fungi, both directly and through fungal products, influence human activities, domestic animals, and even humans themselves.

During warm, dry periods, several of the aspergilli increase rapidly in association with crops. These include aspergilli in the *Aspergillus flavus* group. Prior to 1960, interest in the *A. flavus* group resulted both from the use of certain strains in processing of agricultural products in Europe and the Orient (Beuchat, 1978), and from the ability of some strains to parasitize insects. In the early 1960's fungi in the *A. flavus* group were implicated as the producers of aflatoxins ("*Aspergillus flavus* toxins"), the toxins which poisoned thousands of poultry, pigs and trout; in trout these factors were associated with liver cancer (Goldblatt and Stoloff, 1983). It soon became apparent that aflatoxins also occurred in the human diet and that aflatoxins could pass from feed to milk with only slight modification (Goldblatt and Stoloff, 1983). The most common aflatoxin, aflatoxin $B_1$ , was found to be a potent hepato-carcinogen in rats and trout; carcinomas were induced at rates below $1 \mu g kg^{-1}$ body weight (Robens and Richard, 1992). Aflatoxin content of foods and feeds was eventually regulated in many countries (Stoloff *et al.*, 1991). In some products, such as milk or infant foods, aflatoxin levels below $0.02 \mu g kg^{-1}$ are mandated. Thus, for many, the focus of interest in this diverse and important fungal group became the production of aflatoxins.

There clearly are interactions between agriculture, and both aflatoxins and the fungi in the *A. flavus* group. Some consequences of these interactions are obvious, others are virtually unexplored. The relationship of crop contamination cycles to the life strategies of

*The Genus Aspergillus*, Edited by Keith A. Powell *et al.*,
Plenum Press, New York, 1994

*A. flavus* group fungi is uncertain. The role agriculture plays in structuring *A. flavus* populations and their toxigenic potential is also uncertain. This chapter will address some aspects of the interactions of *A. flavus* with humans and human activities; it includes suggestions on how these interactions may be altered to reduce human exposure to aflatoxins and other detrimental fungal traits.

## INFLUENCES OF THE *ASPERGILLUS FLAVUS* GROUP

### Effects of Aflatoxins on Humans and Domestic Animals

Although aflatoxins are most often noted for ability to induce liver cancer at very low doses, they can cause several problems of economic importance during animal production. The presence of relatively high levels of aflatoxins in feeds can lead to animal death; rabbits, ducks and swine are particularly susceptible ($LD_{50}$= 0.30, 0.35, and 0.62 $mgkg^{-1}$, respectively; Pier, 1992). However, at much lower concentrations, aflatoxins have other effects on domestic animals including immunosuppression and reduced productivity (Pier, 1992; Robens and Richard, 1992). Once consumed, aflatoxins are also readily converted to aflatoxin $M_1$ which occurs in milk and can thus cause both human exposure and sickness in animal offspring (Pier, 1992; Robens and Richard, 1992).

**Incidence of Health Effects due to Contaminated Foods.** In many developed countries, regulations combined with both an enforcement policy and an abundant food supply can prevent exposure of human populations, in most cases, to significant aflatoxin ingestion (Stoloff *et al.*, 1991). However, in countries where either food is insufficient or regulations are not adequately enforced, routine ingestion of aflatoxins may occur (Hendrickse and Maxwell, 1989; Zarba, *et al.*, 1992). In populations with relatively high exposure, a role for aflatoxins as a risk factor for primary liver cancer in humans has repeatedly been suggested, but is still not clear (Robens and Richard, 1992). However, aflatoxins cause a variety of effects on animal development, the immune system and a variety of vital organs. Exposure to aflatoxins, particularly in staples (*i.e.* corn or peanuts) of people dependent upon relatively few nutrient sources, must be considered a serious detriment. The relationship between aflatoxins and kwashiorkor may be only one reflection of this detriment (Hendrickse and Maxwell, 1989).

**Effects of Aflatoxins on Agricultural Enterprise.** Controversies regarding the possible role of aflatoxins in primary liver cancer of humans are moot in the contemporary international marketplace. Brokers and producers of agricultural commodities have found aflatoxins increasingly costly as careful monitoring of aflatoxins limits the use and value of contaminated products (Cappuccio, 1989). Regulations in most developed countries and even many less developed countries restrict the import of contaminated foods and feeds (van Egmond, 1991; Stoloff *et al.*, 1991). Assessing the aflatoxin content of crops is a routine aspect of brokering and often a prerequisite of shipping. Contamination is highly variable and allowable concentrations are at such low levels (some below 1 $\mu gkg^{-1}$), that analysis prior to shipping cannot always ensure acceptable levels upon receipt, even if no increases occur during transit (Horwitz *et al.*, 1993). This increases commodity costs and can decrease competitiveness of imported products. Regulations applied more rigorously to imported than domestic products or set at zero, where the limit of detection determines the enforcement level, can serve as barriers to trade which again increase the cost of products. These increased costs may be the primary effect of aflatoxins felt by most consumers in developed nations.

**Effects of Aflatoxins on Health of Agricultural Workers.** Labourers engaged in production and processing of commodities may be exposed to aflatoxins through inhalation

(Shotwell, 1991). Crops grown under conditions favouring aflatoxin contamination often become covered with large quantities of *A. flavus* propagules. Furthermore, air in areas where contaminated crops are produced may contain thousands of propagules per cubic meter (Lee *et al.*, 1986). These propagules, which are mostly conidia, remain associated with the crops through harvest and processing. Conidia contain large quantities of aflatoxins (over 100 $mgkg^{-1}$ in some strains; Wicklow and Shotwell, 1982). Since most contamination occurs in damaged crop components, fines and dust generated during crop processing have much higher toxin contents than the crop as a whole (Lee *et al.*, 1983). The conidia, fines, and dust, may be inhaled and thus pose an avenue of exposure to aflatoxins; this exposure has been quantified in certain cases (Shotwell, 1991). Recently, occupational exposure to aflatoxins through the handling and processing of contaminated agricultural products has been associated with increased risk of both primary liver cancer and other cancers (Alavanja *et al.*, 1987; Olsen *et al.*, 1988).

## *Aspergillus flavus* group Fungi as Allergens and Animal Pathogens

Several allergic and infective conditions of humans and certain other vertebrates are caused by *Aspergillus* species (Rinaldi, 1983; St. Georgiev, 1992; Wardlaw and Gedes, 1992). These include allergic bronchiopulmonary aspergillosis and invasive pulmonary aspergillosis. The most common cause of most of these conditions is *Aspergillus fumigatus* (Rinaldi, 1983; St. Georgiev, 1992; Wardlaw and Gedes, 1992). However, other aspergilli, including members of the *A. flavus* group, are also often implicated.

**Insect Pathogen.** During epidemics of aflatoxin contamination, high concentrations of *A. flavus* group propagules are associated with most objects resident in fields, including insects; thus insects may serve as vectors (Stephenson and Russell, 1974; Widstrom, 1979). *A. flavus* readily grows and multiplies on insect damaged crops, insect frass and on insects themselves both as dead debris and as parasitized hosts (Sussman, 1951,1952; Stephenson and Russell, 1974; Goto *et al.*, 1988). Many insects typically carry *A. flavus* group isolates internally and many insects are hosts of at least certain strains (Stephenson and Russell, 1974; Widstrom, 1979; Goto *et al.*, 1988). Domesticated insects are included among the hosts of the *A. flavus* group. Domesticated insect diseases include Stonebrood, a rare disease of the honey bee which is of minor importance to bee keepers (Gilliam and Vandenberg, 1990) and koji kabi disease of cultivated silkworm larvae (Ohtomo *et al.*, 1975; Goto *et al.*, 1988).

## Benefits of *Aspergillus flavus* group Fungi

**Industry.** Fungi in this group have had a long history in processing to increase product utility and value. *A. flavus* group strains are used to produce enzymes for food processing and other industrial uses and even to produce therapeutic products such as urate oxidase and lactoferrin (Chavalet *et al.*, 1992; van den Hondel *et al.*, 1992; Ward *et al.*, 1992). A variety of traditional fermented food products have been made with fungi in the *A. flavus* group for centuries (Beuchat, 1978).

**Ecological Benefits.** Although *A. flavus* group fungi are not commonly recognised as beneficial, these ubiquitous organisms become dominant members of the microflora under certain circumstances and exert multiple influences on both biota and environment. These fungi are important degraders of crop debris and may play roles in solubilising and recycling crop and soil nutrients (Ashworth *et al.*, 1969; Griffin and Garren, 1976). *A. flavus* can even degrade lignin (Betts and Dart, 1989). As insect pathogens, these fungi may serve to limit pest populations (Wadhwani and Srivastava, 1985) and have even been considered potential agents to replace chemical insecticides ( Roberts and Yendol, 1971).

## Contamination Cycles

**Contaminated components.** *A. flavus* causes a variety of plant diseases typical of largely saprotrophic "weak" plant pathogens (Widstrom, 1992). These diseases include boll, ear, and pod rots which result in both decreased yield and reduced quality (Shurtleff, 1980; Watkins, 1981). However, crop infection by *A. flavus* takes on a different importance than infections for which concern might focus on yield and quality loss, or increased free fatty acids. Aflatoxins are compounds regulated in parts per billion; yet, these toxins occur in certain infections at concentrations over 100,000 $ugkg^{-1}$. This situation causes high-toxin-containing components to greatly exceed in cost the value of the same components if not contaminated. Variability among components of crops in aflatoxin content is extreme (Figure 1). Most infected components contain low aflatoxin concentrations (below 50 $\mu gkg^{-1}$). However, a small percent contain very high toxin levels, at times exceeding 500,000 $\mu gkg^{-1}$ (Cucullu *et al.*, 1966; Schade *et al.*, 1975; Lee *et al.*, 1990; Steiner *et al.*, 1992). In many cases, elimination of highly contaminated components (over 1,000 $\mu gkg^{-1}$) would result in a commodity with an acceptable average aflatoxin content (Schade *et al.*, 1975; Steiner *et al.*, 1992).

Crop components damaged by wounding or severe stress are colonised and decayed by a variety of fungi. During hot and dry conditions, fungi in the *A. flavus* group out compete many colonising microbes and become the prominent fungi degrading damaged components. In most crops the majority of contamination occurs in damaged plant parts (Wilson et al., 1977; Lee et al., 1983; Cotty and Lee, 1989). Damaged seed can be sorted from high value crops for less profitable use such as production of vegetable oil. However, crushing contaminated seed to produce oil concentrates aflatoxins in the resulting meal which is used for feed. Such toxic meal caused the first recognised aflatoxin problems; peanut meal caused turkey X disease in England and cottonseed meal caused trout hepatocarcinoma in the United States (Goldblatt and Stoloff, 1983). Such meal must either be detoxified (i.e. through ammoniation) or put to non-feed use (Park *et al.*, 1988).

**Geography determines frequency and severity.** Geographic location greatly influences frequency of contamination. Many agricultural areas at low elevation and between the latitudes 35 N and 35 S have perennial risk of contamination. Countries in this zone (which include many countries with insufficient food supply) may view elimination of aflatoxins from the food supply differently than countries whose major agricultural lands lie out of this zone (*i.e.* developed countries in Europe and North America). Producers of contaminated products may base allowable levels of aflatoxins on toxicological data, whereas consumer nations which rarely produce contaminated products may base allowable levels at the lowest level detectable (Stoloff *et al.*, 1991).

Contamination cycles can be considered perennial, sporadic or infrequent based on locale and crop. In all three situations, populations of *A. flavus* are long term residents. However, populations in different areas differ in magnitude (Figure 2) (Griffen and Garren, 1974; Manabe *et al.*, 1978; Shearer *et al.*, 1992) and possibly in the distribution of both qualitative and quantitative traits (Manabe *et al.*, 1978; Cotty, 1992*b*). During periods not conducive to contamination, perennial areas (*i.e.* the desert valleys of Arizona; Lee *et al.*, 1986) support higher *A. flavus* populations than areas with infrequent contamination, *i.e.* midwest corn producing areas (Shearer *et al.*, 1992). Areas with sporadic contamination may have perennial contamination at low levels but, have less regular exposure to important predisposing factors such as hot, dry conditions, *i.e.* contamination of corn in certain areas of the southeastern United States (Widstrom, 1992) or insect pressure, *i.e.* pink bollworm pressure on cotton in western Arizona (Cotty and Lee, 1989). During periods conducive to contamination, a shift in the microflora occurs and aflatoxin producing fungi become dominant colonisers and decayers.

Processes through which crops become contaminated with aflatoxins are varied and complex (Diener *et al.*, 1987). However, certain generalities might be suggested. Contamination cycles may be divided into three phases: Prebloom, Crop Development, and Post Maturation (Figure 2).

**Prebloom.** Contamination does not occur in the field during the period after crop removal and prior to bloom. However, both the microflora and crop may become predisposed to contamination. During this phase: 1. propagules (conidia, sclerotia, colonised organic matter) are dispersed through cultivation, planting, pruning or other activities of animals (including man) or the environment; 2. *A. flavus* populations fluctuate, first decreasing after crop removal and then, if conditions are favourable, increasing on debris from current and

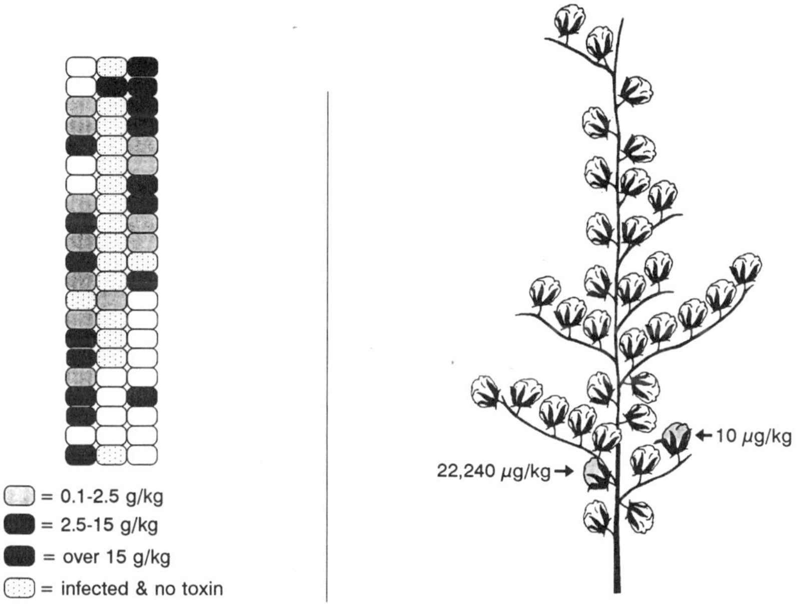


**Figure 1.** Distribution and aflatoxin content of maize kernels on an ear (drawn from data in Lee *et al.*, 1980) and bolls on a plant (Cotty and Lee, 1990). Contamination is highly variable and not all infected seed becomes contaminated.

prior crops (Ashworth *et al.*, 1969; Griffen and Garren, 1974; Lee *et al.*, 1986) 3. The crop may become predisposed by long periods of drought or by luxuriant growth followed by drought (Cole *et al.*, 1982; Deiner *et al.*, 1987; Shearer *et al.*, 1992); 4. Overwintering insects emerge and develop.

**Crop Development.** From flowering to maturation, seeds and fruits are vulnerable to various perturbations. During this phase: 1. If conditions are hot and dry, populations of the *A. flavus* group, in canopy and soil, will outcompete many saprophytic microbes and increase in size. 2. High temperatures and/or drought stress may interfere with crop development and weaken plant defences making the crop more susceptible to infection and contamination (Jones *et al.*, 1981; Cole *et al.*, 1985; Wotton and Strange, 1987; Widstrom,

1992). 3. Wounding of fruits at middle to late stages of development can lead to portions of the crop with very high toxin levels (Lillehoj *et al.*, 1987; Cotty, 1989*b*). In several crops, most aflatoxin is formed during this phase and in certain locations crop predisposal to contamination can be attributed to specific wound types caused by specific insects. Examples are pink bollworm exit holes in cotton in the desert valleys of the western United States (Cotty and Lee, 1989), maize weevil damage in the southern United States (McMillian *et al.*, 1987), navel orange worm damage in nuts in the western United States (Schade *et al.*, 1975; Sommer *et al.*, 1986), and lesser corn stalk borer damage in peanuts in the southern United States (Lynch and Wilson, 1991). In some crops, components prevented from maturing due to stress or early harvest are particularly vulnerable to contamination (Cole *et al.*, 1985; Lynch and Wilson, 1991).

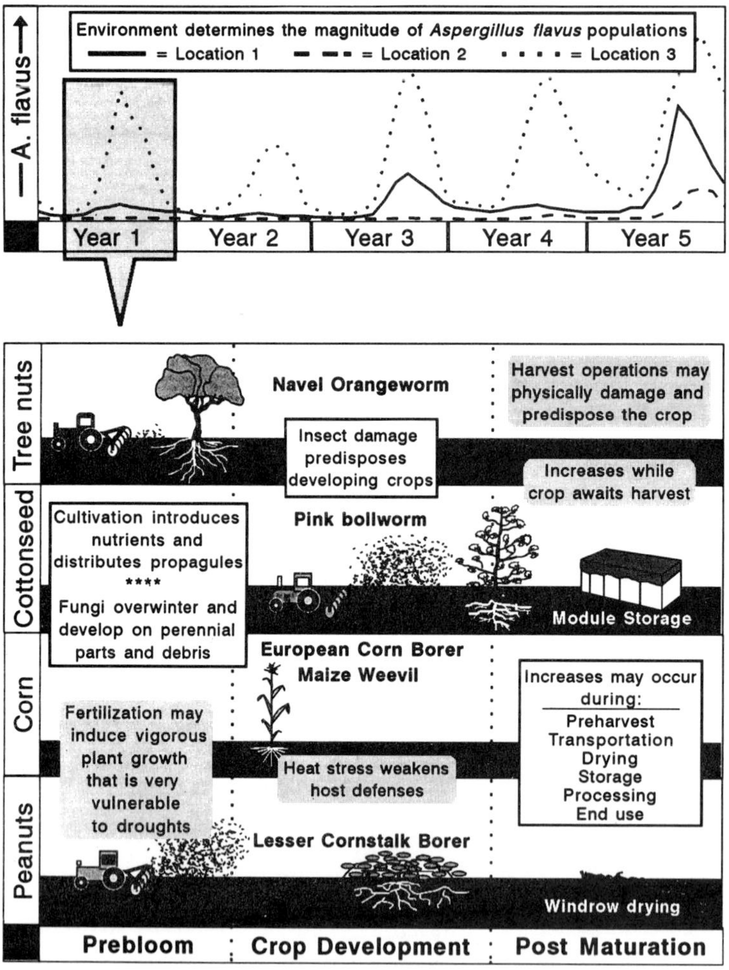


**Figure 2.** Contamination cycles can be divided into three phases. Local conditions determine both the extent of contamination and the magnitude of *A. flavus* populations associated with the crop. Boxed information applies to all crops.

**Post Maturation.** Most crops are susceptible to aflatoxin contamination at maturity and if the crop was grown in an area with perennial contamination or during a period conducive to contamination, the mature crop will be associated with large quantities of *A. flavus* group propagules. These propagules remain associated with the crop as it awaits harvest in the field, during harvest, field storage (i.e. peanuts in windrows, cotton in modules), shipment and processing, and even during storage by the end user. Exposure of the mature crop to periods of wetting and drying under warm conditions may lead to increased contamination. Aflatoxin concentrations are known to be dependent on environmental conditions and competing microflora (see Strain Isolation and Accumulation of Aflatoxins). Mature fruits and seeds are living organisms and factors which compromise seed health, such as wounding or stress, predispose these products to infection and contamination. Harvest operations can simultaneously damage crops and introduce *A. flavus* into wounds (Schroeder and Storey, 1976; Sommer *et al.*, 1986; Siriacha *et al.*, 1989). Insect activity after harvest can disperse aflatoxin-producing fungi and, by increasing host susceptibility, increase aflatoxin levels in a manner similar to insect damage during crop development (Dunkel, 1988). The same insect can affect contamination both prior to and after maturation (ie. the navel orange worm on pistachios).

Post maturation contamination dictates that each handler of the crop be responsible and minimize the potential for aflatoxin increases. Thus dairies which purchase feed with undetectable toxin must still store the feed properly or contaminated milk may occur. With indeterminate crops (*e.g.* cotton) crop development and post maturation phases may occur simultaneously and with all crops the prebloom and post maturation phases occur simultaneously, although at different locations.

Initially, the crop development phase was ignored because all contamination was thought to occur post harvest; recently, most research has been directed at contamination

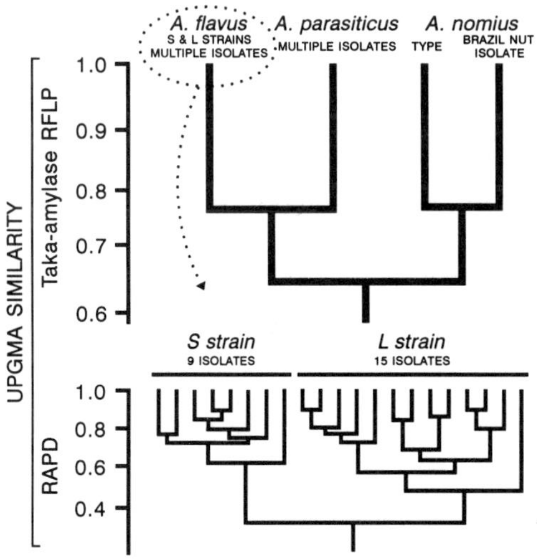


**Figure 3**. Phenograms of *A. flavus* group isolates. Taka-amylase data from Egel and Cotty, (1992); RAPD data from Bayman and Cotty, (1993).

before harvest, sometimes without distinguishing periods of active crop development from periods after maturation (Lillehoj *et al.*, 1976; Goldblatt and Stoloff, 1983). The contamination process can be divided many different ways besides those presented here. However, failure to segregate the contamination process into different phases may result in data that suggests no clear pattern and apparent contradictions. For example, in Arizona, most cottonseed contamination occurs during crop development in cottonbolls damaged by pink bollworms in the absence of rain (Cotty and Lee, 1989). Still, rain on a mature crop awaiting harvest can lead to significant contamination during post maturation, even if developing bolls were not damaged (Cotty, 1991). Similarly, a great fervour occurred about contamination of the midwest U.S. corn crop, in the field, during droughts of 1983 and 1988 (Kilman, 1989; Schmitt and Hurburgh, 1989; Shearer *et al.*, 1992). Yet, in Thailand, contamination typically occurs during the wet season, not during the dry season (Goto *et al.*, 1986). In Thailand's rainy season, contamination occurs during post maturation (Siriacha *et al*, 1989); in the midwestern United States, it typically occurs during crop development (Lillehoj *et al.*, 1976).

## FUNGAL POPULATIONS

### Diversity

**Species of Aflatoxin-Producing Fungi.** There have been a variety of taxonomic schemes used to classify *A. flavus* group strains (Thom and Raper, 1945; Klich and Pitt, 1988; Samson and Frisvad, 1990). Each species represents an assortment of strains which behave as clonal organisms with the exception of occasional parasexuality between members of the same vegetative compatibility group (Papa, 1984, 1986). For the purposes of this discussion we will place all isolates within this group into four species *A. flavus*, *Aspergillus parasiticus*, *A. nomius,* and *Aspergillus tamarii*. Depending on interpretation, these species are supported by clustering algorithms based on DNA polymorphisms (Kurtzman *et al.*, 1987; Moody and Tyler, 1990*a,b*; Egel and Cotty, 1992; Bayman and Cotty, 1993). *A. tamarii* is of minor interest here because no isolates in this species produce aflatoxins. *A. tamarii* isolates apparently have some markedly different adaptations than the remainder of the group and *A. tamarii* is more distantly related to the other three species, than the three are to each other (Kurtzman *et al.*, 1987; Klich and Pitt, 1988). *Aspergillus oryzae* and *Aspergillus sojae* are apparently derived from *A. flavus* and *A. parasiticus*, respectively (Kurtzman *et al.*, 1986) and will be mentioned only in an industrial context. *A. nomius* was named after the genus of alkali bees from which several isolates were obtained (Kurtzman *et al.*, 1987). *A. nomius* comprises a group of strains that are distinct by both physiologic and molecular criteria (Kurtzman *et al.*, 1987; Bayman and Cotty, 1993). The name "*nomius*" may be misleading in associating this species predominantly with the alkali bee when isolates are known from several crops, including wheat (the type isolate) and peanuts (Hesseltine *et al.*, 1970).

**Diversity Within *Aspergillus flavus*.** Within each of the three aflatoxin producing species, there is a great deal of variability among isolates. It may be, that if we sought out all the unusual or atypical isolates within this group and examined them, we would find a continuum as suggested by Thom and Raper (1945). Indeed, based on polymorphisms in the Taka-amylase gene, we have found strains intermediate between *A. flavus* and *A. parasiticus* as well as *A. nomius* isolates almost as different from the *A. nomius* type strain as the *A. parasiticus* type from the *A. flavus* type (Egel and Cotty, 1992; see Brazil nut isolate in Figure 3). Variation among isolates is evident in genetic, physiological and morphological characters. Each of the above species is composed of at least several Vegetative Compatibility Groups (VCGs) and *A. flavus* is composed of many (Papa, 1986; Bayman and Cotty, 1991; P.J. Cotty, unpublished). Physiological and morphological traits are typically much more consistent within a VCG than within the species as a whole (Bayman and Cotty,

1993). Thus, a large portion of the variability perceived within *A. flavus* reflects divergence among VCGs. This divergence has resulted in consistent differences among VCGs in several characters, including enzyme production, plant virulence, sclerotial morphology, and other physiological traits (Cotty, 1989*a*; Cotty *et al.*, 1990*b*; Bayman and Cotty, 1993).

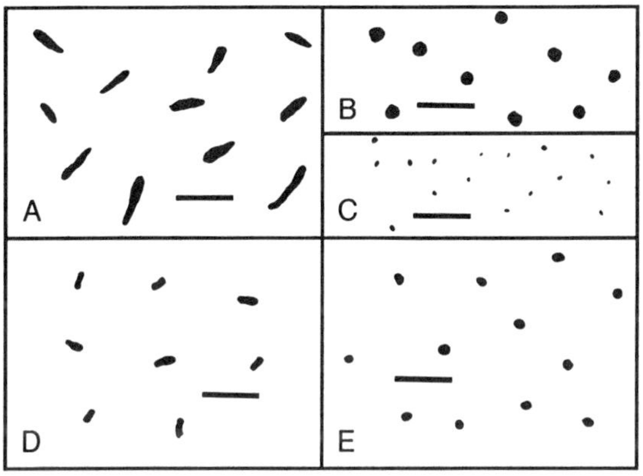


**Figure 4.** Silhouettes of sclerotia produced by aflatoxin producing isolates during 30 days growth at 32 °C on Czapek's agar; A is an unusual isolate of *A. nomius* from a Brazil nut; B is an L strain isolate of *A. flavus*; C is an S strain isolate of *A. flavus*; D is the type isolate of *A. nomius*, E is an isolate of *A. parasiticus*. All bars are 3 mm.

Variability in production of aflatoxins, especially among *A. flavus* isolates, has often been reported and discussed (Joffe, 1969; Davis and Deiner, 1983; Clevstrom and Ljunggren, 1985). *A. flavus* isolates may produce anywhere from no detectable aflatoxins (<1 $\mu gkg^{-1}$) to over 1,000,000 $\mu gkg^{-1}$. *A. parasiticus* and *A. nomius* produce B and G aflatoxins, and *A. parasiticus* produces aflatoxins far more consistently than *A. flavus* (Hesseltine *et al.*, 1970; Dorner *et al.*, 1984; Kurtzman *et al.*, 1987). Too few isolates of *A. nomius* have been examined to discern consistency. *A. flavus* is generally considered to produce only B aflatoxins (Samson and Frisvad, 1990); however, this observation is dependent on how the definition of *A. flavus* is restricted (Saito *et al.*, 1986; Klich and Pitt, 1988). Taxonomy aside, variability in toxin production, and other strain differences indicate divergence and possible differential adaptation. This variability can be a tool for discerning functions of variable traits (Cleveland and Cotty, 1991); it may further be used to develop a better understanding of the ecological niches to which strains are adapted.

On the basis of physiological and morphological criteria, *A. flavus* can be divided into two strains, S and L (Cotty, 1989*a*). Isolates in the S strain of *A. flavus* (actually a collection of strains which belong to numerous VCGs; Cotty, 1989*a*; Bayman and Cotty, 1993; Cotty *et al.*, 1990*b*) produce numerous small sclerotia and fewer conidia than other *A. flavus* isolates (Cotty, 1989*a*; Saito *et al.*, 1986). The L strain is composed of the so called "typical" isolates of *A. flavus* (Saito *et al.*, 1986) which produce larger and fewer sclerotia. Some key differences between the S and L strains are outlined in Table 1.

**Table 1.** Key characteristics of the S and L strains of *Aspergillus flavus*

| Character | L strain | S strain | Reference[1] |
|---|---|---|---|
| Sclerotium size | Average> 300 mm | Average< 300mm | A, B |
| Production of aflatoxins | Variable, Zero to High | Consistent, High | A, B |
| Production of conidia | Heavy | Light | A, B |
| Production of sclerotia | | | |
| on potato dextrose agar | None to many, $10cm^{-2}$ | Many > $50cm^{-2}$ | A |
| on 5% V-8 juice | None to few, < $1cm^{-2}$, | Many $10cm^{-2}$ to $50cm^{-2}$ | A |
| Virulence to cotton | High | Low to High | A |
| Pectinase production | Consistent | Variable | C |
| Primary habitat | Aerial? | Soil? | |

[1]References: A = Cotty, 1989*a*; B = Saito *et al.*, 1986; C = Cotty *et al.*, 1990*b*

**Importance of Infrequent Strains to Contamination.** The etiology of aflatoxin contamination, and the relationship of both the size and structure of *Aspergillus* populations to contamination is complicated by the importance of unusual strain types which occur at low frequency. Aflatoxin contamination is a peculiar and frustrating agricultural problem because less than 1% of the crop may be contaminated with levels high enough to make the average of the entire crop exceed allowable concentrations (Figure 1). During attribution of cause, infrequent but highly toxigenic strains may easily be overlooked or not identified as potential aflatoxin producers. Such may be the case with isolates belonging to the S strain of *A. flavus*. Due to colony and sclerotial appearance (Figure 4) S strain isolates may be passed over in favour of co-occurring "typical" or L strain isolates. Several visitors to our laboratory have been surprised at the identity of S strain isolates and have returned home to discover the occurrence of S strain isolates at their locale. In soils of several areas of the southern United States, the S strain incidence averages around 30% (Cotty, 1992*b*). On average S strain isolates produce much higher aflatoxin levels than L strain isolates, and also more sclerotia and fewer conidia (Saito *et al.*, 1986; Cotty, 1989*a*) (Table 1). Predominance of conidia of L strain isolates on mature crops may at times interfere with attribution of contamination to S strain isolates actually inciting the problem.

Another relatively infrequent aflatoxin-producing fungus is *A. nomius*. *A. nomius* isolates can produce large quantities of aflatoxins but may be misidentified as *A. parasiticus* which produces the same aflatoxins (both B and G) and roughened conidia (Hesseltine *et al.*, 1970; Kurtzman *et al.*, 1987). A case in point is an unusual *A. nomius* isolate from a store-bought brazil nut which contained 8,400 $\mu gkg^{-1}$ total aflatoxins (Figure 4). This isolate produces large quantities of aflatoxins and, based on polymorphisms in the taka-amylase gene, differs almost as much from other *A. nomius* isolates as *A. parasiticus* differs from *A. flavus* (Egel and Cotty, 1992) (Figure 3). This isolate is clearly unusual, but it incited significant contamination in the marketplace. Furthermore, such rare highly contaminated nuts are the primary source of contamination in brazil nuts (Steiner *et al.*, 1992).

**Diversity in Ecological Niches.** Fungi in the *A. flavus* group are broadly adapted to exploit many organic nutrients and to infect a variety of animal and plant hosts. Strains must adapt to compete in ecological niches which provide long term survival. Many strains with diverse adaptations clearly have some success in exploiting crop related resources. However, other niches, which may only support small fungal populations relative to crop associated niches, may have been occupied over long periods by certain strains. Differences among these "minor" niches may drive strain diversification. Similarly, stability of minor niches may stabilise the character of minor strain types. Relatively stable minor niches may have greater long term importance than vast crop resources, because suitability and quantity of crop related resources oscillate widely in response to the environment, insect herbivory, changes in agronomic practice and the crop itself.

Wicklow (1982) showed that strains of the *A. flavus* group used as Koji moulds (moulds used to produce fermented foods) (Beuchat, 1978) germinate faster and have larger spores than wild strains of the species from which these moulds were probably domesticated; thus during domestication, the Koji moulds might have developed traits which favour rapid nutrient capture (and success during intraspecific competition) and lost traits which are not adaptive in the Koji environment, *i.e.* aflatoxin-producing ability (Wicklow, 1982). DNA relatedness among strains of *A. flavus* and *A. parasiticus* and their Koji mould equivalents, *A. oryzae* and *A. sojae,* suggest that the Koji moulds were indeed derived from the wild species (Kurtzman *et al.*, 1986; Egel and Cotty, 1992). However, attributing adaptive value to Koji traits is speculative in the absence of experimental data. Similarly, strain variability suggests multiple adaptations, but our assignment of specific functions to adaptations is largely speculative.

Strains of the *A. flavus* group may not only differ in host or nutrient use, but also in host/nutrient location and strategy to exploit resources. Members of this group are very common both in and above the soil. Although all *A. flavus* group strains contribute to the soil biota, certain strains may be better adapted to capture resources above the soil. Small sclerotia and reduced sporulation among S strain isolates may imply adaptation to infect and capture resources in the soil whereas relatively large sclerotia often facilitates aerial infection and nutrient capture (Garrett, 1960). S strain isolates may have diverged from other *A. flavus* strains through adaptations to the soil environment. There has been a general assumption that sclerotia of this group serve primarily to produce conidia after non-conducive periods (Wicklow and Donahue, 1984). Sclerotia of other fungi can germinate directly to infect hosts or capture resources (Coley-Smith and Cooke, 1971); this may also be an important role for sclerotia in the *A. flavus* group, particularly for S strain isolates which appear to disperse via numerous, small sclerotia.

*A. parasiticus* has also been associated with the soil environment (Davis and Diener, 1983) because in certain locations (e.g. Georgia), it occurs more frequently on peanuts than on corn (Hill *et al.*, 1985). Although relatively few isolates have been compared, in most cases, *A. flavus* isolates are more invasive of crop tissues, even peanut tissues, than *A. parasiticus* (Calvert *et al.*, 1978; Zummo and Scott, 1990; Pitt *et al.*, 1991). Furthermore, based on occurrence of G aflatoxins (produced by *A. parasiticus* and *A. nomius*, but not *A. flavus*) *A. flavus* produces most contamination in peanuts (Hill *et al.*, 1985; Maeda, 1990). Conflicting observations on the association of *A. parasiticus* with peanuts may reflect fungal adaptations to soils or conditions in certain locales where peanuts are a major crop and lack of adaptation to other locales. This is supported by a low frequency of *A. parasiticus* in several agricultural areas and failure of introduced *A. parasitcus* strains to overwinter efficiently at certain locations (Davis and Diener, 1983; Zummo and Scott, 1990; Cotty, 1992*b*).

*A. flavus* strains produce large quantities of extracellular enzymes (van den Hondel *et al.*, 1992) which probably enhance their ability to utilise a broad assortment of organic resources. Enzyme polymorphisms have been used to suggest a role for specific enzymes (elastase and pectinase) in fungal virulence. Certain strains of *A. flavus* have reduced ability to rot cotton bolls and spread between cotton boll locules (Cotty 1989*b*). This reduced ability is associated with failure to produce a specific pectinase isoenzyme, P2C, both in culture and developing cottonbolls (Cleveland and Cotty, 1991; Brown *et al.*, 1992). In one population study, 50% of S strain isolates and no L strain isolates failed to produce P2C. This polymorphism suggests certain S strain clones are not dependent on efficient colonisation of plant hosts. S strain isolates might primarily exploit soil debris and/or insect hosts and thus not require high pectinase. Even though not optimally adapted to exploit plants, pectinase P2C deficient strains do occur on the commercial crop and can cause significant contamination (Cotty, 1989*a*; Cotty *et al.*, 1990b). Therefore, specific adaptation to a crop is not required for strain contribution to contamination. Analogous to P2C variability is variability in production of elastase (an alkaline protease). All *A. flavus* strains isolated from patients suffering from invasive aspergillosis produced elastase whereas strains from other origins produced elastase less frequently (Rhodes *et al.*, 1988). Thus, a role for elastase in human pathogenesis has been suggested (Rhodes *et al.*, 1988), although this role is still controversial (Denning *et al.*, 1992). *A. flavus* is an opportunistic human pathogen and it's unlikely that *A. flavus* elastase evolved to permit infection of mammals. The ecological function of elastase is not clear; however, elastase production may be directed at exploitation of dead mammals or insects (Charnley, 1989; Malanthi and Chakraborty, 1991).

*A. flavus* has an intimate relationship with insects, particularly lepidopterans (Sussman, 1951, 1952). Excretion of large quantities of diverse enzymes, a characteristic of the *A. flavus* group (van den Hondel *et al.*, 1992), may facilitate mutualism as well as parasitism and saprophytism (Martin, 1992). Insect use of fungal excreted enzymes that degrade or detoxify plant products can drive development of fungal-insect mutualisms (Martin, 1992). The *A. flavus* elastase actively degrades multiple enzymes in alkaline environments (Rhodes *et al.*, 1990) and is relatively stable among other proteases (van den Hondel *et al.*, 1992; P.J. Cotty and J.E. Mellon, unpublished). Such activities might ameliorate the lepidopteran gut environment (ie. alkaline and high protease activity) (Martin, 1992) and permit strain establishment and retention. Similarly, aflatoxins may exert influence on insect immune systems (Charnley, 1989) permitting fungal strain retention. *A. flavus*-insect relations meet several predictions of mutualistic relations including fungal asexuality and lack of specificity (Martin, 1992). However, production of a potent insecticide and/or other virulence factors (Sussman, 1952; Ohtomo *et al.*, 1975; Drummond and Pinnock, 1990) within host tissues preclude full mutualism and allows a shift from avirulence to virulence. The associated host death may benefit both saprophytic insect exploitation and movement to plant resources (Bennett, 1981). Speculations about the nature of the relationship aside, diverse arthropods vector *A. flavus* group fungi, predispose crops to aflatoxin contamination and serve as both hosts and predators of many *A. flavus* group strains (Widstrom, 1979). In the latter two roles these animals may exert strong selective pressure on fungal strain character and the fungi may exert considerable pressure on insects (Rodriguez *et al.*, 1979; Wadhwani and Srivastave, 1985).

Specialisation of strains seems not to include pathogen-host specificity, or at least specificity has not been shown. Sussman (1951) showed diverse lepidopterans were infected by the same strain of *A. flavus* and isolates from one crop typically can infect and contaminate other distantly related crops (Schroeder and Hein, 1967; Brown *et al.*, 1991). Similarly, the life strategies of strains causing aspergillosis in poultry and humans are clearly not directed at specifically exploiting those hosts. Different crop associations of *A. parasiticus* and *A. flavus* strains may reflect either as yet undescribed adaptations to specific hosts or other differences in ecological adaptation and life strategy (Moss, 1991) (see above). Many adaptations in this group relate to aggressive saprophytism at elevated temperature and under relatively dry conditions. As pathogens, these fungi generally exploit wounded or stressed hosts and avoid taking on host defenses directly, although *A. flavus* does elicit plant defense mechanisms (e.g. enzyme production and phytoalexins) (Mellon, 1991,1992). Still, healthy and non-compromised hosts (both plants and animals) can be infected (Barbesgaard *et al.*, 1992; Pitt *et al.*, 1992). Infection of healthy plant parts in the absence of symptoms may occur regularly, even if these infections do not include invasion of living host cells. Indeed, through serial isolations, systemic plant infections by *A. flavus* group strains have been observed in corn, peanut and cotton (Klich *et al.*, 1984; Pitt *et al.*, 1991; Mycock *et al.*, 1992).

## Influences of Agriculture on Fungal Populations

**Fungal Population Structure.** Populations of *A. flavus* group fungi are complex. All species within the group may occur on the same crop or in the same field (Schroeder and Boller, 1973; Davis and Diener, 1983; Cotty, 1992*b*). The greatest information on population structure is available for *A. flavus* which is composed of numerous vegetative compatibility groups (Papa, 1986; Bayman and Cotty, 1991) (Figure 5). Populations are complex at every level with multiple strains occupying gram quantities of soil and individual crop pieces. VCG composition of the population infecting a crop does not necessarily reflect the VCG composition of the population within the soil in which the crop is grown (Cotty, 1992*b*). Furthermore, during crop production new resources for *A. flavus* to exploit become available and population composition may change very rapidly (Bayman and Cotty, 1991; Cotty, 1991*b*, 1992*b*); apparently these fluxes in population composition are driven by establishment of relatively rare VCGs on newly available resources. There is little information on *A. flavus* group populations in the absence of agriculture and little information on fungal community responses to agricultural methods (Zak, 1992). However, it is clear that cultivation disturbs and homogenises the soil environment in which these fungi reside and in so doing must disperse conidia, sclerotia and colonised organic matter. At the same time both cultivation and crop development create immense resources for fungi to use. Although disturbance generally results in decreased species richness and heterogeneity, this sudden abundance of resources during environments favouring the *A. flavus* group may permit noncompetitive strain coexistence (Zak, 1992) and a temporary increase in the diversity of strains exploiting particular resources.

**Selection of Fungal Strains.** The *A. flavus* group is broadly distributed but, in the absence of crop cycles, *A. flavus* group populations are generally maintained at relatively low levels (Angle *et al.*, 1982; Shearer *et al.*, 1992) and in the absence of a conducive environment, *A. flavus* populations also maintain low levels on crop resources (Griffin and Garren, 1974; Shearer *et al.*, 1992). Thus, during conducive periods, there is a potential for crops to exert tremendous influence on strain growth and selection. Strains infecting crops are diverse in many characters including type and number of sclerotia, toxin producing ability, VCG, and even virulence to plants. This diversity among infecting strains suggests that agriculture does not aggressively select specific fungal types. However, a lack of requirement for aflatoxin production during crop infection and during fungal increases on crops (Cotty, 1989*a*) may permit disproportionate increases in atoxigenic strains (Bilgrami and Sinha, 1992). Furthermore, the importance of aerial dispersal to spread through a crop may cause the high sporulating L strain of *A. flavus* to outcompete the low sporulating S strain during secondary spread in the canopy. An as yet unknown specific strain-vector association could also permit strain advantage. Cropping process, crop types, geography, and/or climate may select certain strain types (Shroeder and Boller, 1973; Lafont and Lafont, 1977; Wicklow and Cole, 1982; Shearer *et al.*, 1992). However, multiple-year experiments with more rigorous design are needed to reliably establish such selection, if present. Studies should also utilise strain identification methods that are more specific than ability to produce either toxins or sclerotia. Vegetative compatibility analysis has been shown to be useful for monitoring the behaviour of specific strains over both time and space (Cotty, 1991*b*, 1992*c*) and we recently found differences among cotton producing areas in the proportion of *A. flavus* isolates belonging to the S strain (Cotty, 1992*b*).

Crops might exert different influences on populations by exposing strains to either different substrates or resistance factors. Crop components for which contamination is a concern (*i.e.* nuts or kernels) do not exert the only nor often the major influence on populations; other parts (*i.e.* leaves, stems, floral parts, cobs) may play a greater role in forming and maintaining the overall population (Zummo and Scott, 1990; Kumar and Mishra,

1991). Perennial crops (*i.e.* tree nuts) may maintain and select strains on perennial parts and long season crops (*i.e.* cotton) may provide longer periods of increase and shorter periods between crops than short season crops (*i.e.* corn). The nature and magnitude of plant debris and it's successful survival between croppings may be an important determinant of population structure and magnitude (Jones, 1979; Zummo and Scott, 1990). *A. flavus* can colonise very large proportions of plant debris associated with crops and this debris can yield large quantities of conidia (Ashworth *et al.*, 1969; Stephenson and Russell, 1974). Variation among crops in insect microflora may also influence the composition of fungal populations. This phenomenon might occur due to differences in herbivory or variability among insect hosts in both life cycle and susceptibility to fungal strains.

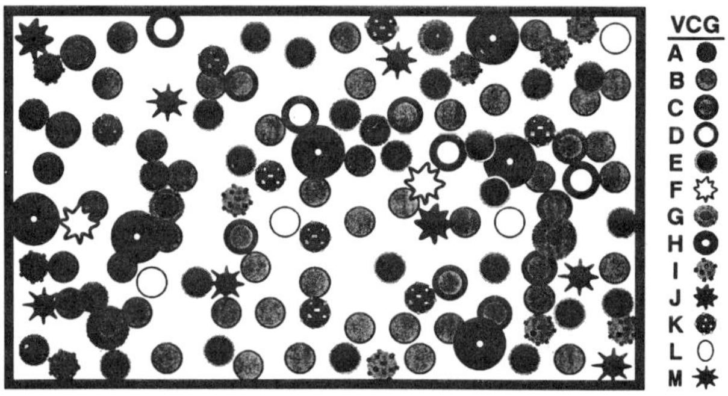


**Figure 5.** Diagrammatic representation of the vegetative compatibility group (VCG) composition of a single agricultural field. Marker frequency indicates the relative incidence of the represented VCG within the field. Marker shape does not reflect morphological differences among VCGs. Only VCGs that make up greater than two percent of the population are represented (Bayman and Cotty, 1991).

## Adaptive Value and Ecological Significance of Aflatoxins

Through their effects on agriculture, aflatoxins prove to be "non-nutritional chemicals controlling the biology of other species in the environment," Torssell's (1983) definition of secondary metabolites. A characteristic of secondary metabolites is a principally unknown function (Torssell, 1983), however, consideration of the adaptive value of aflatoxins is appropriate when discussing interactions among aflatoxins, *Aspergillus,* and agriculture.

**Protection of Survival Structures.** High concentrations (3 $mgkg^{-1}$ to 132 $mgkg^{-1}$ aflatoxin $B_1$) of aflatoxins occur in both conidia and sclerotia of aflatoxin-producing strains (Wicklow and Cole, 1982; Wicklow and Shotwell, 1982; Cotty, 1988). The presence of aflatoxins in sclerotia has received the most attention because sclerotia are long term survival structures and aflatoxins are highly toxic to a variety of predators of fungi, especially insects (Wright *et al.*, 1982; Willetts and Bullock, 1992). Sclerotia of *A. flavus* group fungi typically contain an extensive array of other toxic metabolites in addition to aflatoxins (Wicklow, 1990). Some of these metabolites are not found in other fungal structures and in combination with aflatoxins these toxins may form an elaborate chemical defense system directed at protecting sclerotia from insect predation (Wicklow, 1990; Dowd, 1992). Indeed kojic acid, a metabolite of most *A. flavus* strains, can synergistically increase the toxicity of aflatoxin $B_1$ to caterpillars (Dowd, 1988).

Long term survival for *A. flavus* group propagules requires resistance to degradation by microorganisms during conditions (wet and/or cool) not conducive to successful competition by *A. flavus*. Bacteria are active under these conditions and, although aflatoxins are not very inhibitory to fungi, they do inhibit many bacteria at concentrations present in sclerotia (Burmeister and Hesseltine, 1966, Arai *et al.*, 1967, Angle and Wagner, 1981). Aflatoxins even inhibit certain known bacterial antagonists of *A. flavus* (Kimura and Hirano, 1988). However, pure aflatoxin $B_1$ is rapidly degraded in diverse soils (Angle, 1986), and thus to have long term effects, aflatoxins may themselves have to be shielded from decomposition, a condition possibly provided by the sclerotial rind (Willetts and Bullock, 1992).

*A. flavus* strains which produce sclerotia may either produce aflatoxins or not (Bennett *et al.*, 1979; Cotty, 1989*a*). Furthermore, sclerotia of the same strain with differing aflatoxin content can be produced by growing sclerotia on different substrates (Cotty, 1988). Sclerotia, from multiple sources, with different aflatoxin contents could be evaluated for longevity in field soil, resistance to microbial degradation, and insect predation. If aflatoxins contribute to the defense of sclerotia, some level of correlation between aflatoxin content and sclerotium resistance should occur.

**Association with Sclerotia.** A relationship between sclerotia and aflatoxins has been repeatedly suggested (Mehan and Chohan, 1973; Sanchis *et al.*, 1984). This is not a straightforward relationship because, *in vitro*, certain fungal strains produce aflatoxins but not sclerotia and vice versa (Bennett *et al.*, 1979). The situation is further complicated by attributing quantitative differences in toxin producing ability to the tendency of a strain to produce sclerotia (Mehan and Chohan, 1973; Sanchis *et al.*, 1984). These differences probably reflect differences among phylogenetically diverged groups which may be identified less ambiguously by sclerotial morphology (Cotty, 1989*a*). However, in strains that do produce both sclerotia and aflatoxins, there appears to be an interrelationship between regulation of aflatoxin biosynthesis and regulation of sclerotial morphogenesis (Cotty, 1988). This is suggested by: A) association of increases in aflatoxin production with inhibition of sclerotial maturation when cultures are exposed to either acidic pH or fungicides which inhibit ergosterol biosynthesis (Cotty, 1988; Bayman and Cotty, 1990); B) coincidence of sclerotial maturation with cessation of aflatoxin production (Cotty, 1988); C) high aflatoxin content of sclerotia (Wicklow and Cole; 1982, Wicklow and Shotwell, 1982; Cotty, 1988); D) possible transport of aflatoxins from mycelia into sclerotia (Cotty, 1988; Bayman and Cotty, 1990). These observations suggest that sclerotium maturation is associated with a signal that terminates aflatoxin biosynthesis. Delays in sclerotial maturation may thus delay the termination signal and be associated with increased aflatoxin concentrations.

The interrelationship between sclerotial morphogenesis and aflatoxin biosynthesis is supported by recent advances in our understanding of the molecular biology of aflatoxin production. During characterisation of genes involved in aflatoxin biosynthesis, influences of specific genes on both biosynthesis and morphogenesis has been observed (Skory *et al.*, 1992). However, as Skory *et al.* (1992) point out, it is not clear whether this relationship is a direct influence of either aflatoxins or aflatoxin precursors on sclerotia or a regulatory association. The recent isolation of a putative regulatory element (*apa-2)* that influences both processes (Chang *et al.*, 1993) also corroborates the relationship. The suggestion of Skory *et al.* (1992), that aflatoxins themselves may serve a regulatory role during sclerotial development, is interesting in light of the ability of aflatoxin $B_1$ to directly bind DNA (Muench *et al.*, 1983); a regulatory role for aflatoxins in mature sclerotia is also possible.

**Accumulation of Aflatoxins in Substrates.** Aflatoxins are a concern in agriculture because large quantities can accumulate in certain plant materials. This accumulation may be a survival adaptation directed at either preventing ingestion of infested seed or inhibiting competition (Janzen, 1977; Bilgrami and Sinha, 1992). However, large quantities of aflatoxins are not accumulated in many plant parts in which the fungus increases and is maintained (Griffin and Garren, 1976; Takahashi *et al.*, 1986), and *A. flavus* is not very efficient at either degrading aflatoxins or converting them to use (Doyle and Marth, 1978). Many strains of *A. flavus* do not produce large quantities of aflatoxins (Davis and Deiner, 1983) and when grown with other microbes (which is common in nature) toxin production is greatly curtailed (see Strain Isolation and Accumulation of Aflatoxins). Indeed, it might be argued that most materials in which *A. flavus* grows and is maintained, are not contaminated with large quantities of aflatoxins. If accumulation of aflatoxins in plant substrates is a directed fungal strategy, it is a very inefficient one. Accumulation may be inadvertent, caused by interference with sclerotial morphogenesis (Cotty, 1988). Export of aflatoxins from producing cells (Shih and Marth, 1973) might be directed at creating accumulations in the sclerotial rind (Willetts and Bullock, 1992). Accumulation of aflatoxins intracellularly and in dead cells of the sclerotial rind is testable by histological techniques.

**Microbial Interactions and Aflatoxin Biosynthesis.** Aflatoxin biosynthesis is readily inhibited by microbial competition. Many microbes interfere with aflatoxin production in culture ( Kimura and Hirano, 1988; Roy and Chourasia, 1990) and in crops (Ashworth *et al.*, 1965; Ehrlich *et al.*, 1985). Even *A. flavus* and *A. parasiticus* strains and/or mutants which do not produce aflatoxins can interfere with aflatoxin production and/or crop contamination (Ehrlich, 1987; Cotty, 1990; Brown *et al.*, 1991). Interference apparently occurs through competitive exclusion (Cotty *et al.*, 1990*a*), production of interfering compounds (Shantha *et al.*, 1990) and/or competition for nutrients (Cotty *et al.*, 1990*a*). Isolation of aflatoxin-producing strains from interfering strains is thus prerequisite for accumulation of high aflatoxin concentrations in crops (Bullerman *et al.*, 1975; Roy and Chourasia, 1990). Isolation may occur spatially, as occurs in laboratory tests on sterilised substrates, or physiologically, usually when temperature or substrate moisture or composition favour dominance of *A. flavus*. Strain isolation may be one mechanism through which both wounding during crop development and high temperature favour very high toxin levels; rapid wound colonisation with aggressive invasion of developing tissues may permit such isolation. If isolation of toxigenic strains is inadequate, poor aflatoxin production will result even though conditions favour growth and reproduction of the inciting fungus (Bullerman *et al.*, 1975).

**Interactions with Hosts.** Aflatoxins are toxic and exert several physiologic effects on most hosts of the *A. flavus* group including plants, insects and mammals (Roberts and Yendol, 1971; McLean *et al.*, 1992; Robens and Richard, 1992). Aflatoxins may thus mediate pathogenesis either as a determinant of strain pathogenicity or by increasing strain virulence. Several lines of evidence suggest that this is, at least, not always the case. Even though very high toxin levels have been detected in plants (over 500 $mgkg^{-1}$) (Lee *et al.*, 1990)) and insects (over 13 $mgkg^{-1}$) (Ohtomo *et al.*, 1975). *A. flavus* group isolates from insects, mammals, and plants may either produce aflatoxins or not. Furthermore, isolates which do not produce toxins retain ability to cause disease in the evaluated hosts (Cotty, 1989*a*; Drummond and Pinnock, 1990). In insects, aflatoxins may serve as virulence factors, increasing the rate at which infected insects die (Ohtomo *et al.*, 1975). However, the ability of *A. flavus* to infect and invade insects appears to be more dependent on enzymatic degradation of host proteins and cuticles (Sussman, 1952; Charnley, 1989).

**Conservation of Aflatoxin producing ability.** The *A. flavus* group is a mosaic of numerous strains delimited by a vegetative incompatibility system (Bayman and Cotty, 1991). These strains appear to evolve, at least in general, as clones (groups of identical organisms descended from a single common ancestor by mitosis) (King and Stansfield, 1985). These clones may move rapidly from rare to frequent depending on opportunity (Bayman and Cotty, 1991). The importance of a given trait to strain success may be viewed both by the diversity of strains expressing that trait and the frequency of the expressing strains. Consistent expression of a trait by diverse strains may indicate that trait is beneficial in multiple niches or that it has use in a broad ecological niche. The tendency to produce aflatoxins is highly variable within the overall *A. flavus* group. However, toxin production is more consistent among strains which are more closely related on the basis of morphological, physiological, genetic, or molecular characteristics (Dorner *et al.*, 1984; Saito *et al.*, 1986; Cotty, 1989*a*; Moody and Tyler, 1990*a,b*; Egel and Cotty, 1992; Bayman and Cotty, 1993). Thus, we can associate conservation of toxin production over evolutionary time with certain clusters of strains and loss of toxin producing ability with others. These observations may lead to new insights on potential adaptive values of aflatoxins as we learn more about the basic biology of the various clusters. Aflatoxigenicity is highly conserved among most wildtype strains of *A. parasiticus* and *A. flavus* strain S (Dorner *et al.*, 1984; Cotty, 1989*a*). Aflatoxin-producing ability is readily lost in culture and thus, conservation among field isolates of these evolutionarily diverged clusters implies a strong selective force causing retention of aflatoxin-producing ability. These clusters may share a common use for aflatoxins or may each have different uses. S strain isolates are more closely related to L strain isolates than to *A. parasiticus* (Egel and Cotty, 1992; Bayman and Cotty, 1993). Apparently divergence of the L and S strains is relatively recent compared to divergence of *A. flavus* and *A. parasiticus.* Unstable toxin production (Boller and Schroeder, 1974; Clevstrom and Ljunggren, 1985) and reduced toxin producing ability are characteristics of the L strain (Cotty, 1989*a*), and Bayman and Cotty (1993) found that low toxin producing strains within the L strain are more closely related to atoxigenic strains than to highly toxigenic strains. Thus, atoxigenicity apparently can be a multistep process in this group and for at least one of the ecological niches to which the L strain is adapted, aflatoxins do not confer an important advantage. Aflatoxins do not increase fungal virulence to crops (Cotty, 1989*a*) and atoxigenic *A. flavus* strains have been associated with aerial crop parts (Bilgrami and Sinha, 1992); *A. parasiticus* (Davis and Deiner 1983) and S strain isolates have been associated with a soil habitat (see Diversity of Ecological Niches). Thus, the soil environment may favour conservation of toxin production and the aerial environment may not.

In addition to high aflatoxin production, reduced virulence to plants is also associated with the S strain of *A. flavus* (Cotty, 1989*a*; Cotty *et al.*, 1990*b*). Reduced virulence stems from failure to produce the most active *A. flavus* pectinase (Cotty *et al.*, 1990*b*; Cleveland and Cotty, 1991; Brown *et al.*, 1992). Low virulence, stemming from reduced ability to decay and colonise plant tissues (Brown *et al.*, 1991) may imply adaptation to a niche where such traits are not essential. Thus, high aflatoxin producing ability is associated with strains adapted to ecological niches where infection of crops is probably not essential.

Accumulation of large quantities of G aflatoxins by *A. nomius, A. parasiticus,* and certain S strain isolates (Hesseltine *et al.*, 1970; Dorner *et al.*, 1984; Saito *et al.*, 1986; Kurtzman *et al.*, 1987), but not by other *A. flavus* isolates, provides an additional puzzle. Unique activities have not been associated with G aflatoxins. Therefore, it's difficult to envisage selective advantages conferred by retention of G aflatoxin producing ability. Production of G aflatoxins may merely reflect slight differences in pathway regulation (Bhatnagar *et al.*, 1992) or retention of an ancestral trait.

## SELECTION OF ASPERGILLI ASSOCIATED WITH AGRICULTURE

### Strain Selection in the Past

*A. flavus* group populations have generally been altered by agriculture through disruption of habitat and introduction of nutrients. Production of crops under environmental conditions conferring a competitive advantage to these fungi permits their rapid increase on crop resources (Griffin and Garren, 1976; Lee *et al.*, 1986). After cropping, large quantities of crop remnants and debris are incorporated into field soil; additional remnants remaining after crop processing (from sorting, ginning, paring, shelling, etc.) are also often incorporated. This organic matter may both be superficially associated with large quantities of *A. flavus* group propagules and heavily colonised (Stephenson and Russell, 1974; Griffin and Garren, 1976). Under conditions particularly favourable to *A. flavus*, most organic debris incorporated into the soil can be colonised by *A. flavus* (Ashworth *et al.*, 1969). Strains associated with the crop and debris are diverse and generally not deliberately selected (Bayman and Cotty, 1991). Deliberate selection of specific strains with particular characters has been for production of enzymes for the European baking industry (Barbesgaard *et al.*, 1992) and for production of traditional fermentation products in the orient (Beuchat, 1978).

Fungal selection has reduced strain toxicity (Kurtzman *et al.*, 1986) and increased fungal traits associated with both product quality and efficient fermentation (Wicklow, 1982, 1990). Use of these fungi over centuries has inadvertently resulted in the release of large quantities of spores and colonised organic debris (Wicklow, 1990; Barbesgaard *et al*, 1992). Such strain selection and release may have altered *A. flavus* populations in the vicinity of industries and may partly explain strain distribution (Manabe *et al.*, 1976). The validity of this speculation might be tested with recently developed techniques to characterise and compare structures of *A. flavus* populations (Bayman and Cotty, 1991, 1993).

### The Potential of Strain Selection

There are no methods for preventing aflatoxin contamination that are both reliable and economical. To fully protect crops from contamination, procedures must be active in the field under hot, dry conditions that are not very conducive to crop development but, often are near optimal for *A. flavus* group fungi. Controls must be effective during both the crop development and post-maturation phases of contamination cycles. The procedure must fit within agriculture's economic constraints and for worldwide use, must be effective under suboptimal storage conditions and with low technological input. Furthermore, because most contamination occurs in damaged seed (which for many crops either cannot be sorted out or must be used) controls must prevent contamination of plant parts compromised by either physiological stress or predation. These are difficult requirements for a procedure directed at preventing the relatively rare, highly contaminated seed.

A promising avenue of control, that may meet the above criteria, is the seeding of agricultural fields with atoxigenic *A. flavus* group strains in order to reduce toxigenicities of resident populations (Cotty 1991*b*,1992*a*). *A. flavus* does not require aflatoxins to infect crops and production of large quantities of aflatoxins in crop parts does not increase either strain virulence or strain ability to colonise and utilise crop resources (Cotty, 1989*a*). This led to speculation that applied atoxigenic strains might outcompete toxigenic strains during crop infection and thereby reduce contamination (Cotty, 1989*a*; Cole and Cotty, 1990). Greenhouse and field experiments in which either developing cotton bolls or developing corn ears were wound inoculated with various strain combinations demonstrate the potential of atoxigenic strains to reduce contamination (80 to 90%) during crop development (Cotty, 1990; Brown *et al.*, 1991). Individual crop components are often coinfected by multiple *A. flavus* strains and cottonbolls damaged by pink bollworms are infected by *A. flavus* strains in multiple vegetative compatibility groups at least 50 to 80% of the time (Bayman and Cotty, 1991; P.J. Cotty, unpublished). Pink bollworm-damaged bolls contain the majority of aflatoxins in commercial fields (Cotty and Lee, 1989). Thus, the ability of atoxigenic strains to interfere with contamination in co-infected bolls may be of real practical value. Atoxigenic strains reduce contamination by both spatially excluding toxigenic strains and by competing for resources required for production of aflatoxins (Cotty *et al.* 1990*a*). However, not all atoxigenic strains are capable of reducing contamination during co-infection (Cotty, 1992*a*); thus strain optimisation in co-inoculation tests should be pre-requisite to field evaluation.

In theory, seeding fields with atoxigenic strains relatively early in crop development may permit seeded strains to compete with other resident strains for crop associated resources (Cole and Cotty, 1990). The seeded strains may thus increase in population size along with toxigenic strains when environmental conditions favour aflatoxin contamination (Cole and Cotty, 1990; Cotty, 1992*a,c*). At the same time, the atoxigenic strains may compete for infection sites. In special environmental control plots, Dorner *et al.*. (1992) have demonstrated that *A. parasiticus* strains which accumulate specific aflatoxin precursors (i.e. a native strain that accumulates O-methylsterigmatocystin and a mutant that accumulates versicolorin-A) but not aflatoxins, can interfere with aflatoxin contamination when concentrated propagule suspensions are applied to developing peanuts. Applications resulted in long term (several years) fungal population changes and support the use of atoxigenic *A. flavus* group strains in preventing contamination of peanuts. The most comprehensive field tests, to date, have been performed on cotton grown in Yuma County, Arizona. This area has the most consistent aflatoxin contamination of cottonseed in the United States (Gardner *et al.*, 1974). An atoxigenic strain was seeded on colonised wheat seed (Cotty, 1992*a*) into a field of developing cotton, prior to crop flowering (Cotty, 1991*b*). The distribution of vegetative compatibility groups (VCGs) within this field had been determined in previous years, (Bayman and Cotty, 1990) and a strain in a rare VCG was seeded. Five months later the crop was harvested and the distribution of the applied strain on the crop was determined by mutating isolates to nitrate auxotrophy and assessing VCG. Strain seeding resulted in large and significant reductions in the aflatoxin content of the crop at maturity and aflatoxin content was inversely correlated with the incidence of the seeded VCG (Cotty, 1991*b*). Similar tests, performed in subsequent years, also demonstrate that atoxigenic strains applied early in crop development can partially competitively exclude toxigenic strains and thereby reduce contamination (Cotty, 1992c); this early strain application is associated with neither increased crop infection nor increased *A. flavus* populations on the crop at maturity.

The theoretical advantage of atoxigenic strains of *A. flavus* over other microorganisms that might be used to competitively exclude aflatoxin-producing strains is that atoxigenic strains are apparently adapted to similar environmental conditions as toxigenic strains. Other potential agents, such as bacteria (Kimura and Hirano, 1988; Bowen *et al*, 1992), may be inactive under the hot, dry conditions associated with aflatoxin contamination. The use of atoxigenic strains seeks to limit neither the amount of crop infection by the *A. flavus* group nor the quantity of these fungi associated with the crop. The procedure merely selects which fungi become associated with the crop. Thus, crop quality losses typically associated with fungal infection (*i.e.* increased free fatty acids) will not be ameliorated. Seeding atoxigenic strains might not result in increased crop infection because infection is more heavily dependent on host predisposition and the environment than on the number of propagules of *A. flavus*. Indeed in three years of tests on cotton, seeding has not resulted in increased infection rates (Cotty, 1992*c*). However, under certain circumstances with sufficiently low initial *A. flavus* levels and sufficiently high seeding rates, increased infection rates in treated crops might be expected. However, *A. flavus* typically decays predisposed crop components that, under different environmental conditions, would be infected by other microbes. Thus, these infections probably would not be of a magnitude to cause concern.

Populations of *A. flavus* increase on crops very rapidly under conditions favourable to contamination. The ultimate magnitude of the *A. flavus* group is largely dependent on the available resources and the environment. Thus, even in areas with perennially low aflatoxin contamination, high *A. flavus* populations can rapidly develop during droughts (Shearer *et al.*, 1992); the composition of these rapidly increasing populations might be partially controlled by properly timed seeding.

*A. flavus* group fungi typically become associated with crops in the field during crop development and remain associated with the crop during harvest, storage and processing. Thus, seeding of atoxigenic strains into agricultural fields prior to crop development may provide postharvest protection from contamination by associating the harvested crop with high frequencies of atoxigenic strains. Atoxigenic strains applied both prior to harvest and after harvest have been shown to provide protection from aflatoxin contamination of corn (Brown *et al.*, 1991), even when toxigenic strains are associated with the crop prior to application.

Domestication of *A. flavus* group fungi for seeding into agricultural fields may cause some concern over the pathogenic potential of these fungi to humans (Pore *et al.*, 1970). Although limiting exposure of high risk individuals to aspergilli will reduce infection risk, particularly in hospitals, it might be argued that host predisposal is more important in determining disease incidence than exposure to fungal propagules (Wardlaw and Geddes, 1992; St. Georgiev, 1992; Rinaldi, 1983). In many agricultural industries and communities, workers and residents respire high concentrations of *Aspergillus* spores. Clearly such exposure is undesirable, but such respiration may occur without noticeable disease. This point is particularly clear for fungal strains used to produce koji and baking or brewing enzymes (Barbesgaard *et al.*, 1992). In these industries, generations of workers have been exposed to very high concentrations of spores throughout their working years with a very low incidence of disease (Barbesgaard *et al.*, 1992) . Barbesgaard *et al.* (1992) argues for *A. oryzae* to be classified as "Generally Regarded As Safe" (GRAS), partially on this basis.

Seeding agricultural fields with select fungal isolates can result in *A. flavus* populations with altered composition, but without increased population size (Cotty, 1991*b*, Cotty 1992*c*). Thus, seeding may provide the opportunity to improve the overall safety of fungal populations by reducing human exposure to aflatoxins through both dietary and respiratory routes (see Effects of Aflatoxins on Humans and Domestic Animals). The frequency of fungal traits other than aflatoxin-producing ability might also be altered and, in so doing, fungal virulence to animals might be reduced or fungal sensitivity to therapeutic agents might be increased (Cotty and Egel, 1992). Other fungal traits detrimental to humans or human activities (*i.e.* allergenicity) might also be minimised and beneficial traits (*e.g.* ability to decay crop debris between plantings) might be maximised. The concept of fungal seeding also applies to fungi other than *A. flavus*, particularly to other aspergilli. *A. fumigatus*, a more potent animal pathogen than *A. flavus*, is a very frequent degrader of plant debris (Gandolla and Aragno, 1992). Extremely high concentrations of *A. fumigatus* spores may be associated with composting organic matter (Gandolla and Aragno, 1992). It may be possible to select strains of *A. fumigatus*, in a manner similar to *A. flavus*, in order to optimise both safety and decomposition.

This strategy of seeding fields with select strains of A. flavus has drawn repeated controversy and criticism based on the dangers of *A. flavus* populations (Wicklow, 1993; Kilman, 1993). However, the choice presented is not whether or not there will be fungi. The choice is whether we will determine, through deliberate selection, which strains make up the populations. Current agricultural practice does seed fields with very large quantities of organic matter colonised with *A. flavus* group fungi. This material is in the form of crop remnants, gin trash, corn cobs, etc. It is common practice to incorporate such materials into field soils. This differs from the seeding strategy suggested here in that seeded strains are not selected, the quantity of material incorporated is very large, and incorporation is not timed to give applied strains preferential exposure to the developing crop.

## Ecological Significance

The use of atoxigenic strains of *A. flavus* to control aflatoxin production has been hindered by a lack of information about fungal population biology. In many ways this field lags twenty or thirty years behind comparable studies on animals and plants (Burnett, 1983); the best-known fungus in this respect is probably *Neurospora crassa* (Perkins and Turner, 1988). Dispersal, change in population structure over time, and natural selection are poorly understood in fungal populations. This partly results from difficulty in tracking individuals (McDonald and Martinez, 1991). Interactions between conspecific genetic individuals have not been widely regarded until recently (Rayner, 1991). Furthermore, studies on one group of fungi have often turned out to have limited application to other groups. All these problems are complicated in fungi like the *A. flavus* group by tremendous reproductive and dispersal abilities, the lack of a known sexual stage, parasexuality (Papa, 1984), and mitotic chromosomal rearrangements (Keller *et al.*, 1992).

During the course of experiments discussed here, data has been collected on variation in many characters in many natural isolates. Areas have been sampled repeatedly over several years and known isolates have been introduced into fields and their survival and dispersal followed over the course of years; this has not been done with *Neurospora.* This body of data on how *A. flavus* genetic individuals survive, spread, and interact, may turn out to be as interesting as the biocontrol strategy it was designed to support.

One concept in sustainable agriculture is to "study the forest in order to farm like the forest" (Jackson and Piper, 1989). Understanding distribution, variation, and competition in fungal populations in nature and agriculture may lead to successful use of this principle.

## ACKNOWLEDGMENTS

We are grateful to Mrs. Darlene Downey, technical assistant to P.J. Cotty, for her contributions to the work discussed here.

## REFERENCES

Alavanja, M.C.R., Malker, H. and Hayes, R.B. (1987) Occupational cancer risk associated with the storage and bulk handling of agricultural foodstuff. J. Toxicol. Environ. Health 22, 247-254.

Angle, J.S., Dunn, K.A. and Wagner, G.H. (1982) Effect of cultural practices on the soil population of *Aspergillus flavus* and *Aspergillus parasiticus.* Soil Sci. Soc. Amer. J. 46, 301-304.

Angle, J.S. and Wagner, G.H. (1981) Aflatoxin $B_1$ effects on soil microorganisms. Soil. Biol. Biochem. 13, 381-384.

Angle, J.S. (1986) Aflatoxin decomposition in various soils. J. Environ. Sci. Health 21, 277-288.

Arai, T., Ito, T. and Koyama, Y. (1967) Antimicrobial activity of aflatoxins. J. Bacteriol. 93, 59-64.

Ashworth, L.J, Jr., McMeans, J.L. and Brown, C.M. (1969) Infection of cotton by *Aspergillus flavus,* epidemiology of the disease. J. Stored Prod. Res. 5, 193-202.

Ashworth, L.J.,Jr., Schroeder, H.W. and Langley, B.C. (1965) Aflatoxins: environmental factors governing occurence in spanish peanuts. Science 148, 1228-1229.

Barbesgaard, P., Heldt-Hansen, H.P. and Diderichsen, B. (1992) On the safety of *Aspergillus oryzae*: a review. Appl. Microbiol. Biotechnol. 36, 569-572.

Bayman, P. and Cotty, P.J. (1990) Triadimenol stimulates aflatoxin production by *Aspergillus flavus* in vitro. Mycological Research 94, 1023-1025.

Bayman, P. and Cotty, P.J. (1991) Vegetative compatibility and genetic diversity in the *Aspergillus flavus* population of a single field. Can. J. Bot. 69, 1707-1711.

Bayman, P. and Cotty, P.J. (1993) Genetic diversity in *Aspergillus flavus*: Association with aflatoxin production and morphology. Can. J. Bot. 71:23-31.

Bennett, J.W. (1981) Genetic perspective on polyketides, productivity, parasexuality, protoplasts, and plastids, in "Advances in Biotechnology Volume 3 Fermentation Products" (C. Vezina and K. Singh, eds), pp 409-415. Pergamon Press, Toronto.

Bennett, J.W., Horowitz, P.C. and Lee, L.S. (1979) Production of sclerotia by aflatoxigenic and nonaflatoxigenic strains of *Aspergillus flavus* and *A. parasiticus*. Mycologia 71, 415-422.

Betts, W.B. and Dart, R.K. (1989) Initial reactions in degradation of tri- and tetrameric lignin-related compounds by *Aspergillus flavus*. Mycol. Res. 92, 177-181.

Beuchat, L.R. (1978) Traditional fermented food products, in "Food and Beverage Mycology"(Beuchat, L.R., Ed.), pp. 224-253. AVI, Westport.

Bhatnagar, D., Ehrlich, K.C. and Cleveland, T.E. (1992) Oxidation-reduction reactions in biosynthesis of secondary metabolites, in "Handbook of Applied Mycology Vol. 5, Mycotoxins in Ecological Systems"(Bhatnagar, D., Lillehoj, E. and Arora, D.K., Eds.), pp. 255-286. Dekker, Basel.

Bilgrami, K.S. and Sinha, K.K. (1992) Aflatoxins: their biological effects and ecological significance, in "Handbook of Applied Mycology Vol. 5, Mycotoxins in Ecological Systems"(Bhatnagar, D., Lillehoj, E., and Arora, D.K., Eds.), pp. 59-86. Marcel Dekker, Basel.

Boller, R.A. and Schroeder, H.W. (1974) Production of aflatoxin by cultures derived from conidia stored in the laboratory. Mycologia 66, 61-66.

Bowen, K.L., Kloepper, J.W., Chourasia, H. and Mickler, C.J. (1992) Selection of geocarposphere bacteria as candidate biological control agents for aflatoxigenic fungi and reducing aflatoxin contamination in peanut. Phytopathol. 82, 1121.

Brown, R.L., Cleveland, T.E., Cotty, P.J. and Mellon, J.E. (1992) Spread of *Aspergillus* in cotton bolls, decay of intercarpellary membranes, and production of fungal pectinases. Phytopathology 82, 462-467.

Brown, R.L., Cotty, P.J. and Cleveland, T.E. (1991) Reduction in aflatoxin content of maize by atoxigenic strains of *Aspergillus flavus*. J. Food Protection 54, 623-626.

Bullerman, L.B., Baca, J.M. and Stott, W.T. (1975) An evaluation of potential mycotoxin-producing molds in corn meal. Cereal Foods World 20, 248-253.

Burmeister, H.R. and Hesseltine, C.W. (1966) Survey of the sensitivity of microorganisms to aflatoxin. Appl. Microbiol. 14, 403-404.

Burnett, J.H. (1983) Speciation in fungi. Trans. Brit. Mycol. Soc. 81, 1-14.

Calvert, O.H., Lillehoj, E.B., Kwolek, W.F. and Zuber, M.S. (1978) Aflatoxin B1 and G1 production in developing *Zea mays* kernels from mixed inocula of *Aspergillus flavus* and *A. parasiticus*. Phytopathol. 68, 501-506.

Cappuccio, M. (1989) Effects of new rules on EEC trade. J. Amer. Oil Chem. Soc. 66, 1410-1413.

Chang, P.K., Cary, J., Bhatnagar, D., Cotty, P.J., Cleveland, T.E., Bennett, J.W., Linz, J.E., Woloshuk, C. and Payne, G. (1993) Cloning of the apa-2 gene that regulates aflatoxin biosynthesis in *Aspergillus parasiticus*. Fungal Gen. Newsletter In Press.

Charnley, A.K. (1989) Mycoinsecticides, Present use and future prospects, in "Progress and Prospects in Insect Control", BCPC Mono. No. 43. pp. 165-181.

Chevalet, L., Tiraby, G., Cabane, B. and Loison, G. (1992) Transformation of *Aspergillus flavus*, construction of urate oxidase-deficient mutants by gene disruption. Curr. Genet. 21, 447-453.

Cleveland, T.E. and Cotty, P.J. (1991) Invasiveness of *Aspergillus flavus* isolates in wounded cotton bolls is associated with production of a specific fungal polygalacturonase. Phytopathol. 81, 155-158.

Clevstrom, G. and Ljunggren, H. (1985) Aflatoxin formation and the dual phenomenon in *Aspergillus flavus* Link. Mycopathologia 92, 129-139.

Cole, R.J., Hill, R.A., Blankenship, P.D., Sanders, T.H. and Garren, K.H. (1982) Influence of irrigation and drought stress on invasion of *Aspergillus flavus* of corn kernels and peanut pods. Appl. Environ. Microbiol. 52, 1128-1131.

Cole, R.J., Sanders, T.H., Hill, R.A. and Blankenship, P.D. (1985) Mean geocarposphere temperatures that induce preharvest aflatoxin contamination of peanuts under drought stress. Mycopathologia 91, 41-46.

Cole, R.J. and Cotty, P.J. (1990) Biocontrol of aflatoxin production by using biocompetitive agents, in "A Perspective on Aflatoxin in Field Crops and Animal Food Products in the United States" (Robens, J.R., Ed.) pp. 62-66. Agricultural Research Service, Beltsville.

Coley-Smith, J.R. and Cook, R.C. (1971) Survival and germination of fungal sclerotia. Ann. Rev. Phytopathol. 9, 65-92.

Cotty, P.J. (1988) Aflatoxin and sclerotial production by *Aspergillus flavus*, Influence of pH. Phytopathol. 78, 1250-1253.

Cotty, P.J. (1989*a*) Virulence and cultural characteristics of two *Aspergillus flavus* strains pathogenic on cotton. Phytopathol. 79, 808-814.

Cotty, P.J. (1989*b*) Effects of cultivar and boll age on aflatoxin in cottonseed after inoculation with *Aspergillus flavus* at simulated exit holes of the pink bollworm. Plant Disease 73, 489-492.

Cotty, P.J. (1990) Effect of atoxigenic strains of *Aspergillus flavus* on aflatoxin contamination of developing cottonseed. Plant Dis. 74, 233-235.

Cotty, P.J. (1991*a*) Effect of harvest date on aflatoxin contamination of cottonseed. Plant Dis. 75, 312-314.

Cotty, P.J. (1991*b*) Prevention of aflatoxin contamination of cottonseed by qualitative modification of *Aspergillus flavus* populations. Phytopathol. 81, 1227.

Cotty, P.J. (1992*a*) Use of native *Aspergillus flavus* strains to prevent aflatoxin contamination. United States Patent 5,171,686.

Cotty, P.J. (1992*b*) Soil Populations of *Aspergillus flavus* group fungi in agricultural fields in Alabama, Arizona, Louisiana, and Mississippi. Phytopathol. 82, 1064.

Cotty, P.J. (1992*c*) *Aspergillus flavus*, Wild intruder or domesticated freeloader, in "Aflatoxin Elimination Workshop" (Robens, J.F., Ed.) pp. 28. Agricultural Research Service, Beltsville.

Cotty, P.J. and Lee, L.S (1989) Aflatoxin contamination of cottonseed, Comparison of pink bollworm damaged and undamaged bolls. Trop. Sci. 29, 273-277.

Cotty, P.J. and Lee, L.S. (1990) Position and aflatoxin level of toxin positive bolls on cotton plants, in "Proceedings Beltwide Cotton Production Conference", pp. 34-36. National Cotton Council of America, Memphis.

Cotty, P.J. and Egel, D.S. (1992) Can the safety of fungi associated with crops be managed through fungal domestication? Abst. gen. meet. Amer. Soc. of Microbiol.. p 502.

Cotty, P.J., Bayman, P. and Bhatnagar, D. (1990a) Two potential mechanisms by which atoxigenic strains of *Aspergillus flavus* prevent toxigenic strains from contaminating cottonseed. Phytopathology 80, 944.

Cotty, P.J., Cleveland, T.E., Brown, R.L. and Mellon, J.E. (1990b) Variation in polygalacturonase production among *Aspergillus flavus* isolates. Appl. Environ. Microbiol. 56, 3885-3887.

Cucullu, A.F., Lee, L.S., Mayne, R.Y. and Goldblatt, L.A. (1966) Determination of aflatoxins in individual peanuts and peanut sections. J. Am.Oil Chem. Soc. 52, 448-450.

Davis, N.D. and Diener, U.L. (1983) Biology of *A. flavus* and *A. parasiticus*, some characteristics of toxigenic and nontoxigenic isolates of *Aspergillus flavus* and *Aspergillus parasiticus*, in "Aflatoxin and *Aspergillus flavus* in Corn" (Diener, U.L., Asquith, R.L., and Dickens, J.W., Eds.) pp. 1-5. Auburn University, Auburn.

Denning, D.W., Ward, P.N., Fenelon, L.E. and Benbow, E.W. (1992) Lack of vessel wall elastolysis in human invasive pulmonary aspergillosis. Infect. Immun. 60, 5153-5156.

Diener, U.L., Cole, R.J., Sanders, T.H., Payne, G.A., Lee, L.S. and Klich, M.A. (1987) Epidemiology of aflatoxin formation by *Aspergillus flavus*. Ann. Rev. Phytopathol. 25, 249-270.

Dorner, J.W., Cole, R.J. and Diener, U.L. (1984) The relationship of *Aspergillus flavus* and *Aspergillus parasiticus* with reference to production of aflatoxins and cyclopiazonic acid. Mycopathologia 87, 13-15.

Dorner, J.W., Cole, R.J. and Blankenship, P.D. (1992) Use of a biocompetitive agent to control preharvest aflatoxin in drought stressed peanuts. J. Food Prot. 55, 888-892.

Dowd, P.F. (1988) Synergism of aflatoxin $B_1$ toxicity with the co-occurring fungal metabolite kojic acid to two caterpillars. Entomol. Exper. Appl. 47, 60-71.

Dowd, P.F. (1992) Insect interactions with mycotoxin-producing fungi and their hosts, in "Handbook of Applied Mycology Vol. 5, Mycotoxins in Ecological Systems" (Bhatnagar, D., Lillehoj, E., and Arora, D.K., Eds.), pp.137-155. Marcel Dekker, Basel.

Doyle, M.P. and Marth, E.H. (1978) Aflatoxin is degraded by mycelia from toxigenic and nontoxigenic strains of aspergilli grown on different substrates. Mycopathologia 63, 145-153.

Drummond, J. and Pinnock, D.E. (1990) Aflatoxin production by entomopathogenic isolates of *Aspergillus parasiticus* and *Aspergillus flavus*. J. Invert. Pathol. 55, 332-336.

Dunkel, F.V. (1988) The relationship of insects to the deterioration of stored grain by fungi. Intern. J. Food Microbiol. 7, 227-244.

Egel, D.S. and Cotty, P.J. (1992) Relationships among strains in the *Aspergillus flavus* group which differ in toxin production, morphology, and vegetative compatibility group. in "Aflatoxin Elimination Workshop" (Robens, J.F., Ed.) pp. 27. Agricultural Research Service, Beltsville.

Ehrlich, K.C. (1987) Effect on aflatoxin production of competition between wildtype and mutant strains of *Aspergillus parasiticus*. Mycopathologia 97, 93-96.

Ehrlich, K.C., Ciegler, A., Klich, M. and Lee, L. (1985) Fungal competition and mycotoxin production on corn. Experientia 41, 691-693.

Gandola, M. and Aragno, M. (1992) The importance of microbiology in waste management. Experentia 48, 362-366.

Gardner, D.E., McMeans, J.L., Brown, C.M., Bilbrey, R.M. and Parker, L.L. (1974) Geographical localization and lint fluorescence in relation to aflatoxin production in *Aspergillus flavus*-infected cottonseed. Phytopathol. 64, 452-455.

Garrett, S.D. (1960) "Biology of Root-Infecting Fungi," pp. 179-186. University Press, Cambridge.

Gilliam, M. and Vandenberg, J.D. (1990) Fungi, in "Honey Bee Pests, Predators, and Diseases"(Morse,

R.A. and Nowogradzki, R. , Eds.) pp. 64-90. Cornell University Press, Ithaca.
Goldblatt, L.A. and Stoloff, L. (1983) History and natural occurrence of aflatoxins, in "Proceedings of the International Symposium on Mycotoxins" (Naguib, K., Naguib, M.M., Park, D.L. and Pohland, A.E. Eds.), pp. 33-46. The Gen. Organ. for Govern. Printing Offices, Cairo.
Goto, T., Kawasugi, S., Tsuruta, O., Okazaki, H., Siriacha, P., Buangsuwon, D. and Manabe, M. (1986) Aflatoxin contamination of maize in Thailand 2. Aflatoxin contamination of maize harvested in the rainy seasons of 1984 and 1985. Proc. Jpn. Assoc. Mycotoxicol. 24, 53-56.
Goto, T., Tanaka, K., Tsuruta, O. and Manabe, M. (1988) Presence of aflatoxin-producing *Aspergillus* in Japan. Proc. Japan. Assoc. Mycotoxicol. Supp. 1. 179-182.
Griffin, G.J. and Garren, K.H. (1974) Population levels of *Aspergillus flavus* and the *A. niger* group in Virginia peanut field soils. Phytopathol. 64, 322-325.
Griffin, G.J. and Garren, K.H. (1976) Colonization of rye green manure and peanut fruit debris by *Aspergillus flavus* and *Aspergillus niger* group in field soils. Appl. Environ. Microbiol. 32, 28-32.
Hendrickse, R.G. and Maxwell, S.M. (1989) Aflatoxins and child health in the tropics. J. Toxicol. Toxin Rev. 8, 30-48.
Hesseltine, C.W., Shotwell, O.L., Smith, M., Ellis, J.J., Vandegraft, E. and Shannon, G. (1970) Production of various aflatoxins by strains of the *Aspergillus flavus* series. in "Proceedings of the First U.S.-Japan Conference on Toxic Microorganisms" (Herzberg, M., Ed.), pp. 202-210. U.S. Govern. Printing Office, Washington.
Hill, R.A., Wilson, D.M., McMillian, W.W., Widstrom, N.W., Cole, R.J., Sanders, T.H. and Blankenship, P.D. (1985) Ecology of the *Aspergillus flavus* group and aflatoxin formation in maize and groundnut, in "Trichothecenes and Other Mycotoxins" (Lacey, J., Ed.) pp. 79-95. John Wiley & Sons, New York.
Horwitz, W., Albert, R. and Nesheim, S. (1993) Reliability of mycotoxin assays--an updata. J. Assoc. Off. Anal. Chem. 76:461-491.
Jackson, W. and Piper, J. (1989) The necessary marriage between ecology and agriculture. Ecol. 70, 1591-1593.
Janzen, D.H. (1977) Why fruits rot, seeds mold, and meat spoils. Amer. Natur. 111, 691-713.
Joffe, A.Z. (1969) Aflatoxin produced by 1,626 isolates of *Aspergillus flavus* from ground-nut kernels and soils in Israel. Nature 221, 492.
Jones, R.K. (1979) The epidemiology and management of aflatoxins and other mycotoxins, in "Plant Disease, An Advanced Treatise. Vol. 4." (Horsfall, J.G., and Cowling, E.B., Eds), pp. 381-392. Academic Press, New York.
Jones, R.K., Duncan, H.E. and Hamilton, P.B. (1981) Planting date, harvest date, and irrigation effects on infection and aflatoxin production by *Aspergillus flavus* in field corn. Phytopathol. 71, 810-816.
Keller, N.P., Cleveland, T.E. and Bhatnagar, D. (1992) Variable electrophoretic karyotypes of members of *Aspergillus flavus* section *Flavi*. Curr. Genet. 21:371-375.
Kilman, S. (1989) Fungus in corn crop, a potent carcinogen invades food supplies. The Wall Street Journal, February 23.
Kilman, S. (1993) Food-safety strategy pits germ vs. germ. The Wall Street Journal, March 16.
Kimura, N. and Hirano, H. (1988) Inhibitory strains of *Bacillus subtilis* for growth and aflatoxin production of aflatoxigenic fungi. Agric. Biol. Chem. 52, 1173-1179.
King, R.C. and Stansfield, W.D. (1985) "A Dictionary of Genetics," pp. 76. Oxford University Press, New York.
Klich, M.A., Thomas, S.H. and Mellon, J.E. (1984) Field studies on the mode of entry of *Aspergillus flavus* into cotton seeds. Mycologia 76, 665-669.
Klich, M.A. and Pitt, J.I. (1988) Differentiation of *Aspergillus flavus* from *A. parasiticus* and other closely related species. Trans. Brit. Mycol. Soc. 91, 99-108.
Kumar, R.N. and Mishra, R.R. (1991) Effect of pollen on the saprophytic and pathogenic mycoflora of the phylloplane of paddy. Acta Botanica Indica 19, 131-135.
Kurtzman, C.P., Smiley, M.J., Robnett, C.J. and Wicklow, D.T. (1986) DNA relatedness among wild and domesticated species in the *Aspergillus flavus* group. Mycologia 78, 955-959.
Kurtzman, C.P., Horn, B.W. and Hesseltine, C.W. (1987) *Aspergillus nomius*, a new aflatoxin-producing species related to *Aspergillus flavus* and *Aspergillus tamarii*. Anton. Leeuwen. 53, 147-158.
Lafont, P. and Lafont, J. (1977) Toxigenesis of *Aspergillus flavus* isolated from groundnut fields. Mycopathologia 62, 183-185.
Lee, L.S., Lillehoj, E.B. and Kwolek, W.F. (1980) Aflatoxin distribution in individual corn kernels from intact ears. Cereal Chem. 57, 340-343.
Lee, L.S., Koltun, S.P. and Buco, S. (1983) Aflatoxin distribution in fines and meats from decorticated cottonseed. J. Am. Oil Chem. Soc. 60, 1548-1549.

Lee, L.S., Lee, L.V. and Russell, T.E. (1986) Aflatoxin in Arizona cottonseed, field inoculation of bolls by *Aspergillus flavus* spores in wind-driven soil. J. Amer. Oil Chem. Soc. 63, 530-532.

Lee, L.S., Klich, M.A., Cotty, P.J. and Zeringue, H.J. (1989) Aflatoxin in Arizona cottonseed, Increase in toxin formation during field drying of bolls. Arch. Environ. Contam. Toxicol. 18, 416-420.

Lee, L.S., Wall, J.H., Cotty, P.J. and Bayman, P. (1990) Integration of ELISA with conventional chromatographic procedures for quantitation of aflatoxin in individual cotton bolls, seeds, and seed sections. J. Assoc. Off. Anal. Chem. 73, 581-584.

Lillehoj, E.B., Fennel, D.I. and Kwolek, W.F. (1976) *Aspergillus flavus* and aflatoxin contamination in Iowa corn before harvest. Science 193, 485-496.

Lillehoj, E. B., Wall, J.H. and Bowers, E.J. (1987) Preharvest aflatoxin contamination: Effect of moisture and substrate variation in developing cottonseed and corn kernels. Appl. Environ. Microbiol. 53, 584-586.

Lynch, R.E. and Wilson, D.M. (1991) Enhanced infection of peanut, *Arachis hypogaea* L., seeds with *Aspergillus flavus* group fungi due to external scarification of peanut pods by the lesser cornstalk borer, *Elasmopalpus lignosellus* (Zeller). Peanut Sci. 18, 110-116.

Maeda, K. (1990) Incidence and level of aflatoxin contamination in imported foods which were inspected by the official method of Japan. Proc. Japn. Assoc. Mycotoxicol. 31, 7-17.

Malathi, S. and Chakraborty, R. (1991) Production of alkaline protease by a new *Aspergillus flavus* isolate under solid-substrate fermentation conditions for use as a depilation agent. Appl. Environ. Microbiol. 57, 712-716.

Manabe, M., Tsuruta, O., Tanaka, K. and Matsuura, S. (1978) Distribution of aflatoxin-producing fungi in Japan. Trans. Mycol. Soc. Japan 17, 436-444.

Martin, M.M. (1992) The evolution of insect-fungus associations, from contact to stable symbiosis. Amer. Zool. 32, 593-605.

McDonald, B.A. and Martinez, J.P. (1991) DNA fingerprinting of the plant pathogenic fungus *Mycosphaerella graminicola* (anamorph *Septoria tritici*). Exp. Mycol. 15, 146-158.

McLean, M., Berjak, P., Watt, M.P. and Dutton, M.F. (1992) The effects of aflatoxin $B_1$ on immature germinating maize (*Zea mays*) embryos. Mycopathologia 119, 181-190.

McMillian, W.W., Widstrom, N.W. and Wilson, D.M. (1987) Impact of husk type and species of infesting insects on aflatoxin contamination in preharvest corn at Tifton, Georgia. J. Entomol. Sci. 22, 307-312.

Mehan, V.K. and Chohan, J.S. (1973) Relative performance of selected toxigenic and non-toxigenic isolates of *Aspergillus flavus* Link ex Fries on different culture media. Indian J. Exp. Biol. 11, 191-193.

Mellon, J.E. (1991) Purification and characterization of isoperoxidases elicited by *Aspergillus flavus* in cotton ovule cultures. Plant Physiol. 95, 14-20

Mellon, J.E. (1992) Inhibition of aflatoxin production in *Aspergillus flavus* by cotton ovule extracts. J. Am. Oil Chem. Soc. 69, 945-947.

Moody, S.F. and Tyler, B.M. (1990*a*) Restriction enzyme analysis of mitochondrial DNA of the *Aspergillus flavus* group, *A. flavus*, *A. parasiticus*, and *A. nomius*. Appl. Environ. Microbiol. 56, 2441-2452.

Moody, S.F. and Tyler, B.M. (1990*b*) Use of nuclear DNA restriction fragment length polymorphisms to analyze the diversity of the *Aspergillus flavus* group, *A. flavus*, *A. parasiticus*, and *A. nomius*. Appl. Environ. Microbiol. 56, 2453-2461.

Moss, M.O. (1991) The environmental factors controlling mycotoxin formation, in "Mycotoxins and Animal Feeds" (Smith, J.E. and Henderson, R.S., Ed.) pp. 37-56. CRC Press, London.

Muench, K.G., Misra, R.P. and Humayum, M.Z. (1983) Sequence specificity in aflatoxin $B_1$-DNA interaction. Proc. Natl. Acad. Sci. U.S.A. 80, 6-9.

Mycock, D.J., Rijkenberg, F.H. and Berjak, P. (1992) Systemic transmission of *Aspergillus flavus* var. *columnaris* from one maize seed generation to the next. Seed Sci. and Technol. 20, 1-13.

Ohtomo, T., Murakoshi, S., Sugiyama, J. and Kurata, H. (1975) Detection of aflatoxin $B_1$ in silkworm larvae attacked by an *Aspergillus flavus* isolate from a sericultural farm. Appl. Microbiol. 30, 1034-1035.

Olsen, J.H., Dragsted, L. and Autrup, H. (1988) Cancer risk and occupational exposure to aflatoxins in Denmark. Br. J. Cancer 58, 392-396.

Papa, K.E. (1984) Genetics of *Aspergillus flavus*, linkage of aflatoxin mutants. Can. J. Microbiol. 30, 68-73.

Papa, K.E. (1986) Heterokaryon incompatibility in *Aspergillus flavus*. Mycologia 78, 98-101.

Park, D.L., Lee, L.S., Price, R.L. and Pohland, A.E. (1988) Review of the decontamination of aflatoxins by ammoniation: Current status and regulation. J. Assoc. Off. Anal. Chem. 71, 685-703.

Perkins, D.D. and Turner, B.C. (1988) *Neurospora* from natural populations, Toward the population biology of a haploid eukaryote. Exp. Mycol. 12, 91-131.

Pier, A.C. (1992) Major biological consequences of aflatoxicosis in animal production. J. Anim. Sci. 70, 3964-3967.

Pitt, J.I., Dyer, S.K. and McCammon, S. (1991) Systemic invasion of developing peanut plants by *Aspergillus flavus*. Lett. Appl. Microbiol. 13, 16-20.

Pore, R.S., Goodman, N.L. and Larsh, H.W. (1970) Pathogenic Potential of Fungal Insecticides Am. Rev. Respiratory Dis. 101, 627-628.

Rayner, A.D.M. (1991) The challenge of the individualistic mycelium. Mycologia 83, 48-71.

Rhodes, J.C., Bode, R.B. and McCuan-Kirsch, C.M. (1988) Elastase production in clinical isolates of *Aspergillus*. Diagn. Microbiol. Infect. Dis. 10, 165-170.

Rhodes, J.C., Amlung, T.W. and Miller, M.S. (1990) Isolation and characterization of an elastinolytic proteinase from *Aspergillus flavus*. Infect. Immun. 58, 2529-2534.

Rinaldi, M.G. (1983) Invasive aspergillosis. Rev. Infect. Dis. 5, 1061-1077.

Robens, J.F. and Richard, J.L. (1992) Aflatoxins in animal and human health. Rev. Environ. Contam. and Tox. 127, 69-94

Roberts, D.W. and Yendol, W.G. (1971) Use of fungi for microbial control of insects, in "Microbial Control of Insects and Mites" (Burges, H.D., and Burges, N.W., Eds.) pp. 125-149. Academic Press, New York.

Rodriguez, J.G., Potts, M. and Rodriguez, L.D. (1979) Survival and reproduction of two species of stored product beetles on selected fungi. J. Invert. Pathol. 33, 115-117.

Roy, A.K. and Chourasia, H.K. (1990) Inhibition of aflatoxins production by microbial interaction. J. Gen. Appl. Microbiol. 36, 59-62.

Saito, M., Tsuruta, O., Siriacha, P., Kawasugi, S., Manabe, M. and Buangsuwon, D. (1986) Distribution and aflatoxin productivity of the atypical strains of *Aspergillus flavus* isolated from soils in Thailand. Proc. Jpn. Assoc. Mycotoxicol. 24, 41-46.

Samson, R.A. and Frisvad, J.C. (1990) Taxonomic species concepts of hyphomycetes related to mycotoxin production. Proc. Jpn. Assoc. Mycotoxicol. 32, 3-10.

Sanchis, V., Vinas, I., Jimenez, M. and Hernandez, E. (1984) Diferencias morfologicas y enzimaticas entre cepas de *Aspergillus flavus* productoras y no productoras de aflatoxinas. An. Biol. Spec. Sect. 1, 109-114.

Schade, J.E., McGreevy, K., King, A.D., Jr., Mackey, B. and Fuller, G. (1975) Incidence of aflatoxin in California almonds. Appl. Microbiol. 29, 48-53.

Schmitt, S.G. and Hurburgh, C.R., Jr. (1989) Distribution and measurement of aflatoxin in 1983 Iowa corn. Cer. Chem. 66, 165-168.

Schroeder, H.W. and Hein, H., Jr. (1967) Aflatoxins, production of the toxins in vitro in relation to temperature. Appl. Microbiol 15, 441-445.

Schroeder, H.W. and Boller, R.A. (1973) Aflatoxin production of species and strains of the *Aspergillus flavus* group isolated from field crops. Appl. Microbiol. 25, 885-889.

Schroeder, H.W. and Storey, J.B. (1976) Development of aflatoxin in 'Stuart' pecans as affected by shell integrity. Hort. Science 11, 53-54.

Shantha, T., Rati, E.R. and Shankar, T.N.B. (1990) Behaviour of *Aspergillus flavus* in presence of *Aspergillus niger* during biosynthesis of aflatoxin $B_1$. Anton. Leeuwen. 58:121-127.

Shearer, J.F., Sweets, L.E., Baker, N.K. and Tiffany, L.H. (1992) A study of *Aspergillus flavus/parasiticus* in Iowa crop fields, 1988-1990. Plant Dis. 76, 19-22.

Shih, C.N. and Marth, E.H. (1973) Release of aflatoxin from the mycelium of *Aspergillus parasiticus* into liquid media. Z. Lebensm. Unters.-Forsch. 152, 336-339.

Shotwell, O.L. (1991) Mycotoxins in grain dusts, health implications, in "Mycotoxins and Animal Foods" (Smith, J.E. and Henderson, R.S., Eds.), pp. 415-422. CRC Press, London.

Shurtleff, M.C. (1980) "Compendium of Corn Diseases, Second Edition", pp. 51-60. The Amer. Phytopathol. Soc. St. Paul.

Siriacha, P., Kawashima, K., Kawasugi, S., Saito, M. and Tonboon-ek, P. (1989) Postharvest contamination of Thai corn with *Aspergillus flavus*. Cer. Chem. 66, 445-448.

Skory, C.D., Chang, P.K., Cary, J. and Linz, J.E. (1992) Isolation and characterization of a gene from *Aspergillus parasiticus* associated with the conversion of versicolorin A to sterigmatocystin in aflatoxin biosynthesis. Appl. Envir. Microbiol. 58, 3527-3537.

Sommer, N.F., Buchanan, J.R. and Fortlage, R.J. (1986) Relation of early splitting and tattering of pistachio nuts to aflatoxin in the orchard. Phytopathology 76, 692-694.

St. Georgiev, V. (1992) Treatment and developmental therapeutics in aspergillosis 1. Amphotericin B and its derivatives. Respiration 59, 291-302.

Steiner, W.E., Brunschweiler, K., Leimbacher, E. and Schneider, R. (1992) Aflatoxins and fluorescence in brazil nuts and pistachio nuts. J. Agric. Food Chem. 40, 2453-2457.

Stephenson, L.W. and Russell, T.E. (1974) The association of *Aspergillus flavus* with hemipterous and other insects infesting cotton bracts and foliage. Phytopathol. 64, 1502-1506.

Stoloff, L., van Egmond, H.P. and Park, D.L. (1991) Rationales for the establishment of limits and regulations for mycotoxins. Food Add. and Contam. 8, 213-222.

Sussman, A.S. (1951) Studies of an insect mycosis. II. Host and pathogen ranges. Mycologia 43, 423-429.

Sussman, A.S. (1952) Studies of an insect mycosis. III. Histopathology of an aspergillosis of *Platysamia cecropia* L. Ann. Entomo. Soc. Am. 45, 233-245.

Takahashi, T., Onoue, Y. and Mori, M. (1986) Contamination by moulds and inhibitory effect of hay cube on aflatoxin production by *Aspergillus flavus*. Proc. Jpn. Assoc. Mycotoxicol. 23:15-22.

Thom, C. and Raper, K.B. (1945) "A Manual of the Aspergilli." Williams and Wilkins, Baltimore.

Torssell, K.B.G. (1983) "Natural Product Chemistry, A Mechanistic and Biosynthetic Approach to Secondary Metabolism," pp. 3-18. John Wiley & Sons, New York.

van Egmond, H.P. (1991) Limits and regulations for mycotoxins in raw materials and animal feeds, in "Mycotoxins and Animal Feeds" (Smith, J.E. and Henderson, R.S., Ed.) pp. 423-436. CRC Press, London.

van den Hondel, C.A.M.J.J., Punt, P.J. and van Gorcom, R.F.M. (1992) Production of extracellular proteins by the filamentous fungus *Aspergillus*. Anton. Leeuwen. 61, 153-160.

Wadhwani, K. and Srivastava, M. (1985) *Aspergillus flavus* Link as an antagonist against aphids of crucifers. Acta Botanica Indica 13, 281-282.

Ward, P.P., Lo, J.Y., Duke, M., May, G.S., Headon, D.R. and Conneely, O.M. (1992) Production of biologically active recombinant human lactoferrin in *Aspergillus oryzae*. Bio/Technol. 10, 784-789.

Wardlaw, A. and Geddes, D.M. (1992) Allergic bronchopulmonary aspergillosis, a review. J. Royal Soc. Med. 85, 747-751

Watkins, G.M. (1981) "Compendium of Cotton Diseases, "pp. 20-24. The Amer. Phytopathol. Soc. St. Paul.

Wicklow, D.T. (1982) Conidium germination rate in wild and domesticated yellow-green aspergilli. Appl. Environ. Microbiol. 47, 299-300.

Wicklow, D.T. (1990) Adaptation in *Aspergillus flavus*. Trans. Mycol. Soc. Japan 31, 511-523.

Wicklow, D.T. (1993) The mycology of stored grain, an ecological perspective, in "Stored Grain Ecosystems"(Jayas, D.S., and White, N.D.G., Eds). Marcel Dekker, Basel. In Press.

Wicklow, D.T. and Cole, R.J. (1982) Tremorgenic indole metabolites and aflatoxins in sclerotia of *Aspergillus flavus*, an evolutionary perspective. Can. J. Bot. 60, 525-528.

Wicklow, D.T. and Donahue, J.E. (1984) Sporogenic germination of sclerotia in *Aspergillus flavus* and *A.. parasiticus*. Trans. Br. Mycol. Soc. 82:621-624.

Wicklow, D.T. and Shotwell, O.L. (1982) Intrafungal distribution of aflatoxins among conidia and sclerotia of *Aspergillus flavus* and *Aspergillus parasiticus*. Can. J. Microbiol. 29, 1-5.

Widstrom, N.W. (1979) The role of insects and other plant pests in aflatoxin contamination of corn, cotton, and peanuts: a review. J. Environ. Qual. 8, 5-11.

Widstrom, N.W. (1992) Aflatoxin in developing maize, Interactions among involved biota and pertinent econiche factors, in "Handbook of Applied Mycology Vol. 5, Mycotoxins in Ecological Systems"(Bhatnagar, D., Lillehoj, E. and Arora, D.K., Eds.), pp. 23-58. Marcel Dekker, Basel.

Willetts, H.J. and Bullock, S. (1992) Developmental biology of sclerotia. Mycol. Res. 96, 801-816.

Wilson, D.M., Mixon, A.C. and Troeger, J.M. (1977) Aflatoxin contamination of peanuts resistant to seed invasion by *Aspergillus flavus*. Phytopathol. 67, 922-924.

Wotton, H.R. and Strange, R.N. (1987) Increased susceptibility and reduced phytoalexin accumulation in drought-stressed peanut kernels challenged with *Aspergillus flavus*. Appl. Environ. Microbiol. 53, 270-273.

Wright, V.F., Vesonder, R.F. and Ciegler, A. (1982) Mycotoxins and other fungal metabolites as insecticides , in "Microbial and Viral Pesticides," (Kurstak, E., Ed.), pp 559-583. Marcel Dekker, New York.

Zak, J.C. (1992) Response of soil fungal communities to disturbance, in "The Fungal Community Its Organization and Role in the Ecosystem" (Carroll, G.C., and Wicklow, D.T., Eds.) pp. 403-425. Marcel Dekker, New York.

Zarba, A., Wild, C.P., Hall, A.J., Montesano, R., Hudson, G.J. and Groopman, J.D. (1992) Aflatoxin M1 in human breast milk from The Gambia, West Africa, quantified by monoclonal antibody immunoaffinity chromatography and HPLC. Carcinogenesis 13, 891-894.

Zummo, N. and Scott, G.E. (1990) Relative aggressiveness of *Aspergillus flavus* and *A. parasiticus* on maize in Mississippi. Plant Dis. 74, 978-981.

# BIOSYNTHESIS OF *ASPERGILLUS* TOXINS - NON-AFLATOXINS

Maurice O. Moss

School of Biological Sciences
University of Surrey
Guildford
Surrey, GU2 5XH

## INTRODUCTION

Mycotoxins are a sub-set of the vast array of natural products referred to as secondary metabolites. The evolutionary origin and the role of secondary metabolites for the producing organisms are still areas of vigorous debate but there is a growing concensus that secondary metabolism will benefit the organism and this benefit will usually reflect the nature of the metabolites themselves (Vining, 1992).

Fungal secondary metabolites are relatively small molecules characterised, not only by their structural diversity (Turner and Aldridge, 1983), but also by their diversity of biological activity which includes antibiotic activity, phytotoxicity, animal toxicity and a bewildering array of physiological and pharmacological activities in the mammalian system. There is also a diversity in the species specificity of their production, some being produced by a limited number of isolates of a single species whereas others may be produced by several species. The specificity in production may be such as to suggest the use of secondary metabolites for chemotaxonomic purposes (Frisvad and Samson, 1990) but there are examples of secondary metabolites produced by different organisms which span huge taxonomic boundaries. Thus at least ten of the gibberellins produced by higher plants are produced by *Fusarium moniliforme* Sheldon and one of these, gibberellin $A_4$, is produced by the unrelated mould *Sphaceloma manihoticola* Bitancourt & Jenkins (Rademacher and Graebe, 1979). Another important plant metabolite, abscisic acid, is also produced by the mould *Cercospora rosicola* Passerini (Assante, Merlini and Nasini, 1977). These, and other examples, suggest that lateral gene transfer may provide one of the processes for the acquisition of the biosynthetic pathways for some secondary metabolites by some fungi.

The biosynthesis of secondary metabolites involves the same small pool of important precursor molecules required for the biosynthesis of primary metabolites associated with growth (Table 1). Acetyl coenzyme A plays an especially important role being both an earlier precursor of mevalonate as well as being involved in the biosynthesis of several amino acids. In identifying the amino acid pool as an important group of precursors for both primary and secondary metabolism it should be recalled that the biosynthesis of the aromatic amino acids involves the shikimate pathway and this pathway may also be diverted into the biosynthesis of secondary metabolites. Such

*The Genus Aspergillus*, Edited by Keith A. Powell *et al.*,
Plenum Press, New York, 1994

**Table 1.** Precursors for primary and secondary metabolism

| Precursor | 1° metabolism | 2° metabolism |
|---|---|---|
| Acetyl coenzyme A | Fatty acids | Polyketides |
| Mevalonate | Sterols | Sesquiterpenes<br>Diterpenes |
| Amino acids | Proteins | Diketopiperazines<br>Cyclic peptides |

metabolites occur widely in higher plants and in the basidiomycetes but are not so widespread amongst the group of fungi associated with the study of mycotoxins.

## ASPERGILLUS MYCOTOXINS

The diversity of toxic metabolites produced by *Aspergillus* was reviewed by Moss (1977) who attempted to relate this diversity in chemical structure to the taxonomic diversity within the genus. Even at that time the anamorph genus *Aspergillus* contained species associated with at least eleven teleomorph genera within the Eurotiales (Fennell, 1977), as well as species for which no teleomorph connection had been made. The most intensively studied group of *Aspergillus* mycotoxins is still the aflatoxins produced by *Aspergillus flavus* Link, *A. parasiticus* Speare and, a more recently described species, *Aspergillus nomius* (Kurtzman, Horn and Hesseltine 1987). The biosynthesis of this group will be dealt with in a subsequent chapter and Table 2 summarises the range of other toxic metabolites produced by species of *Aspergillus* indicating in general terms the likely precursors from the metabolic pool referred to in Table 1. Their structures are given in Figures 1 - 24.

**Table 2.** Mycotoxins, other than aflatoxins, produced by *Aspergillus*

| Toxic Metabolite | Fig. | Precursors | Producing Organism |
|---|---|---|---|
| Aflatrem | 1 | amino acid/mevalonate | *A. flavus* |
| Aspochalasin | 2 | amino acid/polyketide | *A. microcysticus* Sappa |
| Austdiol | 3 | polyketide | *A. ustus* (Bain.)Thom & Church |
| Austin | 4 | mevalonate/polyketide | *A. ustus* |
| Austocystins | 5 | polyketide | *A. ustus* |
| Citreoviridin | 6 | polyketide | *A. terreus* Thom |
| Citrinin | 7 | polyketide | *A. candidus* Link<br>*A. carneus* (v. Tieghem) Blochwitz<br>*A. flavipes* (Bain. & Sart.) Thom & Church<br>*A. terreus* |
| Cyclopiazonic acid | 8 | amino acid/mevalonate/ acetate | *A. flavus*<br>*A. oryzae* (Ahlburg) Cohn<br>*A. versicolor* (Vuill.) Tiraboschi |
| Cytochalasin E | 9 | amino acid/polyketide | *A. clavatus* Desm. |
| Fumagillin | 10 | mevalonate/polyketide | *A. fumigatus* Fres. |
| Fumitremorgins | 11 | amino acid/mevalonate | *A. fumigatus* |

**Table 2. (Cont.)**

| Toxic Metabolite | Fig. | Precursors | Producing Organism |
|---|---|---|---|
| Gliotoxin | 12 | amino acids | *A. chevalieri* Thom & Church<br>*A. fumigatus*<br>*A. terreus* |
| Malformins | 13 | amino acids | *A. niger* v. Tieghem |
| Maltoryzine | 14 | polyketide | *A. oryzae* |
| Ochratoxin | 15 | amino acid/polyketide | *A. alliaceus* Thom & Church<br>*A. alutaceus* Berk. & Curt.<br>(= *A. ochraceus* Wilhelm)<br>*A. fresenii* Subram.<br>(=*A. sulphureus* (Fres.) Weh.<br>*A. melleus* Yukawa<br>*A. ostianus* Wehmer<br>*A. petrakii* Vöròs<br>*A. sclerotiorum* Huber |
| Patulin | 16 | polyketide | *A. clavatus*<br>*A. giganteus* Wehmer<br>*A. terreus* |
| Territrems | 17 | polyketide/mevalonate | *A. terreus* |
| Sphingofungins | 18 | polyketide/amino acid | *A. fumigatus* |
| Sterigmatocystin | 19 | polyketide | *A. nidulans* (Eidam) Winter<br>*A. variecolor* (Berk. & Broome) Thom & Raper<br>(=*A. stellatus* Curzi)<br>*A. versicolor* (Vuill.) Tiraboschi |
| Tryptoquivaline | 20 | amino acids/anthranilate | *A. clavatus*<br>*A. fumigatus* |
| Viomellein | 21 | polyketide | *A. fresenii*<br>*A. melleus* |
| Viriditoxin | 22 | polyketide | *A. brevipes* G. Sm.<br>*A. viridinutans* Ducker & Thrower |
| Xanthocillin | 23 | amino acid | *A. chevalieri* |
| Xanthomegnin | 24 | polyketide | *A. fresenii*<br>*A. melleus* |

It can be seen from Table 2 that there are examples amongst the mycotoxins of *Aspergillus* derived from all possible combinations of the pool of precursors. This is illustrated diagrammatically in Figure 25 in which an attempt has also been made to indicate the two phases which occur in most biosynthetic pathways i.e. polymerisation followed by modification.

## Secondary metabolism

Before dealing with the detailed biosynthetic pathways of a selection of the mycotoxins produced by different species of *Aspergillus*, it may be useful to make some general observations about the phenomenon of secondary metabolism in fungi. It has

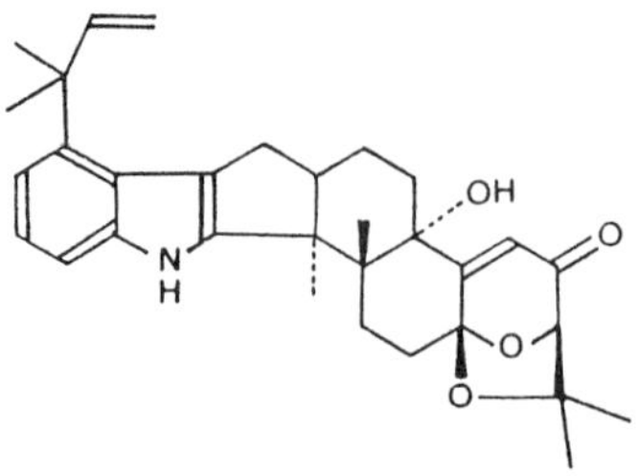


**Figure 1.** Aflatrem

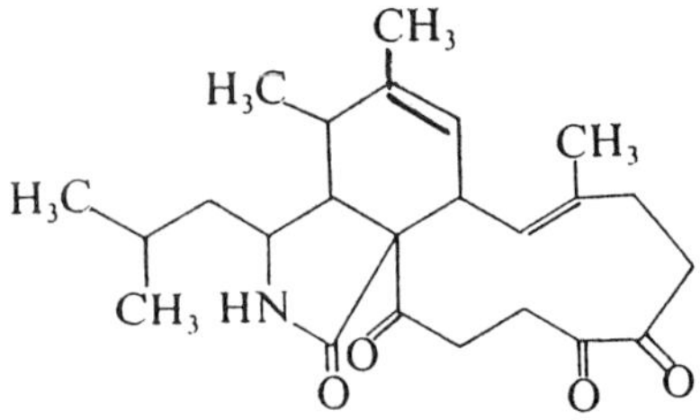


**Figure 2.** Aspochalasin A

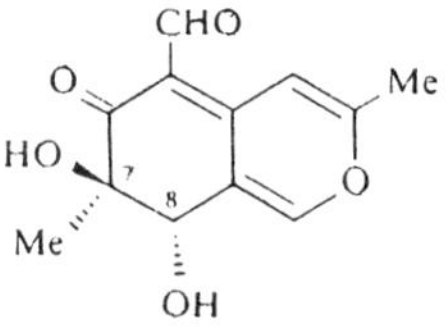


**Figure 3.** Austdiol

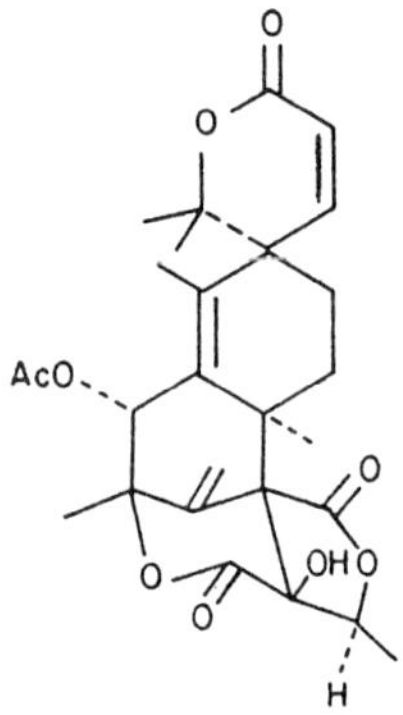


**Figure 4.** Austin

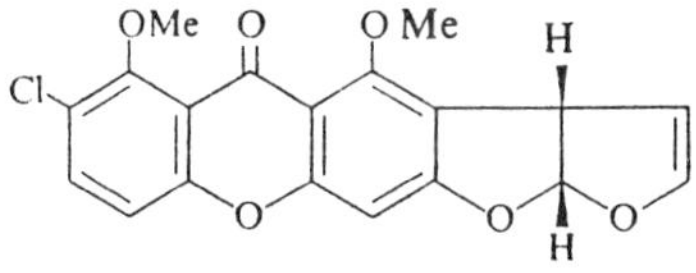


**Figure 5.** Austocystin A

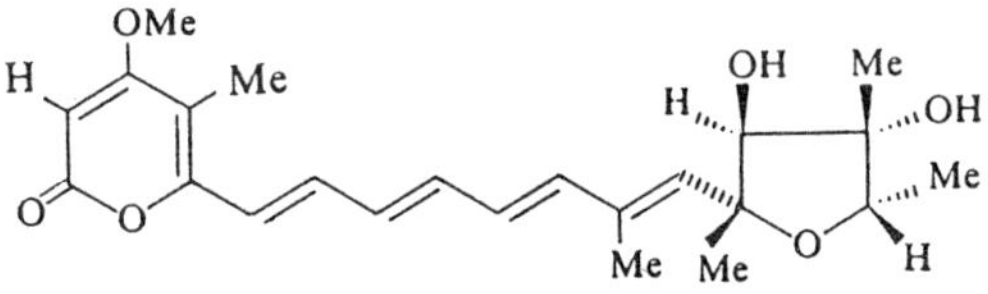


**Figure 6.** Citeoviridin

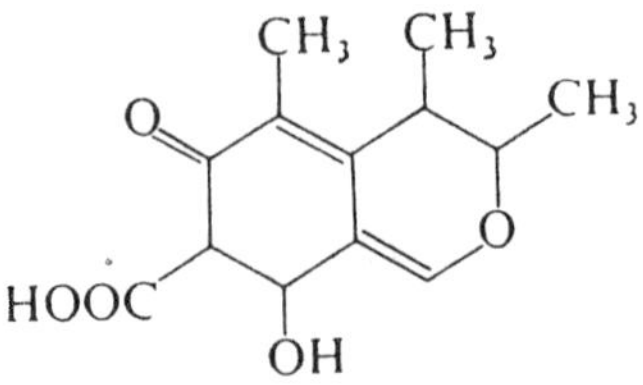


**Figure 7.** Citrinin

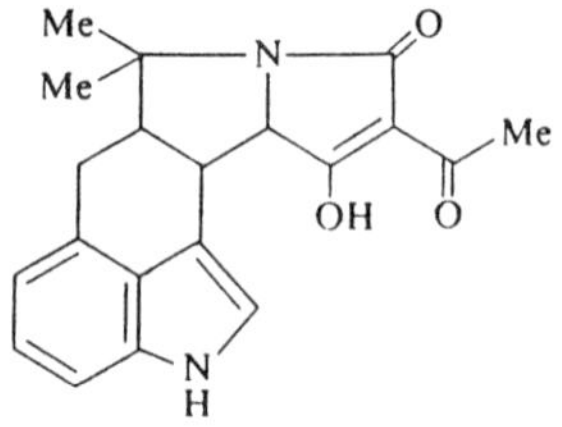


**Figure 8.** Cyclopiazonic acid

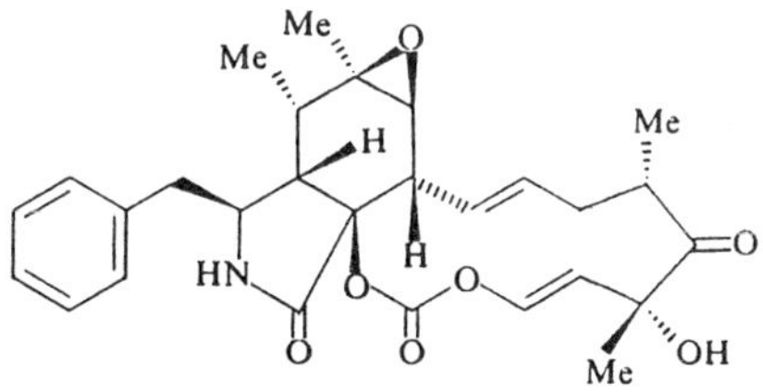


**Figure 9.** Cytochalasin E

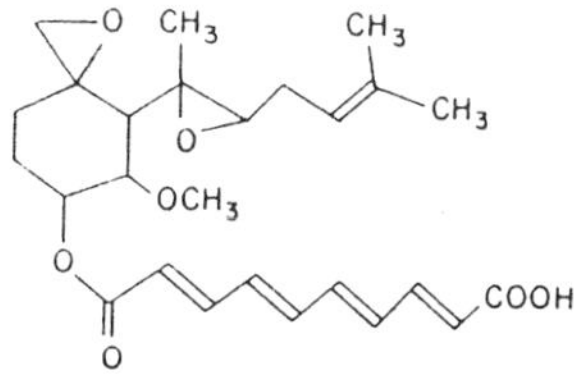


**Figure 10.** Fumagillin

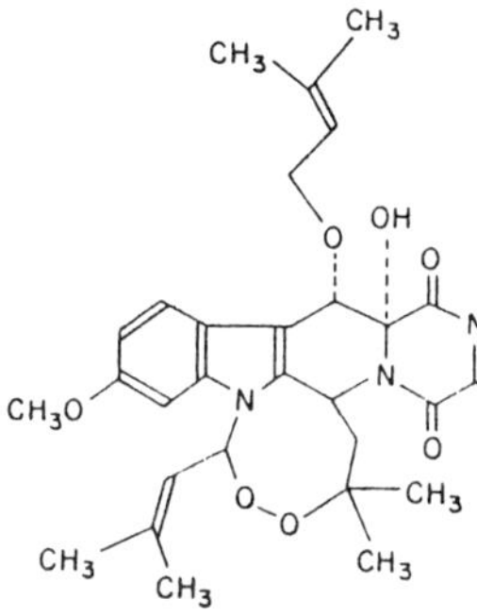


**Figure 11.** Fumitremorgin A

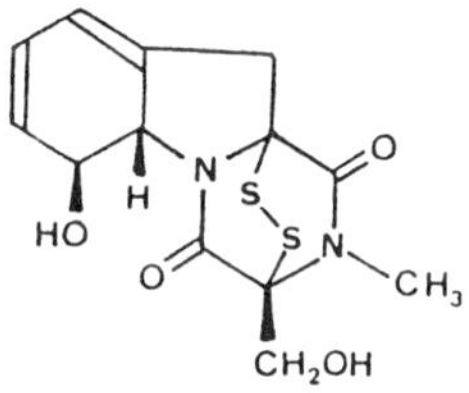


**Figure 12.** Gliotoxin

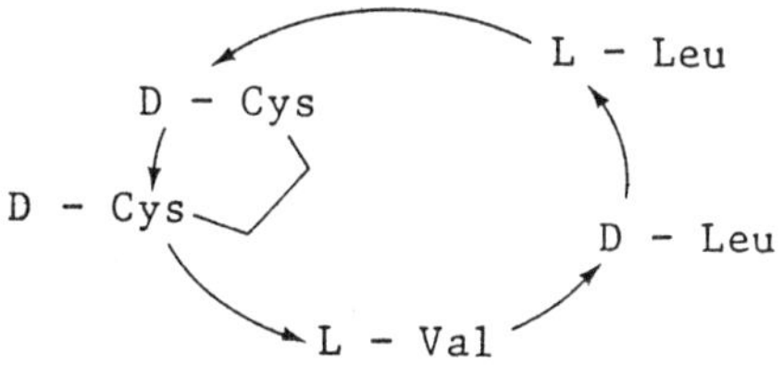


**Figure 13.** Malformin C

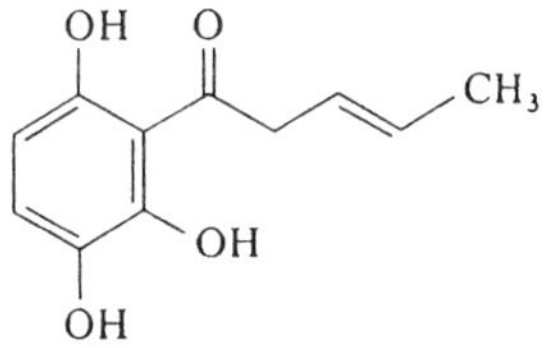


**Figure 14.** Maltoryzine

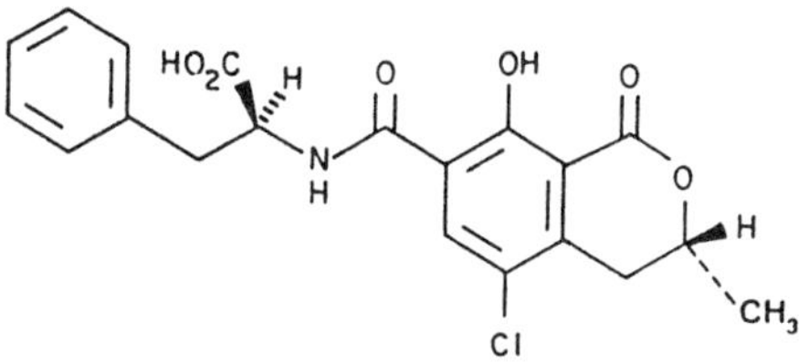


**Figure 15.** Ochratoxin A

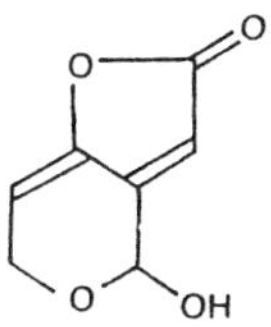


**Figure 16.** Patulin

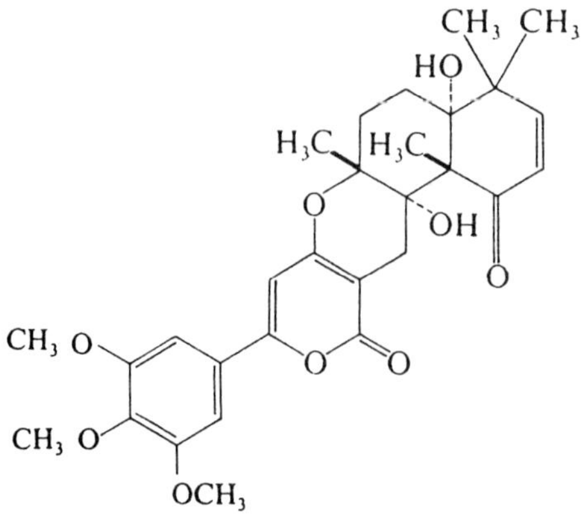

**Figure 17.** Territrem B

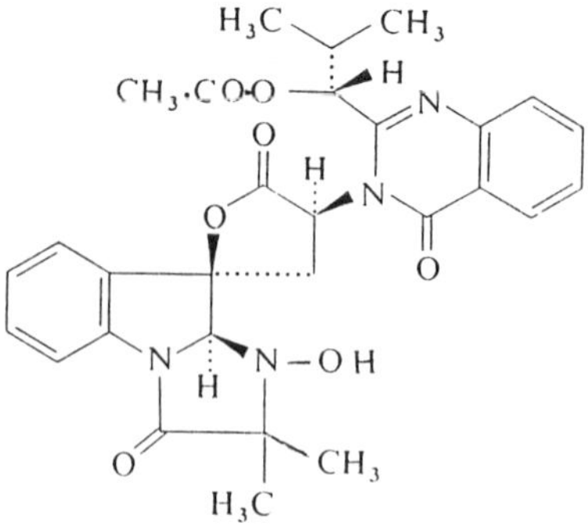

**Figure 20.** Tryptoquivaline

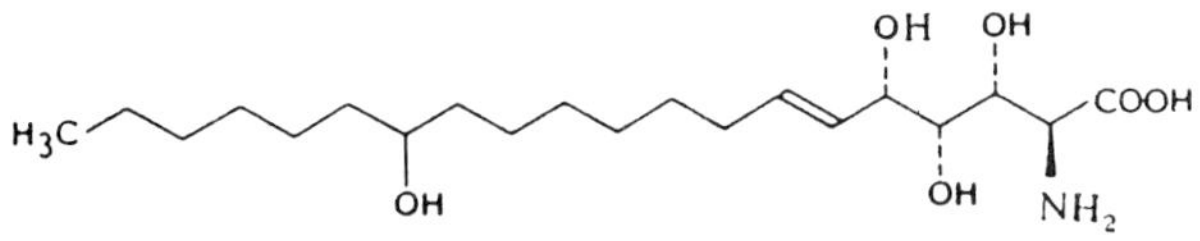

**Figure 18.** Sphingofungin B

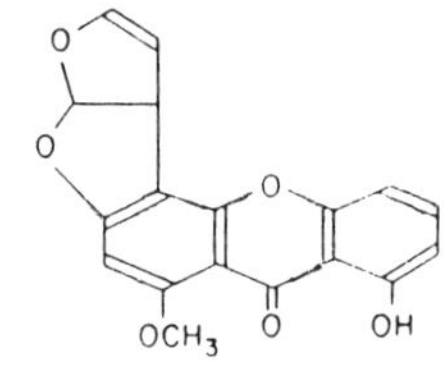

**Figure 19.** Sterigmatocystin

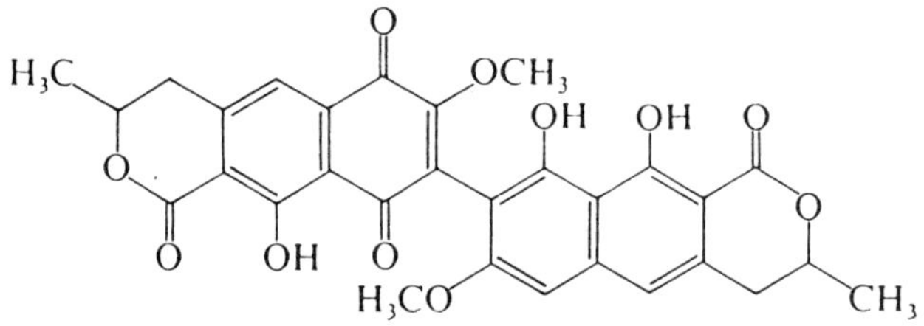

**Figure 21.** Viomellein

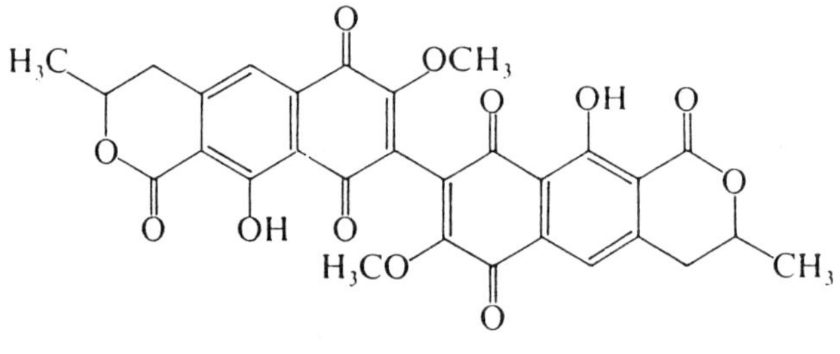

**Figure 22.** Viriditoxin

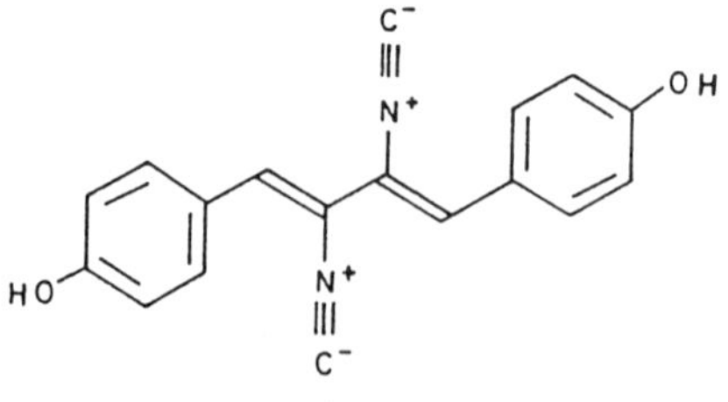

**Figure 23.** Xanthocillin

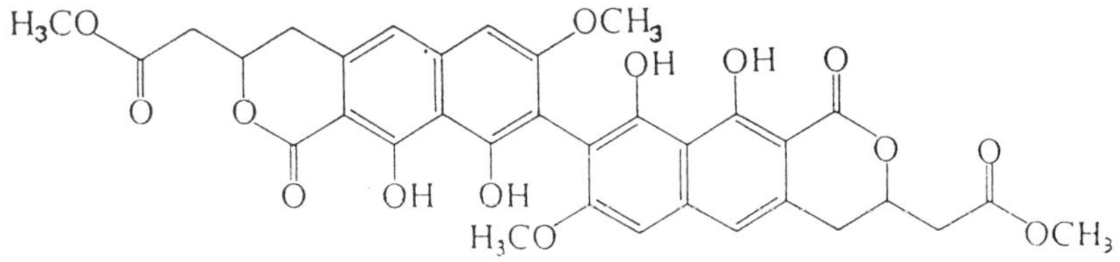

**Figure 24.** Xanthomegnin

been observed that their production is especially diverse amongst the terrestrial filamentous fungi and their prokaryotic counterparts , the actinomycetes especially of the genus *Streptomyces* (Moss, 1984). It is possible that secondary metabolism represents a form of biochemical differentiation associated with changes in the nutritional, physiological and ecological status of an organism with the highly polarised growth form of the hypha extending across and into solid substrates. It is often possible to observe a direct link between the onset of secondary metabolism and processes of morphological differentiation leading to the formation of fruiting structures and spores or resting structures such as sclerotia (Campbell, 1983).

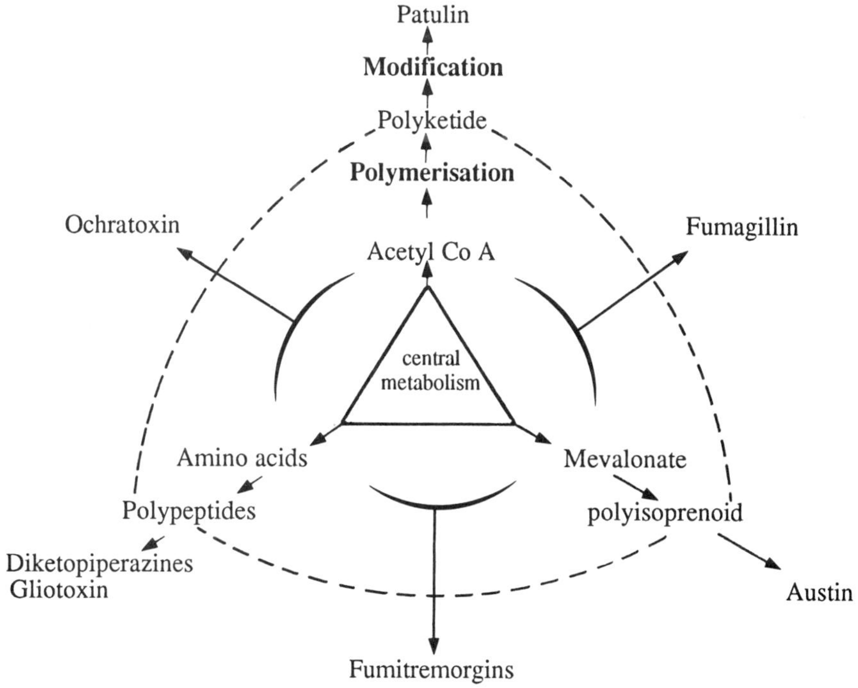


**Figure 25.** The relationship of central metabolism to the production of mycotoxins showing an example from each of the major pools of intermediates.

It is no longer tenable to accept that secondary metabolites, including mycotoxins, are simply waste products of an otherwise essential process (secondary metabolism) or chance phenomena which happen to be neutral in their influence on evolution. It may not be possible to identify the precise role of each secondary metabolite in the physiology and ecology of the producing organism but two important observations suggest that such a role exists. An increasing knowledge and deeper understanding of the enzymology and genetics of the production of some secondary metabolites has shown that the producing organism has a significant investment in their production and the control of that production. Secondly, in the case of some antibiotics, it has been clearly demonstrated that the producing organism may have very sophisticated mechanisms, controlled at the genetic level, of resistance to their own antibiotics (Cundliff, 1992). On the basis of detailed studies of mammalian toxicology, Riley and Goeger (1992) speculate that the indole tetramic acid cyclopiazonic acid may play an important role in the control of metabolic processes associated with differentiation.

The biosynthetic pathways of many secondary metabolites, including mycotoxins, involves two distinct phases. During the first phase the precursor molecules are linked together by enzyme complexes such as polyketide synthases (the polymerisation phase). This group of enzymes may be present but inhibited during the active growth phase of the mould. The products of the polymerisation phase are then modified by enzymes the formation of which is usually repressed during vegetative growth. This second group of inducible enzymes are frequently cytochrome $P_{450}$ oxygenases and may be involved in hydroxylations, oxidative cleavages and rearrangements leading to the remarkable diversity of secondary metabolites characteristic of the filamentous fungi. The following sections will deal with some examples of each of the major pathways indicated in Figure 25

## POLYKETIDES

The head to tail coupling of acetate units, first proposed by Birch and Donovan (1953) and now known to occur *via* malonyl coenzyme A, is brought about by polyketide synthase. The process is entirely analogous to fatty acid synthesis involving the fatty acid synthase complex but without the cycle of reductive and dehydration steps of the latter. It is tempting to think that the evolution of polyketide synthases could have occurred by a process of gene duplication of the complex of genes coding for fatty acid synthase making it possible for the copy to undergo deletions and other modifications leading to the polyketide synthases. However, the sequencing of the genes for 6-methyl salicylic acid synthase (a tetraketide synthase involved in the biosynthesis of patulin) of *Penicillium griseofulvum* Dierckx has shown that it more closely resembles vertebrate fatty acid synthases than the fatty acid synthase of *P. griseofulvum* itself (Beck *et al.*, 1990). Recent developments in the understanding of the molecular genetics of polyketide biosynthesis have been lucidly reviewed by Hopwood and Khosha (1992).

### Patulin

A recent reassessment of the toxicological data available for patulin has led to an increased awareness of its occurrence and significance in fresh apple juice. Although it is produced by a number of species of *Aspergillus*, *Penicillium* and *Paecilomyces*, it is the production by *Penicillium expansum* Link, associated with a brown rot of apples, which is especially important. The production of patulin by *Aspergillus clavatus* growing on spent malted barley has been implicated in the poisoning of cattle when this material has been used as a feed additive. Although a relatively small molecule the biosynthesis of patulin is a clear example of the two stages in the production of a secondary metabolite. The polymerisation of three molecules of malonyl coenzyme A with a molecule of acetyl coenzyme A, during which a reduction and dehydration occur, leads to a tetraketide derivative which readily cyclises to 6-methyl salicylic acid (see Figure 26). This particular metabolite has a special place in the history of our understanding of polyketide biosynthesis, being the first to be produced from [1-$^{14}$C] - labelled acetate, and subsequently degraded to demonstrate where the $^{14}$C had gone in the elegant experiments of Birch, Massy-Westrop and Moye (1955).

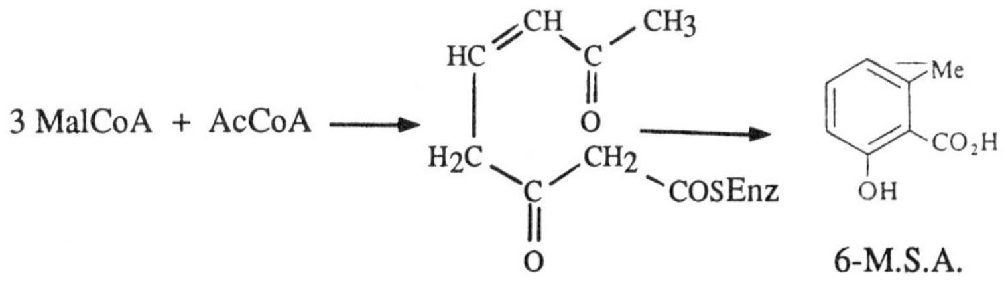


**Figure 26.** The production of 6-methylsalicylic acid from a tetraketide precursor

6-Methylsalicylic acid is the substrate for a complex sequence of oxidations, cleavage and rearrangements which form the second stage in the biosynthesis of patulin. There is no doubt that gentisaldehyde is the substrate for the oxidative cleavage reactions and the arguments for the involvement of the epoxy-quinone, phyllostine, have been set out in detail by Zamir (1980). The major steps, which are actually part of a grid of metabolites, are outlined in Figure 27.

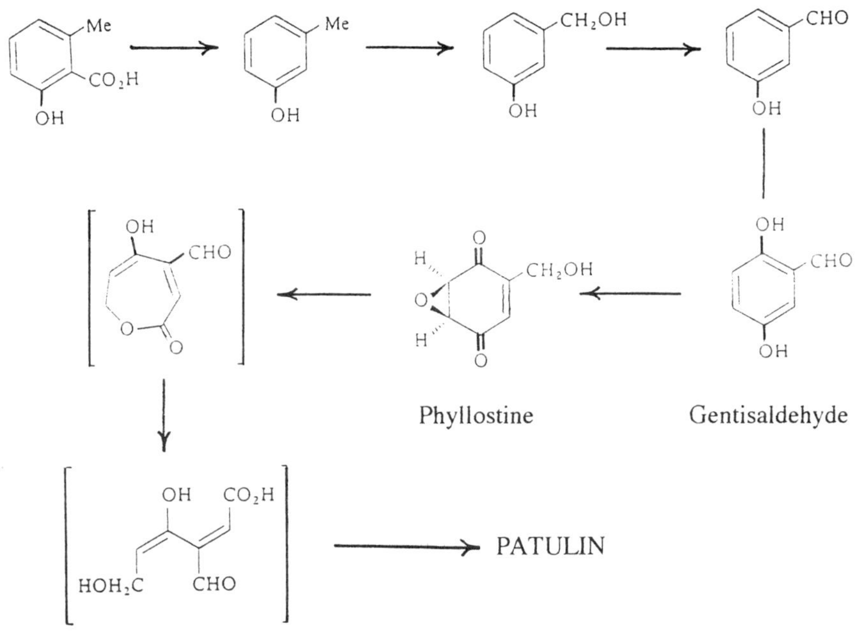


**Figure 27.** The major stages in the biosynthesis of patulin from 6-methylsalicylic acid.

## Citrinin

During the building of a polyketide chain a number of additional groups can be added, especially to the reactive methylene groups, providing another source of diversity in structure. Citrinin is a good example in which three additional $C_1$ groups have been added to the pentaketide chain from the biochemically active one carbon pool involving methionine (see Figure 28).

It should be noted that the common occurrence of C-methyl groups in polyketide derived metabolites produced by prokaryotes such as actinomycetes is often due to the incorporation of methylmalonyl coenzyme A in place of malonyl coenzyme A. In the moulds, although these additional $C_1$ groups are usually initially incorporated as methyl groups, they are frequently further oxidised. Thus, one of the two additional branched carbons in austdiol is present as an aldehyde and one of the three in citrinin is oxidised to the carboxylic acid group.

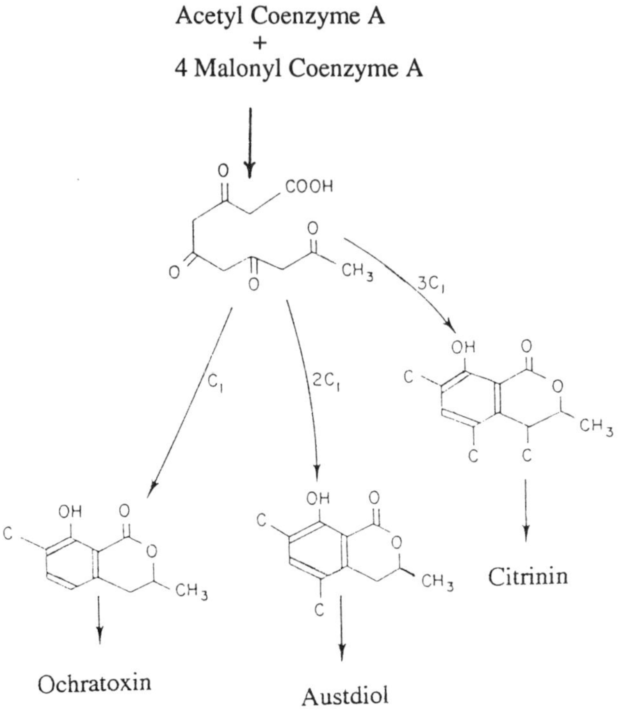


**Figure 28.** The interaction of a pentaketide precursor with the one carbon pool leading to the biosynthesis of ochratoxin, austdiol and citrinin.

## Sterigmatocystin and the austocystins

Sterigmatocystin is an intermediate in the biosynthetic pathway leading to the aflatoxins, and will be dealt with in detail elsewhere, but there are several additional features in the production of these difurano-xanthones which are worth highlighting in this more general discourse. Although derived from $C_{18}$ difurano-anthroquinones, the original polyketide precursor of this nested family of compounds (the versicolorins, sterigmatocystins and finally aflatoxins) is not a nonaketide but a decaketide, such as norsolorinic acid, which is probably derived from hexanoyl coenzyme A with the subsequent addition of seven more malonyl coenzyme A units (Townsend *et al.*, 1984). The rearrangements leading to the unusual fused difuran group result in the loss of an acetate. Once formed this fused difuran ring system is sufficiently stable to remain intact through all the oxidative cleavages and rearrangements leading to aflatoxin $B_1$, aflatoxin $G_1$ and eventually parasiticol.

The austocystins form a complex family of linear difurano-xanthones, some of which are chlorinated, and some have additional mevalonate-derived sidechains (see Figure 29). They are essentially substituted derivatives of sterigmatin produced, along with sterigmatocystin and the versicolorins, by *Aspergillus versicolor* (Hamasaki *et al.*, 1973).

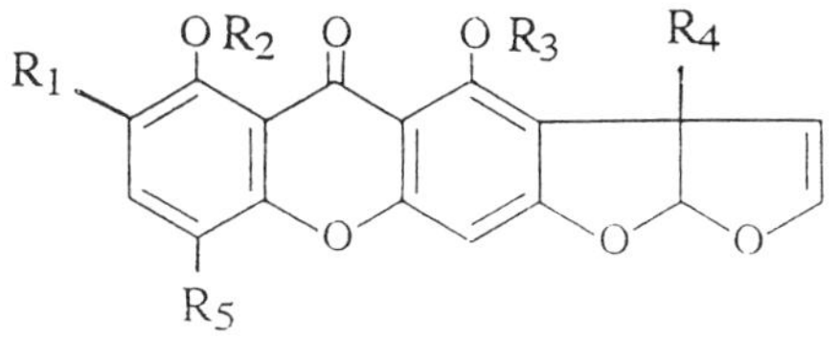

| | | R1 | R2 | R3 | R4 | R5 |
|---|---|---|---|---|---|---|
| Austocystin | A | Cl | $CH_3$ | $CH_3$ | H | H |
| | B | H | H | H | H | $CH_2.CH_2.C(OH).(CH_3)_2$ |
| | C | H | H | $CH_3$ | H | $CH_2.CH_2.C(OH).(CH_3)_2$ |
| | D | H | H | H | OH | $CH_2.CH_2.C(OH).(CH_3)_2$ |
| | E | H | H | $CH_3$ | OH | $CH_2.CH_2.C(OH).(CH_3)_2$ |
| | F | H | H | H | OH | H |
| | G | Cl | $CH_3$ | H | OH | H |
| | H | H | H | H | OH | $CH_2.CH = C.(CH_3)_2$ |
| | I | H | H | $CH_3$ | OH | H |
| Sterigmatin | | H | H | H | H | H |

**Figure 29.** The austocystins and their relationship to sterigmatin

### Mevinolin

Although mevinolin may not to be considered as a mycotoxin, it has a potent pharmacological effect through inhibition of hydroxymethyl glutaryl coenzyme A reductase and is thus an inhibitor of sterol biosynthesis. Indeed mevinolin, a metabolite of *Aspergillus terreus,* is marketed for treating hypercholesterolemia having been shown to lower serum cholesterol in man. It is of interest here as a further demonstration of the diversity of biosynthetic pathways in the production of polyketides by *Aspergillus.* Figure 30 shows the way in which it is formed from two distinct polyketide chains, each of which is methylated *via* methionine (Moore *et al.*, 1985).

## POLYMEVALONATE METABOLITES

Although Turner and Aldridge (1983) list a number of sesquiterpenes and diterpenes produced by *Aspergillus*, those mevalonate derived secondary metabolites of the genus recognised as mycotoxins also involve components derived from acetate (polyketide) or amino acids.

### Austin and terretonin

An interesting complex of metabolites produced by *Aspergillus ustus* and *A. terreus* involves the preliminary coupling of the sesquiterpene precursor farnesyl pyrophosphate and the methylated polyketide, 2,4-dihydroxy-3,5,6-trimethyl benzoic acid, produced from orsellinic acid by the addition of two methyl groups from methionine. The rearrangements of this hypothetical intermediate are so complex that the biosynthetic pathways proposed by Simpson and Stenzel (1981) for austin and McIntyre and Simpson (1981) for terretonin required careful labelling experiments and the tools of modern structural organic chemistry to unravel them (see Figure 31) (Vederas, 1986).

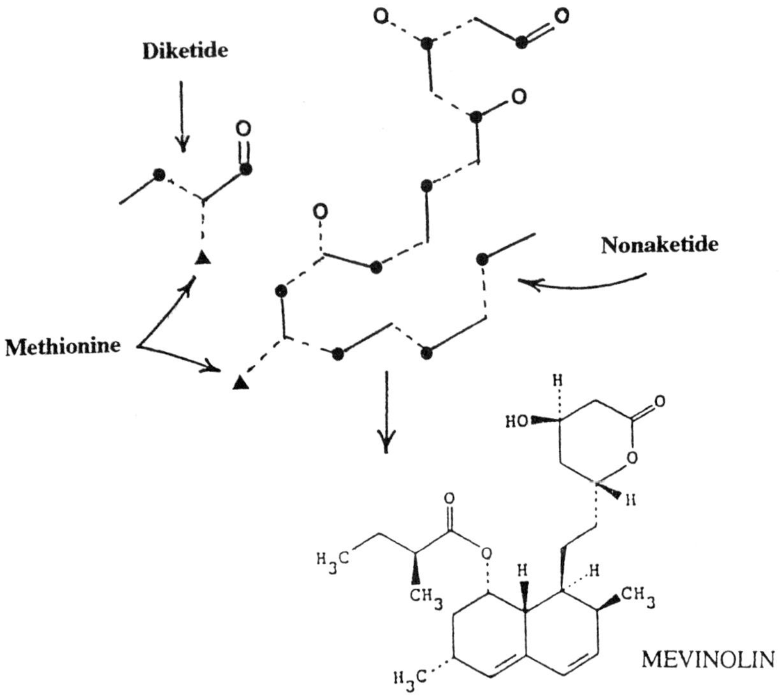


**Figure 30.** The biosynthesis of mevinolin by *Aspergillus terreus.*

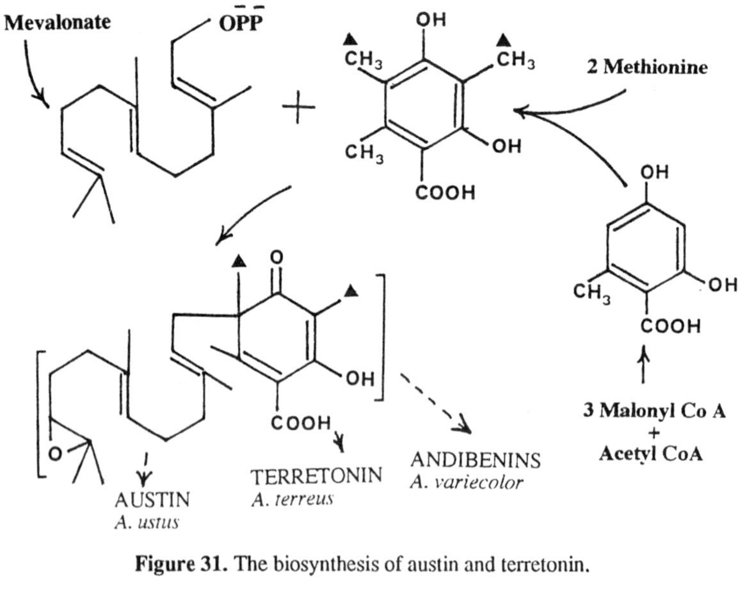


**Figure 31.** The biosynthesis of austin and terretonin.

## Fumagillin

The acute toxicity of fumagillin, produced by *Aspergillus fumigatus,* is sufficiently low that it is a potentially useful antibiotic and has been used in the control of amoebic dysentry. It is none the less an interesting molecule being a very clear example of a secondary metabolite assembled from two distinct pools of precursor intermediates. It is also an opportunity to pay further tribute to the perception of A. J. Birch and his colleagues in understanding the biosynthetic pathways of complex molecules. Birch and Hussain (1969) showed that the decantetraendioic acid moiety has a polyketide origin and that the terminal carboxyl group represents the original carboxyl end of the polyketide. This is possibly protected (as the coenzyme A ester?) while the methyl derived end of the polyketide undergoes ω-oxidation and esterification to the sesquiterpene based nucleus.

Birch and Hussain (1969) also proposed an elegant hypothesis for the origin of the unusual nucleus of fumagillin from farnesyl pyrophosphate *via* a bergamotene intermediate (see Figure 32), an hypothesis which was subsequently supported by the isolation of β-trans bergamotene itself from *A. fumigatus* (Nozoe *et al.*, 1976) and the detailed studies of Cane and Levin (1976) on the biosynthesis of the closely related metabolite of *Pseudeurotium ovalis* Stolk known as ovalicin. This latter mould, which is said to have a *Sporotrichum* anamorph, is a member of the Eurotiales and was originally isolated from nematode cysts. Ovalicin has antibiotic, immunosuppressive and antitumour activities.

**Farnesyl pyrophosphate**

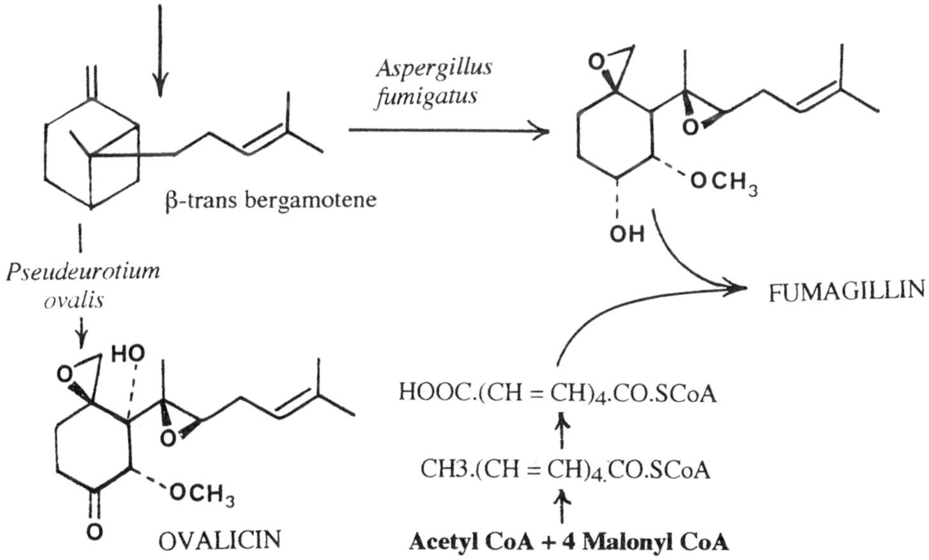


**Figure 32.** Biosynthesis of fumagillin and its relationship with ovalicin.

## The territrems

This family of metabolites, produced by a strain of *A. terreus*, is unusual amongst tremorgenic mycotoxins in not containing nitrogen. First isolated by Ling *et al.* (1979) further members of the family have been described and characterised in a series of papers (Ling *et al.*, 1984; Peng *et al.*, 1985; Ling *et al.*, 1986). It is possible to recognise a sesquiterpene origin for one half of the molecule but one possible route to the other half requires the loss of a carbon atom (see Figure 33).

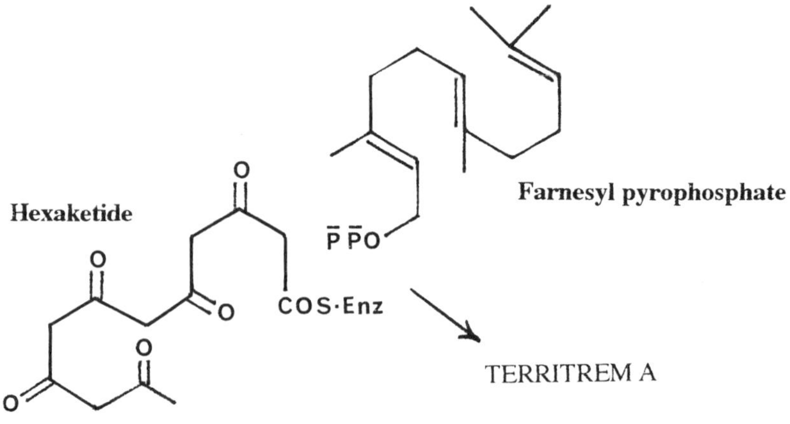


**Figure 33.** A possible route to the biosynthesis of the territrems.

## AMINO ACID DERIVED MYCOTOXINS

Amino acids are the precursors of several distinct families of secondary metabolites, cyclic polypeptides, diketopiperazines, polycyclic compounds with anthranilic acid and as part of complex molecules with polyketide or mevalonate components.

### Malformins

Members of this family of cyclic peptides were first studied because of their effects on plants but, with an $LD_{50}$ (I.P.) of 0.9 mg $kg^{-1}$ in the rat, malformin C is also very toxic to mammals. It is produced by strains of *Aspergillus niger* and related species (Anderegg *et al.*, 1976) and, as is so often the case with biologically active cyclic polypeptides, several of the amino acid residues have the D-configuration.

### Gliotoxin

The biosynthesis of gliotoxin is dealt with in some detail by Kirby and Robins (1980). It is the first of the mould metabolites to be recognised as belonging to the diketopiperazines with polysulphur bridges, a group which includes the aranotins, chaetomin, sirodesmins and the sporidesmins. Although the structure of gliotoxin contains a reduced indole group this is not derived from tryptophan but from phenylalanine and serine (see Figure 34) (see Bu'Lock and Ryles, 1970).

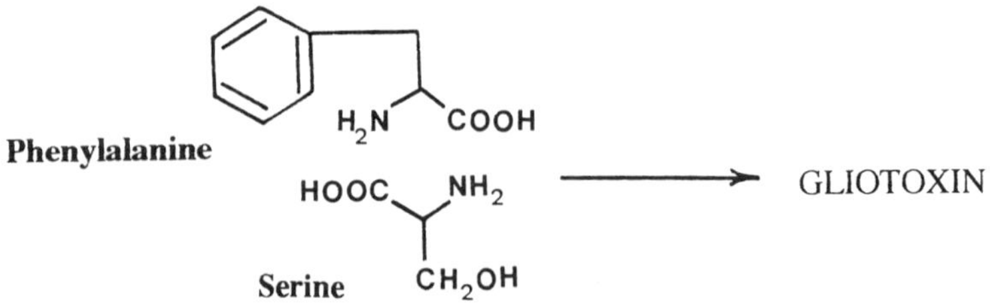


**Figure 34.** The origin of the skeleton of gliotoxin from phenylalanine and serine.

Although originally described as an antifungal metabolite of *Trichoderma lignorum* (Tode) Harz (=*T. viride* Pers.), the producing organism was subsequently identified as *Gliocladium fimbriatum* Gilman & Abbott (= *Myrothecium verrucaria* (Alb. & Schwein.) Ditmar) (Webster and Lomas, 1964). Gliotoxin is also produced by several species of *Penicillium* and *Aspergillus,* the most important of which is perhaps *A. fumigatus,* although Frisvad and Samson (1990) did not detect gliotoxin in any of 26 isolates of *A. fumigatus* and related taxa studied by them.

## Tryptoquivalines

A further reflection of the biochemical potential of *A. fumigatus* is the production of a large number of tryptoquivaline derivatives some of which are also produced by *A. clavatus*. Clardy *et al.* (1975) first isolated the tremorgenic mycotoxins tryptoquivaline itself and tryptoquivalone from a toxigenic strain of *A. clavatus* isolated from a sample of rice implicated in the death of a child in Thailand. The complex molecule which forms the basis of this group of metabolites is probably assembled from the amino acids tryptophan, valine, alanine and anthranilic acid, the latter being also an intermediate in the shikimate pathway to tryptophan (see Figure 35).

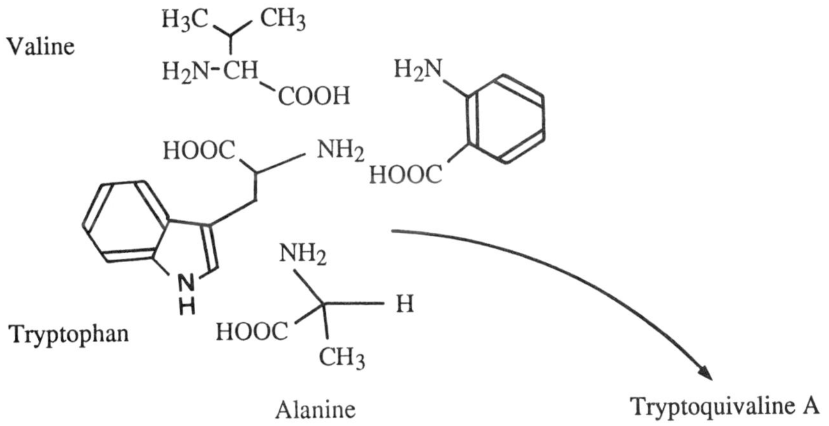


**Figure 35.** The origin of components in the structure of tryptoquivaline A

## AMINO ACID/MEVALONATE DERIVED MYCOTOXINS

The indole nucleus of tryptophan is a common feature in a wide range of fungal metabolites, especially tremorgenic mycotoxins. The rest of what are frequently complex polycyclic structures may be derived from mevalonate, or/and mevalonate provides branched $C_5$ units which are added peripherally. An example of the former is aflatrem and of the latter fumitremorgin.

### Aflatrem

Aflatrem has a special place in the study of mycotoxins being the first of the tremorgenic metabolites to be recognised. Tremorgenic activity was extracted from the mycelium and sclerotia of a strain of *A. flavus* by Wilson and Wilson (1964) and it must be recognised that the toxicology of this and related species is potentially very complex if aflatrem, cyclopiazonic acid and the aflatoxins are all produced together. Although discovered nearly thirty years ago, the structure of aflatrem was not elucidated until 14 years later (Gallagher and Wilson, 1978), a reflection of its complexity and also possibly because its importance as a metabolite of *A. flavus* was overshadowed by the concern over aflatoxin.

Aflatrem is almost certainly formed from tryptophan, geranylgeranyl pyrophosphate and mevalonate, with the loss of a methyl group during the formation of the polycyclic structure (see Figure 36). The structural elucidation (Gallagher, Clardy and Wilson, 1980) was considerably aided by the recognition that aflatrem is very closely related to the tremorgenic compound paspalinine, isolated from *Claviceps paspali* Stevens and Hall (Gallagher *et al.*, 1980). Indeed paspalinine, which could be a precursor of aflatrem, has itself been isolated from an aflatrem-producing isolate of *A. flavus*, along with the very unusual indole-mevalonate metabolites aflavinine and dihydroxyaflavinine (Cole *et al.*, 1981).

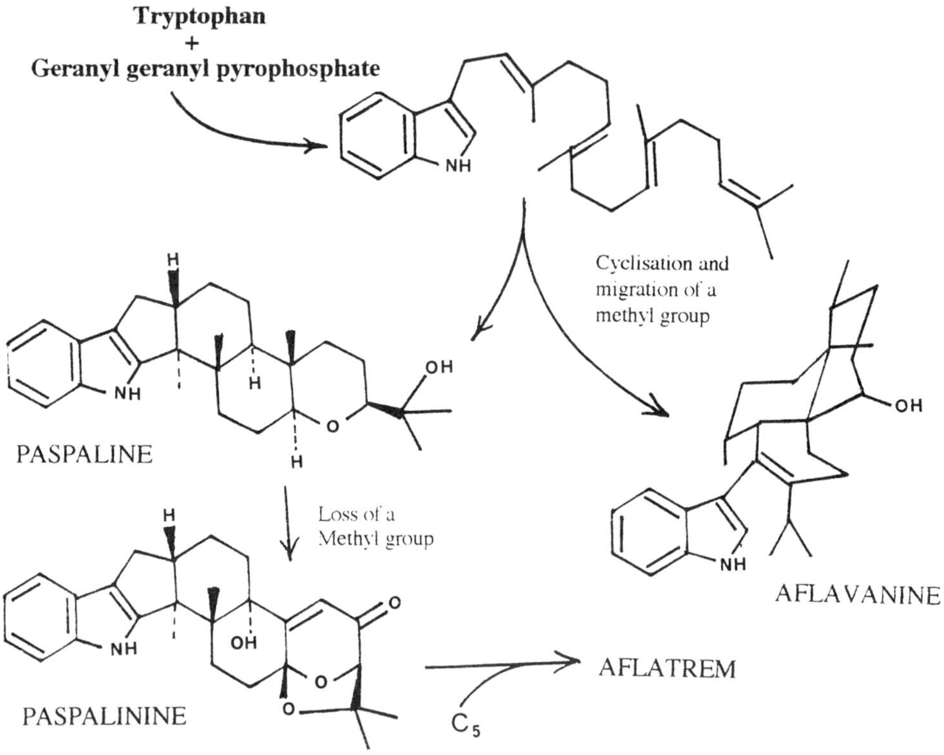


**Figure 36.** The biosynthesis of aflatrem and its relationship to paspalinine and aflavinine

## Cyclopiazonic acid

Although first isolated from *Penicillium cyclopium* Westling (= *P. aurantiogriseum* Dierckx), cyclopiazonic acid is also produced by strains of *Aspergillus flavus*, *A. oryzae* and *A. versicolor*, as well as several other species of *Penicillium*. The molecule is assembled in a step-wise manner from tryptophan and a diketide to yield α-acetyl-γ-(β-indolyl) methyltetramic acid to which is added a $C_5$ dimethylallyl group with subsequent cyclisation to yield cyclopiazonic acid (see Figure 37).

Two of the enzymes involved in the later steps of this pathway have been isolated and partially characterised (Steenkamp, Schabort and Perreira, 1973) and the stereochemistry of the cyclisation has been thoroughly studied (de Jesus *et al.*, 1981).

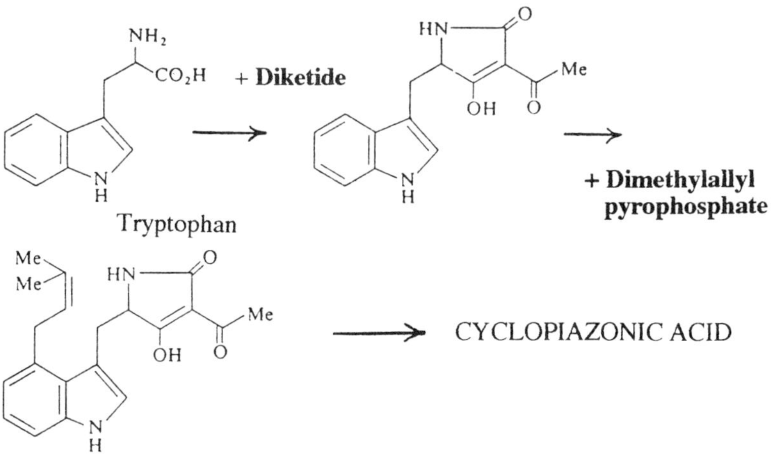


**Figure 37.** Steps in the biosynthetic pathway to cyclopiazonic acid

Of particular interest from a toxicological point of view is the possibility mentioned earlier of isolates producing several mycotoxins simultaneously. Of 56 isolates of *A. flavus* studied by Richard and Gallagher (1979), a single isolate (NRRL 3251) produced aflatrem, cyclopiazonic acid and aflatoxins whereas a further two isolates produced aflatrem and aflatoxin together but no cyclopiazonic acid. In a related study 14 of 54 isolates produced cyclopiazonic acid and aflatoxin concurrently (four produced only aflatoxin and 14 produced cyclopiazonic acid only) (Gallagher *et al.*, 1978).

### Fumitremorgins

These tremorgenic metabolites of certain strains of *A. fumigatus* are a family of compounds derived from a diketopiperazine, formed from the two amino acids tryptophan and proline, to which up to three separate mevalonate derived $C_5$ units have been added. Their discovery and structural characterisation have been concisely reviewed by Yamazaki (1980). Thus, fumitremorgin C has a single dimethylallyl group linked to carbon and nitrogen forming part of the polycyclic structure, fumitremorgin B has a second such group linked to nitrogen and fumitremorgin A has a third dimethyl allyl group linked to oxygen (see Figure 38).

## AMINO ACID/POLYKETIDE DERIVED MYCOTOXINS

Ochratoxin is an especially important nephrotoxic metabolite of a number of species of *Aspergillus* and *Penicillium*. It is a straight forward pentaketide modified by the addition of a single methyl group from the $C_1$ pool. This additional methyl group is oxidised to a carboxyl group through which L-phenylalanine is linked by an amide bond (see Figure 28).

The cytochalasan family of compounds (which includes the aspochalasins produced by *A. microcysticus* and described by Keller-Schierlein and Kupfer, 1979) are remarkable in the diversity of fungi producing them (*Phoma, Helminthosporium, Zygosporium, Metarrhizium, Chaetomium, Rosellinia* and *Aspergillus*) as well as the nature of their biological activity which includes the inhibition of movement and

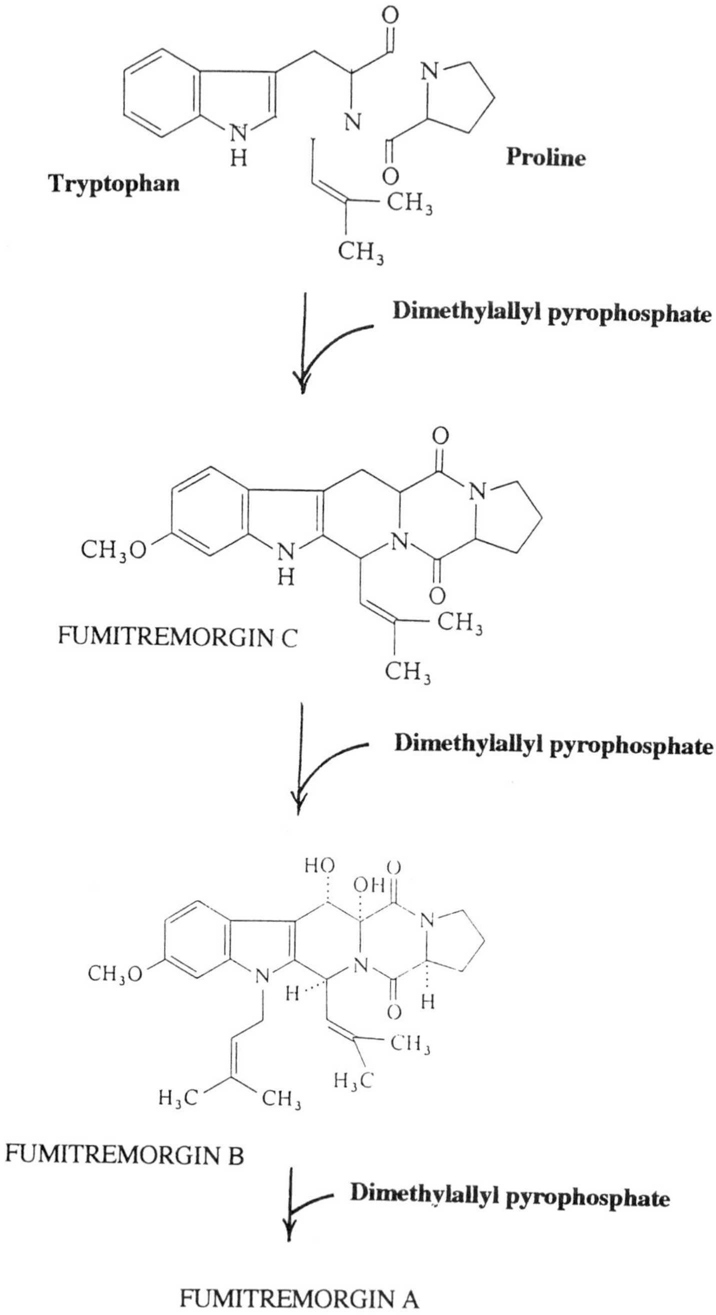


**Figure 38.** The biosynthesis and relationship between the fumitremorgins.

cytoplasmic cleavage of mammalian cells. Much of the biological activity can be accounted for by an interaction with the contractile microfilament system of eukaryote cells.

A detailed account of the biosynthetic pathways leading to the cytochalasans has been provided by Tamm (1980) and cytochalasin E, produced by *A. clavatus* (Büchi *et al.*, 1973), is produced from phenylalanine linked to an octaketide derived macrocyclic structure to which three methionine derived methyl groups have been added (see Figure 39).

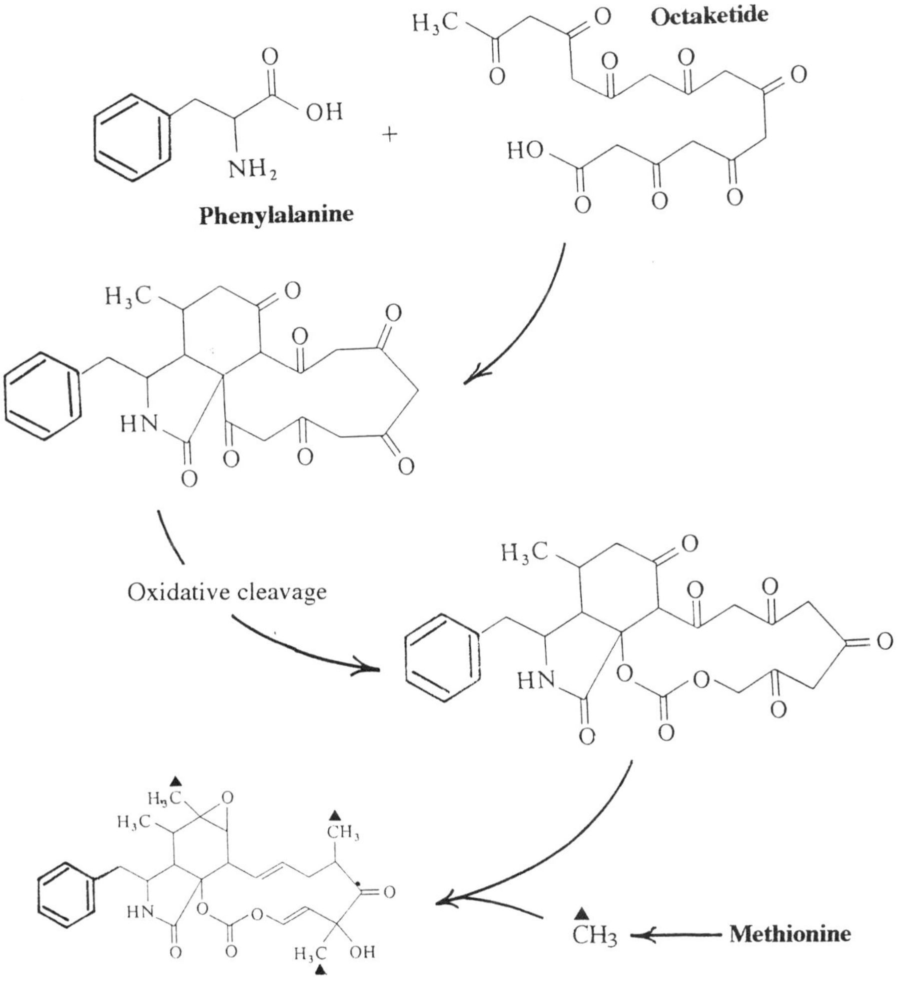


**Figure 39.** The biosynthetic pathway to cytochalasin E

## Sphingofungin

One of the most significant developments in mycotoxicology during the past few years has been the isolation and characterisation of the fumonisins from *Fusarium moniliforme* Sheldon (Bezuidenhout *et al.*, 1988). These remarkable compounds are responsible for equine encephalomalacia (a major outbreak of which occurred in the United States as recently as 1990), liver cancer in rats, pulmonary oedema in pigs and possibly oesophageal carcinoma in man. The fumonisins are potent inhibitors of the incorporation of serine into sphingosine.

Even more recently a group of simpler but similar compounds have been isolated from *A. fumigatus* because of their broad spectrum anifungal activity. Known as sphingofungins (Van Middlesworth *et al.*, 1992a) they have been shown to be specific inhibitors of serine palmitoyltransferase, an enzyme involved in an early step in the biosynthesis of sphingolipids which play an important role in such functions of the outer cytoplasmic membrane as cell recognition and response to virus infection (Zweerink *et al.*, 1992).

The determination of the absolute configuration of the carbon atoms at the carboxyl end of the molecule (Van Middlesworth *et al.*, 1992b) show it to be related to an L-amino acid suggesting that a portion of that end of the molecule may be derived from an amino acid, the rest of the molecule being a reduced polyketide.

## POSTSCRIPT

The genus *Aspergillus* has provided natural products chemists with many challenges in the diversity of complex secondary metabolites many of which have overt biological activity including mammalian toxicity. The aflatoxins are perhaps still the most important mycotoxins in terms of their impact on human health in the tropics and on the economy of international trade in food and animal feeds throughout the world. The fused difuran ring system, present in the aflatoxins, is known to occur in a number of mould metabolites produced by a diverse range of genera but, in all cases until recently this structure is fused onto another ring. Twenty years after the first report of aflatoxin, there is a report of another mycotoxin, also produced by a species of *Aspergillus*, which has a fused difuran ring system which is simply attached to an open chain and not forming part of a more complex ring system. Asteltoxin (see Figure 40) is a toxic metabolite of *A.spergillus stellatus* Curzi with features in common with citreoviridin and the aurovertins (Kruger *et al.*, 1979) but with the unique structural feature of a "naked" bisfuran ring system! The genus *Aspergillus* is going to continue to challenge natural products chemists, and biochemists interested in biosynthetic pathways,but a more important challenge now is to understand the molecular biology of secondary metabolite production and the natural history of their role in the biology of the producing organism.

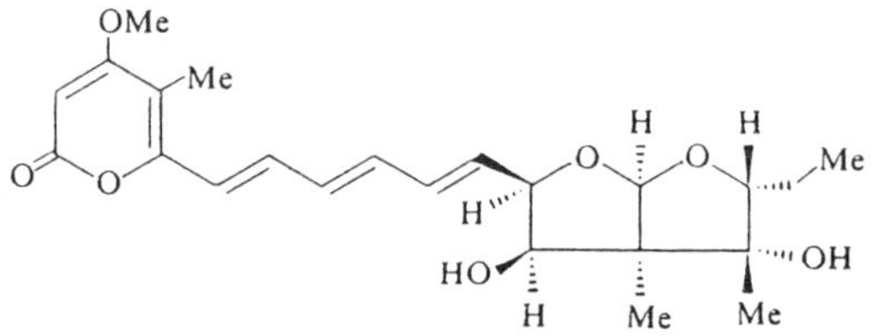


**Figure 40.** Asteltoxin

## REFERENCES


Anderegg, R.J., Biemann, K., Büchi, G. and Cushman, M. (1976) Malformin C, a new metabolite of *Aspergillus niger*. J. Amer. Chem. Soc. 98, 3365-3370.

Assante, G., Merlini, L. and Nasini, G. (1977) (+)-Abscisic acid, a metabolite of the fungus *Cercospora rosicola*. Experientia 33, 1556-1557.

Beck, J., Ripka, S. Siegner, A., Schiltz, E and Schweize, E. (1990) The multifunctional 6-methyl-salicylic acid synthase gene of *Penicillium patulum*: its gene structure relative to that of other polyketide synthases. Eur. J. Biochem. 192, 487-498.

Bezuidenhout, S.C., Gelderblom, W.C.A., Gorst-Allman, C.P., Horak, R.M., Marasas, W.F.O., Spiteller, G. and Vleggaar, R. (1988) Structure elucidation of the fumonisins, mycotoxins from *Fusarium moniliforme*. J. Chem. Soc., Chem. Comm., 743-745.

Birch, A.J. and Donovan, F.W. (1953) Studies in relation to biosynthesis I. Some possible routes to derivatives of orcinol and phloroglucinol. Aust. J. Chem. 6, 360-368.

Birch, A.J. and Hussain, S.F. (1969) Studies in relation to biosynthesis. Part XXXVIII. A preliminary study of fumagillin. J. Chem Soc (C), 1473-1474.

Birch, A.J., Massey-Westrop, R.A. and Moye, C.J. (1955) Studies in relation to biosynthesis VII. 2-hydroxy-6-methyl benzoic acid in *Penicillium griseofulvum* Dierckx. Aust. J. Chem. 8, 539-544.

Büchi, G., Kitaura, Y., Yuan, S.-S., Wright, H.E., Clardy, J., Demain, A.L., Glinsukar, T., Hunt, N., and Wogan, G.N. (1973) Structure of cytochalasin E, a toxic metabolite of *Aspergillus clavatus*. J. Amer. Chem. Soc. 95, 5423-5425.

Bu'Lock, J.D. and Ryles, A.P. (1970) The biogenesis of the fungal toxin gliotoxin: the origin of the "extra" hydrogens as established by heavy isotope labelling and mass spectrometry. J. Chem. Soc. Chem. Comm. 1404-1406

Campbell, I.M. (1983) Correlation of secondary metabolism and differentiation, in "Secondary Metabolism and Differentiation in Fungi" (Bennett, J.W. and Ciegler, A., Eds.) pp. 55-72. Marcel Dekker, New York.

Cane, D.E. and Levin, R.H. (1976) Application of Carbon-13 magnetic resonance to isoprenoid biosynthesis. II Ovalicin and the use of doubly labelled mevalonate. J. Amer. Chem. Soc. 98, 1183-1188.

Clardy, J., Springer, J.P., Büchi, G., Matsuo, K. and Wightman, R. (1975). Tryptoquivaline and tryptoquivalone, two tremorgenic metabolites of *Aspergillus clavatus*. J. Amer. Chem. Soc. 97, 663-665.

Cole, R.J., Dorner, J.W., Springer, J.P. and Cox, R.H. (1981) Indole metabolites from a strain of *Aspergillus flavus*. J. Agric. Fd. Chem. 29, 293-295.

Cundliffe, E. (1992) Self-protection mechanisms in antibiotic producers, in "Secondary Metabolites: Their Function and Evolution, Ciba Foundation Symposium 171", (Chadwick, D.J. and Whelan, J.E., Eds.), pp. 198-208. John Wiley, Chichester.

Fennell, D.I. (1977) *Aspergillus* taxonomy, in "Genetics and Physiology of *Aspergillus*" (Smith, J.E. and Pateman, J.A., Eds.), pp 1-22. Academic Press, London.

Frisvad, J.C. and Samson, .A. (1990) Chemotaxonomy and morphology of *Aspergillus fumigatus* and related taxa, in "Modern Concepts in *Penicillium* and *Aspergillus* Classification" (Samson, R.A. and Pitt, J.I., Eds.) pp 201-208. Plenum Press, New York.

Gallagher, R.T., Clardy, J. and Wilson, B.J. (1980) Aflatrem, a tremorgenic toxin from *Aspergillus flavus*. Tet. Lett. 21, 239-242.

Gallagher, R.T., Finer, J., Clardy, J., Leutwiler, A., Weibel, F., Acklin, W. and Arigoni, D. (1980) Paspalinine, a tremorgenic metabolite from *Claviceps paspali* Stevens and Hall. Tet. Lett. 21, 235-238.

Gallagher, R.T., Richard, J.L., Stahr, H.M. and Cole, R.J. (1978) Cyclopiazonic acid production by aflatoxigenic and non-aflatoxigenic strains of *Aspergillus flavus*. Mycopathologia 66, 31-36.

Gallagher, R.T. and Wilson, B.J. (1978) Aflatrem, the tremorgenic mycotoxin from *Aspergillus flavus*. Mycopathologia 66, 183-185.

Hamasaki, T., Matsui, K., Isano, K. and Hatsuda, Y. (1973). A new metabolite from *Aspergillus versicolor*. Agr. Biol. Chem. 37, 1769-1770.

Hopwood, D.A. and Khosha, C. (1992) Genes for polyketide secondary metabolic pathways in microorganisms and plants, in"Secondary Metabolites: Their Function and Evolution, Ciba Foundation Symposium 171", (Chadwick, D.J. and Whelan, J.E., Eds.), pp. 88-106. John Wiley, Chichester.

de Jesus, A.E., Steyn, P.S., Vleggaar, R., Kirby, G.W., Varley, M.J. and Ferreira, N.P. (1981) Biosynthesis of α-cyclopiazonic acid. Steric course of proton removal during cyclisation of β-cyclopiazonic acid in *Penicillium griseofulvum*. J. Chem. Soc. Perkin I, 3292-3294.

Keller-Schierlein, W. and Kupfer, E. (1979) Stoffwechselprodukte von Mikroorganismen. Uber die Aspochalasine A, B, C und D. Helv. Chim. Acta 62, 1501-1524.

Kirby, G.W. and Robins, D.J. (1980) The biosynthesis of gliotoxin and related epipolythiodioxo piperazines, in "The Biosynthesis of Mycotoxins: a Study in Secondary Metabolism" (Steyn, P.S., Ed.) pp 301-326. Academic Press, New York.

Kruger, G.J., Steyn, P.S., Vleggaar, R. and Rabie, C.J. (1979) X-ray crystal structure of asteltoxin, a novel mycotoxin from *Aspergillus stellatus* Curzi. J. Chem. Soc. Chem. Comm. 441-442.

Kurtzman, C.P., Horn, B.W. and Hesseltine, C.W. (1987) *Aspergillus nomius*, a new aflatoxin-producing species related to *Aspergillus flavus* and *Aspergillus tamarii*. Ant. van Leeuwenhoek 53, 147-158.

Ling, K.H., Liou, H.-H., Yang, C.-M. and Yang, C.-K. (1984) Isolation, chemical structure, acute toxicity, and some physicochemical properties of territrem C from *Aspergillus terreus*. Appl. Environ. Microbiol. 47, 98-100.

Ling, K.-H., Peng, F.-C., Chen, B.J., Wang, Y. and Lee, G.H. (1986) Isolation, physicochemical properties and toxicities of territrems A' and B'. Saengyak Hakhoechi 17, 153-160.

Ling, K.H., Yang, C.-K. and Peng, F.-T. (1979) Territrems, tremorgenic mycotoxins of *Aspergillus terreus*. Appl. Environ. Microbiol. 37, 355-357.

McIntyre, C.R. and Simpson, T.J. (1981) Biosynthesis of terretonin, a polyketide-terpenoid metabolite of *Aspergillus terreus*. J.Chem.Soc. Chem. Comm. 1043-1044.

Moore, R.N., Bigam, G., Chan, J.K., Hogg, A.M., Nakashima, T.T. and Vederas, J.C. (1985) Biosynthesis of the hypocholesterolemic agent mevinolin by *Aspergillus terreus*. Determination of the origin of carbon, hydrogen and oxygen atoms by carbon-13 NMR and mass spectrometry. J. Amer. Chem. Soc. 107, 3694-3671.

Moss, M.O. (1977) *Aspergillus* mycotoxins, in "Genetics and Physiology of *Aspergillus*" (Smith, J.E. and Pateman, J.A., Eds.), pp 499-524. Academic Press, London.

Moss, M.O. (1984) The mycelial habit and secondary metabolite production, in "The Ecology and Physiology of the Fungal Mycelium" (Jennings, D.H. and Rayner, A.D.M., Eds.) pp 127-142. C.U.P., Cambridge.

Nozoe, S., Kobayashi, H. and Marisaki, N. (1976) Isolation of β-trans bergamotene from *Aspergillus fumigatus*, a fumagillin producing fungus. Tet. Lett., 4625-4626.

Peng, F.-C., Ling, K.H., Wang, Y. and Lee, G.-H. (1985) Isolation, chemical structure, acute toxicity, and some physicochemical properties of territrem B' from *Aspergillus terreus*. Appl. Environ. Microbiol 49, 721-723.

Rademacher, W. and Graebe, J.E. (1979) Gibberellin A4 produced by *Sphaceloma manihoticola*, the cause of superelongation disease of cassava. Biochem. Biophys. Res. Comm. 91, 35-40.

Richard, J.L. and Gallagher, R.T. (1979) Multiple toxin production by an isolate of *Aspergillus flavus*. Mycopathologia 67, 161-163.

Riley, R.T. and Goeger, D.E. (1992) Cyclopiazonic acid: speculation on its function in fungi, in "Mycotoxins in Ecological Systems" (Bhatnagar, D., Lillehoj, E.B. and Arora, D.K., Eds.) pp 385-402. Marcel Dekker, New York.

Simpson, T.J. and Stenzel, D.J. (1981) Biosynthesis of austin, a polyketide-terpenoid metabolite of *Aspergillus ustus*. J. Chem. Soc. Chem. Comm. 1042-1043.

Steenkamp, D.J., Schabort, J.C. and Ferreira, N.P. (1973) β-cyclopiazonate oxidocyclase from *Penicillium cyclopium* III Preliminary studies on the mechanism of action. Biochim. et Biophys. acta 309, 440-456.

Tamm, Ch. (1980) The biosynthesis of the cytochalasins, in "The Biosynthesis of Mycotoxins: a Study in Secondary Metabolism" (Steyn, P.S., Ed.) pp 269-299. Academic Press, New York.

Townsend, C.A., Christensen, S.B. and Trautwein, K. (1984) Hexanoate as a starter unit in polyketide biosynthesis. J. Amer. Chem. Soc. 106, 3868-3869.

Turner, W.B. and Aldridge, D.C. (1983) "Fungal metabolites II," pp 1-631. Academic Press, London.

Van Middlesworth, F., Dufresne, C., Wincoth, F.E., Mosley, R.T. and Wilson, K.E. (1992b) Determination of the relative and absolute stereochemistry of sphingofungins A, B, C and D. Tet. Lett. 33, 297-300.

Van Middlesworth, F., Giacoble, R.A., Lopez, M., Garrity, G., Bland, J.A., Bartizal, K., Fromtling, R.A., Polshook, J., Zweerink, M., Edison, A.M., Rozdilsky, W., Wilson, K.E. and Monaghan, R.L. (1992a) Sphingofungins A, B, C and D; a new family of antifungal agents 1. Fermentation, isolation and biological activity. J. Antibiot. 45, 861-867.

Vederas, J.C. (1986) Biosynthetic studies on mycotoxins using multiple stable isotope labelling and NMR spectroscopy, in "Mycotoxins and Phycotoxins" (Steyn, P.S. and Vleggaar, R., Eds.) pp 97-108. Elsevier, Amsterdam.

Vining, L.C. (1992) Role of secondary metabolites from microbes, in "Secondary Metabolites: Their Function and Evolution, Ciba Foundation Symposium 171", (Chadwick, D.J. and Whelan, J.E., Eds.), pp. 184-194. John Wiley, Chichester.

Webster, J. and Lomas, L. (1964) Does *Trichoderma viride* produce gliotoxin and viridin? Trans. Brit. Mycol. Soc. 47, 535-540.

Wilson, B.J. and Wilson, C.H. (1964) Toxin from *Aspergillus flavus*: production on food materials of a substance causing tremors in mice. Science 144, 177-178.

Yamazaki, M. (1980) The biosynthesis of neurotropic mycotoxins, in "The Biosynthesis of Mycotoxins: a Study in Secondary Metabolism" (Steyn, P.S., Ed.) pp 193-222. Academic Press, New York.

Zamir, L.O. (1980) The biosynthesis of patulin and penicillic acid, in "The Biosynthesis of Mycotoxins: a Study in Secondary Metabolism" (Steyn, P.S., Ed.) pp 223-268. Academic Press, New York.

Zweerink, M.M., Edison, A.M., Wells, G.B., Pinto, W. and Lester, R.L. (1992) Characterisation of a novel, potent, and specific inhibitor of serine palmitoyltransferase. J. Biol. Chem. 267, 25032-25038.

# THE MOLECULAR GENETICS OF AFLATOXIN BIOSYNTHESIS

J.W. Bennett,[1] D. Bhatnagar,[2] and P.K. Chang[2]

[1]Department of Cell and Molecular Biology
Tulane University, New Orleans, LA 70118 USA
[2]Southern Regional Research Center
P.O. Box 19687 New Orleans, LA 70179 USA

## INTRODUCTION

Mycotoxins are toxic compounds produced *by* fungi, in contrast, for example, to phytotoxins, compounds which are toxic *to* plants. Chemically, mycotoxins are a heterogeneous group of secondary metabolites with a diverse array of adverse pharmacological effects in human beings and many domesticated animal species.

Mycotoxins arise through a variety of biosynthetic pathways, especially along the polyketide route, the amino acid route, the terpene route, or through a combination of these pathways. Mycotoxin biosynthetic studies have benefited from the decades of research in antibiotic chemistry: appropriate strategies have been adopted for delivering isotopically labelled precursors, culturing filamentous micro-organisms, and utilising blocked mutants and metabolic inhibitors (Steyn, 1980). In terms of biochemistry and genetics, mycotoxin studies also suffer the same handicaps as antibiotics research: the multiple enzymatic steps interact in complex ways to produce taxonomically specific compounds; individual enzymes of the pathways are difficult to isolate; secondary metabolites are not direct gene products so molecular genetic manipulations at the level of proteins or nucleic acids require sophisticated strategies.

The best known mycotoxins are the aflatoxins which are produced by certain strains of *Aspergillus flavus* Link and *Aspergillus parasiticus* Speare. This family of secondary metabolites has attracted considerable research because of their extreme toxicity, mutagenicity, and carcinogenicity. The producing mould species are both common contaminants of foods and feeds, posing a threat to human and domestic animal health. Since their discovery in 1960, these mycotoxins have been actively studied by basic chemists and biologists, as well as a wide array of applied scientists in agriculture, veterinary medicine, epidemiology and related fields. For reviews see Betina (1989), Bray and Ryan (1991), Bhatnagar *et al.* (1992a) and the citations therein.

Statistical correlations between contaminated food supplies and high frequencies of hepatocarcinoma in Africa and Asia have long implicated aflatoxins as risk factors in human liver cancer. A molecular "hot spot" in the p53 gene, a transversion at the third base position (a guanine) of codon 249, has now been identified in independent studies on patients from Qidong, China (Hsu *et. al.*, 1991) and from sub-Saharan Africa (Bressac *et al.*, 1991). These studies constitute among the most convincing demonstrations that a natural product in the environment can induce a specific form of cancer.

In addition to being useful models for studying chemical carcinogenesis, aflatoxins have been used to elucidate new aspects of polyketide biosynthesis, and most recently, have provided a system for analysing the molecular biology of a fungal secondary metabolite.

*The Genus Aspergillus*, Edited by Keith A. Powell *et al.*,
Plenum Press, New York, 1994

## CLASSICAL GENETICS

As imperfect (anamorphic) species, neither *A. flavus* nor *A. parasiticus* can bc crossed by sexual means. Prior to the advent of recombinant DNA research, the only genetic resource available was the parasexual cycle. Utilizing auxotrophic, spore colour, and blocked aflatoxin markers, the parasexual cycle was demonstrated in aflatoxigenic strains of both *A. flavus* and *A. parasiticus*. Because both species contain multinucleate conidiospores, and because nonrandom recovery of segregants from diploids is common for many genotypes from both species, the parasexual cycle is a slow, tedious, and difficult system for doing genetic studies on toxin inheritance. Nonetheless, rudimentary linkage maps are available for *A. flavus* and certain genetic linkages are known in *A. parasiticus*. See Bennett and Deutsch (1986) and Bennett and Papa (1988) for reviews of this rather esoteric research.

These classical genetic studies, largely conducted during the 1970's, are most important in that they provide the raw material for speeding up other research. Mutants defective in aflatoxin synthesis, isolated on the basis of altered fluorescence, have been a mainstay of studies on aflatoxin biosynthesis. More recently, various nutritionally deficient and colour mutants have been employed for aflatoxin molecular genetics. The advent of electrophoretic karyotyping utilising pulsed field gel electrophoresis promises a way of doing direct physical mapping of genes to chromosomes (Keller *et al.*, 1992a), further building on the foundation grounded in the old genetics.

## AFLATOXIN BIOSYNTHESIS

### History

For several reasons, aflatoxin biosynthesis is one of the best understood fungal polyketide pathways. Government agencies have been relatively generous in funding research because of the widespread occurrence of the producing fungi and the biological potency of the toxin itself. In addition, many aflatoxin pathway intermediates are brightly coloured anthraquinones, which facilitates purification, the isolation of blocked mutants, and the design of complementation tests.

The four major naturally occuring aflatoxins are called $B_1$, $B_2$, $G_1$, and $G_2$ based on their characteritistic blue or blue-green fluoresence and relative chromatographic mobility. Amounts of the four major aflatoxins vary with both the genetic constitution of the strain and with the cultural conditions of fungal growth. Almost all of the research has focused on biosynthesis of the B aflatoxins, which are the main congeners produced by *A. flavus*, and the most abundant ones produced by *A. parasiticus*.

Aflatoxins are formed by head-to-tail condensation of acetyl units to form a cyclised polyketide which is enzymatically altered through a series of intermediates. Mutant blocks have been identified at norsolorinic acid, averantin, averufin, hydroxyversicolorin, versicolorin A, and O-methyl sterigmatocysin. The presence of the esterase inhibitor, dichlorvos, leads to the accumulation of versiconal hemiacetal acetate. The overall, accepted scheme for aflatoxin $B_1$ biosynthesis is: acetate + malonate --> linearised polyketide chain --> anthraquinones (e.g., norsolorinic acid, averantin, averufin, versiconals, and versicolorins) --> xanthones (sterigmatocystins) --> aflatoxin (See Figure 1.). Details of the scheme, with citations to relevant experimental papers and reviews are given by Steyn (1980), Bennett and Christensen (1983), and Bhatnagar *et al.* (1992b).

### Polyketide synthase

Polyketides constitute one of the largest families of secondary metabolites. Polyketide synthesis shares many features with fatty acid synthesis, and it has long been theorised that the two pathways share a common evolutionary origin. The enzyme systems that catalyse the reactions leading to fatty acid production are called fatty acid synthases. Analogously, the enzyme systems in polyketide biosynthesis are called polyketide synthases. Since fatty acids are primary metabolites, found universally in living systems, there has been far more research on these pathways than on polyketide synthases, where the end product secondary metabolites display sporadic, often species-specific, distribution (Turner, 1976; Hopwood and Sherman, 1990; Hopwood and Khosla, 1992).

Molecular techniques have greatly enhanced our understanding of microbial polyketide synthases, which were first cloned and sequenced from streptomycete species. In many cases, the sequential order of catalytic domains is similar between fatty acid and polyketide synthase (Hopwood and Khosla, 1992).

The best studied fungal secondary metabolites are not polyketides, but rather the β-lactam antibiotics such as the penicillins and the cephalosporins. Several peptide synthases in the β-lactam pathway show a high degree of sequence similarity between the genes of streptomycetes and fungi, leading some workers to hypothesise a horizontal transfer of this secondary pathway between prokaryotes and eukaryotes (Turner, 1992). In hopes that a similar horizontal transfer might have occurred for polyketide synthase genes, our lab used an *actI/actIII* heterologous probe from *Streptomyces coelicolor* in an unsuccessful attempt to find a related gene in genomic DNA of *A. parasiticus* by Southern blotting (Kale, S. and Bennett, J.W., unpublished).

The simple tetraketide 6-methylsalicylic acid has long been a model compound in polyketide research, so it is perhaps fitting that the first fungal polyketide synthase gene cloned is 6-methylsalicylic acid polyketide synthase from *Penicillium patulum* and *Penicillium urticae* (Beck *et al.*, 1990; Wang *et al.*, 1992). This 6-methylsalicylic acid polyketide synthase gene shows more homology with vertebrate fatty acid synthases than with fungal fatty acid synthases, raising interesting questions about evolutionary origins of the respective pathways. More practically, this gene sequence may now be used as a probe to aid in the isolation of related genes from other fungi.

Preliminary cross hybridisation to *A. parasticus* genomic digests with the *P. urticae* probe revealed some weak cross hybridising bands (Wang, I. and Gaucher, G.M., unpublished). Numerous moulds accumulate mycelial anthraquinones similar or identical to early aflatoxin pathway intermediates (Turner and Aldridge, 1983), which would imply that the polyketide synthase gene family that yields these $C_{20}$ metabolites is widely distributed in filamentous fungi. Moreover, several tantalising recent studies indicate that enzymes involved in producing asexual and sexual spore colour pigments in *A. parasiticus* and *A. nidulans* are also polyketide synthases (Mayorga and Timberlake, 1992; Brown, *et al.*, 1993; Brown, D.W. and Salvo, J., unpublished). The next few years promise to be a fruitful time for research on polyketide-generating enzyme systems in fungi.

**Origin of bisfuran and parallel pathways to aflatoxins $B_1$ and $B_2$**

Townsend's group at Johns Hopkins University in Baltimore, Maryland, has pioneered investigations on the series of oxidative steps involved in the rearrangement of averufin to versiconal and then the formation of the characteristic bisfuan moiety in either the dihydrobisfuran found in versicolorin A, sterigmatocystin, O-methyl sterigmatocystin and aflatoxin $B_1$ or the tetrahyrobisfuran found in versicolorin B, dihydrosterigmatocystin, dihydro-O-methylsterigmatocystin and aflatoxin $B_2$ (McGuire *et al.*, 1989; Townsend *et al.*, 1992). Earlier work using cell free cultures, by Anderson's group at Texas Tech University in Lubbock, Texas (Anderson and Chung, 1990; Anderson *et al.*, 1990), and Hsieh's group at The University of California, Davis, California (Wan and Hsieh, 1980; Hsieh *et al.*, 1989) had shown indirectly that the steps involved in forming the bisfuran moiety involve a number of intermediates not identified in blocked mutants. Work by Townsend's group clearly identified versicolorin B (a racemate of versicolorin C) as a key branching point in the pathway (Mcguire *et al.*, 1989; Townsend *et al.*, 1992), and a desaturase catalysed transformation of versicolorin B to versicolorin A was demonstrated by Yabe *et al.*, (1991).

Simultanous work, utilising precursor feeding studies into mutants blocked at several late stages of the pathway, has been conducted by Bhatnagar and Cleveland's group at the Southern Regional Research Center, New Orleans, Louisiana (Bhatnagar *et al.*, 1991; Bhatnagar *et al.*, 1992b and references cited therein) and Yabe's group at the National Institute of Animal Health, Tsukuba City, Japan (Yabe *et al.*, 1988, 1989). Independently and in mutually reinforcing experiments, these groups have shown that aflatoxins $B_1$ and $B_2$ are produced by parallel pathways. The same methyltransferase catalyses the dihydrobisfuran reaction (sterigmatocystin to O-methylsterigmatocystin) as catalyses the tetrahydrobisfuran reaction (dihydrosterigmatocystin to dihydro-O-methylsterigmatocystin). Similarly, the same oxidoreductase activity transforms both O-

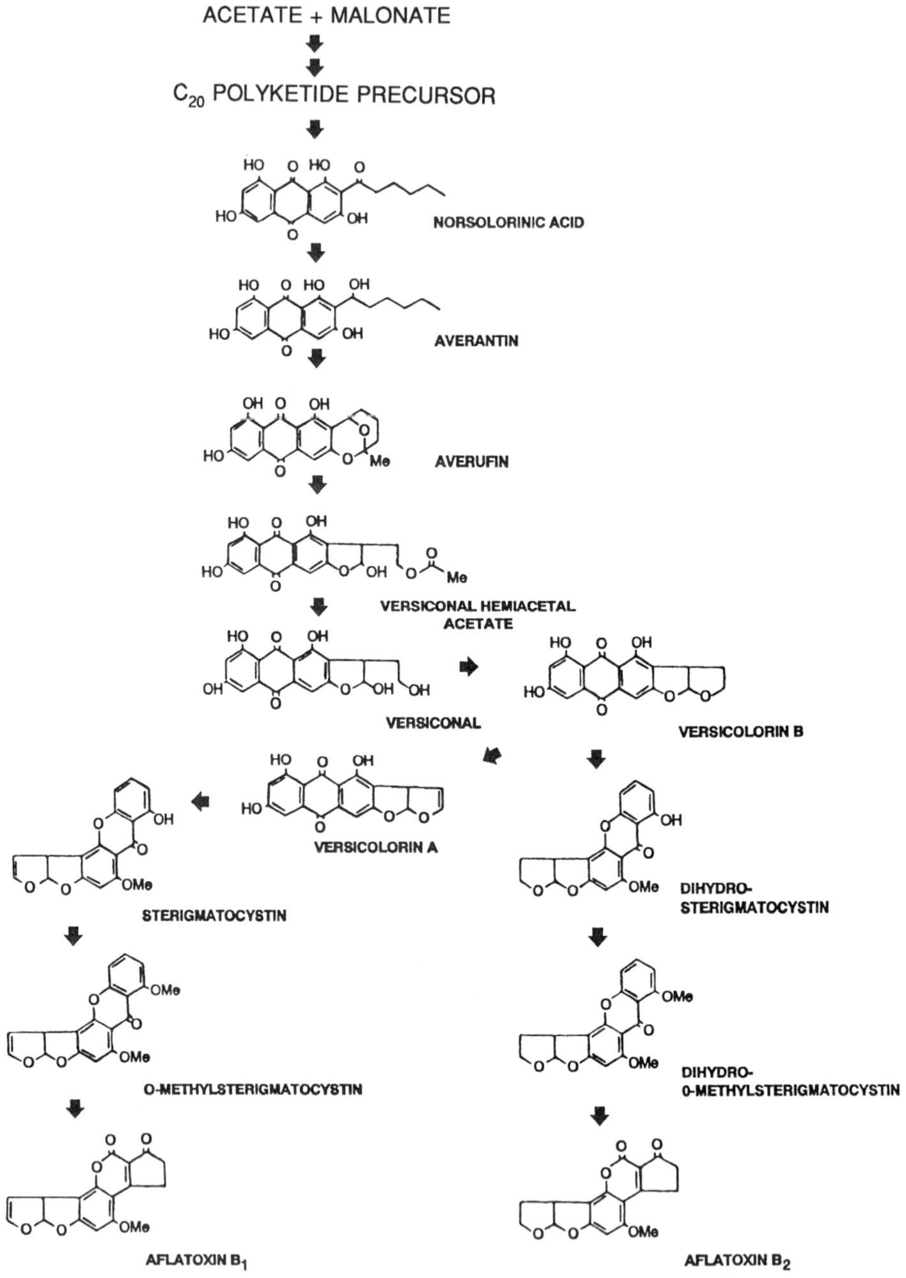


**Figure 1.** Aflatoxin biosynthetic pathway.

methylsterigmatocystin into aflatoxin $B_1$ and dihydro-O-methylsterigmatocystin into aflatoxin $B_2$. In other words, the fidelity of the enzymes is not absolute; one set of enzymes yields more than one end product, with the quantity and structure of the substrates determining the quantity and structure of the products. Many secondary metabolites other than aflatoxins are produced in similar families of congeners, but there are few for which the experimental data supporting parallel pathways are so clear.

## MOLECULAR BIOLOGY

A long time goal in aflatoxin research has been the cloning and sequencing of the relevant pathway genes. The development of transformation systems for both *A. flavus* (Woloshuk *et al.*, 1989) and *A. parasiticus* (Skory *et al.*, 1990) are important landmarks. With the help of appropriate selectable markers, exogenous DNA can be introduced into these species. However, development of efficient gene transfer systems is only one of the tools needed for taking advantage of the power of molecular genetic analysis.

Genes can be isolated in many ways, and fungal geneticists have adopted virtually all of these methods. See Leong and Berka (1991), Bennett and Lasure (1991), and Peberdy *et al.* (1992) for general reviews. Depending on the resources available, strategies at the DNA level might include the use of heterologous sequences to probe libraries or the use of PCR primers to pull out desired sequences. RNA based strategies involve the construction of cDNA expression libraries or substractive hybridisation studies. Oligonucleotide probes can be derived from purified proteins or proteins can be prepared for antibody detection in expression libraries. Complementation tests are useful in a variety of situations where appropriate selectable markers are available, and for clustered genes, chromosome walking can be employed. Virtually all of these approaches have been suggested and tried with respect to the genes of aflatoxin biosynthesis (Keller *et al.*, 1992b; Feng *et al.*, 1992).

Polyketide synthase research has been described above, with the caveat that this field is moving so fast that any review is immediately dated. The rest of this discussion will focus on the genes of later steps of aflatoxin biosynthesis, of which several have now been identified and cloned.

The enzymes of secondary metabolism are notoriously difficult to purify. Cell free extracts are often used *in lieu* of purified enzymes and, indeed, research on the aflatoxin biosynthetic pathway has relied heavily on biotransformation studies in which putative precursors are fed to blocked mutants and metabolic products monitored by chromatography (Dutton, 1988). Although it had been predicted that known enzyme sequences would be used to produce probes for isolating aflatoxin pathway genes, and although several enzymes of the aflatoxin pathway have now been purified to homogeneity, the first aflatoxin pathway genes to be cloned were all isolated by complementation.

Perhaps fittingly, the first gene of the aflatoxin pathway to be cloned concerns norsolorinic acid, which is the first stable cyclisation product detected in aflatoxin biosynthesis. A cosmid library was constructed by inserting wild type *A. parasiticus* genomic DNA into a vector containing a nitrate reductase gene as a selectable marker. A sibling selection strategy was used to recover an aflatoxin-positive cosmid which complemented a norsolorinic acid-producing (=aflatoxin defective) nitrate-nonutilising strain (Chang *et al.*, 1992). The gene has been called both *nar-1* and *nor-1*. A similar strategy, using a cosmid genomic library with the homologous gene (*pyrG*) encoding orotidine monophosphate decarboxylase for the selection of transformants, was used to isolate a gene that encodes an activity associated with the conversation of versicolorin A to sterigmatocystin termed *ver-1* (Skory *et al.*, 1992). The predicted amino acid sequence of the *ver-1* protein shows high sequence similarity with many reductases and dehydrogenases in the EMBL and GenBank data bases. Interestingly, the $pyrG^+$ $Afl^+$ transformants of the versicolorin-deficient recipient strain produce abundant levels of sclerotia (Skory *et al.*, 1992).

A gene involved in aflatoxin biosynthesis has also been cloned from *A. flavus* (Payne *et al.*, 1993). Using blocked aflatoxin strains from the culture collection of the late K. E. Papa and a genomic DNA library from aflatoxigenic *A. flavus* in a cosmid vector containing the *pyr-4* gene of *Neurospora crassa* a cosmid containing the *afl-2* gene was isolated by sib selection. Further complementation tests of the mutated *afl-2* gene in a double mutant containing *afl-2* and a lesion in the gene for norsolorinic acid suggested that the product of the gene works at a step in the pathway before norsolorinic acid. However, the gene seems to be a regulatory sequence rather than part of the aflatoxin pathway polyketide synthase (Payne *et al.*, 1993). Quite recently, our group has cloned a regulatory gene from *A. parasiticus* by transformation of strains with a cosmid containing both *nor-l* and *ver-l*. This gene, termed *apa-2*, shows greater than 95% DNA homology with the *A. flavus afl-2* sequence (Chang *et al.*, 1993). Transformation with *apa-2* of several strains containing aflatoxin biosynthetic genes resulted in overproduction of pathway intermediates. Moreover, the *A. parasiticus apa-2* gene complemented the *A. flavus afl-2* mutant to aflatoxin production, suggesting that the genes function in the same or similar ways in the two aflatoxigenic species (Chang *et al.*, 1993).

A methyltransferase which catalyses the conversion of sterigmatocystin (dihydrosterigmatocystin) to O-methylsterigmatocystin (dihydro-O-methylsterigmatocystin) has been purified from an averantin-accumulating strain of *A. parasiticus*. This S-adenosylmethionine-dependent 40-kilodalton enzyme was purifed to >90% homogeneity and the N-terminal sequence was determined (Keller *et al.*, 1993). A polyclonal antibody was raised against the 40-kilodalton methyltransferase and used to screen a cDNA library from wild type aflatoxigenic *A. parasiticus* (Yu *et al.*, 1993). The deduced amino acid sequence of a gene isolated, *omt-1*, matched a 19 amino acid region of the 22-N-terminl amino acid sequence obtained by direct sequencing of the mature protein (Keller *et al.*, 1993).

What does the future hold? Virtually all of the labs that have been working in aflatoxin biosynthetic research are still active and now using molecular approaches. The clustering of known pathway genes increases the likelihood that during the next few years most of the relevant pathway enzymes will be cloned and sequenced. This will allow comparisons to be made with other polyketide pathways and extend our basic understanding of secondary metabolism. It is also possible that the breakthroughs in molecular genetics will lead to rational methods for controlling the production of this family of potent fungal toxins.

## SUMMARY

Aflatoxins are a serious hazard to food safety and a beguiling challenge to scientists interested in the subtleties of fungal metabolism. It is anticipated that a combination of classical and molecular techniques will ultimately provide practical insights into both the regulation and evolution of aflatoxins. Since aflatoxins are economically important, funds have been available to support fundamental studies in biosynthesis and molecular genetics. For a quarter century, aflatoxin research has been a model system for chemical carcinogenesis and polyketide biosynthesis. The recent isolation of the *afl-2* and *apa-2* genes indicates that the aflatoxin system now may also become a model for studying regulation of complex pathways in filamentous fungi.

## ACKNOWLEDGEMENTS

We thank John Anderson, Maurice Gaucher, and Joseph Salvo for providing reprints and unpublished material; Ed Cleveland, Michael Dutton, Dennis Hsieh, Nancy Keller Louise Lee, John Linz, Gary Payne, Craig Townsend, and their students and postdoctoral associates for their long term collaborations; and John Drwiega for manuscript preparation. Portions of this work were sponsored by a Collaborative Agreement between the U. S. Department of Agriculture and Tulane University (532243).

## REFERENCES

Anderson, J.A. and Chung, C.H. (1990) Conversion of versiconal acetate to versiconal and versicolorin C in extracts from *Aspergillus parasiticus*. Mycopathologia 110, 31-35.

Anderson, J.A., Chung, C.H. and Cho, S. (1990) Versicolorin A hemiacetal, hydroxydihydrosterigmatocystin, and aflatoxin $G_{2a}$ reductase activity in extracts from *Aspergillus parasiticus*. Mycopathologia 111, 39-45.

Beck, J., Ripka, S., Siegner, A., Sclitz, E. and Schweizer E. (1990) The multifunctional 6-methyl-salicylic acid synthase gene of *Penicillium patulum:* its gene structure relative to that of other polyketide synthases. Eur. J. Biochem. 192, 487-498.
Bennett, J.W. and Christensen, S.B. (1983) New perspectives on aflatoxin biosynthesis. Adv. Appl. Microbiol. 19, 53-92.
Bennett, J. W. and Deutsch, E. (1986) Genetics of mycotoxin biosynthesis, in "Mycotoxins and Phycotoxins" (Steyn, P. S. and Vlegger, R., Eds.), pp. 22-25. Elsevier, Amsterdam.
Bennett, J. W. and Lasure, L. S. (1992) "More Gene Manipulations in Fungi." Academic Press, San Diego.
Bennett, J. W. and Papa, K. E. (1988) The aflatoxigenic *Aspergillus* spp. Adv. Plant Path. 6, 263-280.
Betina, V. (1989) "Mycotoxins: Chemical, Biological and Environmental Aspects." Elsevier, Amsterdam.
Bhatnagar, D., Cleveland, T. E. and Kingston, D. G. I. (1991) Enzymological evidence for separate pathways for aflatoxin $B_1$ and $B_2$ biosynthesis. Biochem. 30, 4343- 4350.
Bhatnager, D., Lillehoj, E. B. and Arora, D. K. (1992a) "Handbook of Applied Mycology." Marcel Dekker, New York.
Bhatnagar, D., Ehrlich, K.C. and Cleveland, T.E. (1992b) Oxidation-reduction reactions in biosynthesis of secondary metabolites, in "Handbook of Applied Mycology, Volume 5: Mycotoxins in Ecological Systems" (Bhatnagar, D., Lillehoj, E.B. and Arora, D.K., Eds.), pp. 255-286. Marcel Dekker, New York.
Bray, G.A. and Ryan, D.H. (1991) "Mycotoxins, Cancer, and Health: Vol. 1." Louisiana State University Press, Baton Rouge.
Bressac, B., Kew, M., Wands, W. and Ozturk, M. (1991) Selective G to T mutations of p53 gene in hepatocellular carcinoma from southern Africa. Nature 350, 429-431.
Brown, D. W. , Hauser, F. M., Tommasi, R., Corlett, S. and Salvo, J. J. (1993) Structural elucidation of a putative conidial pigment intermediate in *Aspergillus parasiticus*. Tet. Lett. 34, 419-422.
Chang, P., Skory, C.D. and Linz, J.E. (1992) Cloning of a gene associated with aflatoxin $B_1$ biosynthesis in *Aspergillus parasiticus*. Curr. Genet. 21, 231-233.
Chang, P., Cary, J. W., Bhatnagar, D., Cleveland, T.E., Bennett, J.W., Linz, J.E., Woloshuk, C.P. and Payne, G.A. (1993) Cloning of the *Aspergillus parasiticus apa-2* gene associated with the regulation of aflatoxin biosynthesis. Appl. Environ. Microbiol. (Accepted).
Dutton, M.F. (1988) Enzymes and aflatoxin biosynthesis. Microbiol. Rev. 52, 274-295.
Feng, G. H., Chu, F. C. and Leonard, T.J., (1992) Molecular cloning of genes related to aflatoxin biosynthesis by differential screening. Appl. Environ. Microbiol. 58, 455-460.
Hopwood, D. A. and Sherman, D. H. (1990) Molecular genetics of polyketides and its comparison to fatty acid biosynthesis. Ann. Rev. Genet. 24, 37-66.
Hopwood, D.A. and Khosla, C. (1992) Genes for polyketide secondary metabolic pathways in microorganisms and plants, in "Secondary Metabolites: Their Function and Evolution" (Ciba Foundation Symposium 171), pp. 88-112. Wiley, Chichester.
Hsieh, D.P.H., Wan, C.C. and Billington, J.A. (1989) A versiconal hemiacetal acetate converting enzyme in aflatoxin biosynthesis. Mycopathologia 107, 121-126.
Hsu, I.C., Metcalf, R. A., Sun, T., Welsh, J.A., Wang, N.J. and Harris, C.C. (1991) Mutational hotspot in the p53 gene in human hepatocellular carcinomas. Nature 350, 427-428.
Keller, N.P., Cleveland, T.E. and Bhatnagar, D. (1992a) Variable electrophoretic karyotypes of members of *Aspergillus* section *flavi*. Curr. Genet. 21, 371-375.
Keller, N.P., Cleveland, T.E. and Bhatnagar, D. (1992b) A molecular approach towards understanding aflatoxin production, in "Handbook of Applied Mycology, Volume 5: Mycotoxins in Ecological Systems" (Bhatnagar, D., Lillehoj, E.B. and Arora, D.K., Eds.), pp. 287-310. Marcel Dekker, New York.
Keller, N.P., Dischinger, J.R., Bhatnagar, D., Cleveland, T. E. and Ullah, A. H. J. (1993) Purification of a 40-kilodalton methyltransferase active in the aflatoxin biosynthetic pathway. Appl. Environ. Microbiol. 59, 479-484.
Leong, S. A. and Berka, R. M. (1991) "Molecular Industrial Mycology. Systems and Applications for Filamentous Fungi." Marcel Dekker, New York.
Mayorga, M.E. and Timberlake, W.B. (1992) The develomentally regulated *Aspergillus nidulans wA* gene encodes a polypeptide homologous to polyketide and fatty acid synthases. Mol. Gen. Genet. 235, 205-212.
McGuire, S.M., Brobst, S.W., Graybill, T.L., Pal, K. and Townsend, C.A. (1989) Partitioning of tetrahydo- and dihydrobisfuran formation in aflatoxin biosynthesis defined by cell-free and direct incorporation experiments. J. Am. Chem. Soc. 111, 8308-8309.
Payne, G.A., Nystrom, G.J., Bhatnagar, D., Cleveland, T.E. and Woloshuk, C.P. (1993) Cloning of the *afl-2* gene involved in aflatoxin biosynthesis from *Aspergillus flavus*. Appl. Environ. Microbiol. 59, 156-162.
Peberdy, J. F., Caten, C. E., Ogden, J. E. and Bennett, J. W. (1991) "Applied Molecular Genetics of Fungi." Cambridge University Press, Cambridge.
Skory, C. D., Horng, J. S., Pestka, J. J. and Linz, J. E., (1990) Transformation of *Aspergillus parasiticus* with a homologous gene (pyG) involved in pyrimidine biosynthesis. Appl. Environ. Microbiol. 56, 3315-3320.
Skory, C.D., Chang, P., Cary, J. and Linz, J.E. (1992) Isolation and characterization of a gene from

*Aspergillus parasiticus* associated with the conversion of versicolorin A to sterigmatocystin in aflatoxin biosynthesis. Appl. Environ. Microbiol. 58, 3527-3537.

Steyn, P. S. (1980) "The Biosynthesis of Mycotoxins: A Study in Secondary Metabolism." Academic Press, New York.

Townsend, C.A., McGuire, S.M., Brobst, S.W., Graybill, T.L., Pal, K. and Barry, III, C.E. (1992) Examination of tetrahydro- and dihydrobisfuran formation in aflatoxin biosynthesis: from whole cells to purified enzymes, in "Secondary-Metabolite Biosynthesis and Metabolism" (Petroski, R.J. and McCormick, S.P., Eds.), pp. 141-154. Plenum Press, New York.

Turner, G. (1992) Genes for the biosynthesis of β-lactam compounds in microorganisms, in "Secondary Metabolites: Their Function and Evolution" (Ciba Foundation Symposium 171), pp. 113-128. Wiley, Chichester.

Turner, W.B. (1976) Polyketides and related metabolites, in "The Filamentous Fungi: Biosynthesis and Metabolism, Vol. 2" (Smith, J.E. and Berry, D.R., Eds.), pp. 445-459. Arnold, London.

Turner, W.B. and Aldridge, D.C. (1983) "Fungal Metabolites, Vol. 2." Academic, London.

Wan, N.C. and Hsieh, D.P.H. (1980) Enzymatic formation of the bisfuran structure in aflatoxin biosynthesis. Appl. Environ. Microbiol. 39, 109-112.

Wang, I., Reeves, C. and Gaucher, G.M. (1990) Isolation and sequencing of a genomic DNA clone containing the 3' terminus of the 6-methylsalicylic acid polyketide synthetase gene of *Penicillium urticae*. Can. J. Microbiol. 37, 86-95.

Woloshuk, C. P., Seip, E. R., Payne, G. A. and Adkins, C. R. (1989) Genetic transformation system for the aflatoxin-producing fungus *Aspergillus flavus*. Appl. Environ. Microbiol. 55, 86-90.

Yabe, K., Ando, Y. and Hamasaki, T. (1988) Biosynthetic relationship among aflatoxins $B_1$, $B_2$, $G_1$, and $G_2$. Appl. Environ. Microbiol. 54, 2101-2106.

Yabe, K., Ando, Y., Hashimoto, J. and Hamasaki, T. (1989) Two distinct O-methyltransferases in aflatoxin biosynthesis. Appl. Environ. Microbiol. 55, 2172-2177.

Yabe, K., Ando, Y. and Hamasaki, T. (1991) Desaturase activity in the branching step between aflatoxins $B_1$ and $G_1$ and aflatoxins $B_2$ and $G_2$. Agric. Biol. Chem. 55, 1907-1911.

Yu, J., Cary, J.W., Bhatnagar, D., Cleveland, T.E., Keller, N.P. and Chu, F.S. (1993) Cloning and characterization of a gene for an O-methyltransferase involved in the conversion of sterigmatocystin to O-methylsterigmatocystin in aflatoxin biosynthesis from *Aspergillus parasiticus*. Appl. Environ. Microbiol. (Accepted).

# *ASPERGILLUS* TOXINS IN FOOD AND ANIMAL FEEDINGSTUFFS

Keith A. Scudamore

Central Science Laboratory
Ministry of Agriculture, Fisheries and Food
London Road
Slough, SL3 7HJ
Berkshire, UK

## INTRODUCTION

When moulds grow, many different metabolites may be produced. The number of these recorded in the literature runs into thousands and many more undoubtedly await identification and investigation. Secondary fungal metabolites can be regarded as chemicals produced which are not necessary for the growth of the producing organism. They are a group of sustances with a diverse range of structures, and chemical and physical properties. Amongst them are compounds of considerable current benefit to man and others with still unrecognised potential. In contrast, some are extremely toxic and their physiological effects are as wide in action as their properties. These latter substances are known as mycotoxins.

A number of common genera of fungi are rich sources of mycotoxins and most of the best known and studied are elucidated by species of *Aspergillus*, *Penicillium*, and *Fusarium*. However many other species including, *Claviceps*, *Alternaria, Paecilomyces*, *Pithomyces*, *Cladosporium, Wallemia* may also produce mycotoxins.

Considerable research has been carried out into the production of secondary metabolites in laboratory culture and a large number of scientific papers published describing these findings. However even the most toxic of chemicals produced under such conditions have little toxicological implication for man or animal unless they can be shown to occur in food, water or the atmosphere. This review will consider primarily those mycotoxins of the genus *Aspergillus* known to occur in food and in the environment and which are a known or potential threat to human and animal health.

## TOXIGENIC ASPERGILLI

Many *Aspergillus* spp. have been reported as producing mycotoxins (Kozakiewicz, 1989) although in some instances it is difficult to confirm the findings in these reports. The complex nature of the genus *Aspergillus* has often resulted in confusion and mis-identification in the past as has similarly occurred with the genus *Penicillium,* and it remains certain that some of the mycotoxigenic fungi reported do not in practice produce mycotoxins. For instance there is still considerable doubt about whether or not *Aspergillus candidus* produces citrinin. The current author, for example, has examined a number of isolates of this species and failed to find any citrinin produced in culture. Table 1 lists some of the species which are probably responsible for the main mycotoxins currently known to occur in human and animal foods. However, not all strains of a particular species will produce mycotoxins. Even the plant source which hosts the strains may influence the ability to produce a particular toxin. When a number of isolates of *Aspergillus flavus* were cultured under identical conditions, 74-100% of those isolated from groundnuts and cottonseed produced aflatoxins while only 20-58% of those from rice did so (Shroeder and Boller, 1973). In addition, it is not uncommon for proven mycotoxigenic strains to lose their ability to produce toxins when continuously cultured and sub-cultured in the laboratory.

**Table 1.** Toxigenic species of *Aspergillus* responsible for mycotoxins in food and animal feeds

| fungus | mycotoxins |
|---|---|
| *A. flavus* | aflatoxins B1, B2, cyclopiazonic acid |
| *A. parasiticus* | aflatoxins B1, B2, G1, G2 |
| *A. versicolor* | sterigmatocystin |
| *A. nidulans* | sterigmatocystin |
| *A. ochraceus* | ochratoxin A, citrinin, xanthomegnin, viomellein, penicillic acid |
| *A. clavatus* | patulin, other neurotoxins |
| *A. fumigatus* | gliotoxin, tremorgenic mycotoxins |

It is apparent from this list that the number of *Aspergillus* mycotoxins shown to occur in food is relatively small. By far the most widely known and almost certainly the most toxic of the mycotoxins found most frequently are the aflatoxins.

## TOXICITY OF MYCOTOXINS

Table 2 lists some of the principal toxic effects attributed to the more important mycotoxins of *Aspergillus*. Several, including aflatoxin B1, sterigmatocystin, ochratoxin A and patulin, are proven carcinogens in animals and aflatoxin B1 and sterigmatocystin are strongly suspected to be human carcinogens. Studies carried out in the UK (Garner, 1992) show that aflatoxin B1 can interact with DNA and form long lasting adducts in human tissues that relate to an individuals lifetime exposure to this mycotoxin. The implication of such findings is that the population is challenged by very low levels of this mycotoxin in the diet.

**Table 2.** Toxic effects of mycotoxins

| mycotoxin | effect |
|---|---|
| aflatoxins B1 | carcinogenic, hepatotoxic, DNA damage |
| cyclopiazonic acid | neurotoxic |
| sterigmatocystin | carginogenic |
| ochratoxin A | nephrotoxic, teratogenic, immunosuppressent |
| citrinin | nephrotoxic |
| xanthomegnin, viomellein | photosensitisation, liver and kidney damage? |
| patulin | carcinogenic |

Aflatoxin B1 is one of the most potent tumour producing chemicals known to man and using published data (Fishbein, 1979), Table 3 shows its potency relative to that of a number of other better known carcinogens. Hence to produce tumours in 50% of test rodents over a lifetime would require 30,000 times the amount of beta-naphthalamine compared to that required by aflatoxin B1 to produce the same effect. On the same basis, 1,000 times the amount of beta-naphthalamine would be required to produce a similar effect to that caused by sterigmatocystin.

**Table 3.** Range of potency of carcinogens in test animals

| compound | dose* | relative potency |
|---|---|---|
| trichloroethylene | 3 | 1 |
| beta-naphthylamine | 0.03 | 100 |
| carbon tetrachloride | 0.02 | 150 |
| benzidine | 0.005 | 600 |
| dimethyl nitrosamine | 0.0005 | 6000 |
| sterigmatocystin | 0.00003 | 100000 |
| aflatoxin B1 | 0.000001 | 3000000 |

* = $gkg^{-1}day^{-1}$ to produce tumours in 50% of test animals over a lifetime

## ANALYTICAL METHODS

Any study of the occurrence of mycotoxins in food is dependent on the availability of sensitive, reliable and easy to use analytical methods. In practice such methods are only available for a few mycotoxins, but this does include the aflatoxins and ochratoxin A. Surveys to determine the incidence of such compounds as cyclopiazonic acid, sterigmatocystin and the mycotoxins formed by *A. fumigatus* in food and animal feeds have rarely been carried out because of the non availability of suitable methods. In contrast to many of the man-made contaminants of food such as pesticides, mycotoxins are often very similar in chemical and physical properties to many of the natural components present in food. Hence complete separation of any mycotoxins from other constituents extracted by the solvents used in the analytical procedure, can be difficult and in some cases, almost impossible. Until recently, many of these analyses have been lengthy, labour intensive and insensitive and as a result perhaps no more than 10-15 of all the potential mycotoxins have been the subject of surveillance or control. However, in the past few years, highly specific immunological methods based on anti-bodies raised against a number of mycotoxins have been developed and some marketed commercially. For instance, binding of these antibodies to inert materials has enabled immunoaffinity columns to be marketed for aflatoxins and ochratoxin A (Figure 1).

The enormous impact such developments have had on aflatoxin analysis is illustrated in Figure 2. Chromatogram A was obtained after clean-up of extract solutions using a method developed for aflatoxin, ochratoxin A and zearalenone (Howell, 1981), chromatogram B following gel permeation clean-up (Hetmanski and Scudamore, 1989) and chromatogram C, following a single stage clean-up with an immunoaffinity column. In the last case, no peaks other than those for aflatoxins are present. Unfortunately the development of similar products for other mycotoxins is only likely to occur where an economically viable market is clearly perceived to exist.

## MYCOTOXINS FOUND IN FOODS AND ANIMAL FEEDS

Identification of fungal metabolites produced in culture does not necessarily imply that these compounds will be detected in foods. Hence those mycotoxins listed in Table 1 represent only a small fraction of those capable of being produced in fungal cultures. Study and monitoring of foods in many countries over a number of years has enabled a picture of the principle mycotoxins that occur to be built up. In the UK, the Ministry of Agriculture, Fisheries and Food have regularly published data giving the results of surveillance for mycotoxins in food for many years ( MAFF,1980; 1987; 1993).

A number of the mycotoxins produced by *Aspergillus* spp. which have been shown to occur in foodstuffs are also produced by some species of *Penicillium*. It is thus not always clear whether a mycotoxin when found is due to the presence of *Aspergillus* or *Penicillium,* although the type of commodity and its geographical origin may enable an informed assessment to be made. Table 4 lists those mycotoxins known to be common to both genera.

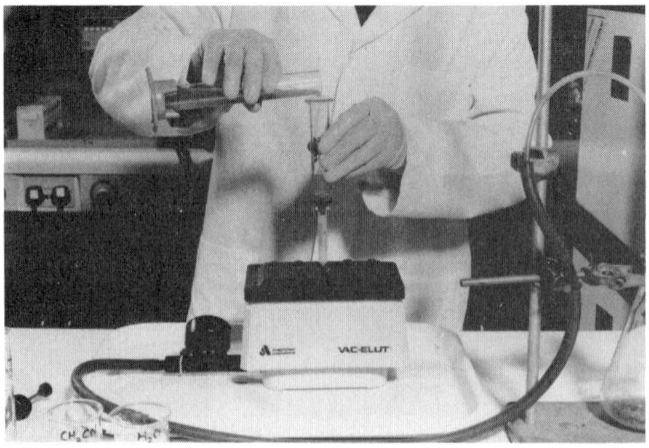


**Figure 1.** Use of immunoaffinity columns in aflatoxin analysis

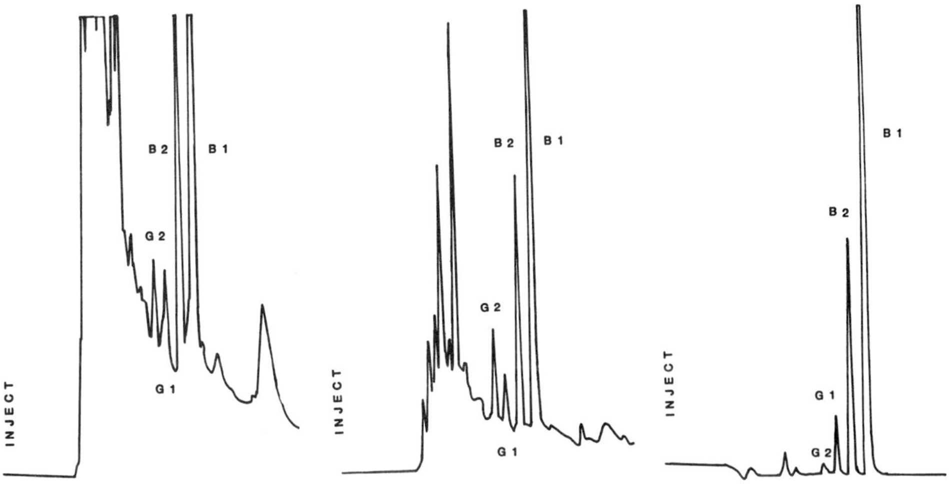


**Figure 2.** HPLC of a compound animal feed showing improvements in analytical methodology. A) solid phase clean-up, B) gel permeation clean-up and C) immunoaffinity column clean-up.

**Table 4.** Mycotoxins formed by both *Aspergillus* and *Penicillium*

| mycotoxin | fungus |
|---|---|
| cyclopiazonic acid | *A. flavus*<br>*Penicillium* spp. |
| ochratoxin A, citrinin | *A. ochraceus*<br>*P. verrucosum* |
| xanthomegnin, viomellein | *A. ochraceus*<br>*P. aurantiogriseum* |
| patulin | *A. clavatus*<br>*P. expansum* |
| penicillic acid | *A. ochraceus*<br>*P. aurantiogriseum* |

Patulin found in apples and pears, for example, is almost certainly due to infection of stored fruit with *Penicillium expansum* in contrast to the formation of patulin during malting of barley which can usually be attributed to the presence of *A. clavatus*. Similarly, the common occurrence of ochratoxin A and citrinin in poorly stored grain in temperate areas of the world is almost certainly usually due to *P. verrucosum,* while the presence of these mycotoxins in products such as cocoa and coffee beans grown in tropical or semi-tropical areas of the world is more likely, but not exclusively so, to be due to *A. ochraceus* or other *Aspergillus* spp.

## AFLATOXINS

*A. flavus.* and *A. parasiticus* can infect almost any foodstuff and have been found on most agricultural products. There is thus a potential for the occurrence of aflatoxins in a wide variety of foods. It is usual for a mixture of aflatoxins B1 and B2 perhaps together with aflatoxins G1 and G2 to be present. The 'B's' usually occur in higher amounts than the 'G's' and the '1's' in higher amounts than the '2's'. Once formed, aflatoxins are quite stable and are likely to persist during storage and processing. However under alkaline or acid conditions, degradation will occur and a number of decontamination procedures have been proposed such as ammonia treatment (Dollear *et al.*, 1968; Park *et al.*, 1988) or the use of other agents such as formaldehyde (Mann *et al.*, 1970) or sodium hydroxide (Dollear *et al.*, 1968).

The widespread incidence of aflatoxins in staple foods in the less developed parts of the world, makes a major contribution to disease, ill health and in extreme cases, death. Groundnuts and many other nuts, palm kernels, maize, cotton seed, spices, rice and figs are commonly contaminated although soya, wheat and barley appear less susceptible. Many countries have regulations controlling the amounts of aflatoxins and other mycotoxins in human and animal food and these were reviewed (van Egmond, 1989). Legislation for aflatoxins in the UK up to 1992, only existed for animal feedingstuffs (MAFF, 1991). However new legislation for human foods was introduced at the end of 1992, which proscribes limits of 10 $ugkg^{-1}$ of total aflatoxins for food which is to be processed further and at 4 $ugkg^{-1}$ in retail products (MAFF, 1992). In those countries where regulations exist, these may specify either aflatoxin B1 or total aflatoxins and thus it is essential to be clear which system is in operation. Monitoring of aflatoxins has shown in general that levels and incidence of this group of mycotoxins in the UK is low. In recent years, problems in susceptible products such as figs, peanuts, maize and animal feedingstuffs have been identified, targeted and largely controlled. However reports on finding aflatoxin in foods sometimes appear in the media (Smith, 1992), or as happened also with figs in the UK (Isherwood and Brown, 1988). Following this problem, a surveillance and control scheme was set up and results from this were subsequently published (Sharman *et al.*, 1991). Turkish figs were strongly implicated and confirmed the findings of earlier studies from Turkey (Boyacioglu and Gonul, 1990) that the amount of aflatoxin A was strongly related to the quality of the figs.

There are many published surveys on aflatoxin occurring in foods. Three recent examples that contrast the situation in different areas of the world are those carried out in Bihar state, India during 1987-1989 (Sinha and Sinha, 1991), that for Foods and Foostuffs in Tokyo, 1986-1990, (Tabata *et al.*, 1993) and a survey of domestic and imported foods in the USSR (Tutelyan *et al.*, 1989).

In Bihar, 162 out of 416 samples of wheat, gram and corn flours were aflatoxin positive, mostly at levels above 20 $ugKg^{-1}$, the maximum level being 2048 $ugKg^{-1}$ in one sample of maize flour. However there was a considerable year to year variation.

The survey in Japan examined 3054 samples of a wide range of commodities. In most products, both the incidence and levels of aflatoxins were low but they were found in rice products, adlay, corn, crude sugar, peanut products, pistachio nuts, brazil nuts, sesame, butter beans, pepper, paprika, nutmeg and mixed spices. A summary of some of these results is given in Table 5.

Between the years 1985-1987, 4532 samples of cereals, legumes, coffee, cocoa beans, tea, cotton seed and nuts were examined in the USSR. The samples most often examined were peanuts and wheat. 788 out of 1559 samples of peanuts contained aflatoxin, 420 at levels above the regulatory limit of 5 $ugKg^{-1}$ with a level of 3650 $ugKg^{-1}$ in one sample. Aflatoxin in wheat occurred much less frequently, only 3.1% of 579 imported samples and 1.5% of 1100 home produced wheat samples were positive. Aflatoxin was found in some cocoa and coffee beans but not in tea.

Peanuts and peanut products are very frequently contaminated with aflatoxin and this was recognised as early as 1960 when groundnut meal was found to be the main cause of turkey-X disease (Allcroft and Carnaghan, 1962). Although much surveillance has been carried out on peanuts, a survey of peanut butters obtained from Health Food outlets in the UK, using an enzyme- linked immunosorbent assay (Mortimer *et al.*, 1988) is of interest in illustrating that this type of food is just as likely to be contaminated with aflatoxin as peanuts produced and stored with the aid of agricultural chemicals.

**Table 5.** Aflatoxins found in foods in Tokyo 1986-1990.

| commodity $ugKg^{-1}$ | No. samples | No. positive | aflatoxin B1(max.value) |
|---|---|---|---|
| rice | 74 | 2 | 2.7 |
| wheat | 168 | 0 | - |
| corn | 181 | 4 | 0.4 |
| crude sugar | 9 | 3 | 1.5 |
| peanut/products | 149 | 11 | 21.7 |
| cashew nuts | 56 | 0 | - |
| sesame | 19 | 2 | 2.4 |
| beans (for bean jam) | 293 | 4 | 52 |
| black pepper | 46 | 0 | - |
| white and red pepper | 94 | 13 | 9.1 |
| nutmeg | 108 | 84 | 13.4 |

When ruminants eat feedingstuffs containing aflatoxins, the toxins are metabolised and excreted as aflatoxin M1 (from B1) and aflatoxin M2 (from B2) in the milk. Humans are thus potentially exposed to these metabolites in milk, cheese and other dairy products. In the developed areas of the world where the amounts of aflatoxins is strictly controlled in animal feed, levels of these toxins are extremely small. Many European and other countries carry out regular surveillance of milk and some of of these results are published (eg. MAFF, 1980; 1987). In 1984 and 1985, in Italy, 313 and 276 samples of imported milk and 159 and 416 samples of cheese respectively were checked for aflatoxin M1. In general, levels in milk were very low (maximum level 23 $ngl^{-1}$) but some aflatoxin M1 was detectected in between 20 and 50% of samples. There was some difference between French, German and Dutch products but levels in 1985 were generally lower overall than in 1984 (Piva *et al.*, 1988). In an earlier survey in 1979-1981 in Brazil, 100 samples of commercially available milk were collected in the state of Sao Paulo. Aflatoxin M1 was detected in only one sample of commercially available cows milk but in milk samples taken directly from farms, 9 out of 50 were positive at levels between 100 $ngl^{-1}$ and 1680 $ngl^{-1}$ (Sabino *et al.*, 1989). However the limit of detection was 100 $ngl^{-1}$ ie 4 times the level of the highest result found from the Italian study reported above.

Although the conversion of aflatoxin B1 to M1 in milk is well recognised, much less has been carried out on the fate of aflatoxin in other animals and whether these residues can be passed on in human food either as the free compound or in a bound form. Studies in Italy have reported a low level of contamination in eggs and that approximately 30% of this residue is in the form of water soluble conjugates ( Micco *et al.*, 1987).

## OCHRATOXIN A AND CITRININ

Current knowledge suggests that without doubt, aflatoxin is the most widespread and important mycotoxin which contaminates food world-wide. Most other familiar mycotoxins produced by *Aspergillus* are of sporadic occurrence or at probably toxicologically unimportant levels. Important mycotoxins such as ochratoxin A and citrinin occurring in cereals are more often likely due to *Penicillium* fungi rather than *Aspergillus*. In view of the uncertainty that may exist in many instances over the fungal origin of these two important mycotoxins in foods, these mycotoxins are briefly reviewed here.

When a mycotoxin is relatively stable in human or animal tissues and body fluids, examination of samples taken from a representative sample of a population should reflect the overall exposure of that population to the mycotoxin from all food sources and may prove a more cost effective way of monitoring. However it will not indicate the original source of the contamination. This approach has been recently adopted more widely eg. in Sweden (Breitholtz *et al.*, 1991) and Germany (Bauer et al., 1986). The data presented by the former workers showed that the mean plasma concentration of ochratoxin A in human blood in Sweden was 0.10 $ngml^{-1}$. This value was then compared with the total daily intake value (TDI), calculated from carcinogenicity data, as proposed (Kuiper-Goodman and Scott, 1989). When this was done, the mean value found was below the TDI, but, the TDI was exceeded for some individuals in the regions where the highest blood levels were found.

Monitoring of animal feed over a number of years in the UK showed that after aflatoxin, ochratoxin A was the second most commonly found mycotoxin often together with citrinin (Buckle, 1983). In this and many other studies, the source of the mycotoxins would most likely be *Penicillium* moulds. In a survey in Czechoslovakia on the occurrence of ochratoxin A in cereals, feedstuffs and porcine kidneys using a radioimmunoassay, levels were found exceeding 20 $ugKg^{-1}$ in 8.8%, 26.7 and 0% of the samples respectively (Fukal, 1990). However, approximately 30% of the porcine kidneys contained amounts between 1 and 20 $ugKg^{-1}$. In the UK, 303 pig kidneys unsuitable for human consumption were examined (Morgan *et al.*, 1986). Of these, 79.8% contained less than 1 $ugKg^{-1}$ and only 2.7% contained more than 5$ugKg^{-1}$ of ochratoxin A.

Of twenty nine samples of green coffee collected in Italy, 58% were found positive in amounts ranging from 0.2 to 15 $ugKg^{-1}$ ochratoxin A. Following roasting, between 50 and 100% of the ochratoxin A had been destroyed and none could be detected in the final beverage (Micco *et al.*, 1989).

## PATULIN AND *ASPERGILLUS CLAVATUS*

The occurrence of patulin in apple and pear juices and products is most likely due to *Penicillium expansum* (Pitt and Hocking, 1985). However, *A. clavatus* is known to produce patulin in culture and appears to be well suited for growth during malting and hydroponic production of leafage for animal feed. This fungus can produce a number of toxic compounds in culture, such as ascladiol, patulin, tryptoquivoline, tryptoquivolone and cytochalasins E and K. There are many cases where neurological symptoms in animals have been linked to the presence of *A. clavatus* in cereals. However the source of such toxicity remains unknown. A comprehensive review of the occurrence of this fungus in cereals and problems related to its presence has been published (Flannigan,1986).

## STERIGMATOCYSTIN

Although sterigmatocystin is readily produced in culture by a number of different fungi, evidence for its occurrence in foodstuffs is scarce. Several surveys of feedingstuffs and cereals in which a relatively large number of samples have been examined for sterigmatocystin have been reported.

In Canada, over 2000 samples of animal feed in Ontario between 1972-77 were examined and only found one sample to be positive for this toxin (Funnell, 1979).

In the UK, of 523 samples examined by the Agricultural Development and Advisory Service of MAFF, 17 were found positive for sterigmatocystin (Buckle, 1983). However this was not a random survey as many of the samples tendered for examination were suspected of being responsible for feed-related problems in animals or of being of inferior quality. In a survey of mouldy cereals in the UK, 7 out of 11 samples were positive for sterigmatocystin (Scudamore *et al.*, 1986) .

In Brazil, a survey of 296 samples of raw and processed food as sold to consumers was carried out (Valente Soares and Rodriquiz-Amaya, 1989). Samples were corn, cassava flour, rice and dried beans. No sterigmatocystin was detected, although nine of the samples were positive for aflatoxin.

A survey of 167 corn samples collected in 1986 in Turkey were examined for moulds and mycotoxins (Ozay and Heperkan, 1989). Sterigmatocystin was detected in ten samples, although in each case this was close to the detection limit of 20 $ugKg^{-1}$. While sterigmatocystin is regarded as a potent carcinogen, the limited data available suggests that its occurrence is at the worst only occasional although many species of *Aspergillus,* which have the potential to produce sterigmatocystin, can frequently be isolated from a wide variety of foods.

## CYCLOPIAZONIC ACID

The study of mycotoxins received a major impetus in 1960 with the occurrence of turkey-X disease in which about 100,000 turkeys and ducklings died. The cause was quickly tracked down to the use of groundnut meal heavily contaminated with aflatoxins (Allcroft and Carnaghan, 1962). However more recently a paper was published which discussed the possibility that the groundnut meal had also contained high levels of cyclopiazonic acid on the basis that many of the affected birds exhibited an unusual postering movement which has been noted as characteristic of cyclopiazonic acid toxication (Cole, 1986). It is well established that *Aspergillus flavus* can produce both groups of mycotoxins.

The detection of cyclopiazonic acid in animal feed is analytically considerably more difficult than for aflatoxin and even today surveys rarely include this mycotoxin. While its natural occurrence is firmly proven, only limited surveillance data exists. However, in 1990 a survey of 50 samples of peanuts and 45 samples of maize was conducted in Georgia, USA, to investigate the possible co-occurrence of aflatoxins and cyclopiazonic acid (Takashi *et al.*, 1992). Over 90% of peanuts and 50% of the maize contained cyclopiazonic acid at similar levels to that found for aflatoxin, Table 3.

**Table 6.** Co-occurrence of aflatoxins and cyclopiazonic acid

| sample | total taken | aflatoxins $ugKg^{-1}$ positive | range | (cyclopiazonic acid $ugKg^{-1}$) positive | range |
|---|---|---|---|---|---|
| peanut | 50 | 45 | <50-2900 | 50 | 3-22000 |
| maize | 45 | 23 | <25-2800 | 39 | 1-2300 |

## *ASPERGILLUS FUMIGATUS* MYCOTOXINS

There is little direct evidence for the natural occurrence of the toxic metabolites attributed to *A. fumigatus*. However the fungus is widely distributed and is often found in stored products which have heated and spoiled. In this laboratory during the course of investigating a number of incidents involving the ill health or death of livestock, *A. fumigatus* has commonly been isolated from feed samples, sometimes at exceedingly high levels and on one occasion was found to be the only fungus present in significant amounts. However examination of the feed samples for some of the toxic metabolites reported for this fungus has to date failed to detect any compound which could be responsible for the symptoms observed. Such investigations are hampered by unsatisfactory analytical methods or non availability of analytical standards. While it would probably appear unlikely that human food would be significantly at risk from *A. fumigatus* mycotoxins, the implication that these mycotoxins may play an important part in animal health remains.

## VIOMELLEIN AND XANTHOMEGNIN

The natural occurrence of viomellein in barley was first reported in Denmark (Hald *et al.*, 1983). Subsequent studies in the United Kingdom (Scudamore *et al.*, 1986; 1986) showed that viomellein and xanthomegnin together with a related compound, vioxanthin, could be found quite often in poorly stored cereals and animal feedingstuffs. However it is almost certain that the origin of these mycotoxins in Denmark and the United Kingdom was *Penicillium* fungi as none of the species of *Aspergillus* known to produce these compounds was isolated from the samples examined.

## OTHER *ASPERGILLUS* TOXINS

Approximately 100 toxic fungal metabolites in addition to those currently accepted as mycotoxins were identified in a review (Watson, 1985). Those attributed to *Aspergillus* were aflatrem, ascadiol, asp-hemolysin, aversin, chrysophanic acid, cytochalasin B, emodin, fumitremorgens A, B and C, gliotoxin, malformin C, 5-methoxy and 5,6-dimethoxy sterigmatocystin, naphtho-g-pyrones, oxalic acid, paspalinine, physchion, secalonic acid, terreic acid, territrems, verrucologen, TR-2 toxin tryptoquivaline, tryptoquivalone, xanthoascin, and xanthocillins. However only one reference could be found to the occurrence of any of these metabolites in foodstuffs viz naphtho-Y-pyrones in mango fruit, pulp and skin (Ghosal *et al.*, 1979 ).

The principal reason that so few *Aspergillus* mycotoxins have been found in foodstuffs is probably that the combination of factors required for the fungi to elucidate the appropriate metabolites does not often occur. Hence a combination of temperature, water activity and balanced nutrient source is never achieved or not maintained for sufficient time to enable the metabolic processes necessary for mycotoxin production to be achieved. However, apart from aflatoxin and ochratoxin A, few other *Aspergillus* toxins are normally actively sought. Most published methods for sterigmatocystin, xanthomegnin and viomellein are insensitive, those for cyclopiazonic acid and citrinin in general are not very satisfactory and methods suitable for detection of most other mycotoxins in foods are not available or not developed sufficiently well to be used in surveillance.

Two main factors are required to initiate and fund surveys for lesser studied mycotoxins: sufficient evidence to suggest the likely occurrence of a particular mycotoxin in foods and sufficient toxicological evidence to raise concern over any possible natural occurrence. However until a suitable analytical method was developed it would be difficult to obtain this evidence for the occurrence of additional toxins. In the current climate where funds for research are limited, it is likely that confirmation of the natural occurrence of further mycotoxins will be slow.

## CONCLUSIONS

Although many toxic metabolites can be produced in culture by *Aspergillus* moulds, the number proven to occur in foodstuffs is very small. The most widespread and the most toxic of these compounds are the aflatoxins which can occur in a very wide range of products. Ochratoxin A probably ranks second to aflatoxins and often co-occurs with citrinin, although the lack of reliable methods for this latter mycotoxin means that monitoring for it is rarely carried out. Patulin is occasionally detected in apple and pear juices or processed fruit products, but the cause is most likely infection of the stored fruit by *Penicillium expansum.* Other *Aspergillus* toxins are sought less frequently, but cyclopiazonic acid and sterigmatocystin are occasionally found. However, many other contaminants with the potential to be formed by *Aspergillus,* are rarely or never sought and thus their natural occurrence in food products remains a matter for speculation.

## ACKNOWLEDGEMENT

The mycotoxin studies at the Central Science Laboratories are funded by the Pesticide Safety and Chemical Safety of Food Divisions of the UK, Ministry of Agriculture, Fisheries and Food.

## REFERENCES


Allcroft, R. and Carnaghan, R.B.A. (1962) Groundnut toxicity - *Aspergillus flavus* toxin (aflatoxin) in animal products: preliminary communication. The Veterinary Record 74, 863.

Anon (1989) Poisoned birdseed warning to pet owners. Reading Herald and Post 16 November, 4.

Bauer, J., Gareis, M. and Gedek, B. (1986) Incidence of ochratoxin A in blood serum and kidneys of man and animals, In Proceedings of the 2nd World Congress on Foodborne Infections and Intoxications, Berlin, 26-30 May 1986, 907.

Boyacioglu, D. and Gonul, M. (1990) Survey of aflatoxin contamination of dried figs grown in Turkey in 1986. Food Additives and Contaminants 7, 235.

Breitholtz, A., Olsen, M., Dahlback, A. and Hult, K. (1991) Plasma ochratoxin A levels inthree Swedish populations surveyed using ion-pair HPLC technique. Food Additives and Contaminants 8, 183.

Buckle, A.E.(1983) The occurrence of mycotoxins in cereals and animal feedstuffs. Vet. Res.Commun. 7, 171.

Cole, R.J. (1986) Etiology of Turkey'X' disease in retrospect: a case for the involvement of cyclopiazonic acid. Mycotoxin Research 2, 3.

Dollear, F.G., Mann, G.E., Codifer, L.P., Jr., Gardner, H.G., Jr., Koltun, S.P. and Vix, H.L.E. (1968) Elimination of aflatoxin from peanut meal. J. Am.Oil Chem. Soc. 45, 862.

Fishbein, L.(1979) Range of potency of carcinogens in animals. In "Potential Industrial Carcinogens and Mutagens" Elsevier Scientific Publishing Company, Amsterdam, Oxford and New York: 1.

Flannigan, B. (1986) *Aspergillus clavatus* - an allergenic, toxigenic, deteriogen of cereals and cereal products, Proceedings of the Biodeterioration Society Meeting on spoilage and mycotoxins of cereals and animal feedstuffs. Int. Biodet. 22 (supplement), 79.

Fukal, L. (1990) A survey of cereals, cereal products feedstuffs and porcine kidneys for ochratoxin A by radioimmunoassay. Food Additives and Contaminants 7, 253.

Funnell, H.S. (1979) Mycotoxins in animal feedstuffs in Ontario. J. Am. Oil Chem. Soc. 43, 243.

Garner, R.C. (1992) Aflatoxin- a cancer problem that refuses to go away. In " Food Safety and Quality Assurance Applications of Immunoassay Systems"( Morgan, M.R.A., Smith, C.J.and Williams, P.A.,Eds) , Elsevier Science Publishers Ltd. 93, 102.

Ghosal, S., Biswas, K. and Chakrabarti, D.K. (1979) Toxic naphtho-g-pyrones from *A. niger*. J. Agric. Food Chem. 27, 1347.

Hald, B., Christensen, D.H. and Krogh, P. (1983) Natural occurrence of the mycotoxin viomellein in barley and the associated quinone producing penicillia. Appl. Envir, Microbiol. 46, 1311.

Hetmanski, M.T. and Scudamore, K.A. (1989) A simple, quantitative HPLC method for determination of aflatoxins in cereals and animal feeding stuffs using Gel Permeation Chromatography clean-up. Food Additives and Contaminants 6,35.

Howell, M.V. and Taylor, P.W. (1981) Determination of aflatoxin, ochratoxin A and zearalenone in mixed feeds with detection by thin layer chromatography or high performance liquid chromatography. J. Assoc. Off. Anal. Chem. 69, 1356.

Isherwood, J. and Brown, G. (1988) Liver cancer toxin fear in figs. Daily Telegraph, 1 December.

Kozakiewicz, Z. (1989) *Aspergillus* Species on Stored Products, International Mycological Institute, Mycological paper No. 161, CAB International, Wallingford, Oxon, UK.

Kuiper-Goodman, T., Scott, P.M. (1989) Risk assessment of the mycotoxin ochratoxin A. Biomed. and Environ. Sc. 2, 179.

Mann, G.E., Codifer, L.P., Jr., Gardner, H.G., Jr., Koltun, S.P. and Dollear, F.G. (1970) Chemical deactivation of aflatoxins in peanuts and cottonseed meals. J. Am.Oil Chem. Soc. 47,173.

Micco, C., Grossi, M., Miraglia, M. and Brera, C. (1989) A study of the contamination by ochratoxin A of green and roasted coffee beans. Food Additives and Contaminants. 6, 333.

Micco, C., Brera, C., Miraglia, M. and Onori, R. (1987) HPLC determination of the total content of aflatoxins in naturally contaminated eggs in free and conjugate forms. Food Additives and Contaminants. 4, 407.

Ministry of Agriculture, Fisheries and Food (1980) Surveillance of mycotoxins in the United Kingdom. The fourth report of the Steering Group on Food Surveillance The Working Party on Mycotoxins. Food Surveillance Paper No 4, London, HMSO, 35 pages,

Ministry of Agriculture, Fisheries and Food (1987) Mycotoxins. The eighteenth report of the Steering Group on Food Surveillance The Working Party on Naturally Occurring Toxicants in Food: Sub-Group on Mycotoxins. Food SurveillancePaper No 18, London, HMSO, 44 pages.

Ministry of Agriculture, Fisheries and Food (1991) The Feeding Stuffs Regulations. Statutory Instruments 1991 No. 2840, London, HMSO, 80 pages.

Ministry of Agriculture, Fisheries and Food (1993) Mycotoxins: Third Report. The thirty fourth report of the Steering Group on Chemical Aspects of Food Surveillance Sub-Group on Mycotoxins. Food Surveillance Paper No 34, London, HMSO,

Ministry of Agriculture, Fisheries and Food (1992) Aflatoxins in nuts, nut products,dried figs and dried fig products Regulations 1992 Statutory Instruments 1992 No. 3236, London, HMSO, 12 pages.

Morgan, M.R.A., McNerney, R., Chan, H.W.S. and Anderson, P. H. (1986) Ochratoxin A in pig kidney determined by enzyme-linked immunosorbent assay (ELISA). J. Sci. Food Agric. 37, 475.

Mortimer, D.N., Shepherd, M.J., Gilbert, J. and Morgan M.R.A. (1988) Survey of the occurrence of aflatoxin B1 in peanut butters by enzyme linked immunosorbent assay. Food Additives and Contaminants 5, 127.

Ozay, G. and Heperkan, D. (1989) Mould and mycotoxin contamination of stored corn in Turkey. Mycotoxin Research 5, 81.

Park, D.L., Lee, L.S., Price, R.L. and Pohland, A.E. (1988) Review of the decontamination of aflatoxins by ammoniation. J. Assoc. Off. Anal. Chem. 71, 685.

Pitt, J.I. and Hocking, A.D. (1985) Fungi and food spoilage. Academic Press.

Piva, G., Pietri, A., Galazzi, L. and Curto, O. (1988) Aflatoxin M1 occurrence in Dairy products marketed in Italy. Food Additives and Contaminants. 5, 133.

Sabino, M., Purchio, A. and Zorzetto, M.A.P. (1989) Variation in the levels of aflatoxin in cows milk consumed in the city of Sao Paulo, Brazil. Food Additives and Contaminants. 6, 321

Scudamore, K.A., Atkin, P.M. and Buckle, A.E. (1986) Natural occurrence of the naphthoquinone mycotoxins, xanthomegnin, viomellein and vioxanthin in cereals and animal feedstuffs. J. Stored Prod. Res. 22, 81.

Scudamore, K.A., Clarke, J.H. and Atkin, P.M. (1986) Natural occurrence of fungal naphthoquinones in cereals and animal feedstuffs. Proceedings of the Biodeterioration Society Meeting on spoilage and mycotoxins of cereals and animal feedstuffs. Int. Biodet. 22 (supplement), 71

Sharman, M., Patey, A.L., Bloomfield, D.A. and Gilbert, J. (1991) Surveillance and control of aflatoxin contamination of dried figs and fig paste imported into the United Kingdom. Food Additives and Contaminants. 8, 299.

Shroeder, H.W. and Boller, R.A. (1973) Aflatoxin production in species and strains of the *Aspergillus flavus* group isolated from field crops. Appl. Microbiol. 25, 885.

Sinha, K.K. and Sinha, A.K. (1991) Monitoring and identification of aflatoxins in wheat, gram and maize flours in Bahir state (India). Food Additives and Contaminants. 8, 453-457.

Smith, S. (1989) Aflatoxin alert, killer fungus found in US maize. Farming News 10 March: 1.

Tabata, S., Kamimura, H., Ibe, A., Hashimoto, H., Iida, M., Tamura, Y. and Nishima, T. (1993) Aflatoxin contamination in foods and foodstuffs in Tokyo. J. Assoc. Off. Anal. Chem. 76, 32-35.

Takashi, U., Trucksess, M.W., Beaver, R.W., Wilson, D.M., Dorner, J.W. and Dowell, F.E. (1992) Co-occurrence of cyclopiazonic acid and aflatoxin in corn and peanuts. J. Assoc. Off. Anal. Chem. 75, 838.

Tutelyan, V.A., Eller, K.I. and Sobolev (1989) A survey using normal phase high performance liquid chromatography of aflatoxins in domestic and imported foods in the USSR. Food Additives and Contaminants 6, 459.

Valente Soares, L.M. and Rodriguiz-Amaya, D.B. (1989) Survey of aflatoxin, ochratoxin A, zearalenone and sterigmatocystin in some Brazilian foods by using multi-toxin Thin Layer Chromatographic Method. J. Assoc. Off. Anal. Chem. 72, 22.

Van Egmond, H.D. (1989) Current situation on regulations for mycotoxins. Overview of tolerances and status of standard methods for sampling and analysis. Food Additives and Contaminants 6, 139.

Watson, D.H. (1985) Toxic fungal metabolites in food. CRC Critical Reviews in Food Science and Nutrition 22, 177.

# ASPERGILLI IN FEEDS AND SEEDS

J. Lacey

AFRC Institute of Arable Crops Research
Rothamsted Experimental Station
Harpenden
Herts AL5 2JQ
UK

## INTRODUCTION

Species with anamorphs representing most sections of the genus *Aspergillus*, some with teleomorphs in the genera *Eurotium*, *Emericella* and *Fennellia* are among the most characteristic fungi of stored feeds and seeds, even when there is little available water. The species recorded from these substrates cover most environmental conditions likely to be encountered, with few geographical and substrate differences in incidence. The species present can indicate the storage history of the substrate. They are important not only because of their effects on quality of feeds and seeds but also because some species produce toxic metabolites while others have been implicated in asthma and allergic alveolitis.

Fungi have been classified into field and storage species but some aspergilli show the limitations of this classification since species characteristic of storage in temperate regions are common and often abundant out of doors in humid tropical climates. Thus, in considering the ecology of aspergilli in stored feeds and seeds, it is necessary to consider first their occurrence on growing and ripening crops and during postharvest handling as well as their development during storage. In the context of this chapter, species will be referred to by their anamorph name.

## *ASPERGILLUS* SPECIES IN FEEDS AND SEEDS

More than 40 *Aspergillus* species have been recorded in stored feeds and seeds (Table 1). Most are classified within the Sections Aspergillus (12 species), Nidulantes and Flavi (six species each). However, not all species are common components of the microflora. The most common are probably *A. vitis, A. reptans, A. chevalieri, A. intermedius, A. rubrobrunneus, A. restrictus, A. fumigatus, A. clavatus, A. nidulans,*

*The Genus Aspergillus*, Edited by Keith A. Powell *et al.*,
Plenum Press, New York, 1994

**Table 1.** *Aspergillus* species recorded in feeds and seeds

**Subgenus ASPERGILLUS**

**Section ASPERGILLUS**

- *A. glaucus* Link (*Eu. herbariorum* (Wiggers: Fr.) Link)
- *A. vitis* Novobr. (*Eu. amstelodami* Mangin)
- *A. chevalieri* Mangin (*Eu. chevalieri* Mangin)
- *A. intermedius* Blaser (*Eu. intermedium* Blaser)
- *A. reptans* Samson and W. Gams (*Eu. repens* de Bary)
- *A. rubrobrunneus* Samson and W. Gams (*Eu. rubrum* König, Spiekermann and Bremer)
- *A. brunneus* Delacr. (*Eu. echinulatum*) (Delacr.) Thom and Church
- *A. halophilicus* Christensen, Papavizas and Benjamin (*Eu. halophilicum* Christensen *et al.*)
- *A. heterocaryoticus* Christensen, Lopez and Benjamin (*Eu. heterocaryoticum* Christensen *et al.*)
- *A. medius* Meissner (*Eu. medium* Meissner)
- *A. atheciellus* Samson and W. Gams (*Edyuillia athecia* Raper and Fennell)

**Section RESTRICTI**

- *A. restrictus* G. Smith
- *A. conicus* Blochwitz
- *A. penicillioides* Speg.
- *A. gracilis* Bainier
- *A. caesiellus* Saito

**Subgenus FUMIGATI**

**Section FUMIGATI**

- *A. fumigatus* Fres.

**Section CERVINI**

- *A. parvulus* G. Smith

**Section CLAVATI**

- *A. clavatus* Desm.
- *A. giganteus* Wehmer
- *A. clavato-nanicus* Batista, Maia and Alecrim

**Subgenus NIDULANTES**

**Section NIDULANTES**

- *A. nidulans* (Eidam) Winter (*Em. nidulans* (Eidam) Vuill.)
- *A. aurantiobrunneus* (Atkins, Hindson and Russell) Raper and Fennell (*Em. aurantiobrunnea* (Atkins *et al.*) Malloch and Cain)
- *A. egyptiacus* Moubasher and Moustafa
- *A. stellifer* Samson and W. Gams (*Em. variecolor* Berk. and Br.)
- *A.tetrazonus* Samson and W. Gams (*Em. quadrilineata* (Thom and Raper) C.R. Benjamin
- *A. rugulovalvus* Samson and W. Gams (*Em. rugulosa* (Thom and Raper) C.R. Benjamin)

**Section VERSICOLORES**

- *A. sydowii* (Bainier and Sartory) Thom and Church
- *A. versicolor* (Vuill.) Tiraboschi
- *A. spelunceus* Raper and Fennell

**Section USTI**

- *A. ustus* (Bainier) Thom and Church

**Section TERREI**

- *A. terreus* Thom and Church

**Section FLAVIPEDES**

- *A. flavipes* (Bainier and Sartory) Thom and Church (*Fennellia flavipes* Wiley and Simmons)
- *A. niveus* Blochwitz (*Fennellia nivea* (Wiley and Simmons) Samson
- *A. carneus* Blochwitz

**Subgenus CIRCUMDATI**

**Section CIRCUMDATI**

- *A. ochraceus* K. Wilh.
- *A. ostianus* Wehmer
- *A. sclerotiorum* Hüber
- *A. sulphureus* (Fres.) Thom and Church
- *A. alliaceus* Thom and Church (*Petromyces alliaceus* Malloch and Cain

**Section WENTII**

- *A. wentii* Wehmer
- *A. terricola* Marchal

**Section FLAVI**

- *A. flavus* Link
- *A. nomius* Kurtzmann *et al.*
- *A. oryzae* (Ahlb.) Cohn
- *A. parasiticus* Speare
- *A. tamarii* Kita
- *A. avenaceus* G. Smith
- *A. zonatus* Kwon and Fennell apud Raper and Fennell

**Section NIGRI**

- *A. niger* v. Tieghem
- *A. japonicus* Saito
- *A. carbonarius* (Bainier) Thom
- *A. awamori* Nakazawa

**Section CANDIDI**

- *A. candidus* Link

Sources: Raper and Fennell, 1965; Wallace 1973; Abdel Kader *et al.*, 1979; Lacey, 1988 and unpublished results; Kozakiewicz, 1989.

*A. versicolor, A. sydowii, A.terreus, A. ochraceus, A. flavus, A. wentii, A. niger and A. candidus*. Some species occur rarely and have sometimes been isolated only once. Such rare occurrences include *A. spelunceus* from barley grain (Lacey, 1971) and rapeseed (Lacey, unpublished) and *A. medius, A. egyptiacus* and *A. aurantiobrunneus* from Iranian wheat (Lacey, 1988) although records of colonies composed only of hülle cells (Abdel Kader *et al.*, 1979) could also be *A. aurantiobrunneus*. *A. clavatus* can be abundant on poorly germinating malting barley and on malt sprouts, causing allergic alveolitis in maltworkers, through its numerous spores, and mycotoxicosis in cattle (Flannigan, 1993).

## COLONISATION OF DEVELOPING SEEDS

Before harvest in temperate regions, few *Aspergillus* spp. are present on feeds and seeds, often below the limits of detection methods. However, sufficient inoculum of the most frequent species is present to enable 500 g dry hay or 2.5 kg cereal grains of a given water activity ($a_w$) to mould predictably (Festenstein *et al.*, 1965; Hill *et al.*, 1983a). Airborne inoculum can also be detected over hay fields (Lacey, 1975). However, in tropical climates, *Aspergillus* spp. are common in soil, on vegetation and in the air out of doors and crops are often contaminated early in their growth.

All developing groundnut crops studied in India were contaminated with *A. flavus* and *A. niger* even at flowering (pod development stage 1 (PDS 1; Usha *et al.*, 1991) Table 2). *A. flavus* was most numerous at PDS 2-3 (gynophore elongation) but its incidence decreased near harvest when 5-8% of seeds were infected. At harvest, it was less frequently isolated from shells than from kernels whereas the reverse was true for *A. niger*. *A. niger* was most abundant at PDS 6-9 (pod enlargement to start of lignification) contaminating 40-70% of kernels but then decreased as the pods ripened to infect only 9-22% by harvest *Aspergillus terreus* occurred erratically from PDS 4 (soil penetration), mostly on kernels, but could not be detected towards the end of seed development. In some years, aflatoxin $B_1$ ($AFB_1$) could already be detected in kernels at PDS 5 (Ovary horizontal in soil) in concentrations of up to 8.75 ng $AFB_1$ $g^{-1}$ tissue but then decreased to 2.6 ng $g^{-1}$ at PDS 10 (empty spaces appearing between nuts and shells) before increasing to 3.5 to 60.0 ng $g^{-1}$ at harvest.

Pods can be infected from the soil, by airborne spores or systemically, and often through insect or mechanical damage and split pods (Mehan *et al.*, 1991; Pitt *et al.*, 1991). Water and temperature stress together are important in determining amounts of *A. flavus* infection and $AFB_1$ contamination of undamaged kernels although neither alone stimulated aflatoxin production. Infection and heavy $AFB_1$ contamination is favoured at 25-31°C with 0.85-0.95 $a_w$ in the pods during drought periods lasting 20-30 days during the later stages of pod development (Cole *et al.*, 1985; Sanders *et al.*,

**Table 2.** Mycoflora of groundnut at different stages of development in 1991.

| Species PDS[1] | 1 | 2 | 3 | 4 | 5 | 6 | 7 | 8 | 9 | 10 |
|---|---|---|---|---|---|---|---|---|---|---|
| *A. flavus* | 1.8 | 16.8 | 4.3 | 0.8 | 0.8 | 7.0 | 7.7 | 1.5 | 2.3 | 2.3 |
| *A. niger* | 16.3 | 53.8 | 43.0 | 46.6 | 50.0 | 71.0 | 37.0 | 25.3 | 27.5 | 22.0 |
| *A. terreus* | 0 | 0 | 1.3 | 0 | 0.8 | 0 | 0 | 0 | 3.3 | 0.8 |

1, Pod development stages (PDS) as defined by Usha *et al.* (1991).

1985). Most edible grade groundnuts were infected when drought stressed plants were grown with geocarposphere temperatures close to 30°C but larger, more mature groundnut kernels required longer periods of drought stress than smaller, more immature kernels (Sanders *et al.*, 1985). No $AFB_1$ occurred in undamaged irrigated crops, irrespective of temperature, nor in drought stressed crops when temperatures were less than 25°C or more than 32°C during this period of susceptibility, even though *A. flavus* was present (Sanders *et al.*, 1983; Cole *et al.*, 1985). Small temperature differences could markedly alter aflatoxin contamination. Lack of aflatoxin production in irrigated plots at high geocarposphere temperatures (mean 34.5°C) may either be due to the irrigation (Hill *et al.*, 1983b) or to the high temperature preventing aflatoxin synthesis (Cole *et al.*, 1985). Drought stress could decrease metabolic activity in the kernels, temperature stress could affect the development of microbial competitors in the geocarposphere and both could, perhaps, affect phytoalexin production (Cole *et al.*, 1989; Mehan *et al.*, 1991).

In maize, fungal colonisation is restricted by the husks that enclose the ears so that kernels remain sterile for up to three weeks after anthesis. However, after eight weeks some grains in all ears carry some fungi (Hesseltine and Bothast, 1977). Colonisation of maize in the field probably occurs from spores produced on overwintered sclerotia, left in the soil at harvest. Contamination with aflatoxins can again occur before harvest. Insect damaged ears are more likely to be infected with *A. flavus* and to contain aflatoxin than undamaged ears. Aflatoxin was found in crops where 21% of ears were damaged by ear worms but not when only 5% were damaged (Lillehoj and Hesseltine, 1977). However, in south-eastern U.S.A., only 10% of *A. flavus* infection was associated with insect damage. Maize is most susceptible to *A. flavus* infection at GS 77-87 (late milk to dough) and high temperatures and drought at this stage increase fungal incidence and aflatoxin contamination before harvest. Consequently, crops sown on different dates may differ in the time and amount of aflatoxin production depending on whether drought and/or temperatures above 30°C occur when the crop is most susceptible (Hill *et al.*, 1985).

*Aspergillus flavus* could usually be isolated from both flag leaves and florets of Indian rice crops from GS 47 but it was infrequent (up to 9% grains infected) and could not be isolated at threshing. Most $AFB_1$, 20.2 ng $g^{-1}$, was found at GS 61 but few samples were contaminated at harvest, with up to 7.5 ng $AFB_1$ $g^{-1}$. *A. niger* was most frequently isolated at GS 57 (21% of grains infected) but its incidence differed between samples and it also was not found after harvest. *A. candidus* and *A. versicolor* were found in small numbers at GS 47.

A range of *Aspergillus* spp. were found on sorghum grain before harvest (Table 3) but the incidence of *A. flavus* differed greatly between farms. It was isolated from both flag leaves and florets but although incidence remained constant to GS 85 on one group of three farms, it then increased greatly to GS 91 on one farm but then decreased to harvest, increased on another to GS 92 while on the third it changed little to harvest but then increased greatly afterwards when threshing was delayed by 7 days. $AFB_1$ was first detected at GS 71 but then increased to GS 75 before declining to harvest. Contamination was greater in cv CSH-5 than in cv local.

Cottonseed in high temperature growing areas of the U.S.A. is often infected by *A. flavus*, especially following bollworm infestation. A 30-45 day period following the opening of the first bolls, when storms disturb the desert raising "sand devils" that may drive spores with soil into the bolls, is critical for toxin formation. Six days is necessary for infection to penetrate from simulated insect damage of the carpel walls to the seed but drying is then necessary for the fungus to penetrate the seed (Lee *et al.*, 1986, 1987).

**Table 3.** Colonisation of sorghum grain during ripening in 1990.

| Species | Percent grains infected | | | | | | | | | |
|---|---|---|---|---|---|---|---|---|---|---|
| G.S. | 45 - 49 | 51 | 59 | 61 | 69 | 71 | 75 | 85 | 91 | 92 |
| *Aspergillus flavus* | 0.8 | 1.0 | 1.0 | 2.5 | 9.7 | 9.5 | 7.3 | 5.2 | 4.5 | 9.0 |
| *A. niger* | 0.2 | 0.8 | 0.3 | 0.5 | 3.2 | 2.7 | 4.3 | 6.0 | 4.3 | 10.8 |
| *A. ochraceus* | - | - | - | - | - | 0.2 | - | - | - | - |
| *A. oryzae* | - | - | 2.8 | 1.2 | 2.8 | 3.5 | 1.3 | 0.5 | 0.7 | 0.5 |
| *A. versicolor* | 0.3 | 0.5 | 0.2 | 0.2 | 0.2 | - | 0.8 | 0.7 | 0.3 | 1.6 |
| *Eurotium* spp. | 0.7 | 0.3 | 0.5 | 0.2 | 0.3 | 0.3 | - | - | 5.3 | 13.0 |
| | $AFB_1$ (ng $g^{-1}$) | | | | | | | | | |
| cv CSH-5 | N/D | N/D | N/D | N/D | N/D | 23.3 | 35.8 | 31.6 | 20.8 | 10.0 |
| cv Local | N/D | N/D | N/D | N/D | N/D | 8.3 | 11.6 | 7.9 | 3.4 | 2.6 |

N/D, not detected.

## POSTHARVEST DEVELOPMENT

There is a profound change in the grain ecosystem at harvest, from the environmental extremes of the field to the comparatively equable environment of the store. It also provides an opportunity for the redistribution of fungal inoculum in the grain and for the introduction of further inoculum. However, although inoculum of *Penicillium* spp. was increased more than 250 times during harvesting by contamination with soil and residues from the harvester, inoculum of Section Aspergillus species was little changed (Flannigan, 1978).

In humid tropical regions, especially on small farms, freshly harvested grain may not be dried immediately. In the Brunca and Huetar Regions of Costa Rica, it may be 5 days before maize grain is dried. Both Regions receive 3.5-3.9 m rain annually yet they differ greatly in the amount of aflatoxin contamination. This appeared to be caused by differences in the handling the crop after harvest. In one region, cobs were shelled immediately after harvest while in the other the grain was kept on the cob until just before drying. Experiments compared both methods of postharvest treatment in both regions. *A. flavus* was present in nearly all samples and increased markedly in some to infect up to 90% of grains after 5 days of harvest and 98% after 8 days. *A. flavus* was isolated from twice as many shelled grains as from kernels stored on the cob. Up to 18.5% (mean 7.0%) of shelled kernels were damaged but only 0.26% of those stored on the cob. Heaviest *A. flavus* infection was found on the most heavily damaged kernels but neither amounts of damage or water content showed a close relationship with infection. Aflatoxin contamination increased with shelling, water content and time before drying, especially if this extended beyond 5 days (Fig. 1). After a six day delay, aflatoxin contamination could increase during the next 12 h from about 50 ng $g^{-1}$ by three to ten times. Differences induced by postharvest treatment were maintained through storage. Populations of *A. flavus* and *Fusarium* spp. in the maize appeared to be inversely related. *A. flavus* populations were small when there was much *Fusarium* infection while numbers of *Fusarium* spp. declined when *A. flavus* infected more than 40-60% of the grains. *Aspergillus parasiticus* and *A. flavus* may also be important in Indian rice when drying is delayed after wet harvests (Sahay and Gangopadhyay 1985).

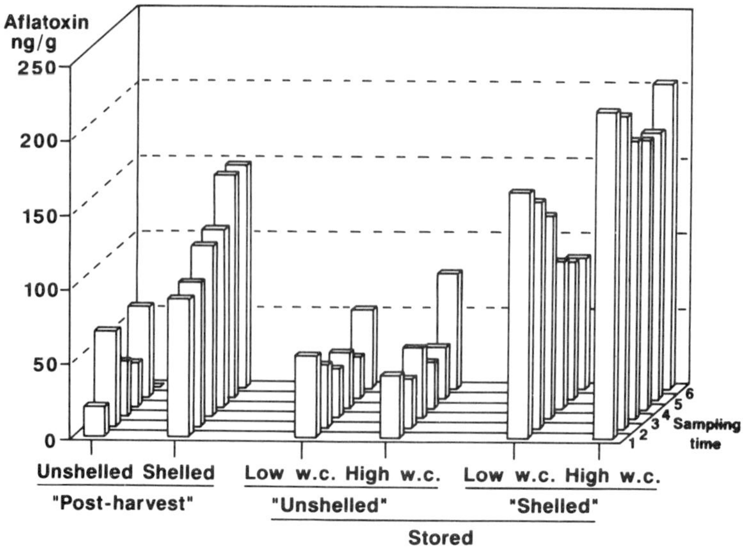


**Figure 1.** Mean aflatoxin contamination of maize during post-harvest treatments and storage. (Sampling was daily during the post-harvest period and monthly during storage; shelled and unshelled refer to postharvest handling).

Postharvest treatment also affects *A. flavus* and aflatoxin contamination of groundnuts. Kernels contain 20-50% water at lifting, depending on soil conditions, and are most susceptible to *A. flavus* infection while they contain 12-30% water. Infection during field drying depends on the extent of mechanical damage at lifting, the attitude of the lifted plants, weather conditions and the rate of drying. Slow drying during wet seasons, when rain falls at night, especially on partially dried pods, or plants are left in random windrows allows more infection than in dry seasons, when rain falls early in the day or when plants in the windrow are all inverted (Mehan *et al.*, 1991). When the pods are separated from the haulms and dried in thin layers with protection from rain, drying is rapid and infection slight (Burrell *et al.*, 1964). Nevertheless, even in periods of rain, the risk of *A. flavus* infection and aflatoxin contamination is likely to be less in pods on windrowed plants than in combined seeds held in baskets or in poorly ventilated driers (Dickens, 1977).

## DEVELOPMENT IN STORAGE

### Temperature and water availability

*Aspergillus* spp differ widely in their temperature and water requirements. Each has a characteristic temperature range and minimum and optimum water activities ($a_w$) for germination, growth and sporulation (Tables 4, 5). Usually, growth requires higher $a_w$ than germination but lower than sporulation (Magan and Lacey, 1984a). Also

**Table 4.** Temperature and water requirements of *Aspergillus* species

| | Temperature (°C) | | Water activity for growth ($a_w$) | |
|---|---|---|---|---|
| | Range | Optimum | Minimum | Optimum |
| *A. rubrobrunneus* | 5-40 | 25-27 | 0.70 $a_w$ | 0.93 $a_w$ |
| *A. restrictus* | 9-40 | 30 | 0.71 $a_w$ | 0.96 $a_w$ |
| *A. chevalieri* | 5-43 | 30-35 | 0.71 $a_w$ | 0.93-0.95 $a_w$ |
| *A. vitis* | 5-46 | 33-35 | 0.71-0.73 $a_w$ | 0.93-0.96 $a_w$ |
| *A. reptans* | 7-40 | 25-27 | 0.75 $a_w$ | 0.90-0.95 $a_w$ |
| *A. candidus* | 3-44 | 25-32 | 0.75 $a_w$ | 0.90-0.98 $a_w$ |
| *A. ochraceus* | 8-37 | 24-31 | 0.77 $a_w$ | 0.95-0.98 $a_w$ |
| *A. versicolor* | 4-39 | 25-30 | 0.78 $a_w$ | 0.95 $a_w$ |
| *A. flavus* | 6-45 | 35-37 | 0.78 $a_w$ | 0.95 $a_w$ |
| *A. fumigatus* | 10-55 | 40-42 | 0.85 $a_w$ | 0.98-0.99 $a_w$ |

**Table 5.** Minimum water potentials for germination, growth and sporulation of *Aspergillus* spp. at 25°C (after Magan and Lacey, 1984a)

| | Water activity ($a_w$) | | |
|---|---|---|---|
| Species | Germination | Linear growth | Sporulation (anamorph) |
| *A. candidus* | 0.78 | 30.7 | 25.6 |
| *A. fumigatus* | 0.94 | 0.94 | 0.95 |
| *A. vitis* | 0.72 | 0.73 | 0.78 |
| *A. nidulans* | 25.6 | 30.7 | 30.7 |
| *A. reptans* | 0.72 | 0.75 | 0.78 |
| *A. versicolor* | 37.8 | 0.78 | 30.7 |

teleomorph production may require higher $a_w$ than anamorph. Thus in Section Aspergillus, conidia can form down to 0.75-0.78 $a_w$ and ascospores to 0.77-0.86 $a_w$ while the corresponding figures for *A. nidulans* are, respectively, 0.85 and 0.95 $a_w$ (Pitt, 1975; Lacey, 1986). In the more xerophilic species, growth is greatest, not when water is freely available, but with 0.90-0.95 $a_w$, depending on incubation temperature, pH, solutes in the medium and, possibly, on whether growth is measured as biomass in liquid medium or as colony diameters on agar (Avari and Allsopp, 1983; Magan and Lacey, 1984a)

Temperature and $a_w$ together determine which species can grow, their relative growth rates (Fig. 2) and their abundance (Table 6). $A_w$ and temperatures favouring colonisation by many species with *Aspergillus* anamorphs are well known. The predominant species can thus indicate previous storage conditions. Many of those fungi numerous before harvest, e.g., *Alternaria*, persist while grain is stored well with no moulding or heating but rapidly disappear if these occur. The most xerophilic

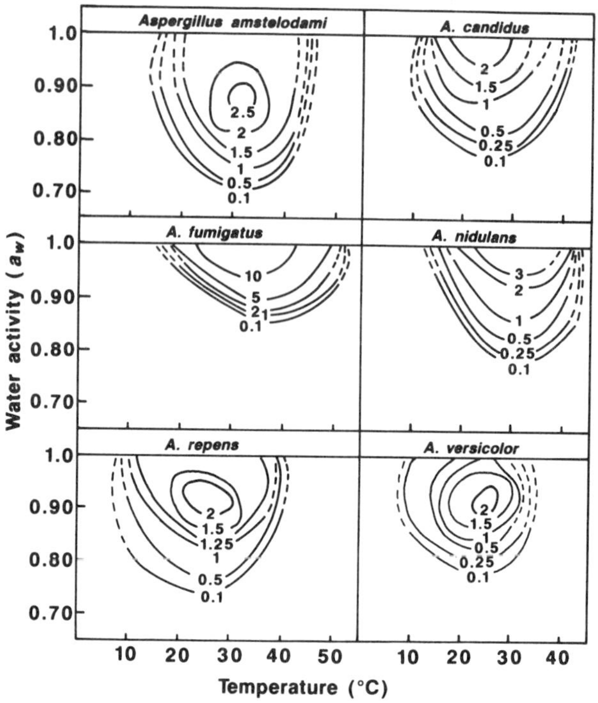


**Fig 2.** Effects of temperature and water potential on growth rates of *Aspergillus* species. (Numbers on the isopleths indicate growth rates in mm day$^{-1}$). (Reprinted from Magan, N. and Lacey, J. (1984 a) Effect of temperature and pH on water relations of field and storage fungi. Trans. Br. Mycol. Soc. 82, 71-81.)

aspergilli in stored grain are species of the Section Aspergillus and *Aspergillus restrictus* which can grow down to 0.70 $a_w$. However, their slow growth may prevent detection before 3 months. As water becomes more readily available, more species are able to grow and spontaneous heating occurs, through energy being released by respiration faster than it can be dissipated. High temperatures favour species, like *A. fumigatus*. However, the conditions at which individual species are most numerous may be those at which they survive or compete better than others rather than those at which they grow best in pure culture but. Thus, Section Aspergillus species are most abundant in grain at 30°C and 0.70 $a_w$ but grow best at 30°C and 0.90 $a_w$. Flooding in Canada resulted in strata of grain with different microflora (Mills and Abramson, 1981). Putrid submerged grain was covered by a layer of damp germinated grain, colonized by *A. flavus*, *A. versicolor* and *Penicillium* spp., with species in Section Aspergillus only in the drier surface grain. Many of the fungi found in maize, rice, sorghum and other cereals grown in the tropics are the same as those in temperate grains, apart from a few additional species and, perhaps, a greater frequency of *Aspergillus* and fewer or different *Penicillium* spp. in tropical grains.

Heating of grain during drying is unlikely to kill most *Aspergillus* spp. since conidia of *A. candidus* will survive 10 min at 50°C while 18-25% of *Eurotium (A.) chevalieri* ascospores survive 10 min at 70°C and a decimal reduction time for this species at 80°C ($D_{80°C}$) has been calculated as 3.3 min.

Chilling, in low ambient temperatures or by ventilation with cool air, increases

**Table 6** Effects of water availability and spontaneous heating on colonisation of cereal grains by *Aspergillus* spp. and other fungi.

| $a_w$ | Max. temp. (°C) | Germination (%) | Predominant fungi | |
|---|---|---|---|---|
| | | | *Aspergillus* | Other fungi |
| <0.60 | Ambient | 90 - 100 | | Field fungi |
| 0.75 | Ambient | 75 - 90 | *A. restrictus* | |
| 0.85 | Ambient | 45 - 76 | *A. vitis*<br>*A. reptans*<br>*A. rubrobrunneus* | |
| 0.87 | Ambient | nd | | *P. brevicompactum* |
| 0.88 | Ambient | nd | | *P. verrucosum* |
| 0.89 | Ambient | nd | | *P. expansum* |
| 0.90 | 25 | 15 - 45 | *A. versicolor* | *P. aurantiogriseum*<br>*P. hordei* |
| 0.92 | 30 | nd | | *P. capsulatum* |
| 0.93 | 35 | nd | | *P. piceum* |
| 0.95 | 50 | 0 - 15 | *A. candidus*<br>*A. flavus*<br>*A. niger*<br>*A. ochraceus*<br>*A. terreus*<br>*A. nidulans* | *P. purpurogenum*<br>*Absidia corymbifera* |
| >0.95 | 60 | 0 | *A. fumigatus* | *Rhizomucor* spp.<br>*Thermomyces* spp. |

Key: *A.*, *Aspergillus*; *P.*, *Penicillium*; nd, Not determined.

the predominance of *Penicillium* spp., especially at high $a_w$. However, *Aspergillus flavus* and *A. versicolor* could still become abundant in cold Canadian winters when heating from -5° to 0°C was initiated by *P. aurantiogriseum*. However, it was killed by heating to 64°C (Sinha and Wallace, 1965).

Fungi, including *Aspergillus* spp., may penetrate the aleurone layer of wheat grains to the sub-aleurone and endosperm cells (Chełkowski and Cierniewski, 1983) and contribute greatly to loss of viability (Tuite and Christensen, 1955). Storage period and water availability accounted for most of the total variability in Canadian stored wheat grain. In the first principal component they contributed to an increase in *Aspergillus* spp., including *A. versicolor*, and to a decrease in field fungi and germinability (Wallace and Sinha, 1981). Damage to maize kernels, especially near the embryo, increased susceptibility to invasion and increased the likelihood of visible molding (Perez *et al.*, 1982; Tuite *et al.*, 1985) but mould damage in barley appeared independent of the degree of injury (Welling, 1968).

Growth, sporulation and toxin production by *A. flavus* may be governed by total heat input (Stutz and Krumperman, 1976) although the relationship may be complex (Park and Bullerman, 1981, 1983). At least 208 degree hours per day were required for growth and 270 degree hours per day for sporulation and toxin production. Below the optimum, 772 degree hours per day, temperature cycling decreased the time required for sporulation and toxin formation. Growth, sporulation and toxin production by *A. flavus* were faster when temperatures alternated every 12 h between 5 and 25°C than at 15°C but slower than at 18 or 25°C but with *A. parasiticus*, production was greater with cycling temperatures than at all three temperatures.

*Aspergillus* spp. in moulding hay are similar to those in grain stored at the same $a_w$ and temperatures (Gregory *et al.*, 1963; Festenstein *et al.*, 1965). Section Aspergillus species, especially *A. vitis* and *A. reptans*, predominate in hays following storage at about 0.90 $a_w$ with heating to 35°C, *A. versicolor* at 0.95 $a_w$ and 40°C, *A. nidulans* at 0.97 $a_w$ and 50°C and *A. fumigatus* at more than 0.98 $a_w$ with heating to 60-65°C. Methods of hay-making which restrict heating may modify colonisation as in Finland, where hay dried on racks over a long period becomes colonised by *A. rubrobrunneus* (*A.umbrosus*) (Terho and Lacey, 1979).

**Gas composition**

Fungi are usually considered to be aerobic but their tolerance of high $CO_2$ concentrations is often under-estimated and some species may grow anaerobically while they efficiently scavenge $O_2$ and can often grow in atmospheres containing <1% $O_2$ (Magan and Lacey, 1984b). Spores usually retain their viability in 50-80% $CO_2$, at least at low $a_w$ (Peterson *et al.*, 1956; Wells and Uota, 1970), but anaerobiosis may sharply decrease survival, especially in *Aspergillus versicolor*, *A. fumigatus* and *A. flavus* (Richard-Molard *et al.*, 1980). $N_2$ probably affects fungi only through its low oxygen content but several species, including *A. reptans*, *A. niger* and *A. fumigatus*, have been shown to grow well in 100% $N_2$, sometimes at rates close to those obtained in air (Hocking, 1990) especially if certain growth factors, such as vitamins, oxygen donors and higher oxidation forms of certain elements are present (Tabak and Cook 1968).

Changing the proportions of $CO_2$ and $O_2$ in the atmosphere can affect spore germination, growth rate and speed of sporulation. Time taken for spore germination (lag phase) correlates negatively with $O_2$ and positively with $CO_2$ concentrations (Magan and Lacey, 1984b). At 23°C/0.98 $a_w$, decreasing $O_2$ from 21% to 0.14% increased the lag phase in *A. candidus* from ≤ 1 day to 4 days and while at 0.85 $a_w$/23°C in 1% $O_2$ it was 13 days. Inhibition by $CO_2$ is much greater if concentrations of $O_2$ are low than when it is abundant. *Aspergillus candidus*, *A. fumigatus* and *A. reptans* failed to grow in even 10% $CO_2$ if oxygen was absent. Some species are stimulated by low concentrations of $CO_2$ but *A. reptans* showed only slight stimulation in 5% $CO_2$ at 23°C and 0.95 $a_w$. Otherwise, growth rates declined with increasing $CO_2$ concentration. Concentrations of $CO_2$ and $O_2$ at which growth rates were halved ($LC_{50}\ CO_2$, $LC_{50}\ O_2$) are shown in Table 7. *Aspergillus fumigatus*, *A. nidulans* and *A. versicolor* were among the most sensitive of 14 species to $CO_2$ concentration (Magan and Lacey, 1984b). Growth of some *Aspergillus* spp. was only affected with <5% $O_2$. The effects of $CO_2$ and $O_2$ on growth were enhanced by low temperatures.

Fungi sometimes appear more tolerant of $CO_2$ in stored products than in culture. Thus, Section Aspergillus species survived and grew in 50% $CO_2$/21% $O_2$ and 79% $CO_2$/21% $O_2$ in grain (Petersen *et al.*, 1956; Wells and Uota, 1970) although not in 85% $CO_2$/3% $O_2$ (Hocking, 1990) even though growth in agar showed little evidence of such tolerance. Their numbers were little affected during 84 days storage of wheat in bins with 6 or 14-day half-lives, with $CO_2$ concentrations maintained at 15% but increased to 50% every 7 days during the first month and with $O_2$ concentrations never below 11% (White and Jayas, 1991). By contrast, *A. chevalieri*, in maize stacks enclosed in PVC sheet, was decreased significantly following treatment with 2.4 kg $CO_2$/ton (Dharmaputra *et al.*, 1991). Maize stored at 0.75-1.00 $a_w$ and 25-45°C in

**Table 7.** Concentrations of carbon dioxide ($LC_{50}\ CO_2$) and oxygen ($LC_{50}\ O_2$) required to halve linear growth of field and storage fungi at 23 and 14°C.

| Species | Temperature | | | | | | |
|---|---|---|---|---|---|---|---|
| | 23°C | | | | 14°C | | |
| $a_w$ | 0.98 | 0.95 | 0.90 | 0.85 | 0.98 | 0.95 | 0.90 |
| | Carbon dioxide | | | | | | |
| *A. candidus* | >15.0 | >15.0 | >15.0 | 10.0 | 4.5 | 13.0 | 4.0 |
| *A. fumigatus* | >15.0 | 5.2 | 12.5 | N.G. | N.T. | - | - |
| *A. versicolor* | 12.0 | >15.0 | 14.5 | N.G. | N.T. | - | - |
| *A. reptans* | >15.0 | >15.0 | >15.0 | 13.0 | >15.0 | 9.0 | >15.0 |
| | Oxygen | | | | | | |
| *A. candidus* | 0.45 | 1.00 | 0.45 | 5.00 | 5.80 | <0.17 | 9.40 |
| *A. fumigatus* | 3.40 | 5.40 | 6.20 | N.G. | N.T. | - | - |
| *A. versicolor* | 6.40 | 0.80 | 4.50 | N.T. | N.T. | - | - |
| *A. reptans* | 0.60 | 3.00 | 5.00 | 10.20 | 0.85 | 0.90 | 4.00 |

*A.*, *Aspergillus*; N.G. no growth; N.T., not tested.

atmospheres containing 0.1% $O_2$ developed significantly less mould than in air (Bottomley *et al.*, 1950) while only *Aspergillus candidus* grew well in <0.2% $O_2$ in wheat containing 17.4% water (Shejbal and Di Maggio, 1976; Di Maggio *et al.*, 1976). Barley, maize, rice, sorghum and groundnut were colonised less by *Aspergillus* spp., including *A. flavus, A. terreus* and Section Aspergillus species, in 10% $CO_2$/10% $O_2$, 15% $CO_2$/15 or 18% $O_2$ or 20% $CO_2$/ 20% $O_2$ atmospheres than in air. Growth and sporulation of *A. flavus* on groundnut decreased with each 20% increment of $CO_2$ until growth was totally inhibited above 80% $CO_2$ (Landers *et al.*, 1967; Sanders *et al.*, 1968). Growth in <5% $O_2$ was much less than in air and was almost completely inhibited with <1% $O_2$. However, populations of *A. flavus* in shelled (5% water content) or unshelled (7.5% water content) ground-nuts stored in 3% $O_2$ or 82% $CO_2$ did not differ from those in sinilar groundnuts stored in air (Jackson and Press, 1967).

Storage of maize or barley containing 19-40% water (about 0.90-1.0 $a_w$) in sealed or unsealed silos relies on respiration replacing $O_2$ with $CO_2$. Concentrations can reach 60->90% $CO_2$ and <10% $O_2$ within three weeks of filling but may decline to only 15% $CO_2$, even in sealed silos, through imperfect sealing, daily temperature and pressure fluctuations and removal of grain. When nearly empty, respiration is often insufficient to maintain an inhibitory atmosphere and moulding may occur, in the later stages when spontaneous heating occurs, with *A. nidulans* and *A. fumigatus*. The microflora in unsealed silos is similar but heating and moulding is controlled by the rate of unloading at the exposed surface (Burmeister and Hartman, 1966; Lacey, 1971).

Mycotoxin production is generally confined within narrower limits of gas concentration, $a_w$ and temperature than growth and can be inhibited before growth is affected. Conditions for mycotoxin production differ with species and toxin. Thus, aflatoxin was produced at 25-35°C in both air and a controlled atmosphere of 10% $CO_2$/1.8% $O_2$/88.2% $N_2$ but no toxin was produced at 15°C although growth was not affected while weak growth still occurred at 12 °C but only in air (Epstein *et al.*, 1970).

Although B vitamins could enhance aflatoxin production in $N_2$ atmospheres 15-fold, they failed to reverse $CO_2$ inhibition (Clevström *et al.*, 1983).

In groundnut, increasing $CO_2$ concentration from 0.03% to 20% did not affect on growth and sporulation of *A. flavus* visibly but aflatoxin production was decreased to 25% of that in air (Table 8). Increasing $CO_2$ concentration further decreased aflatoxin production up to 80% $CO_2$. Aflatoxin production declined with concentration from 15% $O_2$ although inhibition was most marked below 5% $O_2$ when growth and spore production were first affected (Table 8). Decreasing $O_2$ as $CO_2$ was increased affected aflatoxin production more severely than when either was altered alone (Table 8). Only 11% of the aflatoxin formed in air was produced with 20% $CO_2$ and 20% $O_2$, 0.6% with 20% $CO_2$ and 5% $O_2$ and none with larger $CO_2$ concentrations (Landers *et al.*, 1967; Sanders *et al.*, 1968; Diener and Davis, 1977).

**Table 8.** Aflatoxin production in groundnut kernels at 0.99 $a_w$ after 42 days in different concentrations of $CO_2$ and $O_2$, as percentages of production in air (after Diener and Davis 1977).

| $CO_2$ concentration (%) | $O_2$ concentration (%) | | | | | |
|---|---|---|---|---|---|---|
| | 20 | 15 | 10 | 5 | 1 | 0.1 |
| 0 | 100 [1,2] | 101[2] | 61.8[2] | 30.1[2] | 1.2[2] | <0.1[2] |
| 20 | 24.9[2] | | | | | |
| | 10.9[3] | | | 0.6[3] | | |
| 40 | 11.8[2] | | | 0[3] | | |
| 60 | 6.6[2] | | | <0.1[3] | | |
| 80 | <0.1[2] | | | 0[3] | | |
| 100 | 0[2] | | | | | |

1, air, 0.03% $CO_2$/21% $O_2$; all other treatments $CO_2$ or $O_2$ with balance of $N_2$;
2, incubation at 30°C; 3, incubation at 15°C.

Water activity and temperature modify interactions with the gaseous components of the environment. Aflatoxin production in 20% $CO_2$ decreased as $a_w$ decreased from 0.99 to 0.86 $a_w$ but little or none was produced with 40 or 60% $CO_2$ at even 0.92 $a_w$ (Table 9). Decreasing temperatures from 25°C to 17°C decreased aflatoxin production in all combinations of $CO_2$, $O_2$ and $a_w$ tested. Aflatoxin production and groundnuts could also be prevented with carbon monoxide in controlled atmospheres. Large amounts of aflatoxin accumulated in in air or 13.6% $CO_2$/0.1% CO/1.5% $O_2$/83.9% $N_2$ but with <1% $O_2$ or >60%$CO_2$, aflatoxin production was largely inhibited, although visible fungal growth was not prevented (Wilson and Jay, 1975, 1976). Atmospheres containing 2% $O_2$/10% CO inhibited aflatoxin production by 98% of that in 2% $O_2$ or in air (Buchanan *et al.*, 1985).

Ochratoxin production by *Aspergillus ochraceus* in semi-synthetic media at 16°C was completely inhibited in atmospheres containing 30% $CO_2$ or more, regardless of $O_2$ concentration (Paster *et al.*, 1983). Growth was only inhibited when $CO_2$ concentrations exceeded 60% and was completely stopped by 80% $CO_2$. With atmospheres containing 10 or 20% $CO_2$, ochratoxin production was decreased if there was <20% $O_2$ but enhanced by 40 or 60% $O_2$. Ochratoxin production with 1 or 5% $O_2$ was as good as in air in the absence of $CO_2$.

**Table 9.** Effects of $CO_2$ concentration and relative humidity on aflatoxin formation in groundnut kernels at 25°C and different RH after 14 days (Diener and Davis 1977).

| Gas composition (%) | | Water activity ($a_w$) | Kernel water content (%) | Aflatoxin content ($\mu g\ g^{-1}$) |
|---|---|---|---|---|
| $CO_2$ | $O_2$ | | | |
| 0.03[1] | 21 | 0.99 | 22.4 | 206.3 |
| 0.03 | 21 | 0.92 | 26.5 | 185.2 |
| 0.03 | 21 | 0.86 | 15.0 | 72.1 |
| 40 | 20 | 0.99 | 29.2 | 3.8 |
| 40 | 20 | 0.92 | 24.8 | 0.3 |
| 40 | 20 | 0.86 | 17.4 | 0 |
| 60 | 20 | 0.99 | 24.5 | 0.2 |
| 60 | 20 | 0.92 | 20.0 | <0.1 |
| 60 | 20 | 0.86 | 11.8 | 0 |
| Untreated | | | 6.0 | 0 |

1, balance made up of $N_2$

## Fungal Interactions

Feeds and seeds are usually contaminated by a range of different fungi during growth, harvesting and storage. Propagules of different species may be close to one another or widely dispersed through the bulk. When environmental conditions are suitable, they germinate, grow and interact with one another, competing for substrate with other species. Interspecific interactions have often been studied in artificial media but these differ chemically and physically from natural substrates. Interaction effects reported among aspergilli include inhibition of growth and of mycotoxin production.

Magan and Lacey (1984c) gave numerical values to different types of interaction in agar culture and used these to calculate indices of dominance ($I_D$) for different pre-harvest, *Aspergillus* and *Penicillium* spp. competing with one another. *A. candidus, A. fumigatus* and *A. nidulans* gave high scores at 25°C and 0.98 $a_w$, although lower than *Penicillium brevicompactum*, but became less competitive as $a_w$ decreased, while *A. reptans* and *A. versicolor* were more competitive at 0.95 $a_w$ than at 0.98 or 0.90$a_w$. Some species became more competitive fungi as temperature was increased, from 15°C to 25°C for *A. candidus* and *A. nidulans* and to 30°C for *A. fumigatus*. However, *Aspergillus* spp. competed poorly against the preharvest species, *Epicoccum purpurascens* and *Fusarium culmorum*. On autoclaved wheat grain inoculated simultaneously with five *Aspergillus* spp., patterns of colonisation were observed that it was difficult to equate with Petri dish studies of competitiveness (Magan and Lacey, 1985). At 15°C and 0.95$a_w$, *A.versicolor* and *A. reptans* increased first followed by *A. candidus*. At 25°C, *A. versicolor* and *A. candidus* predominated and, at 30°C, *A.nidulans* and *A. versicolor* with *A. fumigatus*. At 0.90 $a_w$, populations increased more slowly but the species isolated were similar, except that *A. fumigatus* could not be isolated at 30°C. These results were obtained by dilution plating but sometimes directly plated grains yielded species not found on dilution plates, e.g., *A. reptans* at 30°C and 0.90 $a_w$.

Field studies suggested that interactions between *A. flavus* and other fungi might lead to the inhibition of aflatoxin production. Positive associations have been noted between *A. flavus* and *A. niger* and between *A. niger* and *P. funiculosum* on maize ears (Wicklow, 1988) which could affect aflatoxin production. Aflatoxin was formed in maize grain when the ratio of propagules of *A. flavus* to *A. niger* exceeded 19:1 but not when it was less than 9:1 (Hill *et al.*, 1983b). *A. niger* is reported to decrease aflatoxin contamin-ation (Wicklow *et al.*, 1980; Horn and Wicklow, 1983) while *Penicillium rubrum* may both enhance and inhibit and rubratoxin and T-2 toxin enhance aflatoxin production by *A. parasiticus* in culture (Fabbri *et al.*, 1984; Moss and Frank, 1985). When inoculated simultaneously into autoclaved grain, *A. flavus* and *A. niger* each covered half the grain but no *A. flavus* was seen with *T. viride* (Wicklow *et al.*, 1980). Aflatoxin production was greatly decreased in both pairings. The effect of *A. niger* has been attributed to lowering of the substrate pH, an inhibitory substance produced by the fungus or to degradation of aflatoxin (Boller and Schroeder, 1974). Aflatoxin production by *A. parasiticus* was also inhibited by *A. chevalieri* and *A. candidus* when *A. parasiticus* could still be isolated from many grains (Boller and Schroeder, 1973, 1974). The bacteria, *Brevibacterium linens* and *Streptococcus lactis*, have both been reported to inhibit and *Acetobacter aceti* to enhance aflatoxin production in *A. parasiticus* or *A. flavus* (Weckbach and Marth, 1977; Wiseman and Marth, 1981).

Cuero *et al.* (1987, 1988) used irradiated maize grain to avoid the large physical changes that occur when grain is autoclaved and incubated inoculated grain in microporous bags to maintain close environmental control. Zearalenone production by *Fusarium graminearum* was generally inhibited by the presence of *A. flavus* but enhanced toxin production was found in pairings of *A. flavus* with the bacterium, *Bacillus amyloliquefaciens* and the yeast, *Hyphopichia burtonii*, at 0.90-0.98 $a_w$ and 16 or 25°C. The filamentous fungi, *A. niger*, *Penicillium viridicatum* and *Fusarium graminearum* inhibited aflatoxin production at 0.98 $a_w$ but stimulated it at 0.95 and 0.90 $a_w$. However, Ramakrishna (1990) found that the effect on mycotoxin production was not only a function of $a_w$, temperature and the species paired but also of period of incubation. Enhancement was often only temporary and could be followed by a decline in mycotoxin production.

### Chemicals

Chemicals, especially volatile fatty acids, have been used as preservatives for stored feeds but inadequate treatment can have undesirable effects. Propionic acid is effective when uniformly applied, the dose is adequate for the water content of the substrate, about 1.3 g propionic acid per 100 ml water, and allowance is made for application losses and variations in swath density and water content (Lacey and Lord, 1977). Larger doses of formic acid are required to give equal mould control. Some fungi, including Section Aspergillus species and *A. flavus*, can tolerate concentrations of organic acids sufficient to control other fungi, metabolise them and allow less tolerant species to grow. *A. flavus* is tolerant of both propionic and formic acids and both can stimulate aflatoxin production barley grain or hay are inadequately treated (Hacking and Biggs, 1979; Clevström *et al.*, 1981). Tolerance of a range of *Aspergillus* species decreased with decreasing $a_w$ (Magan and Lacey, 1986). *A. flavus* and *A. nidulans* were more tolerant than *A. fumigatus, A. niger* and *A. versicolor* at 0.95 $a_w$ and *A. rubrobrunneus* was more tolerant than *A. vitis* and *A. reptans* at 0.85$a_w$. Section Aspergillus species tended to predominate in propionic acid treated hay and rice grain

(Lacey *et al.*, 1983). Oil seeds may require larger doses of propionic acid than cereal grains since colonisation of wheat grain (30% water content) inoculated with *A. parasiticus* required 3 mg $g^{-1}$ for inhibition of growth and $AFB_1$ production for at least 14 days at 30°C and sunflower seeds 4 mg $g^{-1}$ and with 20% water content, respectively, 2 and 3 mg $g^{-1}$.

Treatment of wheat seed (20% water content), inoculated with *A. parasiticus*, with 1 mg sorbic acid $g^{-1}$ was insufficient to prevent moulding and aflatoxin formation. Aflatoxin production was enhanced throughout 21 days storage of seeds treated with 0.5 mg sorbic acid $g^{-1}$ but only after 21 days with 1 mg $g^{-1}$. The sterol biosynthesis inhibitors, itraconazole and isoconazole, at $10^{-6}$, $10^{-5}$ and $10^{-4}$ M, delayed both growth of *A. parasiticus* and aflatoxin production in inoculated wheat and sunflower seeds (20% water content). Ketoconazole was less effective.

Essential oils of spices differed in their effectiveness in preventing moulding and species differed in their sensitivity. Both *A. flavus* and *A. niger* persisted in rice grain treated with up to 6 μl cinnamon oil $g^{-1}$ but *A. versicolor* predominated, especially at 0.90 $a_w$. *A.flavus* could not be isolated from rice grain treated with 8 μl $g^{-1}$ 0.85 $a_w$. Section Aspergillus species were most abundant after 60 days storage and then declined as *A. versicolor* increased. *A. candidus, A. terreus* were also isolated from treated grain. Treatment of sorghum with 4 μl cinnamon oil $g^{-1}$ completely inhibited Section Aspergillus species. Neem, pongamia and mustard oils failed to inhibit growth of *A. parasiticus*. However, almond and cardamom oils decreased *A. flavus* in rice grain but Section Aspergillus species and *A. versicolor* both increased.

Fumigants have chiefly been used for the control of insects in stored seeds. However, some may inhibit fungi although much larger doses are required than to kill insects. All spores of *Aspergillus ochraceus* and *A. flavus* were killed by 120 mg methyl bromide $l^{-1}$ for 4 h (ctp 480 gh $m^{-3}$) at 25°C but only 60% of *A. niger* spores. which required 40 mg $l^{-1}$ for 24 h (ctp 960 gh $m^{-3}$). The maximum dose used to kill insects in France is 100 gh methyl bromide $m^{-3}$ (Pierrot, 1988). A given ctp was more effective if given as a high dose over a short period of time than as a smaller dose over a longer period. Treatment with 0.1-0.3% sulphur dioxide controlled most moulds in maize. Spore germination in *A. flavus, A. ochraceus* and *A. terreus* was inhibited by 50 mg $SO_2$ $l^{-1}$ but not in *A. niger* (Magan, unpublished data). Non fungicidal doses of phosphine (0.3 mg $l^{-1}$ for 21 days, ctp 150 gh $m^{-3}$) decreased growth and aflatoxin production by *A. flavus* group isolates (Leitao *et al.*, 1987) and storage fungi, especially *Aspergillus parasiticus*, developed less rapidly in inoculated freshly harvested paddy rice treated with 0.1 mg phosphine $m^{-3}$ for 14 or 28 days (ctp 33.6 and 67.2 gh $m^{-3}$) (Hocking and Banks, 1991). After four weeks phosphine treatment, populations of *A. parasiticus* and *A. chevalieri* were two orders of magnitude smaller and aflatoxin production was halved but larger doses of phosphine are needed for periods of storage longer than a few days.

**Irradiation**

*Aspergillus* spp. in cereal grains were mostly inhibited by 4 kGy γ-irradiation but *A. flavus* in rice required 6 kGy. Larger doses were required to inhibit fungi in groundnut, 8 kGy for *A. flavus* and *A. niger* and 10 kGy for *A. versicolor*. However, barley grain treated with only 2 kGy was recolonised by *A. vitis* within 2 weeks and rice grain by *A. versicolor* after 30 days at 27°C/0.90 $a_w$. Section Aspergillus species also colonised sorghum treated with up to 8 kGy after 30 days although it had not been evident before irradiation.

## CONCLUSION

*Aspergillus* spp. are widespread and versatile storage fungi. They can tolerate and grow in a wide range of environmental conditions and it is difficult to entirely eliminate them using conventional drying and storage methods, especially in developing countries where harvesting and drying conditions frequently mean that feeds and seeds must be stored wetter than is desirable and where contamination with aspergilli and mycotoxins may occur before harvest. Prevention of mould and mycotoxin contamination of stored grain therefore requires an integrated approach to ensure that feeds and seeds for storage are free from mycotoxins before being placed in store, damage is minimal, drying is carried out as far as possible and the material is then kept dry and other available treatments are applied to prevent moulding. Tests of mould control methods have generally emphasised one control strategy, neglecting possible synergism between different methods of control. For instance, fungi generally tolerate lower water activities when temperatures are optimal than above or below the optimum (Northolt *et al.*, 1979; Magan and Lacey, 1984a) and interactions between $CO_2$ and $O_2$ concentrations have been described above. Busta *et al.* (1980) concluded that three way interactions between $a_w$, temperature and atmosphere were likely. Recently, interactions between physical and chemical treatments have been utilised to extend the period of storage without moulding (Paster *et al.*, 1992). Treatment with 0.2% propionic acid, 2 kGy $\gamma$-irradiation and modified atmospheres with 40 or 60% $CO_2$ and 20% $O_2$ prevented moulding better than any of the component treatments and allowed storage of maize with 18% water content for up to 45 days with little fungal development. In developing countries, modified atmospheres may be provided by the fermentation of waste materials (Paster *et al.*, 1990). Such methods would control *Aspergillus* spp. together with other fungi.

## ACKNOWLEDGEMENTS

Part of the work reported here was funded by the the Commission of the European Community under contract TS2.0164 M(H) and the collaboration of Dr N. Paster (ARO Volcani Institute, Israel), Professor C. Fanelli (University of Rome "La Sapienza", Italy), Professor H.S. Shetty (University of Mysore, India) and Dr M. Mora (University of Costa Rica) is gratefully acknowledged. The contributions of Drs R.A. Hill, N. Magan, R.G. Cuero and N. Ramakrishna are also acknowledged.

## REFERENCES

Abdel - Kader, M.I.A., Moubasher, A.H. and Abdel - Hafez, S.I.I. (1979) Survey of the mycoflora of barley grains in Egypt. Mycopathologia 68, 143 - 147.

Avari, G.P. and Allsopp, D. (1983) The combined effects of pH, solutes, and water activity ($a_w$) on the growth of some xerophilic *Aspergillus* species. In "Biodeterioration 5" (Oxley, T.A. and Barry, S. Eds.) pp. 548-556. John Wiley, Chichester.

Boller, R.A., and Schroeder, H.W. (1973) Influence of *Aspergillus chevalieri* on production of aflatoxin in rice by *Aspergillus parasiticus*. Phytopathology 63, 1507-1510.

Boller, R.A., and Schroeder, H.W. (1974) Influence of *Aspergillus candidus* on production of aflatoxin in rice by *Aspergillus parasiticus*. Phytopathology 64, 121-123.

Bottomley, R.A., Christensen, C.H. and Geddes, W.F. (1950) Grain storage studies IX. The influence of various temperatures, humidities and oxygen concentration on mould growth and biochemical changes in stored yellow corn. Cereal Chem. 27, 271-296.

Buchanan, J., Sommer, N.F. and Fortlage, R.J. (1985) Aflatoxin suppression by modified atmospheres containing carbon monoxide. J. Am. Soc. Hort. Sci. 110, 638-641.

Burmeister, H.R. and Hartman, P.A. (1966) Yeasts in ensiled high moisture corn. *Appl. Microbiol.* 14, 35-38.
Burrell, N.J., Grundey, J.K. and Harkness, C. (1964) Growth of *Aspergillus flavus* and production of aflatoxin in groundnuts. Part V. Trop. Sci. 6, 74-90.
Busta, F.F., Smith, L.B. and Christensen, C.M. (1980) Microbiology of controlled atmosphere storage of grains - an overview. In "Controlled Atmosphere Storage of Grains" (Shejbal, J., Ed.) pp. 121-132. Elsevier, Amsterdam.
Chełkowski, J. and Cierniewski, A. (1983) Mycotoxins in cereal grain. Part X. Invasion of fungal mycelium into wheat and barley kernels. Nährung 6, 533 - 536.
Clevström, G., Göransson, B., Hlödversson, and Pettersson, H. (1981) Aflatoxin formation in hay treated with formic acid and in isolated strains of *Aspergillus flavus*. J. stored Prod. Res. 20, 71-82.
Clevström, G., Ljunggren, H., Tegelstrom, S. and Tideman, K. (1983) Production of aflatoxin by an *Aspergillus flavus* isolate cultured under limited oxygen supply. Appl. environ. Microbiol. 46, 400-405.
Cole, R.J., Sanders, T.H., Hill, R.A. and Blankenship, P.D. (1985) Mean geocarposphere temperatures that induce preharvest aflatoxin contamination of peanuts under drought stress. Mycopathologia 91, 41-46.
Cole, R.J., Sanders, T.H., Dorner, J.W. and Blankenship, P.D. (1989) Environmental conditions required to induce preharvest aflatoxin contamination of groundnuts: summary of six years' research. In "Aflatoxin contamination of groundnut: Proceedings of the Sixth International Workshop, 6-9 October, 1987", ICRISAT Center, India. ICRISAT, Patancheru, India.
Cuero, R.G., Smith, J.E. and Lacey, J. (1987). Stimulation by *Hyphopichia burtonii* and *Bacillus amyloliquefaciens* of aflatoxin production by *Aspergillus flavus* in irradiated maize and rice grains. Appl. Environ. Microbiol. 53, 1142 - 1146.
Cuero, R., Smith, J.E. and Lacey, J. (1988) Mycotoxin formation by *Aspergillus flavus* and *Fusarium graminearum* in irradiated maize grains in the presence of other fungi. J. Food Prot. 51, 452-456.
Dharmaputra, O.S., Tjidtrosomo, H.S.S., Aryani, F. and Sidik, M. (1991) The effects of carbon dioxide on storage fungi of maize. In "Proceedings of the 5th International Working Conference on Stored Product Protection" (Fleurat-Lessard F., and Ducom, P., Eds.) pp. 291-301.
Di Maggio, D., Shejbal, J. and Rambelli, A. (1976) Studio della sisterriactica della flora fungina in frumentoa diversa umidita conservato in atmosfera controlla. Inform. Fitopatol. 26, 11-18.
Dickens, J.W. (1977) Aflatoxin occurrence and control during growth, harvest and storage of peanuts. In "Mycotoxins in Human and Animal Health" (Rodricks, J., Hesseltine, C.W. and Mehlman, M.A., Eds.) pp. 99-105. Pathotox Publishers, Park Forest South, Ill.
Diener, U.L. and Davis, N.D. (1977) Aflatoxin formation in peanuts by *Aspergillus flavus*. Bull. Agric. Exp. Sta., Auburn Univ. No. 493.
Epstein, E., Steinberg, M.P., Nelson, A.I. and Wei, L.S. (1970) Aflatoxin production as affected by environmental conditions. J. Food Sci. 35, 389-391.
Fabbri, A.A., Panfili, G., Fanelli, C. and Visconti, A. (1984) Effect of T-2 toxin on aflatoxin production. Trans. Br. mycol. Soc. 83, 150 - 152.
Festenstein, G.N., Lacey, J., Skinner, F.A., Jenkins, P.A. & Pepys, J. (1965) Self-heating of hay and grain in Dewar flasks and the development of farmer's lung hay antigens. J. gen. Microbiol. 41, 389 - 407.
Flannigan, B. (1978) Primary contamination of barley and wheat grain by storage fungi. Trans. Br. mycol. Soc. 71, 37 - 42.
Flannigan, B. (1993) In "*Aspergillus*" (Powell, K.A., Peberdy, J. and Renwick, A., Eds.) pp. - . Plenum, New York.
Gregory, P.H., Lacey, M.E. (1963) Microbial and biochemical changes during the moulding of hay. J. gen. Microbiol. 33, 147-174.
Hacking, A. and Biggs, N.R. (1979) Aflatoxin $B_1$ in barley. Nature, Lond., 282, 128.
Hesseltine, C.W. and Bothast, R.J. (1977) Mold development in ears of corn from tasseling to harvest. Mycologia 69, 328-340.
Hill, R.A. & Lacey, J. (1983a) Factors determining the microflora of stored barley grain. Ann. appl. Biol. 102, 467 - 483.
Hill, R.A., Blankenship, P.D., Cole, R.J. and Sanders, T.H. (1983b) Effects of soil moisture and temperature on preharvest invasion of peanuts by the *Aspergillus flavus* group and subsequent aflatoxin development. Appl. Environ. Microbiol. 45, 628-633.
Hill, R.A., Wilson, D.M., McMillian, W.W., Widstrom, N.W., Cole, R.J., Sanders, and T.H. Blankenship, P.D. (1985) Ecology of the *Aspergillus flavus* group and aflatoxin formation in maize and groundnut. In "Trichothecenes and other Mycotoxins" (Lacey, J., Ed.) pp. 79-95. John Wiley, Chichester.

Hocking, A.D. (1990) Responses of fungi to modified atmospheres. In "Fumigation and controlled atmosphere storage of grain: Proceedings of an International Conference, Singapore, 14-18 February 1989" (Champ, B.R. and Highley, E., Eds.), ACIAR Proceedings No. 25, 70-82.

Hocking, A.D. and Banks, H.J. (1991) Effects of phosphine on the development of storage microflora in paddy rice. In "Proceedings of the 5th International Working Conference on Stored Product Protection" (Fleurat-Lessard, F. and Ducom, P., Eds.), pp. 823-832.

Horn, B.W., and Wicklow, D.T. (1983) Factors influencing the inhibition of aflatoxin production in corn by *Aspergillus niger*. Can. J. Microbiol. 29, 1087-1091.

Jackson, C.R. and Press, A.F. (1967) Changes in the mycoflora of peanuts stored at temperatures in air or high concentrations of nitrogen or carbon dioxide. Oleagineux 22, 165-168.

Kozakiewicz, Z. (1989) *Aspergillus* species on Stored Products. Mycol. Pap. 161.

Lacey, J. (1971) The microbiology of moist barley storage in unsealed silos. Ann. appl. Biol. 69, 187-212.

Lacey, J. (1975) Airborne spores in pastures. Trans. Br. mycol. Soc. 64, 265-281.

Lacey, J. (1986). Water availability and fungal reproduction, patterns of spore production, liberation and dispersal. In "Water, Fungi and Plants" (Ayres, P.G. and Boddy, L., Eds.) (B.M.S. Symposium No. 11), pp. 65 - 86. Cambridge University Press, Cambridge.

Lacey, J. (1988) The microbiology of cereal grains from areas of Iran with a high incidence of oesophageal cancer. J. Stored Prod. Res. 24, 39 - 50.

Lacey, J. and Lord, K.A. (1977). Methods for testing chemical additives to prevent moulding of hay. Ann. appl. Biol. 87, 327-335.

Lacey, J., Lord, K.A., Cayley, G.R., Holden, G.R. and Sneath, R.W. (1983) Problems of testing novel chemicals for the preservation of damp hay. Anim. Feed Sci. Technol. 8, 283-301.

Landers, K.E., Davis, N.D. and Diener, U.L. (1967) Influence of atmospheric gases on aflatoxin production by Aspergillus flavus in peanuts. Phytopathology 57, 1086-1090.

Leitao, J., De Saint-Blanquat, G. and Bailly, J.-R. (1987) Action of phosphine on production of aflatoxins by various *Aspergillus* strains isolated from foodstuffs. Appl. environ. Microbiol. 53, 2328-2331.

Lee, L.S., Lee, L.V. and Russell, T.E. (1986). Aflatoxin in Arizona cottonseed: field inoculation of bolls by *Aspergillus flavus* spores in wind driven soil. J. Am. Oil Chem. Soc. 63, 530-532.

Lee, L.S., Lacey, P.E. and Goynes, W.R. (1987) Aflatoxin in Arizona cottonseed: a model study of insect-vectored entry of cotton bolls by *Aspergillus flavus*. Plant Dis. 71, 997-1001.

Lillehoj, E. and Hesseltine, C.W. (1977) Aflatoxin control during plant growth and harvest of corn. In "Mycotoxins in Human and Animal Health" (Rodricks, J., Hesseltine, C.W. and Mehlman, M.A., Eds.), pp. 107-119. Pathotox Publishers, Park Forest South, Ill.

Magan, N. and Lacey, J. (1984 a) Effect of temperature and pH on water relations of field and storage fungi. Trans. Br. Mycol. Soc. 82, 71-81.

Magan, N. and Lacey, J. (1984 b) Effect of water activity, temperature and substrate on interactions between field and storage fungi. Trans. Br. mycol. Soc. 82, 83-93.

Magan, N. and Lacey, J. (1984 c) Effects of gas composition and water activity on growth of field and storage fungi and their interactions. Trans. Br. Mycol. Soc. 82, 305-314.

Magan, N. and Lacey, J. (1985) Interactions between field and storage fungi on wheat grain. Trans. Br. mycol. Soc. 85, 29-37.

Magan, N. and Lacey, J. (1986) The effects of two ammonium propionate formulations on growth in vitro of *Aspergillus* species isolated from hay. J. Appl. Bacteriol. 60, 221 - 225.

Mehan, V.K., McDonald, D., Haravu, L.J. and Jayanthi, S. (1991) "The Groundnut Aflatoxin Problem: Review and Literature Database." ICRISAT, Patancheru, India.

Mills, J.T. and Abramson, D. (1981) Microflora and condition of flood-damaged grains in Manitoba, Canada. Mycopathologia 73, 143-152.

Moss, M.O. and Frank, J.M. (1985) The influence on mycotoxin production of interactions between fungi and their environment. In "Trichothecenes and other mycotoxins" (J. Lacey, Ed.) pp. 257-268. Chichester, John Wiley.

Northolt, M.D., van Egmond, H.P. and Paulsch, W.E. (1979) Differences between *Aspergillus flavus* strains in growth and aflatoxin $B_1$ production in relation to water activity and temperature. J. Food Prot. 40, 778-781.

Park, K.Y. and Bullerman, L.B. (1981) Increased aflatoxin production by *Aspergillus parasiticus* under conditions of cycling temperatures. J. Food Sci. 46, 1147-11 .

Park, K.Y. and Bullerman, L.B. (1983) Effect of cycling temperatures on aflatoxin production by *Aspergillus parasiticus* and *Aspergillus flavus* in rice and cheddar cheese. J. Food Sci. 48, 889-896.

Paster, N., Lisker, N. and Chet, I. (1983) Ochratoxin A production by *Aspergillus ochraceus* Wilhelm grown under controlled atmospheres. Appl. environ. Microbiol., 45, 1136-1139.

Paster, N., Calderon, M., Menasherov, M. and Mora, M. (1990) Biogeneration of modified atmospheres in small storage containers using plant wastes. Crop Prot. 9, 235-238.
Paster, N., Menasherov, M., Lacey, J. and Fanelli, C. (1992) Synergism between methods for inhibiting the spoilage of damp maize during storage. Postharvest Biol. Technol. 2, 166-170.
Perez, R.A., Tuite, J. and Baker, K. (1982) Effect of moisture, temperature, and storage time on the subsequent storability of shelled corn. Cereal Chem. 59, 205 - 209.
Petersen, A., Schlegel, V, Humnel, B., Cuendet, L.S., Geddes, W.F. and Christensen, C.M. (1956) Grain storage studies XXII. Influence of oxygen and carbon dioxide concentrations on mold growth and grain deterioration. Cereal Chem. 33, 53-66.
Pierrot, R. (1988) Chemical control of stored insects. In "Preservation and storage of grains, seeds and their by-products" (Multon, J.L., Ed.) pp. 832-842. Lavoisier, New York.
Pitt, J.I. (1975) Xerophilic fungi and the spoilage of foods of plant origin. In "Water relations of food" (Duckworth, R.B., Ed.), pp. 273-307. Academic Press, London.
Pitt, J.I., Dyer, S.K. and McCammon, S. (1991) Systemic invasion of developing peanut plants by *Aspergillus flavus*. Lett. appl. Microbiol. 13, 16-20.
Ramakrishna, N. (1990). Assessment of the effects of fungal interactions on mycotoxin production in barley using monoclonal antibodies. Ph.D. thesis, University of Strathclyde.
Raper, K.B. and Fennell, D.I. (1965) "The Genus *Aspergillus*." Williams and Wilkins, Baltimore.
Richard-Molard, D., Cahagnier, B. and Poisson, J. (1980) Wet grains storages under modified atmospheres. Microbiological aspects. In "Controlled Atmosphere Storage of Grains" (Shejbal, J., Ed.) pp.173-182. Elsevier, Amsterdam.
Sahay, M.N. and Gangopadhyay, S. (1985) Effect of wet harvesting on biodeterioration of rice. Cereal Chem. 62, 80 - 83.
Sanders, T.H., Davis, N.D. and Diener, U.L. (1968) Effect of carbon dioxide, temperature and relative humidity on production of aflatoxin in peanuts. J. Am. Oil Chem. Soc. 10, 683-685.
Sanders, T.H., Cole, R.J., Blankenship, P.D. and Hill, R.A. (1983) Drought soil temperature range for aflatoxin production in preharvest peanuts. Proc. Am. Peanut Res. Ed. Soc. 15, 90.
Sanders, T.H., Cole, R.J., Blankenship, P.D. and Hill, R.A. (1985) Relation of environmental stress duration to *Aspergillus flavus* invasion and aflatoxin production in preharvest peanuts. Peanut Sci. 12, 90-93.
Shejbal, J. and Di Maggio, D. (1976) Preservation of wheat and barley in nitrogen. Meded. Fac. Landbouwwetenschappen Rijksuniversiteit Gent 41, 595-606B.
Sinha, R.N. and Wallace, H.A.H. (1965) Ecology of a fungus induced hot spot in stored grain. Can. J. Pl. Sci. 45, 48-59.
Stutz, H.K. and Krumperman, P.H. (1976) Effect of temperature cycling on the production of aflatoxin by *Aspergillus parasiticus*. Appl. Environ. Microbiol. 32, 327-332.
Tabak, H.H. and Cook, W.B. (1968) Growth and metabolism of fungi in an atmosphere of nitrogen. *Mycologia* 60, 115-140.
Terho, E.O. and Lacey, J. (1979) Microbiological and serological studies of farmer's lung in Finland. Clin. Allergy 9, 43-52.
Tuite, J.F. and Christensen, C.M. (1955) Grain storage studies. XVI. Influence of storage conditions upon the fungus flora of barley seed. Cereal Chem. 32, 1-11.
Tuite, J., Koh - Knox, C., Stroshine, R., Cantone, F.A., and Bauman, L.F. (1985) Effect of physical damage to corn kernels on the development of *Penicillium* species and *Aspergillus glaucus* in storage. Phytopathology, 75, 1137-1140.
Usha, C.M., Patkar, K.L., Shekara Shetty, H. & Lacey, J. (1991) Frequency of occurrence of *Aspergillus flavus* and *Aspergillus niger* in developing groundnut from flowering to harvest. Int. Arachis Newsl. 10, 18-19.
Wallace, H.A.H. (1973) Fungi and other organisms associated with stored grain. In "Grain Storage - Part of a System" (Sinha, R.N. and Muir W.E., Eds.), pp. 71 - 97. Avi Publishing Company, Westport, Conn.
Wallace, H.A.H. and Sinha, R.N. (1981) Causal factors operative in distributional patterns and abundance of fungi: a multivariate study. In "The Fungal Community: its Organisation and Role in the Ecosystem" (Wicklow D.T. and Carrol G.C., Eds.), pp. 233-247. Marcell Dekker, New York.
Weckbach, L.S. and Marth, E.H. (1977) Aflatoxin production by *Aspergillus parasiticus* in a competitive environment. Mycopathologia 62, 39-45.
Welling, B. (1968) The influence of threshdamage on microflora and germination of barley during storage. Tidsskr. PlAvl. 72, 513-519.
Wells, J.M. and Uota, M. (1970) Germination and growth of five fungi in low oxygen and high carbon dioxide. Phytopathology 60, 50 - 53.
White, N.D.G. and Jayas, D.S. (1991) Effects of periodically elevated carbon dioxide on stored wheat

ecosystems at cool temperatures. In "Proceedings of the 5th International Working Conference on Stored Product Protection" (Fleurat-Lessard F. and Ducom, P., Eds.), pp. 925-933.

Wicklow, D.T. 1988. Patterns of fungal association within maize kernels harvested in North Carolina. Pl. Dis. 72, 113-115.

Wicklow, D.T., Hesseltine, C.W., Shotwell, O.L. and Adams, G.L. (1980) Interference competition and aflatoxin levels in corn. Phytopathology 70, 761-764.

Wicklow, D.T., Horn, B.W., Shotwell, O.L., Hesseltine, C.W., and Caldwell, R.W. 1988. Fungal interference with *Aspergillus flavus* infection and aflatoxin contamination of maize grown in a controlled environment. Phytopathology 78, 68-74.

Wilson, D.M. and Jay, E. (1975) Effect of controlled atmosphere storage on aflatoxin production in high moisture peanuts. J. stored Prod. Res. 12, 97-100.

Wilson, D.M., Jay, E., Hale, O.M. and Huang, L.H. (1976) Controlled atmosphere storage of high-moisture corn and peanuts to control aflatoxin production. Ann. Technol. Agric. 26, 339-342.

Wiseman, D.W. and Marth, E.H. (1981) Growth and aflatoxin production by *Aspergillus parasiticus* when grown in the presence of *Streptococcus lactis*. Mycopathologia, 73 49-56.

# ANTIINSECTAN EFFECTS OF *ASPERGILLUS* METABOLITES

[1]D. T. Wicklow, [1]P. F. Dowd and [2] J. B. Gloer

[1]National Center for Agricultural Utilization Research
Agricultural Research Service, U.S.D.A.
Peoria, Illinois, U.S.A.

[2]J. B. Gloer
Department of Chemistry
University of Iowa
Iowa City, Iowa, U.S.A.

## INTRODUCTION

Species of *Aspergillus* have been isolated from living, diseased and dead insects. They colonise discrete nutrient-rich substrates for which insects and fungi may compete (e.g. seeds, fruit, dung, etc.). There are numerous examples of antiinsectan effects attributed to *Aspergillus*, including symptoms of fungal infection (Madelin, 1963; Charnley, 1984; Moore and Erlandson, 1988), observations of insect response to fungal cultures ( Loschiavo and Sinha, 1966; Sinha, 1971; Hill, 1978), moulded grain (Sinha, 1966, van Bronswijk and Sinha, 1971, mould contaminated diets (Wicklow and Dowd, 1989), and toxicity studies with specific fungal secondary metabolites (Wright *et. al.*, 1982; Dowd *et al.*, 1988; Wicklow, 1988; Dowd, 1992). The antiinsectan effects include growth retardation, reduced pupal and adult size, lower fecundity, loss of fertility, mortality, repellency, and genetic changes. In nature, these entomotoxic fermentation products of *Aspergillus* may prevent loss of substrate to insects and/or discourage arthropod fungivores (Janzen, 1977; Wicklow, 1988). Experiments to test the suitability of *Aspergillus* species as food for the grain beetle *Ahasverus advena* revealed that beetle larvae consumed the *Aspergillus* conidial heads of *Eurotium* (syn. *Aspergillus glaucus*) but did not eat the ascomata (Hill, 1978). *Eurotium* ascomata are covered with a felt of sterile aerial hyphae encrusted with yellow, orange, or red granules which, if toxic or unpalatable, could make such fungal structures less attractive to arthropodes.

Moulds such as *Eurotium* spp., *Aspergillus restrictus*, G. Smith, *Aspergillus candidus* Link, *Aspergillus flavus* Link, *Aspergillus ochraceus* Wilhelm (syn. *Aspergillus alutaceus* Berk. and Curt.), and *Aspergillus versicolour* (Vuill) Tiraboschi all invade the germ or embryo of cereal grains (Griffin, 1966; Christensen and Kaufmann, 1974; Tsuruta *et al.*, 1981; Singh *et al.*, 1991), causing deterioration losses of stored grain (Mills, 1983; Sinha, 1992). Several of the sclerotium-forming aspergilli that attack insects also infect seeds

*The Genus Aspergillus*, Edited by Keith A. Powell *et al.*,
Plenum Press, New York, 1994

**Table 1.** Sclerotium-forming aspergilli recorded from insects and or diseased plants.

| Fungal Taxa | Isolated from: Diseased/Dead Insect | Seed/ Cereal Product | Infected Plant Organ |
|---|---|---|---|
| *Aspergillus* | | | |
| *A. alliaceus* | RF | --- | RF |
| *A. alutaceus* (syn. *A. ochraceus*) | RF DGA | RF DGA | --- |
| *A. avenaceus* | --- | RF | --- |
| *A. candidus* | DGA | DGA | --- |
| *A. carbonarius* | --- | --- | RF |
| *A. dybowskii* | --- | SS | --- |
| *A. erythrocephalus* | --- | SS | --- |
| *A. flavus* var. *flavus* | RF DGA | RF DGA | RF |
| *A. flavus* var. *parasiticus* | RF DGA | DGA | --- |
| *A. fresenii* (syn. *A. sulphureus*) | ATC | ATC | --- |
| *A. japonicus* var. *aculeatus* | --- | DGA | DGA |
| *A. melleus* | DGA | DGA ATC | RF DGA |
| *A. nomius* | KHH | KHH | KHH |
| *A. ostianus* | --- | ARS | --- |
| *A. sclerotiorum* | DGA ARS | RF DGA ARS | RF DGA |
| *A. tamarii* | RF | --- | ATC |
| *A. terreus* var. *africanus* | DGA | DGA | DGA |
| *A. togoensis* | --- | SS | --- |

Citation: ATC = Jong and Ganth, 1984; ARS = ARS Culture Collection Records; KHH = Kurtzman *et al.*, 1987; DGA = Domsch *et al.*, 1980; RF = Raper and Fennell, 1965; SS = Samson and Seifert, 1985.

(Table 1). *A. flavus* is a necrotrophic pathogen of maize kernels, deriving nutrients from cells that are killed in advance of the invading hyphae (Smart *et. al.*, 1990), particularly the lipid-rich nucellar tissues (Fennell *et al.*, 1973). In nature, the ability of *Aspergillus* species to be successful seed colonists may depend on their production of specific kinds of toxic secondary metabolites. Antibiosis is generally recognised as the principal mechanism of interference competition by which fungi exclude other microorganisms from resources potentially available to competitors (Wicklow, 1992). Sewell (1959) characterised the "*Penicillium* growth pattern" where a fungus densely colonises small individual substrates with no extension of the mycelium into the surrounding soil and then produces conidia heavily over the surface. This pattern is characteristic of the aspergilli that invade seeds or other discrete, nutrient-rich substrates, enabling these fungi to exploit the entire resource. Fungi displaying the "*Penicillium* growth pattern" are usually not participants in a fungal succession (Sewell, 1959; Cooke and Rayner, 1984). As the dominant or perhaps the sole fungal colonists of a dead insect or seed, *Aspergillus* has the full enzymatic compliment to efficiently utilise these resources. It should not be surprising then that *Aspergillus* species should provide such a wealth of enzymes, many remaining to be discovered and exploited by industry (K. M. Oxenbol, this volume). Janzen (1977) has argued that "seed-eating fungi" are often under strong selection to render seeds as objectionable or unusable to larger seed-eating animals (i.e. insects, mammals, birds) in the shortest possible time, as a means of protecting themselves and the substrate they colonise from predation. Janzen (1977) proposes that fungi achieve this through the production of mycotoxins and metabolic conversions of seeds; the fungus both feeds itself and, at the same time, protects its food. There have been numerous reports showing that *Aspergillus* species known to colonise grain, contaminate the grain with mycotoxins and, when grown on autoclaved grain, will prove toxic when fed to laboratory or farm animals (e.g. Wilson, 1971; Semeniuk *et al.*, 1971; Smith and Moss, 1985; Moss, 1989). Nearly all compounds classified as cytotoxic or neuroactive mycotoxins are entomotoxic when tested against insects at levels up to 250 or 100 ppm dry weight (Wright *et. al.*, 1982).

*A. flavus* makes a suite of secondary metabolites with known toxicity to insects. The *A. flavus* group mycotoxins aflatoxin B1, G1, cyclopiazonic acid, sterigmatocystin, β-nitropropionic acid and kojic acid are toxic to silkworm larvae (*Bombyx mori* L.) (Murakoshi *et al.*, 1973; Yokota *et al.*, 1981). The aflatoxins are implicated as having a role in the entomopathogenicity of *A. flavus* (Ohtomo *et al.*, 1975), although non-aflatoxigenic strains of *A. parasiticus* isolated from naturally infected sugarcane mealybugs *Saccharicoccus sacchari* exhibited a high level of pathogenicity (Drummond and Pinnock, 1990). Becker *et al.* (1969) reported that both aflatoxigenic and non-aflatoxigenic strains of *Aspergillus* destroyed termites, and concluded that other toxins produced by the non-aflatoxigenic strains were responsible for death of the termites. Bennett (1981) proposed that the function of aflatoxins is to kill insect vectors of *A. flavus* "after the fungus is delivered to a food supply"; the insect then "serves as a substrate which the fungus uses to create a large inoculum".

There is much indirect evidence of selection for ability to produce antiinsectan metabolites/mycotoxins involving *A. flavus*; (1) The fungal toxin cyclopiazonic acid was produced primarily by strains of *A. flavus* isolated from warmer latitudes, possibly an evolutionary response to the greater threat of predation (Wicklow and Cole, 1982); (2) In the more southern islands of Japan as well as in the Philippines, Thailand, and Indonesia, numbers of aflatoxin-producing strains isolated from soils were substantially greater than from the northern part of Japan (Manabe and Tsuruta, 1978); (3) The most potent mycotoxins tend to occur towards the end of a long biosynthetic route in which substances of lower toxicity are intermediates (Bu'Lock, 1980). Jarvis *et al.*, (1984) demonstrated this by contrasting the toxicity of aflatoxin G1 and selected biosynthetic precursors (sterigmatocystin, versicolorin A, averufin, and norsolorinic acid) to larvae of the European corn borer *Ostrinia nubilalis*. If toxic polyketides provide a selective advantage against mycophagous arthropods, the evolution of detoxification systems by the arthropods has led to the synthesis of increasingly potent polyketides (Bennett, 1983); (4) Toxin-producing ability can be lost following repeated subculture. Wicklow (1984) theorised that domesticated strains (koji moulds) of yellow-green aspergilli lost their ability to produce toxic metabolites such as aflatoxin because in a koji environment these metabolites are no longer of adaptive value to the fungus. Wicklow and Dowd (1989) fed corn earworm larvae and fall armyworm larvae (*Spodoptera frugiperda*) on diets consisting of maize kernels fermented (8 days; 28°C) with individual strains of "wild" or domesticated yellow-green aspergilli according to Kurtzman et *al.* (1986); e.g. wild isolates = *A. flavus* var. *flavus*, *A. flavus* var. *parasiticus*; domesticated, non-aflatoxigenic koji moulds = *A. flavus* var. *oryzae*, *A. flavus* var. *sojae*). Kernels fermented with individual strains of var. *sojae* were less toxic to larvae than kernels fermented with isolates of the related wild variety *parasiticus*. Likewise, kernels fermented with isolates of the wild variety *flavus* were, with one exception, more toxic to fall armyworm larvae than kernels fermented with isolates of the related domesticated variety *oryzae*. However, the domesticated strains of *oryzae* were no less toxic as a group to corn earworm larvae than the var. *flavus* strains we tested. Wicklow and Dowd (1989) suggest that koji moulds used in food fermentations be screened by means of insect bioassays as a relevant test of the toxigenic potential of individual domesticated strains.

Fungi provide an important nutritional benefit to certain insect pests of stored grain and food products (Sinha, 1971; Hill, 1978; Wright *et al.*, 1980). Another possible advantage for larvae that develop in moulded and toxin-contaminated grain is that they may escape potential competitors and predators (Dowd, 1992). Insects that feed on mouldy grain may be able to metabolise toxic fungal metabolites. For example, adults of the fungus-feeding beetle *Carpophilus hemipterus* L. (Nitidulidae; Coleoptera) are unaffected by 25 ppm aflatoxin B1 (Dowd, 1992) and larvae are ten times more efficient in metabolising the *Fusarium* produced trichothecene 4-monoacetoxyscirpenol than non-fungus-feeding caterpillars, the fall armyworm, *Spodoptera frugiperda* and the corn earworm, *Heliothis zea* (Boddie) (Dowd and Van Middlesworth, 1989). Stored product beetles such as *Tribolium confusum* are generally resistant to mycotoxins, yet are also selective in their feeding behaviours (Wright *et al.*, 1982).

The interpretation of entomotoxic activity within a complex mixture of naturally occurring fungal secondary metabolites is further confounded by the fact that toxins simultaneously produced by *A. flavus* may have synergistic effects. Dowd (1988*a*) has shown that another *A. flavus* metabolite, kojic acid, which is biosynthetically simple to produce, synergised the toxicity of aflatoxin B1 to corn earworm and fall armyworn larvae. In the presence of kojic acid (25 ppm), only one-tenth the amount of aflatoxin B1 (0.25 ppm) was needed to produce levels of toxicity/mortality seen for aflatoxin B1 at 2.5 ppm. Dowd (1988*b*) suggests that kojic acid acts synergistically because it inhibits oxidative enzymes likely to be involved in aflatoxin B1 detoxification. Here the fungus can inhibit the detoxification systems of an insect at comparatively low metabolic cost, thus enabling it to use aflatoxin or other metabolically expensive toxins more efficiently (Dowd, 1988*b*).

## LIFE HISTORY AND ECOLOGY

With the exception of *A. flavus* (Diener *et al.*, 1987; Wicklow, 1991) we have little understanding of the life history and ecology of *Aspergillus* spp. in agricultural fields and almost no information about their ecological role in native plant communities (Raper and Fennell, 1965; Domsch *et al.*, 1980; Christensen and Tuthill, 1985). *A. flavus* produces yellow-green conidia that function in dispersal and as infective inoculum, in addition to sclerotia, long-lived survival structures. Both types of propagules are associated with damaged maize kernels and are dispersed onto the ground during combine harvesting (Wicklow and Horn, 1984; Wicklow *et al.*, 1984). Wicklow and Wilson (1986) observed that sclerotium germination (sporogenic) occurred in maize fields just prior to silking. *A. flavus* is probably like many ascomycetes that made the transition from buffered tropical forest habitats to more seasonal, less predictable situations. According to Evans (1988) these fungi appear to have sacrificed their teleomorph, and thus the ability for genetic recombination, in exchange for the massive production of dry conidia. Evans (1988) suggests that for fungal pathogens of insects, this may be correlated with seasonal population explosions of host insects and the requirement that the fungus survive in the soil during periods of low host density or unfavourable climatic conditions.

Lussenhop and Wicklow (1990) demonstrated the role of nitidulid beetles (Nitidulidae: Coleoptera) in transmitting *A. flavus* infective inoculum to preharvest maize ears in commercial maize fields in the Georgia Coastal Plain, U.S.A. The nitidulids gain entry to the ears through wounds caused by other insects and birds and are capable of entering some ears on their own, particularly loose-husked varieties where kernels near the ear tip are exposed (Connell, 1956). Downed and overwintered maize ears that become infested with *A. flavus* represent an important source of infective inoculum (Lussenhop and Wicklow, 1990). This is because populations of nitidulid beetles also colonise these ears. Nitidulids may preferentially feed on rotted and yeast contaminated plant tissues, using the yeast as food (Miller and Mrak, 1954). Dorsey and Leach (1956) suggest that the association between nitidulid and oak wilt fungus is mutualistic. The vectoring of plant pathogenic fungi by nitidulids provides the beetle with a fungus-enriched plant substrate as food. There is no information on the feeding choices nitidulid beetles make upon encountering kernel rotting moulds or their toxins in nature. Nitidulids are generally more tolerant of individual mycotoxins than non-fungus-feeding insects (Dowd, 1992). The nutritional benefits of fungivory (Martin, 1987) may justify the metabolic cost of detoxification (Dowd, 1992).

## SCLEROTIAL CHEMICAL DEFENSES

The key observation that served as a starting point for our collaboration on the antiinsectan metabolites of *Aspergillus* sclerotia was the fact that sclerotia of *A. flavus* are avoided by the common detritivorous beetle *C. hemipterus*. We were able to demonstrate that aflavinine secondary metabolites distributed exclusively in *A. flavus* sclerotia deter feeding by *C. hemipterus* (Wicklow *et al.*, 1988). The adult beetles will consume the mycelium and conidia of *A. flavus*, but not the intact or ground sclerotia. Dihydroxyaflavinine and three other natural products with the aflavinine ring system, and a novel bicoumarin metabolite, were isolated as major components of *A. flavus* sclerotia with antifeedant activity against *C. hemipterus* when incorporated into a pinto bean diet at 25-100 ppm (Figure 1); (Wicklow *et al.*, 1988; Gloer *et al.*, 1989; TePaske *et al.*, 1992). Naturally occurring levels of each of these compounds were found to be higher than 100 ppm in *A. flavus* sclerotia. (TePaske, 1991). These sclerotial metabolites were not

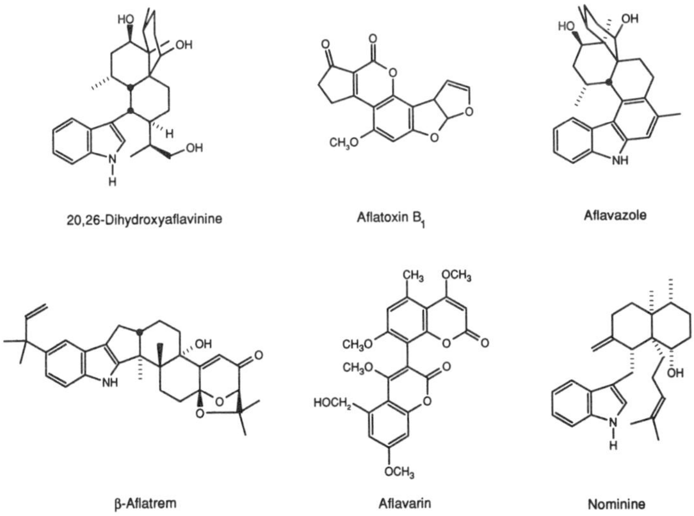


**Figure 1.** Antiinsectan metabolites from sclerotia of *Aspergillus flavus:* dihydroxyaflavinine; aflatoxin B1; aflavazole; aflatrem B; aflavarin; nominine.

detected in the mycelium, conidia, or medium from which the sclerotia were harvested. Thus, it became clear that the sclerotia and ascomata of *Aspergillus* spp. represent an untapped source of potentially novel secondary metabolites with insecticidal/antifeedant activity to insect pests of crops.

Recognising the importance of the fungal sclerotium as a survival structure in the life cycle of *A. flavus*, Wicklow and Cole (1982) and Wicklow and Shotwell (1983) predicted that there should be a greater fungal investment (metabolic cost) in sclerotial chemical defence systems. These authors found that aflatoxins B1, B2, and cyclopiazonic acid were isolated from the sclerotia, conidia, mycelium, and culture medium, whereas aflatoxins G1, G2, aflatrem, and dihydroxyaflavinine were found only in sclerotia harvested from these same cultures. Aflatrem was discovered during a survey of fungal cultures for toxigenicity to mice when grown on autoclaved maize kernels (*Zea mays* L.). (Wilson and Wilson, 1964). It is no coincidence that the aflatrem-producing strain of *A. flavus* QM6738 (=NRRL 500) also produces large numbers of sclerotia on agar culture media. *Aspergillus* and *Claviceps* have many biosynthetic similarities particularly among biologically active indole-derived metabolites (Cole and Cox, 1981; Cole, 1981; Turner and Aldridge, 1983). It is significant that both aflatrem and dihydroxyaflavinine, from the sclerotia of *A. flavus*, are chemically related to paspalinine and other tremorgenic metabolites found in the sclerotia of *Claviceps paspali*, a taxonomically unrelated fungus (Cole, 1981). Wicklow and Cole (1982) theorised that the ergot alkaloids and related indole metabolites found concentrated in the sclerotia of *Claviceps* species (Floss, 1976) probably function in protecting the sclerotium from predation, an ecological role attributed to alkaloids in vascular plant tissues (Robinson, 1979; Clay, 1990). *Claviceps* is a biotrophic pathogen that infects the flowers of grasses and replaces the developing grain (Luttrell, 1981). *Claviceps* conidia (honey dew stage of life cycle) are also dispersed by insects. Ecologists have shown that more insects and larger animals prefer the seed heads of plants due to their rich nitrogen and energy reserves (Fenner, 1985). This may explain why fungal taxa as unrelated as *Claviceps* and seed-infecting, insect-vectored species of *Aspergillus* share so many biosynthetic similarities in their apparent chemical defence systems.

Whittaker and Feeney (1971) and Janzen (1977) stimulated thinking about the potential role of fungal toxins in defending fungal tissues from fungivorous arthropods and vertebrates (Wicklow and Cole, 1982; Wicklow, 1988), and suggested ecological relevance to studies of the secondary metabolite profiles and chemotaxonomy of these species (Frisvad, 1986; Frisvad, 1992). It is intriguing to consider how the presence of such metabolites could impact on sclerotial survival. Sclerotia typically form in fungus-infected plant or insect remains. Many insects and other invertebrates consume fungi (Martin, 1987; Ingham, 1992). Fungal tissue is equal to plant tissue in its nutritional value to insects (Southwood, 1973). A sclerotium would represent a substantial nutrient reward for an insect fungivore, especially since sclerotia possess a much higher nutrient content than the mycelium (Willets, 1972). The soil beneath one square meter of a typical European grassland reportedly contains approximately 100-400 woodlice, 500-1000 beetles and beetle larvae. 200-500 ants, 10,000-40,000 springtails (collembola), 20,000-120,000 mites, and 1.8-120 million nematodes, although these figures are subject to seasonal and locational variation (Kevan, 1965). Thus, countless opportunities for consumption of or damage to a sclerotium could arise during its lengthy dormant period. The dark brown pigmented phenolic materials found concentrated in the outer rind of fungal sclerotia (Bullock *et al.*, 1980) could represent "quantitative" defenses like the phenolic polymers (i.e. tannins) that reduce the digestibility of plant tissues (Rhoades and Cates, 1976; Swain, 1979). Tannins are difficult for herbivores to evolve resistance against, but are also metabolically expensive to produce (Swain, 1979). Distribution of pigmented phenolic polymers in the sclerotium rind would constitute a cost-effective means of defending the sclerotium. Although some sclerotia produce an outer rind, there is evidence which clearly indicates that insects (e.g. fungus gnats, phalacrid beetles) are capable of chewing through sclerotial rind (Steiner, 1984; Anas and Reeleder, 1987). Sclerotia damaged by insect larvae are much more susceptible to microbial decay than are undamaged sclerotia (Baker and Cook, 1974). Surprisingly, few studies have been reported regarding the resistance of sclerotia to predation by insects. If sclerotia commonly contain metabolites which somehow limit feeding by insects, this property could clearly influence the longevity of these important fungal bodies.

## SCLEROTIUM FORMATION IN NATURE AND IN CULTURE

In *Aspergillus*, sclerotia and sclerotioid ascomata are produced by species classified in the Trichocomaceae (=Eurotiaceae): Dichlaenoideae (Malloch, 1985). Many of these species are frequently associated with starchy or oily seeds and fruit. *Aspergillus* sclerotia and sclerotioid ascomata form directly on or within substrates moulded by these fungi. There are few reports of sclerotia or sclerotioid ascomata from nature. Malloch and Cain (1972) discovered that ascospores from ascocarps of *Dichlanea lentisci* germinated to produce an *Aspergillus* conidial state. The ascocarps were discovered on twigs and dung in a wood rat nest near San Francisco, California. In culture, the fungus produced sclerotium-like bodies which exactly duplicated the ascocarps on the natural substrata. *A. flavus* sclerotia were detected by Wicklow *et al.* (1984) in samples of insect-damaged and mouldy maize ears from a field that was left unharvested following the 1981 growing season near Tifton, Georgia. As part of the same study, two *A. flavus* sclerotia were recovered from surface soil (900 g) collected following combine harvesting. Zummo and Scott (1990) report finding *A. flavus* sclerotia in the pith tissues of maize cobs that were dispersed onto the ground at harvest and had overseasoned (1-year) in Mississippi, U.S.A. Additional reports of sclerotigenic aspergilli from diseased or dead insects, infected seeds, moulded cereal products, or infected plant organs other than seeds, have been summarised in Table 1.

Fungal sclerotia and sclerotioid ascomata have been reported from naturally contaminated substrates incubated in laboratory moist chambers: (1) *A. flavus* sclerotia were produced during incubation (30-35°C; >85% RH)) of Arizona, U.S.A. cotton seed (*Gossypium hirsutum* L.) infected with *A. flavus* (Waked and Nouman, 1982); (2) *A. flavus* sclerotia formed during moist-chamber incubation of insect-damaged and fungus-infected corn ears that were hand harvested from fields near Tifton, Georgia, U.S.A. (Wicklow *et al.*, 1982); (3) *Aspergillus sclerotiorum* sclerotia were photographed where they had formed in subterranean cells of the ground-nesting alkali bee (*Nomia melanderi* Ck11.) reared in a laboratory insectary (Batra *et al.*, 1973); (4) *Aspergillus leporis* sclerotia formed on dung from a caged laboratory rabbit given pelletised feed inoculated with *A. leporis* conidia (Wicklow, 1985); (5) *A. flavus* sclerotia developed in wound-inoculated pre-harvest maize ears produced in a plant growth room (Wicklow and Horn, 1984).

### Physiological Requirements of Sclerotium Formation

The nutrient requirements which promote sclerotium-formation in *Aspergillus* are similar to those recorded for plant pathogenic fungi where the greatest number of sclerotia developed on media which also favoured mycelial growth (Townsend, 1957). In general, this occurred when both peptone and glucose (or other carbohydrates) were at relatively high concentration. Rudolph (1962) studied the effect of carbon-nitrogen balance, temperature, pH, and light upon the formation of sclerotia by six species of *Aspergillus*. High nitrate (0.3%-0.6% $NaNO_3$) and high (10-20%) sucrose concentrations were optimal for sclerotium-formation on Czapek's agar. The numbers of sclerotia produced by *A. flavus* NRRL 3517 and *A. parasiticus* NRRL 3145 on Czapek's agar were greatest with 5% sucrose in the medium and 0.5% $NaNO_3$ as contrasted with any combination of these ingredients at lower concentrations (Hesseltine *et al.*, 1970). Fourteen of 20 different carbon compounds tested supported sclerotium-production by *Aspergillus niger* (Agnihotri, 1969). Maximum sclerotial yields were obtained with a standard medium containing 2% fructose, while ribose, mannitol, malonic acid, fumaric acid, and citric acid did not support sclerotium formation. The dry weight of *A. niger* sclerotia varied depending on the source of carbon, nitrogen, or sulphur compounds used in the culture medium (Agnihotri, 1968; 1969). On media where maximum numbers of sclerotia were recorded, average sclerotium size was substantially reduced. For these reasons, it was not possible for Agnihotri to correlate sclerotium yield (mg) with the number of sclerotia produced. Among nine carbohydrates tested by Paster and Chet (1980), the maximal number of *A. ochraceus* NRRL 3174 sclerotia were formed on basal medium containing sucrose (2 x $10^{-1}$ molar concentration). No sclerotia were formed with lactose, galactose, or arabinose as sources of carbon. Maximum sclerotial yields (dry weight) were recorded with glutamic acid ($10^{-2}$ M) added to the basal medium. Among 14 sulphur compounds tested by Agnihotri (1969), $MgSO_4$ yielded the highest number of sclerotia, maximum yields being recorded at a concentration of 0.3% $MgSO_4$.

Rudolph (1962) showed that the pH conditions most favourable for vegetative growth of *Aspergillus* on Czapek's agar (pH 5.0-7.0) are also those most advantageous for the production of sclerotia. *A. ochraceus* sclerotia formed in a defined medium adjusted to initial pH 4.0-9.0, but not at pH 3.0 where there was no growth, or pH 11.0 where there was abundant mycelial growth (Paster and Chet, 1980). *A. niger* produced maximum numbers of sclerotia on a glucose-sodium nitrate medium adjusted to pH 7.0 (Agnihotri, 1969) and on Czapek-Dox agar adjusted to pH 7.5 (Rai *et al.*, 1967). Cotty (1988) reported that sclerotium production by *A. flavus* increased with increased pH (initial pH = 4.5 to 5.5) on citrate-buffered ammonium medium.

Production of *Aspergillus* sclerotia was maximal at or slightly below the optimum temperature for mycelial growth of *Aspergillus* on Czapek's agar (30-34°C) (Rudolph, 1962). Rai *et al.* (1967) recorded similar findings for two sclerotium-forming strains of *A. flavus* grown on Czapek-Dox agar + 0.5% yeast extract. Likewise, Paster and Chet (1980) recorded abundant sclerotium production at 28-32°C, temperatures shown to support maximum growth in *A. ochraceus*. Agnihotri (1969) recorded maximum numbers of *A. niger* sclerotia in cultures grown at 35°C. Rudolph (1962) reported no significant difference in growth, sporulation, or production of sclerotia between *Aspergillus* cultures grown in continuous light and those grown in continuous darkness at 25°C. However, others have shown that light inhibits sclerotium formation in *Aspergillus* (Rai *et al.*, 1967; Bennett *et al.*, 1978; Paster and Chet, 1980).

Paster and Chet (1982) investigated the effect of $O_2/CO_2$ combinations on sclerotium formation in *A. ochraceus*. In this study, sclerotial yields were highest when the fungus culture was grown under normal air. Excellent growth and sclerotium production occurred when $O_2$ levels were as low as 1%. The production of sclerotia decreased with an increase in $O_2$ or $CO_2$. At levels of 20% and above $CO_2$ completely inhibited sclerotium formation, regardless of the $O_2$ concentration. Mycelial growth was inhibited only at a level of 80% $CO_2$. Rai *et al.*, (1967) demonstrated that for *A. niger* light-yellow sclerotial rudiments resembling mycelial knots will form under oxygen-deficient conditions, but fail to mature unless the cultures are subsequently aerated.

### Sclerotium-Production on Autoclaved Corn Kernels

In our investigations of antiinsectan sclerotial metabolites, *Aspergillus* sclerotia were produced by solid substrate fermentation of soaked (@50% moisture) and autoclaved corn kernels distributed in a layer (1-2 kernels deep) covering the bottom of a Fernbach flask (Wicklow *et al.*, 1988). Inoculated kernels were incubated in the dark for 21 days at 28°C. The sclerotium-forming strains of *Aspergillus* that we have examined may differ individually in their ability to produce sclerotia under these fermentation conditions (D. T. Wicklow, unpublished). Even so, the conversion of corn substrate into sclerotial biomass can be a highly efficient process. Many of the *Aspergillus* strains we have cultured on autoclaved corn kernels have yielded more than 30g sclerotia (air-dried at room temperature) for each 100g dry weight of starting substrate.

## ASSAYS FOR ANTIINSECTAN ACTIVITY

Selection of sclerotial extracts for chemical study was based on bioactivity against two insect pests of crops, the corn earworm (*Helicoverpa zea*; syn. *Heliothis zea,* tobacco budworm and cotton bollworm), and the dried fruit beetle (*Carpophilus hemipterus*). *C. hemipterus* was chosen as an assay organism because of its ecological relevance. *C. hemipterus* is a near subtropical to tropical pest of fruit and some vegetables (Hinton, 1945) and may encounter the sclerotia of many fungi while feeding on decaying crop residues in the soil. Our laboratory population of *C. hemipterus* is maintained on a pinto bean diet. Testing of extracts is performed by incorporation of the extract into the diet (Dowd, 1988*a,b*) at levels approximating the concentrations found in the sclerotia. Assays of fractions and pure compounds are performed at levels suggested by the results of the extract tests and at 100 ppm wet weight (400 ppm dry weight) for pure compounds. Samples to be tested are added as an acetone or aqueous solution to a warm, liquified pinto bean diet, which is then blended with a vortex mixer. Traces of solvent are evaporated, and the diet is allowed to cool and solidify. The diet is then cut into ca. 600mg blocks. Each block is placed in a well of a 24-well immunoassay plate. Five beetles are added to each well, and the wells are covered and incubated at 27 ± 1 °C and 40 ± 10% relative humidity, with a 14:10 hour light:dark photoperiod. Four concurrent replications of each test are performed, and separate tests are performed on adults and larvae. Controls consist of solvent blanks. Feeding activity is measured according to a qualitative visual assessment of damage done to each block of diet compared to controls after 7 days (rating scale of 1-4). The concurrent trials are averaged to obtain an approximate percent feeding reduction (%FR). The insects are not given a choice in this assay, and they nearly always cause at least a background level of physical damage to the block of diet even when a potent feeding deterrent is present. Rather than attempt a qualitative "background subtraction", values are conservatively estimated. Thus, no values greater than "75% FR" are assigned in this assay. Additional details of the *C. hemipterus* assay are reported by Wicklow *et al.*, (1988).

The corn earworm *Helicoverpa zea* (syn. *Heliothis zea*, tobacco budworm, cotton bollworm) was chosen as a test organism because of its considerable agricultural importance (Leeper *et al.*, 1986; Kogan *et al.*, 1989). *H. zea* is not generally fungivorous, and is less likely to encounter fungal sclerotia, but sclerotial compounds which interfere with development or cause mortality in *H. zea* could possess practical significance. The colony of *H. zea* is also maintained on a pinto bean-based diet (Dowd, 1988*a,b*). For assay purposes, pinto bean test diets are prepared at the same concentrations described above. In this case, the diet is cut into ca. 250 mg diet block, and each block is placed in a well of a 24-well immunoassay plate (Dowd, 1988*a,b*). A single neonate (newly hatched) larva is added to each well. The plate is then covered, secured, and incubated as described above. Mortality is checked at two, four and seven days, and seven-day survivors are weighed. Test weights and mortalities are compared to controls. Additional details of the *H. zea* bioassay are provided by Dowd (1988*a,b*).

## EXTRACTION, PURIFICATION AND CHARACTERISATION OF ANTIINSECTAN METABOLITES FROM *ASPERGILLUS* SCLEROTIA

Sclerotia are ground with a mortar and pestle or in a Perstorp grinding mill and sequentially extracted with pentane or hexane, ethyl acetate or chloroform, acetone, methanol, and/or water (in some cases) to fractionate the mixture according to polarity. Extracts are then tested for activity against *C. hemipterus* and *H. zea.* Assay results from these studies dictate which strains are subjected to scale-up and detailed examination.

Active fractions are compared with inactive or less active fractions from the same sclerotial sample by thin-layer chromatography (silica gel) in at least two solvent systems. In some cases, the approximate location of active component(s) on the TLC plate can be estimated using these comparisons.

Active extracts are typically subjected to chromatography on silica gel, reversed phase packing (C18), or Sephadex LH-20. Column fractions containing active agents are then subjected to further purification using additional columns or HPLC. Structural studies of compounds we isolate involve the following steps: (1) acquisition of routine IR, UV, NMR, and MS data; (2) dereplication; (3) HRMS to confirm molecular formula; (4) establishment of proton spin-systems by homonuclear decoupling and/or $^{1}H$-$^{1}H$ COSY experiments; (5) determination of one-bond C-H correlations through HMQC; (6) elucidation of long-range C-H correlations using selective INEPT and/or HMBC techniques; and (7) determination of spatial and stereochemical relationships through NOESY experiments.

*Aspergillus* sclerotia have proven to be a rich source of potentially important new natural products. Results of our bioassay-guided studies, from 1987 to date, have led to the isolation of over seventy natural products, forty-five of which possess previously unreported chemical structures, including representitives of eight new ring systems. Some of these compounds possess antiinsectan activity that compares favourably with commercial insecticides. An up-to-date summary of our work on antiinsectan sclerotial metabolites is presented below.

## Sclerotial Metabolites from *Aspergillus flavus* NRRL 6541, and Others (Figure 1)

Sclerotia of *A. flavus* offer an impressive assemblage of antiinsectan compounds. A sclerotium-producing strain of *A. flavus* NRRL 6541, isolated from maize, was initially selected for our study (Wicklow *et al.*, 1988) because it had been shown earlier not to produce aflatoxins. Wicklow and Shotwell (1983) demonstrated that sclerotia produced by some *A. flavus* strains may contain substantial quantities of aflatoxins B1, B2, G1, and G2. We reasoned that because non-aflatoxigenic, sclerotium-producing strains of *A. flavus* are commonly isolated from nature, fungal metabolites other than aflatoxin should be important in protecting the sclerotium from insect predation. Reversed-phase HPLC of the chloroform extract of these sclerotia afforded six major antiinsectan metabolites, five of which are indole diterpenoids. The most potent of the indole diterpenoids, 20, 25-dihydroxyaflavinine, had been previously reported from a whole solid-substrate fermentation extract of *A. flavus*, but had not been associated with any biological activity (Cole *et al.*, 1981). Wicklow and Cole (1982) reported that 20, 25-dihydroxyaflavinine was a product of *A. flavus* sclerotia and not the fungal mycelium or conidia. 20, 25-dihydroxyaflavinine has potent antifeedant activity against adult *C. hemipterus*, causing nearly complete inhibition of feeding when incorporated into the diet at a 25 ppm level (Wicklow *et al.*, 1988). This compound is also toxic to *H. zea* at 25 ppm (Dowd *et. al.*, 1988). Despite this potent antiinsectan activity, 20, 25-dihydroxyaflavinine is nontoxic and nontremorgenic to young chickens at 300 $mgkg^{-1}$ *per os* (Cole *et al.*, 1981). Three of the other indole metabolites are related to 20, 25-dihydroxyaflavinine but have not been previously reported (Gloer *et al.*, 1988). These metabolites (20-hydroxyaflavinine; 24,25-dehydro-10,11-dihydro-20-hydroxyaflavinine;10,11-dihydro-11,12-dehydro-20-hydroxyaflavinine) also show activity against *C. hemipterus* when tested at higher concentrations (i.e. 400-1, 100 ppm) as found in the sclerotia (TePaske, 1991). As stated earlier, 20,25-dihydroxyaflavinine had some moderate activity against *H. zea*, but the other aflavinine derivatives did not. All of the aflavinines characterised by Gloer *et al.*, (1988) also exhibited some antibacterial activity against *Bacillus subtilis*. Furthermore, the aflavinine derivatives were abundant in the sclerotia of eleven different isolates of *A. flavus* and *A. parasiticus* that have been examined (Gloer *et al.*, 1988; TePaske, 1991). In subsequent studies (TePaske *et al.*, 1989*c*) aflavinine itself was isolated as a major hexane-soluble component of the sclerotia of both *A. flavus* and *A. parasiticus*, but it proved inactive in our antiinsectan assays.

Spectral data were significantly different for another major indole metabolite in the chloroform extract and led to our discovery of aflavazole (TePaske *et al.*, 1990). Aflavazole displays significant antifeedant activity against *C. hemipterus* (TePaske *et al.*, 1990). When incorporated into the diet at concentrations found in *A. flavus* sclerotia (200-600 ppm), virtually complete feeding deterrence is observed. Aflavazole had no activity against corn earworm larvae. Aflavazole is closely related to the aflavinines and the tubingensins (see below), but possesses a unique ring system.

Aflavarin, a new bicoumarin, was the remaining antiisectan metabolite from the chloroform extract of *A. flavus* sclerotia, and exhibited potent antifeedant activity against *C. hemipterus* adults (TePaske *et al.*, 1992). Aflavarin is related to the kotanins and the desertorins, which have been isolated from *Aspergillus* spp. that do not produce sclerotia. Kotanin and demethylkotanin were also isolated from *A. flavus* sclerotia but exhibited no activity against *C. hemipterus.*

Studies of the hexane extract of *A. flavus* sclerotia afforded a new antiinsectan metabolite beta-aflatrem, an isomer of aflatrem, in addition to the known metabolites, aflatrem, nominine (see below), paspalinine, and paspaline (TePaske *et al.*, 1992). Of the known compounds, only nominine had not previously been reported as a metabolite of *A. flavus.* Beta-aflatrem causes a significant reduction in the growth rate of the corn earworm *H. zea* but did not interfere with *C. hemipterus* feeding. Beta-aflatrem is also closely related to paspalitrem C, a metabolite of *Claviceps paspali* (Dorner *et al.*, 1984).

## Sclerotial Metabolites from *Aspergillus nomius* NRRL 13137 and NRRL 6552. (Figure 2)

*A. nomius* is closely related to *A. flavus* (Kurtzman *et al.*, 1986). The indeterminate nature of the sclerotia, which are dark brown to black and white-tipped, vertically elongate, and 500-6000 µm in length, provided a ready contrast with the determinant sclerotia formed by *A. flavus*. Nominine, a new indole diterpenoid biogenetically related to the aflavinines, was isolated as the major organic-soluble (pentane and chloroform extracts) component of the sclerotia of *A. nomius* (Gloer *et al.*, 1989). Nominine exhibits

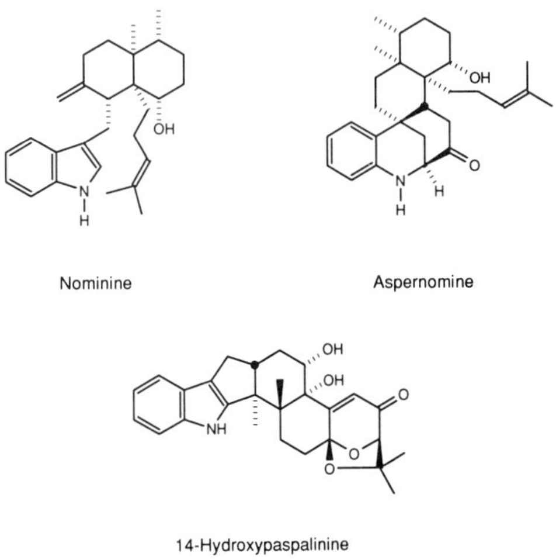


**Figure 2.** Antiinsectan metabolites from sclerotia of *Aspergillus nomius*: nominine; aspernomine; 14-hydroxypaspalinine.

potent activity against *H. zea* in controlled feeding experiments, causing 40% mortality and a 97% reduction in weight relative to the controls when incorporated into the standard test diet at 25 ppm. In our assays, nominine was substantially more effective than rotenone and exhibited activity comparable to that of permethrin. Nominine also elicits an antifeedant response in *C. hemipterus* larvae and exhibited some antibacterial activity against *B. subtilis.* The restriction of nominine to the sclerotia of *A. nomius* parallels the selective allocation of aflavinines (Gloer *et al.*, 1988) and ergot alkaloids (Mantle, 1978) to the sclerotia of *A. flavus* and *Claviceps* spp., respectively. In addition to nominine, pentane and chloroform extracts afforded paspalinine, paspaline and two new related compounds, 14-hydroxypaspalinine and 14-(N,N-dimethylvalyloxy)paspalinine (Staub *et al.*, 1993). Both new compounds exhibit potent activity against *H. zea* causing a 91% reduction in weight gain of the test insects relative to controls at 25 ppm. Interestingly paspalinine is inactive in this assay at the same concentration. These compounds are the first members of the paspalinine/paspaline/penitrem class to possess functionality at the 14 position. 14-hydroxypaspalinine showed moderate antifeedant activity to nitidulids. Aspernomine, isolated from a pentane extract of *A. nomius* sclerotia, caused a 35% reduction in growth rate in *H. zea* when incorporated into the standard test diet at 25 ppm but exhibited no activity to *C. hemipterus* (Staub *et al.*, 1992). Aspernomine contains a previously undescribed ring system, and is the first quinoline-type alkaloid encountered in out studies of sclerotial metabolites.

## Sclerotial Metabolites from *Aspergillus leporis* NRRL 3216 (Figure 3).

*A. leporis* NRRL 3216 was found growing on dung of the white-tailed jackrabbit (*Lepus townsendii* Bachman) collected from a sagebrush grassland community in Wyoming (States and Christensen, 1966), and was observed to produce sclerotia on rabbit dung (Wicklow, 1985). Leporin A, a unique new antiinsectan N-alkoxypyridone, is the most abundant component of the non-polar extracts of *A. leporis* NRRL 3216 sclerotia (TePaske *et al.*, 1991). Leporin A exhibited moderate activity against *H. zea.* Efforts to locate minor analogs or homologs in polar and non-polar extracts by HPLC using photodiode array UV detection provided no evidence for the presence of any related compounds in these mixtures. Some minor compounds isolated from this source were identified as aflavinine derivatives identical with those encountered in *Aspergillus tubingensis* sclerotia (see below).

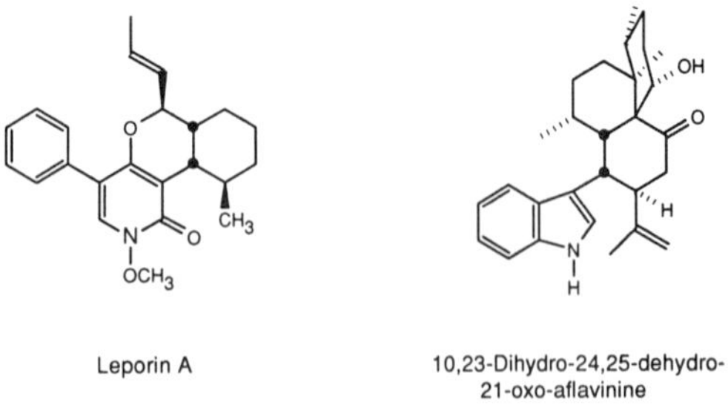


**Figure 3.** Antiinsectan metabolites from sclerotia of *Aspergillus leporis*: leporin A; 10,23-dihydro-24,25-dehydro-21-oxo-aflavinine.

## Sclerotial Metabolites from *Aspergillus tubingensis* NRRL 4700 (Figure 4).

The hexane extract of sclerotia produced by *A. tubingensis* NRRL 4700, a strain isolated from soil in India, exhibited potent activity against *C. hemipterus* (TePaske *et al.*, 1989*a*). Aurasperone A and fonsecinone A are representative examples of a series of *bis*-naphthopyrones that were isolated from extracts of *A. tubingensis* sclerotia, some of which were active against both *C. hemipterus* adults and *H. zea* larvae (TePaske, 1991). This general class of compounds had been previously described, but their effects on insects had not been reported. Quantitative analysis indicated that these compounds are about ten times more concentrated in the sclerotia of *A. tubingensis* than in other tissues of this fungus (TePaske, 1991). Two carbazole alkaloids, tubingensins A and B, exhibited moderate activity against *H. zea*, causing 10-11% mortality when individually incorporated into the standard diet at 125 ppm (TePaske *et al.*, 1989*a*; TePaske *et al.*, 1989*b*). Three new aflavinine derivatives were also encountered in the sclerotia of *A. tubingensis*. These compounds differ from the set of aflavinines produced by *A. flavus* and *A. parasiticus* with regard to the locations and orientations of substituents (TePaske *et al.*, 1989*c*). One of these compounds, 10,23-dihydro-24,25-dehydro-21-oxo-aflavinine, exhibits a 68% reduction in weight gain of *H. zea* and also causes a 38% reduction in feeding rate by *C. hemipterus* when incorporated into the standard test diet at 125 ppm. This compound also exhibits mild antibacterial activity in standard disk assays against *Bacillus subtilis* (TePaske *et al.*, 1989*c*). *A. tubingensis* is a member of the *A. niger* group, and the sclerotia of *A. niger* isolates we have examined contain the same antiinsectan compounds as *A. tubingensis*. Interestingly, we have recently found that *A. tubingensis* also produces nominine (TePaske, 1991).

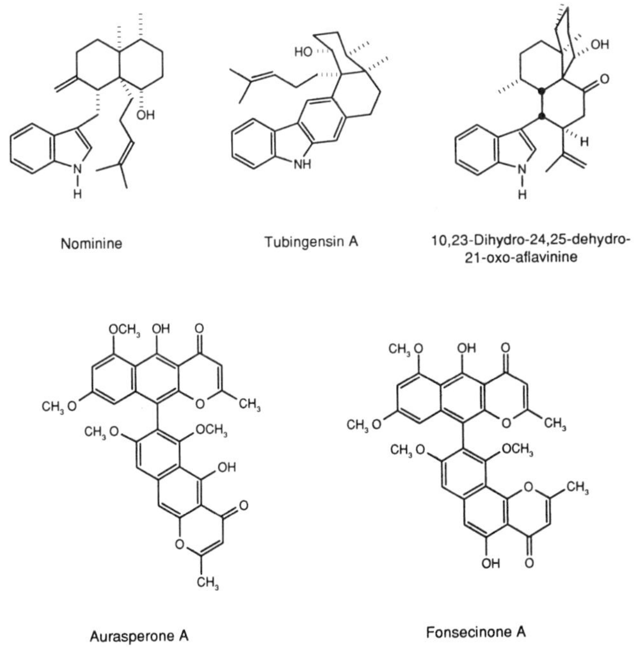


**Figure 4.** Antiinsectan metabolites from sclerotia of *Aspergillus tubingensis* NRRL 4700: tubingensin A; aurasperone A; fonsecinone A; nominine; 10,23-dihydro-24,25-dehydro-21-oxo-aflavinine.

## Sclerotial Metabolites from *Aspergillus sulphureus* NRRL 4077 (Figure 5).

Sclerotia of *A. sulphureus* (Fres.) Thom NRRL 4077, a strain isolated from soil in India, provided a number of new antiinsectan compounds. This strain of *A. sulphureus* NRRL 4077 has been reported to produce ochratoxin A and penicillic acid when grown in liquid culture on a yeast extract-sucrose medium (Ciegler, 1972). The methylene chloride extract of the ground sclerotia exhibited significant antiinsectan activity which led to the discovery of four new indole diterpenoids, Radarins A-D (Laakso *et al.*, 1992*a*). Radarin A, the most potent of these compounds, induces a 53% reduction in weight gain in *H. zea*. The methylene chloride extract also yielded four new antiinsectan compounds of the paspaline/penitrem class including sulpinines A-C and secopenitrem B, all of which exhibit potent activity against *H. zea* (Laakso *et al.*, 1992*b*). Sulpinines A-C could be considered hybrids of the paspalinine and penitrem classes, while radarins A-D have no close literature analogs. Sulpinine A possessed the most potent activity against *H. zea* causing a 96% reduction in weight gain relative to controls. Another major antiinsectan component of this extract was discovered to be penitrem B, a tremorgenic compound originally isolated from *Penicillium palitans* NRRL 3468. Secopenitrem B is a new, ring-opened analog of penitrem B. Penitrem B, sulpinine B, and secopenitrem B brought about similar weight gain reductions (87%). Sulpinine C induced a moderate feeding reduction in *C. hemipterus* adult beetles. A new analog of penitrem B, 10-oxo-11,33-dihydropenitrem B, was discovered in a continued survey of the methylene chloride extract of *A. sulphureus* NRRL 4077 (Laakso *et al.*, 1993). This compound exhibited potent activity in feeding trials against *H. zea* (95% reduction in weight gain relative to controls), and is also moderately active in assays against *C. hemipterus* adults. Although sclerotia are not generally formed in shake cultures of *Aspergillus* spp., standing cultures of *A. sulphureus* formed small quantities of sclerotia on the mycelial surface. Extracts of these sclerotia contained the antiinsectan metabolites as major components. However, none of these compounds were detected in the other fungal tissues or in the medium (Laakso, 1992). Thus, the compounds of interest are indeed heavily concentrated in the sclerotia. Of the sulpinines, only the apparent oxidation product (sulpinine C) lacks potent activity against *H. zea*. This indicates that the indole moiety has a considerable impact on potency, and also suggests a possible biodegradation mechanism for this class of compounds.

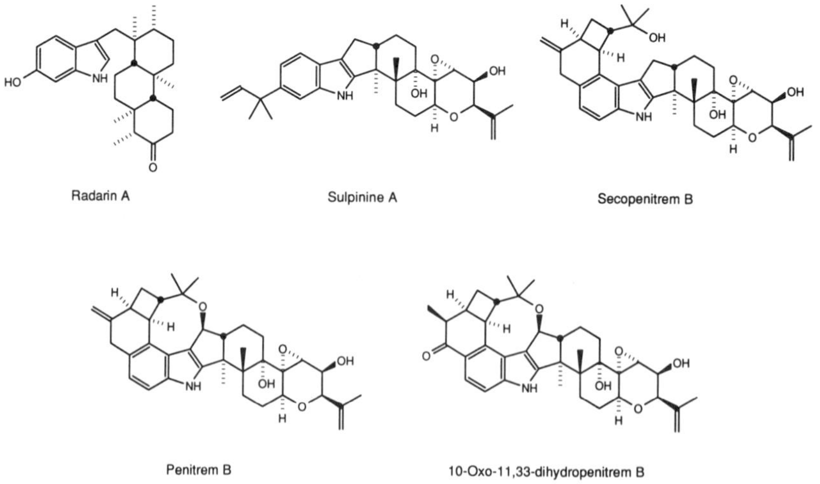


**Figure 5.** Antiinsectan metabolites from sclerotia of *Aspergillus sulphureus* NRRL 4077: radarin A; sulpinine A; secopenitrem B; 10-oxo-11,33-dihydropenitrem.

**Sclerotial Metabolites from *Aspergillus ochraceus* NRRL 3519 (Figure 6).**

Sclerotia produced by a strain of *A. ochraceus* Wilhelm NRRL 3519, received from J. Winitsky, Buenos Aires, and identified by D. I. Fennell after 1965 as *A. ochraceus*, provided three new diketopiperazine-containing metabolites, two of which caused 30% and 33% reductions in weight gain in *H. zea* relative to controls when incorporated into the test diet at 25 ppm (De Guzman *et al.*, 1992). N-methylepiamauromine was the most abundant metabolite from the hexane and methylene chloride extracts of the intact sclerotia, but was less active than epiamauromine or cycloechinulin against *H. zea*.

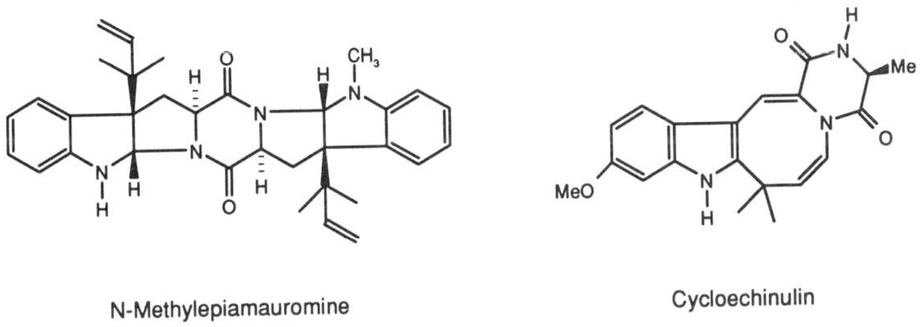


**Figure 6.** Antiinsectan metabolites from sclerotia of *Aspergillus ochraceus* NRRL 3519: N-methylepiamauromine; cycloechinulin.

## TECHNOLOGY TRANSFER

The oral toxicity of some of the compounds we have tested to insects is comparable to the toxicity of the classic "stomach poison" commercial insecticide malathion (compare toxicity of malathion (Dowd, 1988*b*) with nominine (Gloer *et al.*, 1989)). Thus there is potential for direct use of some of the compounds we have discovered as insecticides in their present form, especially in toxic baits that are now used to control such insects as cockroaches and ants. "Attracticidal" baits, where insects are attracted to a toxic bait through use of pheromones, are being developed increasingly and may be an appropriate use of sclerotial insecticides as well. Certainly they could also be used in "organic" insect control programs as are other "botanicals" such as rotenone, pyrethrins, ryania, and sabadilla (primarily as dusts). Unfortunately, the contact toxicity of the most active compounds we have discovered is poor compared to the most active contact insecticides now on the market. For example, none of the compounds we have tested topically at 2 μg/2 mg caterpillar causes any significant mortality, but the commercial insecticide cypermethrin causes approximately 50% mortality at 0.002 μg/2 mg caterpillar (P. F. Dowd, unpublished data). However, one should remember that the natural product that cypermethrin is based on, the pyrethrins, are 100 to 1000-fold less toxic, depending on the insect species. It was only after several years of modification to optimise target site activity and minimise metabolism by the insect (but not non-target vertebrates) that such compounds were developed.

We discovered several analogs of specific classes of sclerotial insecticides that have vastly different activity, and therefore provide some direction for modifications to increase activity. For example, the position and degree of hydroxylation of aflavinines (Dowd *et al.*, 1988; Gloer *et al.*, 1988; Wicklow *et al.*, 1988; TePaske *et al.*, 1989*c*), aurasperones and fonsecinones (TePaske, 1991), radarins (Laakso *et al.*, 1992*a*) and paspalinines (Staub *et al.*, 1993) has a significant effect on the activity against the corn earworm and/or dried fruit beetle. Further derivatives may be obtained by manipulating the growth materials. For example, halogenated forms of penitrems (Mantle *et al.*, 1983) can readily be obtained by adding appropriate salts to the media. Chemical derivatisation of the basic compounds is also possible, and has been used to obtain more active forms of avermectins, which are produced by *Streptomyces avermitilis* (Meinke *et al.*, 1992).

Sclerotial insecticides may have an additional potential use as pharmaceuticals. Insects are complex organisms that share many physiological systems with man and livestock (Dowd, 1992). Thus, target sites present in insects affected by sclerotial compounds may also occur in vertebrates, and be desirable pharmacological targets. For example, we have found penitrem analogs that are active against insects (Dowd *et al.*, 1988; Laakso *et al.*, 1992*a*; 1993). These and several other tremorgenic mycotoxins are thought to interact with the receptor site of the neurotransmitter gamma-amino butyric acid (GABA), a receptor that is found in both insects and vertebrates (Dowd, 1992). Presently compounds with this mode of action are being examined as anti-epileptics (Nilsson *et al.*, 1989).

Prior to our work, the only published chemical studies of fungal sclerotia resulted in the discovery of the pharmaceutically important ergot alkaloids (e.g. ergotamine, ergometrine, ergocornine, ergocristine, ergocryptine, and lysergic acid) (Mantle, 1978; Floss, 1976; Stadler and Giger, 1984). These chemical studies of *Claviceps* were not based on a general interest in sclerotial metabolites, but instead were stimulated by the obvious implication of ergots in poisonings of humans and livestock. It is especially significant that the ergot alkaloids were originally found only in the sclerotia of the fungus (Mantle, 1978). *Claviceps* sclerotia are not formed in liquid fermentation cultures, and many of the fungal ergot alkaloids would not have been discovered through screening liquid cultures alone. The medicinal importance of the ergot alkaloids has led to the gradual development by the pharmaceutical industry of *Claviceps* strains which will produce some of the compounds in liquid fermentations, but even in these cultures, alkaloid production is associated with "sclerotial-like" cells (Mantle, 1978). The pharmaceutical industry apparently has not expanded beyond the ergot fungus to screen the sclerotia of other fungi for their biologically active metabolites. There may be a simple explanation. Sclerotia form on plant/insect hosts or their remains, and may also be produced through solid substrate fermentation. Sclerotia rarely form in liquid shaken culture, which historically has been the industry standard for producing fungal mycelium and fermentation broths, to be extracted for their primary biological screens. There is no reason to believe that production of bioactive metabolites is limited to *Claviceps* sclerotia. Indeed, the results presented above clearly demonstrate that sclerotia from other fungi contain bioactive natural products.

While scale-up for production and harvest of sclerotia is a potential problem, we believe that existing technologies including solid substrate koji fermentation (see B. Wood, K. Aidoo and J. E. Smith, This volume) and corn wet-milling (May 1987) can be readily modified for these purposes. One advantage of producing sclerotia on solid substrates such as corn is that it turns a low value crop material into a high value commodity (the insecticides). Alternatively, liquid culture media could be absorbed into crop byproducts and used for sclerotia production as is done with sugar cane pith and production of ergot alkaloids of *Claviceps fusiformis*, which incidentally afforded a 4-fold greater than for submerged fermentation (Hernandez *et al.*, 1992). Novel chemical cues may also be used to stimulate sclerotium production. For example, 2,4-D has been used to stimulate production of sclerotia by *Rhizoctonia solani* (Sharada *et al.*, 1992).

## SUMMARY

Species of *Aspergillus* have been isolated from living, diseased, and dead insects, and colonise discrete nutrient-rich substrates for which insects and fungi may complete (e.g. seeds, fruit, dung, etc.). There are numerous examples of antiinsectan effects attributed to *Aspergillus*, including symptoms of fungal infection, observations of insect response to fungal cultures, moulded grain, mould contaminated diets, and toxicity studies with specific fungal secondary metabolites. In nature, these entomotoxic fermentation products of *Aspergillus* may prevent loss of substrate to insects and/or discourage arthropod fungivores. Some species of *Aspergillus* produce sclerotia or sclerotioid ascostromata, long-lived resting structures that enable the fungus to survive in the absence of a living host, or to endure seasonal environments unsuitable for fungal growth. We theorised that fungal sclerotia should have superior chemical defences to prevent losses to fungus-feeding insects. Indeed, *Aspergillus* sclerotia contain novel antiinsectan metabolites not produced by the fungal mycelium. Our bioassay system utilises the fungus-feeding sap beetle, *Carpophilus hemipterus* (Nitidulidae) and the corn earworm (*Helicoverpa zea,* Noctuidae), an important insect pest of crops, to identify highly biologically active sclerotial metabolites. Results of our bioassay-guided studies, from 1987 to date, have led to the isolation of over seventy natural products, forty-five of which possess previously unreported chemical structures, including representitives of eight new ring systems. Some of these compounds possess antiinsectan activity that compares favorably with commercial insecticides. These discoveries have been made in spite of the fact that the chemistry of *Aspergillus* spp. has been widely studied for many years. Sclerotia form on plant/insect hosts or their remains, and may also be produced through solid substrate fermentation. Sclerotia rarely form in liquid shaken culture, which historically has been the industry standard for producing fungal mycelium and fermentation broths, to be extracted for biological screening.

## ACKNOWLEDGEMENTS

The work on antiinsectan metabolites from *Aspergillus* sclerotia was conducted under Cooperative Research Agreement No. 58-5114-M-010 between the USDA Agricultural Research Service and the University of Iowa. We thank the National Science Foundation (CHE-8905894 and CHE-9211252) and Biotechnology Research and Development Corporation for financial support.

## REFERENCES

Agnihotri, V.P. (1968) Effect of nitrogenous compounds on sclerotium formation in *Aspergillus niger*. Can. J. Microbiol. 14, 1253-1258.

Agnihotri, V.P. (1969) Some nutritional and environmental factors affecting growth and production of sclerotia by a strain of *Aspergillus niger*. Can. J. Microbiol. 15, 835-840.

Anas, O. and Reeleder, R.D. (1987) Recovery of fungi and arthropods from *Sclerotinia sclerotiorum* sclerotia in Quebec muck soils. Phytopathology 77, 327-331.

Baker, K.F. and Cook, R.J. (1974) Biological Control of Plant Pathogens W. H. Freeman and Co., San Francisco.

Batra, L.R., Batra, S.W.T. and Bohart, G.E. (1973) The mycoflora of domesticated and wild bees (Apoidea). Mycopath. et Mycol. appl. 49, 13-44.

Becker, G., Frank, H.K. and Lenz, M. (1969) Die Giftwirkung *Aspergillus flavus* stampnen auf termitenin Bezeihung zu ihrem von Aflatoxin-gehalt. Z. Angew. Zool. 56, 451-464.
Bennett, J.W. (1981) Genetic perspective on polyketides, productivity, parasexuality, protoplasts, and plasmids, in "Advances in Biotechnology", Vol. III. Fermentation Products. (Vezina, C. and Singh, K., Eds.), pp. 409-415. Pergamon Press, Toronto.
Bennett, J.W. (1983) Differentiation and secondary metabolism in mycelial fungi, in "Secondary Metabolism and Differentiation in Fungi" (Bennett, J.W. and Ciegler, A. Eds.), pp. 1-32. Marcel Dekker, New York.
Bennett, J.W., Fernholz, F.A. and Lee, L.S. (1978) Effect of light on aflatoxins, anthraquinones, and sclerotia in *Aspergillus flavus* and *A. parasiticus*. Mycologia 70, 104-116.
Bullock, S., Willetts, H.J. and Ashford, A.E. (1980) The structure and histochemistry of sclerotia of *Sclerotinia minor* Jagger. I. Light and electron microscope studies on sclerotial development. Protoplasma 104, 315-331.
Bu'Lock, J. D. (1980) Mycotoxins as secondary metabolites, in " The Biosynthesis of Mycotoxins: A Study in Secondary Metabolism" (Steyn, P.S., Ed.), pp. 1-16. Academic Press, New York.
Charnley, A K. (1984) Physiological aspects of destructive pathogenesis in insects by fungi: a speculative review, in "Invertebrate-Microbial Interactions" (Anderson, J.W., Rayner, A.D.M., and Walton, D.W.,Eds.), pp. 229-270. Cambridge Univ. Press, London.
Christensen, C.M. and Kaufmann, H.H. (1974) Microflora, in "Storage of Cereal Grains and their Products" (Christensen, C.M., Ed.), pp. 158-192. Amer. Assn of Cereal Chemists, St. Paul, Minn.
Christensen, M. and Tuthill, D.E. (1985) *Aspergillus*: an overview, in "Advances in *Penicillium* and *Aspergillus* Systematics" (Samson, R.A. and Pitt, J.I., Eds.), pp. 195-209. Plenum Press, New York.
Ciegler, A. (1972) Bioproduction of ochratoxin A and penicillic acid by members of the Aspergillus ochraceus group. Can. J. Microbiol. 18, 631-636.
Clay, K. (1990) Insects, endophytic fungi and plants, in "Pests, Pathogens and Plant Communities" (Burdon, J.J. and Leather, S.R., Eds.), pp. 111-130. Blackwell Scientific Publications, Oxford, U.K.
Cole, R.J. (1981) Fungal tremorgens. J. Food Prot. 44, 715-722.
Cole, R.J. and Cox, R.H. (1981) Handbook of Toxic Fungal Metabolites. Academic Press, New York.
Cole, R.J., Dorner, J.W. Springer, J.P., and Cox, R.H. (1981) Indole metabolites from a strain of *Aspergillus flavus*. J. Agr. Food Chem. 29, 293-295.
Connell, W.A. (1956) Nitidulidae of Delaware. Univ. Del. Agric. Exper. Sta. Bull. 318.
Cooke, R.C. and Rayner, A.D.M. (1984) The Ecology of Saprotrophic Fungi. Butterworths, London.
Cotty, P.J. (1988) Aflatoxin and sclerotial production by *Aspergillus flavus*: Influence of pH. Phytopathology 78, 1250-1253.
De Guzman, F.S., Gloer, J.B., Wicklow, D.T. and Dowd, P.F. (1992) New diketopiperazine metabolites from the sclerotia of *Aspergillus ochraceus*. J. Nat. Prod. 55, 931-939.
Diener, U.L., Cole, R.J., Sanders, T.J., Payne, G.A., Lee, L.S. and Klich, M.A. (1987) Epidemiology of aflatoxin formation by *Aspergillus flavus*. Ann. Rev. Phytopathol. 25, 249-270.
Domsch, K.H., Gams, W. and Anderson, T.H. (1980) A Compendium of Selected Soil Fungi, Vols. I and II. Academic Press, New York 859 pp.
Dorner, J.W., Cole, R.J., Cox, R.H. and Cunfer, B.M. (1984) Paspalitrem C, a new metabolite from sclerotia of *Claviceps paspali*. J. Agric. Food Chem. 32, 1069-1071.
Dorsey, C.K. and Leach, J.G. (1956) The bionomics of certain insects associated with oak wilt, with particular reference to the Nitidulidae. J. Econ. Entomol. 49, 219-230.
Dowd, P.F. (1988*a*) Synergism of aflatoxin B1 toxicity with the co-occurring fungal metabolite kojic acid to two caterpillars. Entomol. Exper. Appl. 47, 69-71.
Dowd, P.F. (1988*b*) Toxicological and biochemical interactions of the fungal metabolites fusaric acid and kojic acid with xenobiotics in *Heliothis zea* (F.) and *Spodoptera frugiperda* (J. E. Smith). Pesticide Biochem. Physiol. 32, 123-134.
Dowd, P.F. (1992) Insect interactions with mycotoxin-producing fungi and their hosts, in "Handbook of Applied Mycology. Vol.5; Mycotoxins in Ecological Systems" (Bhatnagar, D., Lillehoj, E.B. and Arora, D.K., Eds.), pp. 137-155. Marcel Dekker, Inc. New York.
Dowd, P.F. and van Middlesworth, F.L. (1989) *In vitro* metabolism of the trichothecene 4-monoacetoxyscirpenol by fungus- and non-fungus feeding insects. Experientia 45, 393-395.
Dowd, P.F., Cole, R.J. and Vesonder, R.V. (1988) Toxicity of selected tremorgenic mycotoxins and related compounds to *Spodoptera frugiperda* and *Heliothis zea*. J. Antibiotics 41, 1868-1872.
Drummond, J. and Pinnock, D.E. (1990) Aflatoxin production by entomopathogenic isolates of *Aspergillus parasiticus* and *Aspergillus flavus*. J. Invertebrate Pathology 55, 332-336.
Evans, H.C. (1988) Coevolution of entomogenous fungi and their insect hosts in "Coevolution of Fungi

with Plants and Animals" (Pirozynski, K.A. and Hawksworth, D.L., eds.), pp. 149-171. Acad. Press, New York.

Fennell, D.I., Bothast, R.J., Lillehoj, E.B. and Peterson, R.E. (1973) Bright greenish-yellow fluorescence and associated fungi in white corn naturally contaminated with aflatoxin. Cereal Chem. 50, 404-414.

Fenner, M. (1985) Seed Ecology. Chapman and Hall Ltd., London, England. 151 p.

Floss, H.G. (1976) Biosynthesis of ergot alkaloids and related compounds. Tetrahedron 32, 873-912.

Frisvad, J.C. (1986) Taxonomic approaches to mycotoxin identification, in "Modern Methods in the Analysis and Structure Elucidation of Mycotoxins" (Cole, R.J., Ed.), pp. 415-457. Academic Press, New York.

Frisvad, J.C. (1992) Chemometrics and chemotaxonomy: a comparison of multivariate statistical methods for the evaluation of binary fungal secondary metabolite data. Chemometrics and Intelligent Laboratory Systems 14, 253-269.

Gloer, J.B., TePaske, M.R., Sima, J.R., Wicklow, D.T. and Dowd, P.F. (1988) Antiinsectan aflavinine derivatives from the sclerotia of *Aspergillus flavus*. J. Org. Chem. 53, 5457-5460.

Gloer, J.B., Rinderknecht, B.L., Wicklow, D.T. and Dowd, P.F. (1989) Nominine: A new insecticidal indole diterpene from the sclerotia of *Aspergillus nomius*. J. Org. Chem. 54, 2530-2532.

Griffin, D.M. (1966) Fungi attacking seeds in dry seed-beds. Proc. Linnean Soc. N.S.W. 91, 84-89.

Hernandez, M.R.T., Raimbault, M.S. and Lonsane, B.K. (1992) Potential of solid state fermentation for production of ergot alkaloids. Let. Appl. Micro. 15, 156-159.

Hesseltine, C.W., Sorenson, W.G. and Smith, M. (1970) Taxonomic studies of the aflatoxin-producing strains in the *Aspergillus flavus* group. Mycologia 62, 123-132.

Hill, S.T. (1978) Development of *Ahasverus advena* (Coleoptera: Silvanidae) on seven species of *Aspergillus* and on food moulded by two of these. J. Stored Prod. Res. 14, 227-231.

Hinton, H.E. (1945) *Carpophilus hemipterus* (Linnaeus), in "A Monograph of the Beetles Associated with Stored Products", pp. 87-95. Jarrold and Sons, Norwich, England.

Ingham, R.E. (1992) Interactions between invertebrates and fungi: Effects on nutrient availability, in "The Fungal Community", Second Edition, (Carroll, G.C. and Wicklow, D.R., Eds.), pp. 669-690. Marcel Dekker, New York.

Janzen, D.H. (1977) Why fruits rot, seeds mould, and meat spoils. Amer. Nat. 111, 691-713.

Jarvis, J.L., Guthrie, W.D. and Lillehoj, E.B. (1984) Aflatoxin and selected biosynthetic precursors: effects on the European corn borer in the laboratory. J. Agric. Entomol. 1, 354-359.

Jong, S.C. and Ganth, M.J. (1984) ATCC Catalog of Fungi/Yeasts. 16th edition, American Type Culture Collection, Rockville, MD.

Kevan, D.K. (1965) The soil fauna: its nature and biology, in "Ecology of soil-borne plant pathogens" (Baker, K.F. and Snyder, W.C., Eds.), pp. 3-50. University of California Press, Berkeley.

Kogan, M., Helm, C.G., Kogan, J. and Brewer, E. (1989) Distribution and economic importance of *Heliothis virescens* and *Heliothis zea* in North Central, and South America and their natural enemies and host plants in " Proceedings of the Workshop in Biological Control of Heliothis: Increasing the Effectiveness of Natural Enemies". New Delhi, India, November 11-15, 1985, pp. 241-297. Office of International Cooperation and Development, USDA.

Kurtzman, C.P., Smiley, M.J., Robnett, C.J. and Wicklow, D.T. (1986) DNA relatedness among wild and domesticated species in the *Aspergillus flavus* group. Mycologia 78, 955-959.

Laakso, J.A. (1992) Ph. D. Dissertation: New biologically active metabolites from aquatic and sclerotium-producing fungi. Department of Chemistry, University of Iowa, Iowa City, IA.

Laakso, J.A., Gloer, J.B., Wicklow, D.T. and Dowd, P.F. (1992*a*) Radarins A-D: New antiinsectan indole diterpenes from the sclerotia of *Aspergillus sulphureus*. J. Org. Chem. 57, 138-141.

Laakso, J.A., Gloer, J.B., Wicklow, D.T. and Dowd, P.F. (1992*b*) Sulpinines A-C and secopenitrem B: New antiinsectan metabolites from the sclerotia of *Aspergillus sulphureus*. J. Org. Chem. 57, 2066-2071.

Laakso, J.A., Gloer, J.B., Wicklow, D.T. and Down, P.F. (1993) A new penitrem analog with antiinsectan activity from the sclerotia of *Aspergillus sulphureus*. J. Ag. Food Chem. 41, 973 - 975.

Leeper, J.R., Roush, R.T. and Reynolds, H.T. (1986) Preventing or managing resistance in arthropods in "Pesticide Resistance: Strategies and Tactics for Management" (Committee on strategies for the management of pesticide-resistant pest populations, Eds) pp. 335-346. National Academy Press, Washington, D.C.

Loschiavo, S.R. and Sinha, R.N. (1966) Feeding, oviposition, aggregation by the rusty grain beetle *Cryptolestes ferrugineus* (Coleoptera: Cucujidae) on seed-borne fungi. Ann. Ent. Soc. Amer. 59, 578-585.

Lussenhop, J. and Wicklow, D.T. (1990) Nitidulid beetles as a source of *Aspergillus flavus* infective inoculum. Transactions of the Japanese Mycological Society 31, 63-74.
Luttrell, E.S. (1981) Tissue replacement diseases caused by fungi. Annu. Rev. Phytopath. 19, 373-389.
Madelin, M.F. (1963) Diseases caused by hyphomycetous fungi, in "Insect Pathology: An Advanced Treatise" Vol. 2, (Steinhaus, E.A., Ed.), pp. 233-271. Academic Press, New York.
Malloch, D.M. (1985) The Trichocomaceae: Relationships with other Ascomycetes, in "Advances in *Penicillium* and *Aspergillus* Systematics" (Samson, R.A. and Pitt, J.I., Eds.), pp. 365-382. Plenum Press, New York.
Malloch, D. M. and R. F. Cain. (1972). The Trichomataceae: Ascomycetes with *Aspergillus, Paecilomyces*, and *Penicillium* imperfect states. Can. J. Bot. 50: 2613-2628.
Manabe, M. and Tsuruta, O. (1978). Geographical distribution of aflatoxin-producing fungi inhabiting in southeast Asia. Jap. Agric. Res. Quarterly 12: 224-227.
Mantle, P. G. (1978). Industrial exploitation of ergot fungi, in "The Filamentous Fungi, Vol. I. Industrial Mycology" (Smith, J.E. and Berry , D.R., Eds.), pp. 281-300. John Wiley and Sons, New York.
Mantle, P. G. Perera, K. P. W. C., Maishman, N. J. and Mundy, G. R. (1983). Biosynthesis of penitrems and roquefortine by *Penicillium crustosum*. Appl. Environ. Microbiol. 45: 1486-1490.
Martin, M. M. (1987). Invertebrate-Microbial Interactions. Cornell University Press, Ithaca, NY.
May, J. B. (1987). Wet milling: Process and products, pp. 377-397. In: Corn: Chemistry and Technology. S. A. Watson and P. E. Ramstad (Eds.), Am. Assoc. Cereal Chem., St. Paul, MN.
Meinke, P. T., O'Connor, S. P., Mrozik, H. and Fisher, M. H. (1992). Synthesis of ring contracted, 25-nor-6, 5-spiroketal-modified avermectin derivatives. Tetrahedron Lett. 33: 1203-1206.
Miller, M. W. and E. M. Mrak. (1954). Yeasts associated with dried fruit beetles in figs. Appl. Environ. Microbiol. 1: 174-178.
Mills, J. T. (1983). Insect-fungus associations influencing seed deterioration. Phytopathology 73; 330-335.
Moore, K. C. and Erlandson, M. A. (1988). Isolation of *Aspergillus parasiticus* Speare and *Beauveria bassiana* (Bals.) Vuill. from melanopline grasshoppers (Orthoptera: Acrididae) and demonstration of their pathogenicity in *Melanoplus sanguinipes* (Fabricius). Can. Entomol. 120: 989-991.
Moss, M. O. (1989). Mycotoxins of *Aspergillus* and other filamentous fungi. Sym. Ser. Soc. Appl. Bacteriol. Oxford: Blackwell Scientific Publications 18: 695-815.
Murakoshi, S., Ohtomo. T., and Kurata, J. (1973). Toxic effects of various mycotoxins to silkworm (*Bombyx mori* L.) larvae in ad libitum feeding test. J. Food Hyg. Soc. Japan 14: 65-68.
Nilsson, M., Hansson, E. and Ronnback, L. (1989). Uptake of sodium valproate and effects on GABA transport in astroglial primary cultures. Atl. Lab. Anim. A.T.L.A. 16: 244-247.
Ohtomo, T., Murakoshi, S., Sugiyama, J. and Kurata, H. (1975) Detection of aflatoxin B1 in silkworm larvae attacked by an *Aspergillus flavus* isolate from a sericultural farm. Appl. Microbiol. 30, 1034-1035.
Paster, N. and Chet, I. (1980) Effect of environmental factors on growth and sclerotium formation in *Aspergillus ochraceus*. Can. J. Bot. 58, 1844-1850.
Paster, N. and Chet, I. (1982) Influence of controlled atmospheres on formation and ultrastructure of *Aspergillus ochraceus* sclerotia. Trans. Br. Mycol. Soc. 78, 315-322.
Rai, J.N., Tewari, J.P. and Sinha, A.K. (1967) Effect of environmental conditions on sclerotia and cleistothecia production in *Aspergillus*. Mycopathol. Mycol. Appl. 31, 209-224.
Raper, K.B. and Fennell, D.I. (1965) The Genus *Aspergillus*. Williams and Wilkins, Baltimore. 686 pp.
Rhoades, D.F. and Cates, R.G. (1976) A general theory of plant antiherbivore chemistry, in "Biochemical Interaction between Plants and Insects" Wallace, J.W. and Mansell, R.L., Eds.), pp. 168-213. Plenum Press, New York.
Robinson, T. (1979) The evolutionary ecology of alkaloids, in "Herbivores: Their Interaction with Secondary Plant Metabolites" Rosenthal, G.A. and Janzen, D.H., Eds.), pp. 413-448. Academic Press, New York.
Rudolph, E.D. (1962) The effect of some physiological and environmental factors in sclerotial aspergilli. Amer. J. Bot. 49, 71-78.
Samson, R.A. and K.A. Seifert (1985) The ascomycete genus *Penicilliopsis* and its anamorphs, in "Advances in *Penicillium* and *Aspergillus* Systematics" (Samson, R.A. and Pitt, J.I., Eds.), pp. 397-428. Plenum Press, New York.
Semeniuk, G., Harshfield, G.S., Carlson, C.W., Hesseltine, C.W. and Kwolek, W.F. (1971) Mycotoxins in *Aspergillus*. Mycopathol. Mycol. Appl. 43, 137-152.
Sewell, G.W.F. (1959) Studies of fungi in a Calluna-heathland soil. II. Perfect states of some Penicillia. Antonie van Leeuwenhoek 33, 297-314.
Sharada, K., Ikrhsmi, H. and Hyakumachi, M. (1992) 2,4-D induced, c-AMP mediated, sclerotial formation in *Rhizoctonia solani*. Mycol. Res. 96, 863-866.

Singh, K., Frisvad, J.C., Thrane, U. and Mathur, S.B. (1991) An Illustrated Manual on Identification of some Seed-borne Aspergilli, Fusaria, Penicillia and their Mycotoxins. Danish Govt. Inst. of Seed Pathology for Developing Countries, Hellerup, Denmark. 133 pp.

Sinha, R.N. (1966) Association of granary mites and seed-borne fungi in stored grain in outdoor and indoor habitats. Ann. Entomol. Soc. Am. 59, 1170-1181.

Sinha, R.N. (1971) Fungus as food for some stored-product insects. J. Econ. Entomol. 64, 3-6.

Sinha, R.N. (1992) The fungal community in the stored grain ecosystem, in "The Fungal Community", Second Edition, (Carroll, G.C. and Wicklow, D.T., Eds.), pp. 797-815. Marcel Dekker, Inc., New York.

Smart, M.G., Wicklow, D.T. and Caldwell, R.W. (1990) Pathogenesis in *Aspergillus* ear rot of maize: light microscopy of fungal spread from wounds. Phytopathology 80, 1287-1294.

Smith, J.E. and Moss, M.O. (1985) Mycotoxins: formation, analyses, and significance. John Wiley, Chichester.

Southwood, T.R.E. (1973) The insect/plant relationship - an evolutionary perspective, in "Insect/Plant Relationships (van Emden, H.F., Ed.), pp. 3-29. Wiley, New York.

Stadler, P.A. and Giger, R.K.A. (1984) Ergot alkaloids and their derivatives in medicinal chemistry and therapy, in "Natural Products and Drug Development" proceedings of the 20th Alfred Benzon Symposium (Krogsgaard-Larsen, P., Christensen, S.P. and Kofod, H., Eds.), pp. 463-485. Munksgaard, Copenhagen.

States, J. and Christensen, M. (1966) *Aspergillus leporis*, a new species related to *A. flavus*. Mycologia 58, 738-742.

Staub, G.M., Gloer, J.B., Wicklow, D.T. and Dowd, P.F. (1992) Aspernomine: a cytotoxic antiinsectan metabolite with a novel ring system from the sclerotia of *Aspergillus nomius*. J. Amer. Chem. Soc. 114, 1015-1017.

Staub, G.M., Gloer, K.B., Gloer, J.B., Wicklow, D.T. and Dowd, P.F. (1993) New paspalinine derivatives with antiinsectan activity from the sclerotia of *Aspergillus nomius*. Tetrahedron Lett. 34, 2569-2572.

Steiner, W.F. (1984) A review of the biology of Phalacrid beetles, in "Fungus-Insect Relationships" (Wheeler, Q. and Blackwell, M., Eds.), pp. 424-445. Columbia Univ. Press, New York.

Swain, T. (1979) Tannins and lignins, in "Herbivores: their Interaction with secondary plant metabolites" (Rosenthal, G.A. and Janzen, D.H., Eds.), pp. 657-700. Acad. Press, New York.

TePaske, M.R. (1991) Ph.D. Dissertation: Isolation and structure determination of antiinsectan metabolites from the sclerotia of *Aspergillus* species. Department of Chemistry, University of Iowa, Iowa City, IA, 194 p.

TePaske, M.R., Gloer, J.B., Wicklow, D.T. and Dowd, P.F. (1989*a*) Tubingensin A: An antiviral carbazole alkaloid from the sclerotia of *Aspergillus tubingensis*. J. Org. Chem. 54, 4743-4746.

TePaske, M.R., Gloer, J.B., Wicklow, D.T. and Dowd, P.F. (1989*b*) The structure of tubingensin B: A cytotoxic carbazole alkaloid from the sclerotia of *Aspergillus tubingensis*. Tetrahedron Lett. 30, 5965-5968.

TePaske, M.R., Gloer, J.B., Wicklow, D.T. and Dowd, P.F. (1989*c*) Three new aflavinine derivatives from the sclerotia of *Aspergillus tubingensis*. Tetrahedron 45, 4961-4968.

TePaske, M.R., Gloer, J.B., Wicklow, D.T. and Dowd, P.F. (1990) Aflavazole: A new antiinsectan carbazole metabolite from the sclerotia of *Aspergillus flavus*. J. Org. Chem. 55, 5299-5301.

TePaske, M.R., Gloer, J.B., Wicklow, D.T. and Dowd, P.F. (1991) Leporin A: An antiinsectan N-alkoxypyridone from the sclerotia of *Aspergillus leporis*. Tetrahedron Lett. 41, 5687-5690.

TePaske, M.R., Gloer, J.B., Wicklow, D.T. and Down, P.F. (1992) Aflavarin and beta aflatrem: New antiinsectan metabolites from the sclerotia of *Aspergillus flavus*. J. Nat. Prod. 55, 1080-1086.

Townsend, B.B. (1957) Nutritional factors influencing the production of sclerotia by certain fungi. Annals of Botany, N.S. 21, 153-166.

Tsuruta, O., Gohara, S. and Saito, M. (1981) Scanning electron microscopic observations of a fungal invasion of corn kernels. Trans. Mycol.. Soc., Japan 22, 121-126.

Turner, W.B. and Aldridge, D.C. (1983) Fungal Metabolites II. Academic Press, New York. 631 pp.

van Bronswijk, J.E.M.H. and Sinha, R.N. (1971) Interrelations among physical, biological and chemical variates in stored-grain ecosystems: a descriptive and multivariate study. Ann. Ent. Soc. Amer. 64, 789-803.

Waked, M.Y. and Nouman, K.A. (1982) The relationship of sclerotia formation to aflatoxin content of cottonseeds infected with *Aspergillus flavus* Link. Med. Fac. Landbouww. Rijksuniv. Gent, 47, 201-209.

Whittaker, R.H. and Feeney, P. (1971) Allelochemics: chemical interactions between species. Science 171, 757-770.

Wicklow, D.T. (1984) Adaptation in wild and domesticated yellow-green Aspergilli, in "Toxigenic Fungi–

their Toxins and Health Hazard" (Kurata,H. and Ueno, Y., Eds.), pp. 78-86. Elsevier, Amsterdam.

Wicklow, D.T. (1985) *Aspergillus leporis* sclerotia form on rabbit dung. Mycologia 77, 531-534.

Wicklow, D.T. (1988) Metabolites in the coevolution of fungal chemical defence systems, in "Coevolution of Fungi with Animals and Plants" (Pirozynski, K.A. and Hawksworth, D., Eds.), pp. 173-201. Academic Press, London.

Wicklow, D.T. (1991) Epidemiology of *Aspergillus flavus* in corn, in "Aflatoxin in Corn: New Perspectives" (Shotwell, O.L. and Hurburgh, C.R., Jr., Eds.), pp. 315-328. Iowa Agriculture and Home Economics Experiment Station Research Bulletin 599, Ames, Iowa.

Wicklow, D.T. (1992). Interference Competition, in "The Fungal Community - Its Organisation and Role in the Ecosystem" Second Edition (Carroll, G.C. and Wicklow, D.T., Eds.), pp. 265-274. Marcel Dekker, New York, Basel and Hong Kong.

Wicklow, D.T. and Cole, R.J. (1982) Tremorgenic indole metabolites and aflatoxins in sclerotia of *Aspergillus flavus*: An evolutionary perspective. Can. J. Bot. 60, 525-528.

Wicklow, D.T. and Dowd, P.F. (1989) Entomotoxigenic potential of wild and domesticated yellow-green aspergilli, toxicity to corn earworm and fall armyworm larvae. Mycologia 81, 561-566.

Wicklow, D.T. and Horn, B.W. (1984) *Aspergillus flavus* sclerotia form in wound inoculated corn. Mycologia 76 (3), 503-505.

Wicklow, D.T., Horn, B.W. and Cole, R.J. (1982) Sclerotium production by *Aspergillus flavus* on corn kernels. Mycologia 74, 398-403.

Wicklow, D.T., Dowd, P.F., Tepaske, M.R. and Gloer, J.B. (1988) Sclerotial metabolites of *Aspergillus flavus* toxic to a detritivorous maize insect (*Carpophilus hemipterus*, Nitidulidae). Trans. Br. Mycol. Soc. 91, 433-438.

Wicklow, D.T. and Shotwell, O.L. (1983) Intrafungal distribution of aflatoxins among conidia and sclerotia of *Aspergillus flavus* and *Aspergillus parasiticus*. Can. J. Microbiol. 29, 1-5.

Wicklow, D.T. and Wilson, D.M. (1986) Germination of *Aspergillus flavus* sclerotia in a Georgia maize field. Trans. Br. Mycol. Soc. 87, 651-653.

Wicklow, D.T., Horn, B.W., Burg, W.R. and Cole, R.J. (1984) Sclerotium dispersal of *Aspergillus flavus* and *Eupenicillium ochrosalmoneum* from corn during harvest. Trans. Br. Mycol. Soc. 83 (2), 299-303.

Willetts, H.J. (1972) The morphogenesis and possible evolutionary origins of fungal sclerotia. Biol. Rev. 47, 515-536.

Wilson, B. J. (1971) Miscellaneous *Aspergillus* toxins, in "Microbial toxins. VI. Fungal Toxins" (Ciegler, A., Kadis, S. and Ajl, S.J., Eds.), pp. 207-295. Academic Press, New York.

Wilson, B.J. and Wilson, C.H. (1964) Toxin from *Aspergillus flavus*: production on food materials of a substance causing tremors in mice. Science 144, 177-178.

Wright, V.F., Harein, P.K. and Collins, N.A. (1980) Preference of the confused flour beetle for certain *Penicillium* isolates. Environ. Entomol. 9, 213-216.

Wright, V.F., Vesonder, R.F. and Ciegler, A. (1982) Mycotoxins and other fungal metabolites as insecticides, "Microbial and Viral Pesticides" (Kurstak, E., Ed.), pp. 559-583. Marcel Dekker, New York.

Yokota, T., Sakurai, A., Iriuchijima, S. and Takahashi, N. (1981) Isolation and 13 C NMR study of cyclopiazonic acid, a toxic alkaloid produced by muscardine fungi *Aspergillus flavus* and *A. oryzae*. Agric. Biol. Chem. 45, 53-56.

Zummo, N. and Scott, G.E. (1990). Relative aggressiveness of *Aspergillus flavus* and *A. parasiticus* on maize in Mississippi. Plant Disease 74, 978-981.

# *ASPERGILLUS* SPOILAGE: SPOILAGE OF CEREALS AND CEREAL PRODUCTS BY THE HAZARDOUS SPECIES *A. CLAVATUS*

Brian Flannigan and Andrew R. Pearce

Department of Biological Sciences
Heriot-Watt University
Edinburgh, EH14 4AS, UK

## INTRODUCTION

*Aspergillus clavatus* Desm. is one of a small group of species in the Clavati subgenus and section of the genus that is characterised by large blue-green clavate conidial heads. The heads in *A. clavatus* later split into divergent columns of compacted conidial chains. Members of the *A. clavatus* group are common in soils and decomposing materials, their occurrence in soil perhaps being associated with the presence of faecal material (Raper and Fennell, 1965). They often grow on dung of a variety of animals, e.g. chicken and pigeon droppings and stable manure. It has been suggested that the type species, *A. clavatus*, is almost entirely tropical, subtropical and Mediterranean in its distribution (Domsch *et al.*, 1980). However, *A. clavatus* is probably cosmopolitan; in a review (Flannigan, 1986) it was noted that it had been reported present in temperate countries such as Canada, UK, France. Germany, Rumania, Bulgaria and Jugoslavia, and it has also been isolated in Czechoslovakia (Fassatiová *et al.*, 1991). *A. clavatus* has been isolated from cultivated soils under crops of cotton, legumes, potatoes, rice and sugar-cane, and specifically from the rhizosphere of banana, basil, clover, French bean, groundnut, rice and wheat, and appears to be strongly influenced by the growing crop and its residues, the rotational crop sequence and the nitrogen regime. Spore germination and germ-tube growth are stimulated by plant tissue and root exudates (Flannigan, 1986)

## OCCURRENCE OF *A. CLAVATUS* IN STORED PRODUCTS

Although *A. clavatus* may be abundant in soil, it is seldom found as a contaminant of plant organs in the field. For example, Mehrotra and Dwivedi (1977) could not detect it in wheat gathered immediately before harvest and then surface-disinfected, but isolated it from 5.7% of similarly treated stored grains. However, Christensen (1975) considered that it could not be regarded as a storage fungus in the generally accepted sense of the term as it has a high moisture requirement and is usually only isolated in the final stages of decay. Where *A. clavatus* appears in stored products it is associated with either inadequate drying or extremely poor storage of the products. For example, Lindenfelser *et al.* (1978) found that it was only when commercially cured and dried wild rice, *Zizania aquatica*, became

*The Genus Aspergillus*, Edited by Keith A. Powell *et al.*,
Plenum Press, New York, 1994

accidentally damp and mouldy that *A. clavatus* became a predominant member of the mycoflora. Compared with other aspergilli associated with storage, *A. clavatus* is relatively rarc. For example, it was less frequently isolated than other storage fungi from kernels of binned wheat in U.S. by Carter and Young (1950), and did not appear to be involved in the heating of the grain. Also, whilst *A. clavatus* was isolated from roughly 5% of samples of stored wheat surveyed in India (Table 1), more xerophilic storage aspergilli, such as members of the *A. glaucus* group (*Eurotium* spp.), *A. candidus* and *A. flavus*, were much more common (Mehrotra, 1976). In Jugoslavia, it was less common in stored wheat than in maize (Pepeljnjak and Cvetnic, 1984), but in Khazakstan it was among the *Aspergillus* spp. which were prevalent in wheat (Nikov *et al.*, 1977). *A clavatus* has also been isolated from dry-stored barley in Canada (Wallace, 1973), Romania (Stankushev, 1969), Egypt (Abdel-Kader *et al.*, 1979) and India (Mehrotra, 1976), with approaching one-third of samples in the last-named country being contaminated by the species (Table 1). In the U.K., it has been isolated as a minor component of the mycoflora from moist-stored barley (Lacey, 1971).

*A. clavatus* has been found in stored maize (Hassan and Selim, 1982; Pepeljnjak and Cvetnic, 1984) and in meal prepared from mouldy crib-stored maize (Marasas and Smalley, 1972). Wallace and Sinha (1975) found it in both maize and sorghum exported from U.S. to Japan, and Burroughs and Sauer (1971) isolated *A. clavatus* from sorghum with high moisture content in U.S. Various millets have also been found to be contaminated by *A. clavatus* in India (Table 1) and in Khazakstan (Nikov *et al.*, 1977), as has rice. *A. clavatus* has been noted in rice both grown in (Yamazaki, 1971) and imported into Japan (Kurata, 1978), the degree of contamination and damage in some cases being pronounced.

**Table 1.** Occurrence of *A. clavatus* in Indian cereals (after Mehrotra, 1976).

| Cereal | Samples contaminated (%) Grain | Flour |
|---|---|---|
| Wheat | 4.6 | 8.3 |
| Rice | 8.6 | 12.5 |
| Barley | 32.0 | 20.0 |
| Great Millet (*Sorghum bicolor*) | 10.0 | 0 |
| Pearl Millet (*Pennisetum typhoides*) | 23.5 | 25.0 |
| Finger Millet (*Eleusine coracana*) | 18.1 | - |
| Italian Millet (*Setaria italica*) | 20.0 | - |

N.B. No samples of maize grain or flour were found to be contaminated by *A. clavatus*.

In addition to being present in milled wheat fractions (bran, semolina and fine flour) *A. clavatus* was detected in barley, rice and pearl millet flours in India (Table 1), wheat flour and dough products in U.S. (Graves and Hesseltine, 1966), in wheat flour and meal in Khazakstan (Nikov *et al.*, 1977) and wheat and rice flours in Japan (Udagawa *et al.*, 1970).

Although it is with cereals that *A. clavatus* is particularly associated, it has also been isolated from a wide range of other stored commodities (Flannigan, 1986), including a variety of leguminous seeds, black pepper and other spices, pecans and other commercially packed nuts and dried fruits, tomatoes, pomegranates and cocoa beans. Bean paste, dried milk and a range of other foodstuffs have also been found to harbour the organism, and it has been isolated from animal feed pellets composed largely of maize and alfalfa and from poultry feed-mixes containing maize and groundnut meal (Flannigan, 1986).

## *A. CLAVATUS* AS A SPOILAGE ORGANISM

From the substrates which it can utilise and the extracellular enzymes it produces, it can be judged that *A. clavatus* is able to colonise a wide range of organic materials and is therefore able to initiate spoilage. Strains of *A. clavatus* have been shown to utilise pentose and hexose monosaccharides, the disaccharides lactose and sucrose, the trisaccharide raffinose and the polysaccharides dextrin, glycogen, inulin and starch as sole carbon source (Agnihotri, 1963; Varshney, 1981). While Vojtkova-Lepiskova and Kockova-Kratochvilova (1968) found that *A. clavatus* showed lower extracellular activity against starch than a number of yeasts, Ogundero and Osunlaja (1986*a*) reported that starch was broken down rapidly by the organism, which produced a glucoamylase. This enzyme appeared to be produced constitutively, but production was enhanced when glycogen, starch or maltose was used as carbon source. Although there is some doubt as to whether *A. clavatus* is truly cellulolytic (Marsh *et al.*, 1949) or not (Reese and Downing, 1951), fractionation of culture filtrates of *A. clavatus* grown with insoluble cellulose or carboxy methyl cellulose (CMC) as sole carbon source indicated that there was a cellulolytic enzyme system consisting of four components with different specificities (Olutiola, 1977). Pectinolytic activity has been demonstrated (Ragheb and Fabian, 1955), and it has been deduced that it is lignoclastic (Murtuza *et al.*, 1981). *A. clavatus* also appears to be keratinolytic (Reese and Downing, 1951). Cohen (1981) found that, although *A. giganteus* produced an extracellular acid protease, *A. clavatus* was among those *Aspergillus* spp. releasing neutral or alkaline proteases, synthesis of which was controlled by derepression. Ogundero and Osunlaja (1986*b*) found that a purified alkaline protease with maximum activity at 37°C and pH 7.8 rapidly hydrolysed gelatin. Extracellular RNase activity was reported by Bezberodova *et al.* (1969), and has been the subject of continued research in the former USSR. Alkaline and acid phosphatases were first noted by Jacquet *et al.* (1958), and Morozova and Bezberodova (1972) reported production of an acid phosphodiesterase and thermolabile acid monophosphoesterase by *A. clavatus*. Plate and tube tests indicated that lipolytic activity was less common among isolates of *A. clavatus* than amylase, pectinase, cellulase, protease, amidase, RNase, DNase or phosphatase (Flannigan, 1986).

Another fact to be considered in relation to spoilage is that some fungi which grow during malting can affect malt quality or are the cause of taints in the beer or spirit produced. Whether *A. clavatus* does so has not been reported, but some strains have been described as having a strongly foetid odour (Raper and Fennell, 1965). Seifert and King (1982) recorded that aliphatic alcohols and ketones comprised about one-half of the identified compounds in vacuum steam concentrates of broth cultures of the mould. The gas chromatographic pattern was strain-dependent, with the main compounds being 1-octen-3-ol, 4-methylbenzaldehyde, phenylacetaldehyde and 2-methylphenol. It can therefore be supposed that if there is appreciable growth of *A. clavatus* there is likely to be carry-over of these compounds into the wort prepared from the malt and a possibility of taints appearing in the fermented products.

## *A. CLAVATUS* AS A RESPIRATORY HAZARD

The reason for Christensen (1975) stating that the organism is not a 'storage' fungus in the sense that the term is most often used is evident when its water relations are examined. Laboratory studies on pure cultures have shown that the minimum water activity ($a_w$) for its growth appears to be around 0.88 and the optimum $a_w$ approximately 0.98 (Panasenko, 1967). The corresponding values for sporulation are around $a_w$ 0.90 and 0.98-1.00 (Panasenko, 1967) and for spore germination $a_w$ 0.85 and 1.00 (Mehrotra, 1976). Clearly, any substrate must have a very high $a_w$ if it is to be colonised by this species, but colonisation is markedly affected by the nature of the substrate itself and the microflora which it bears. For example, on extruded pasta dough held in an atmosphere of 95% RH ($a_w$ 0.95) *A. clavatus* did not grow in the presence of competing toxigenic aspergilli and penicillia (Stoloff *et al.*, 1978), and on malt contaminated with more xerophilic aspergilli (Flannigan *et al.*, 1986), viable counts of *A. clavatus* only increased when the $a_w$ was >0.966. However, an ecological niche which favours *A clavatus* is created when cereals are germinated in order to produce malt or fodder. In the conditions under which malt for the beverage and food industries is produced, the fungus may come into its own, growing and sporulating profusely and presenting a potential health hazard to those in the malting industry.

That there was a health hazard to workers in the malting industry was recognised some 60 years ago, when chronic respiratory diseases such as bronchitis and emphysema were noted as being a particular problem among malt workers (Bridge, 1932). However, it was only 25 years ago that the importance of *A. clavatus* was appreciated and the occupational extrinisic allergic alveolitis known as malt worker's lung described (Riddle *et al.*, 1968). This serious disease is caused by exposure to high concentrations of *A. clavatus* spores, which have walls particularly rich in allergenic substances (Blyth, 1978). The reasons for the growth of *A. clavatus* and other moulds during malting is evident when the malting process is examined. In traditional floor malting, as in the more modern systems which have replaced it in many places, the barley is steeped until its moisture content (M.C.) is approximately 45% so that it can germinate. To accelerate germination of the grain, steeping is not continuous; the steep is interrupted at least once by draining to give an 'air-rest'. It is finally cast onto the malting floor where it is spread out in a layer 10-15cm deep to germinate. The germinating grain is turned periodically to prevent the temperature of the bed rising above ambient - usually 13-16°C in U.K. This also ensures the uniform aeration needed for even germination and prevents the emerging rootlets becoming matted. The malting barley may be turned manually by shovel and rake, or by a mechanical turner. Germination to produce a 'green malt' with the desired degree of enzyme development and endosperm modification may take upwards of 7 days. To produce a 'finished malt', the green malt is then kilned, firstly at 40-60°C and then finally for a brief period at 70-80°C, depending on the type of malt to be produced. The coleoptiles and rootlets, known as culms, and any fragmented material are then screened off mechanically, leaving malt kernels with a M.C. of 3-4%.

By the time the steeped barley is cast onto the malting floor there will have been some growth of moulds, yeasts and bacteria in the grain (Douglas and Flannigan, 1988), but the major proliferation of micro-organisms occurs during germination of the grain on the floor (Flannigan *et al.*, 1982). Although *Alternaria, Aureobasidium, Fusarium, Geotrichum, Rhizopus* and other fungi proliferate in beds of germinating grain, it is this environment which is evidently ideally suited for the growth and sporulation of *A. clavatus*. It may form blue-green mats on germinating barley (Shlosberg *et al.*, 1991), so that in extreme cases visibility across the malting floor may be affected by dense clouds of spores raised during turning (Riddle *et al.*, 1968). However, it is not just when the germinating grain is being turned that there is an inhalation hazard; loading green malt into kilns and subsequent unloading also result in exposure to large numbers of airborne spores. The spores are readily dispersed, to the extent that, in a distillery, Channell *et al.* (1969) isolated *A clavatus* from the sputum of employees who were not maltsmen and seldom entered the maltings. It should be added here that, although kilning inactivates a high proportion of the inoculum carried on kernels and culms, subsequent handling of these materials may still present a respiratory hazard. For example, maltworkers cleaning culms out of storage bins have been known to show serious respiratory symptoms (Riddle, 1974).

In many maltings, floor malting has been superseded by box (Saladin) maltings, in which a much deeper layer of barley is turned mechanically by helical screws, or by enclosed systems such as germination vessels, rotating drums or continuous malting plant. Although these more modern systems speed up the process, the high level of moisture availability and the presence of solubilised and partially solubilised nutrients still favour microbial growth, although less so than floor malting (Flannigan *et al.*, 1982). As far as malt worker's lung is concerned, Grant *et al.*, (1976) concluded from a survey of Scottish maltings that the incidence of the disease would diminish with mechanisation. The survey revealed that 5.2% of all workers were affected by the disease, with almost as many workers in box maltings (6.2%) as in floor maltings (6.8%) having the disease, but where enclosed systems had been installed only 1.1% (Grant *et al.*, 1976). However, as more recent reports show, floor malting systems are still in use and play host to *A. clavatus* (Rabie and Lübben, 1984; Shlosberg *et al.*, 1991). Furthermore, in southern Africa, where *A. clavatus* is associated with the malting of sorghum, the mould appears to be one of the principal species in the mycoflora of malt produced from sorghum not only in outdoor commercial floor maltings but also in industrial Saladin-type maltings (Rabie and Lübben, 1984). *A clavatus* has also been found in automated malting plant (Flannigan, 1986).

Since *A. clavatus* appears to be rather uncommon on dry grain, it is not clear how malting premises become contaminated by the organism in the first instance. Particular consignments of grain (Channell *et al.*, 1969) and the presence of feral pigeons in grain stores (Riddle *et al.*, 1968) have been suggested, but, whatever the original source, once it is in a maltings *A. clavatus* is difficult to eradicate. Dry grain stored at the maltings becomes contaminated by airborne spores, and the spores on the grain so contaminated are not killed by the addition of hypochlorite to steep water at the concentrations frequently employed (Flannigan *et al.*, 1984). Any dead, cracked, broken or pregerminated kernels (Riddle *et al.*, 1968) are readily colonised centres from which contamination may spread. Clearly, strict standards of hygiene and the use of only top-quality grain have a role to play in controlling the organisms, but the control of temperature is also of particular importance. According to Panasenko (1967), the cardinal temperatures for the growth of *A. clavatus* are: minimum, 5-6°C; optimum, 20-25°C; maximum, 42°C. It has been shown that raising the temperature to 25°C during experimental malting can lead to the viable count of the mould being four orders of magnitude higher than at 16°C (Flannigan *et al.*, 1984). It would not therefore be unexpected that, when the malting temperature in one U.K. maltings was raised well above 16°C in order to shorten the time required for the process (Riddle *et al.*, 1968), problems should have arisen. In some other British maltings where the means of controlling the temperature cannot cope with periods of warm weather (Flannigan *et. al.*, 1984), again problems have arisen, and a fatal mycotoxicosis in Israel has been attributed to profuse growth of *A. clavatus* in consequence of a heat-wave (Shlosberg *et al.*, 1991). The fact that in South Africa the temperature during malting can often be as high as 28°C seems likely to be an important factor in accounting for the prevalence of *A. clavatus* in sorghum malt (Rabie and Lübben, 1984).

## MYCOTOXICOSES ASSOCIATED WITH *A. CLAVATUS*

### Toxic Feeds Containing *A. clavatus*

As culms screened off from kilned malt are either used directly to feed farm animals or are incorporated into feed formulations, the presence of any toxigenic organism in culms may present a serious hazard to animal health. This is instanced in a mycotoxicosis which resulted in the death of most of 250 dairy cows fed on 'malt germs' in Bulgaria (Tomov, 1965), and the culling of the survivors because of decreased lactation or neurological and respiratory abnormalities. *A. clavatus* was shown to be the cause of this episode, as it was in a 'malt germ intoxication' of cattle in Germany (Abadjieff *et al.*,

1966; Fritzsch and Abadjieff, 1967). A method of extracting the toxic principle was subsequently devised by Schultz and Motz (1973), who proposed that culms intended for feed should be assayed for *A. clavatus* and its toxic principle. Adoption of this proposal could well have prevented further episodes, including a recent mycotoxicosis of cattle and sheep in U.K. (Gilmour *et. al.*, 1989). In this case, a supplementary feed containing as the main ingredient culms from a distillery maltings was the cause of death in some cattle, but not sheep. The culms were found to carry 15 x $10^6$ colony forming units $g^{-1}$ culms. However, even more recently, 96% mortality was observed among sheep fed on sprouted barley grains from a factory producing malt extract in Israel (Shlosberg *et al.*, 1991). It is not just culms from maltings which are used as feed, wet residues from sorghum beer production which had been spread out to dry and were found to be contaminated with *A. clavatus* were the cause of fatal tremorgenic disease among cattle (Kellerman *et al.*, 1976). It is not clear from this report whether the mould was present in the residues before spreading out to dry or whether it was adventitiously contaminated when drying, but the practice was considered to be conducive to its growth (Kellerman *et al.*, 1976).

Because of the possibility of casual contamination of by-products, we have examined the possibility of growth of *A. clavatus* on uncontaminated culms which became damp. Firstly, however, several months after the original outbreak we examined a sample of culms involved in the mycotoxicosis reported by Gilmour *et al.*, (1989) and found that *A. clavatus* could still be isolated as the predominant mould, but some penicillia were also present. At that stage, we could not demonstrate toxicity in extracts prepared from the culms after the method of Glinsukon *et al.* (1973) and tested against human epithelial (HEp) cells using the technique of Robb and Norval (1985). However, the extract caused limited inhibition of growth of the yeast *Saccharomyces cerevisiae* (Pearce, 1990), as judged by its effect on conductance change in broth cultures (Figure 1). The highest concentration used, 50 µg extract $ml^{-1}$ culture medium, had an inhibitory effect which was <40% of that of 5 µg patulin $ml^{-1}$.

When uncontaminated dry culms were dry-inoculated with spores of the strain isolated from the culms and stored at a range of different relative humidities (R.H.),

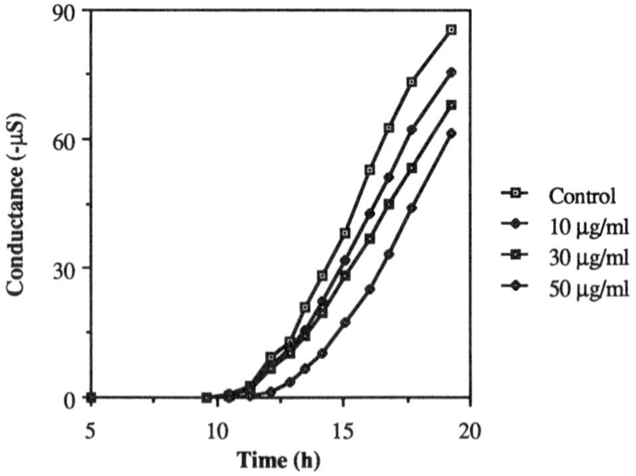


**Figure 1.** Effect of extract prepared from *A. clavatus*-contaminated culms on change in conductance of *Saccharomyces cerevisiae* cultures during growth in modified wort broth at 25°C.

visible mouldering was first evident after 10-12 days at 92.5% R.H., 7 days at 96.6% R.H. and 5-6 days at 98% R.H. and above. Sporulation occurred more rapidly at the higher R.H., and after 21 days >50% of the substratum was covered by sporulating mycelium at 98%, <50% at 96.6% and <25% at 92.5% R.H.. As small quantities of culms were used and dry culms are rather hygroscopic, they would have rapidly come into equilibrium with the storage atmosphere and it is therefore reasonable to assume that culms with a $a_w$ of 0.925 or above will support growth. However, when extracts prepared after this strain was grown on fully hydrated culms at 25° for 21 days were tested, they had no demonstrable effect on yeast growth, although thin layer chromatography (TLC) indicated the presence of traces of patulin in the extracts.

As shown by Kellerman *et al.* (1976) for sorghum malt, there is also the possibility of *A. clavatus* growing on the solid residues remaining after mashing of malt in hot water to produce fermentable wort, i.e. brewer's grains. When brewer's grains (M.C. 76%), were inoculated in our laboratory with a South African strain of *A. clavatus*, known to be highly toxic to ducklings, and incubated at 25°C for 21 days, the resulting extract had a pronounced inhibitory effect on yeast growth (Figure 2). A concentration of 10 µg extract $ml^{-1}$ culture medium extended the time for cultures to show a conductance change of -50 µS by nearly 30% relative to the controls. However, no cytotoxic effect was observed at this concentration when the extract was tested against HEp cells. As *A. clavatus* can clearly grow on this animal feed, which can become adventitiously contaminated by the organism, particularly if a brewery or distillery is adjacent to a maltings in which *A. clavatus* is established, there is a potential hazard if the material is stored for any length of time before it is fed to stock. It has been recommended that, where brewer's grains are to be held for extended periods, they should be ensiled. We have found that in laboratory experiments the application of propionic acid, as used for moist-stored grain, can prevent growth of the mould, but at low dosage rates there is some growth enhancement.

Since the *A. clavatus* grows during the hydroponic phase of malt production, it is to be expected that problems may also arise when cereals are sprouted hydroponically for

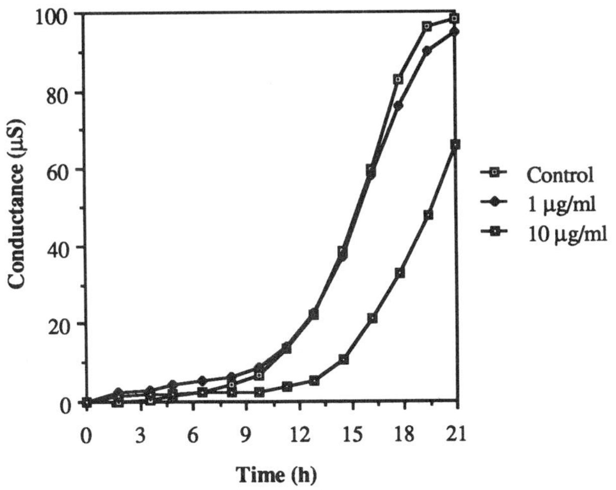


**Figure 2.** Effect of extract prepared from *A. clavatus* culture grown for 21 days at 25°C on brewer's grains on change in conductance of *Saccharomyces cerevisiae* cultures during growth in modified wort broth at 25°C.

use as animal fodder. When Moreau and Moreau (1960*a*) examined one of a number of farms where cattle had died after feeding on barley or wheat forage produced hydroponically because grazing was scarce during an exceptionally dry summer, they found that the mould was present in both the forage and the building in which it was produced. Neurological symptoms were observed among the affected cattle in this case and also in a later mycotoxicosis when cattle were fed on wheat germinated for 6 days to provide winter feed (Jacquet et al., 1963). Nikov et al. (1965) reported mass intoxication of Bulgarian cattle fed on barley sprouts from which this species and other moulds were isolated, and Kellerman et al. (1984) reported that feeding of sprouted maize contaminated with *A. clavatus* had caused fatal tremorgenic mycotoxicosis of dairy cattle in South Africa. *A. clavatus* has also been implicated in a case of mycotoxicoses associated with 'barley grass' in U.K. (Flannigan, 1986).

## Mycotoxins of *A. clavatus* and Their Effects

When the fungal metabolites which could cause disease are considered, it has long been established that *A. clavatus* can produce the toxic unsaturated lactone, patulin, originally named clavacin (Waksman *et al.*, 1942), but it can also produce a range of other mycotoxins (Table 2).

**Table 2.** Mycotoxins produced by *A. clavatus.*

| Mycotoxin | Reference |
|---|---|
| Patulin | Waksman *et al.* (1942) |
| Cytochalasin E | Büchi *et al.* (1973) |
| Cytochalasin K | Steyn *et al.* (1982) |
| Tryptoquivaline, tryptoquivalone | Clardy *et al.* (1975) |
| Nortryptoquivaline, deoxytrytoquivaline, deoxynortryptoquivaline, deoxynortryptoquivalone | Büchi *et al.* (1977), Springer (1979) |
| Ascladiol | Suzuki *et al.* (1971) |
| Kojic acid | Parrish *et al.* (1966) |

Patulin is acutely toxic to laboratory rodents, and was considered by Moreau and Moreau (1960*b*) to be responsible for neurological and other disorders in cattle, and certainly in mice injected intravenously with the toxin (Capitaine and Balouet, 1974) some of the symptoms exhibited by cattle were observed. However, Kellerman *et al.* (1976) were unable to isolate either patulin or the tremorgens tryptoquivaline and tryptoquivalone in the toxic residues which caused fatal tremorgenic disease in cattle. Patulin was obtained by Lindenfelser *et al.* (1978), from naturally contaminated wild rice which had been processed and in which *A. clavatus* was dominant, but it is noteworthy that this toxin shows appreciable instability in grain and grain products (Pohland and Allen, 1970). At high $a_w$, the half-life may only be 1.8-6.8 days, depending on the cereal (Harwig *et al.*, 1977). As indicated by Harwig *et al.* (1977), it is therefore possible that, by the time products which have been responsible for mycotoxicoses are analysed, recoverable patulin may no longer be present. Another factor which may have accounted for failure to detect the toxin is that conditions may not have favoured elaboration of patulin, which is apparently synthesised within much more confined temperature and moisture conditions than those which permit growth of *A. clavatus.* Northolt *et al.* (1978) reported that the optimum temperature for production of patulin in this species was 16°C and that it was not produced at <0.99$a_w$. A second toxin, ascladiol, is structurally similar to patulin, but has only about one-quarter of its acute toxicity to mice (Suzuki *et al.*, 1971).

The tremorgens, tryptoquivaline and tryptoquivalone were found to cause hypersensitivity and persistent tremors, and death after eight days, in intraperitoneally injected rats, but post-mortem examination revealed no specific histopathological changes in the major organs. However, similarly administered cytochalasin E proved fatal within 2-18 h (Glinsukon *et al.*, 1973). Cerebral oedema, severe congestion of the liver and other major organs and pulmonary haemorrhage were observed after either intraperitoneal (Glinsukon *et al.*, 1973) or oral administration of the toxin (Glinsukon *et al.*, 1975*a*, 1975*b*; Glinsukon and Lekutai, 1979), and it was concluded that death resulted from shock caused by effusion of plasma fluid, albumin and globulin through damaged vascular walls. Steyn *et al.* (1982) subsequently indicated that production of cytochalasins E and K accounted for the toxicity of a strain of *A. clavatus* from sorghum malt. Although laboratory studies have shown that kojic acid is toxic, there have been no outbreaks of mycotoxicoses in which it has been perceived to be the cause, so that the fact that some *A. clavatus* strains produce kojic acid (Parrish *et al.*, 1966) is apparently of little significance.

## Symptoms and Pathology of Mycotoxicoses

When more recent outbreaks of *A. clavatus*-associated mycotoxicoses are examined, it can be seen that the clinical signs and neuropathological changes described in stock intoxicated by culms from a distillery maltings in Scotland (Gilmour *et al.*, 1989) were similar to those in cattle in South Africa (Kellerman *et al.*, 1976). The prominent clinical feature in both outbreaks was ataxia, which was evident in the Scottish case as stiffness of the rear limbs and knuckling over, leading to recumbency in sheep and cattle and death in cattle. The frothing from the mouth in the one episode (Gilmour *et al.*, 1989) was probably comparable to the 'excessive salivation' in the other (Kellerman *et al.*, 1976). However, in the South African mycotoxicosis hyperaesthesia and tremors were observed, but not in the outbreak in Scotland. Although acute neural degeneration similar to chromatolysis was not observed in spinal ganglia by Kellerman *et al.*, (1976), it was in the Scottish episode, in which axonal swelling and disruption and intermyelinic oedema were also observed. Otherwise the neuropathological features in the two outbreaks were similar, with microcavitation and degeneration being seen in the spinal white matter.

In feeding trials, Kellerman *et al.*, (1976, 1984) were able to reproduce the clinical symptoms and neuropathological changes in both sheep and cattle, and Gilmour *et al.* (1989) produced neuropathological change, but not clinical symptoms, in feeding trials with two sheep. When two lambs were fed with dried germinated barley associated with the fatal mycotoxicoses of sheep in Israel, neither symptoms nor histopathological changes were observed (Shlosberg *et al.*, 1991). The spinal cord was not examined at necropsy, but swollen neurones showing chromatolysis were observed in the hippocampus and medulla of animals in the main outbreak. On the basis of tremor and other clinical signs, Shlosberg *et al.* (1991) concluded that the outbreak in Israel resembled the episodes in South Africa (Kellerman *et al.*, 1976, 1984), but was probably not the same syndrome as observed by Gilmour *et al.* (1989). They also presumed that the South African strains of *A. clavatus* produced the widest range of toxins. Although there were many similarities between the South African and British episodes, the differences, in particular the hyaline degeneration, necrosis, oedema and petechiation observed in cattle in South Africa, led Gilmour *et. al.* (1989) to consider that differences in dose, or a different toxin or mixture of toxins, might have caused these. They cited an outbreak in China (Jiang *et al.*, 1982) in which cerebrocortical haemorrhage and malacia occurred, i.e. neuropathological lesions substantially different from those observed in Scotland and South Africa, as possible evidence of the involvement of a different toxin. In this instance, more than one-half of 38 dairy cows which received mouldy malt sprouts, in which *A. clavatus* was the dominant fungus, developed acute or subacute symptoms, including hyperaesthesia, tremors, ataxia and posterior paralysis. In addition to neuropathological changes, other lesions included necrotic hepatitis, pulmonary oedema and emphysema.

Among other effects associated with intoxication by *A. clavatus* are hyperkeratosis, first demonstrated in calves by Forgacs *et al.*, (1954); proliferation of tubular epithelial cells, fibroblasts and macrophages in the kidneys (van Rensburg *et al.*, 1971); and interference with mitosis (Saito *et al.*, 1971). Mutagenicity of alkaline hydrolysates of *A. clavatus* has been demonstrated, and in laboratory mice pulmonary tumours developed after nasal administration of spores (Blyth and Hardy, 1982).

It can be seen from the laboratory studies mentioned earlier that patulin, cytochalasins and tremorgens could all have individual roles in such natural mycotoxicoses, but it is almost certain that the syndromes observed result from the synergistic action of components of complex mycotoxin mixtures. It is well-known that there are differences in toxigenic potential in different strains of *A. clavatus*. For example, Vesely *et al.* (1984) observed that only one-third of *A. clavatus* strains produced cytochalasin E. Since there are such strain differences, and differences in the conditions under which the organism grows, it is not surprising that the precise nature of the syndrome which develops should vary from case to case.

## REFERENCES


Abadjieff, W., Bar, H.J., Beyer, J., Ehrentraut, W., Elze, K., Fritzsch, W., Leipnitz, W., Linder, K., Neundorf, R., Panndorf, H., Priboth, W., Schuppel, K.F., Stolzenburg, P., Ulbrich, M., and Voigt, O. (1966) Intoxikationen bei Rindern durch Verfütterung pilzhaltiger Malzkeime Monats. Veterinär. 21, 452-458.

Abdel-Kader, M.I.A., Moubasher, A.H., and Abdel-Hafez, S.I.I. (1979) Survey of the mycoflora of barley grains in Egypt. Mycopathol. 69, 143-147.

Agnihotri, V.P. (1963) Studies on Aspergilli. 10. Utilisation of polysaccharides. Can. J. Microbiol. 9, 703-707.

Bezberodova, S.I., Borodaeva, L.I., Ivanova, G.S., and Morozova, V.G. (1969) Properties of extracellular RNase of *Aspergillus clavatus*. Biokhim. 34, 1129-1136.

Blyth, W. (1978) The occurrence and nature of alveolitis-inducing substances in *Aspergillus clavatus*. Clin. Exp. Immunol. 32, 272-282.

Blyth, W., and Hardy J.C. (1982) Mutagenic and tumourigenic properties of the spores of *Aspergillus clavatus*. Br. J. Canc. 45, 105-117.

Bridge, J. C. (1932) Health, in "Annual Report of the Chief Inspector of Factories", pp. 41-57. London, His Majesty's Stationery Office.

Büchi, G., Kitaura, Y., Yuan, S., Wright. H.E., Clardy, J., Demain, A.L., Glinsukon, T., Hunt, N. and Wogan, G.N. (1973) Structure of cytochalasin E, a toxic metabolite of *Aspergillus clavatus*. J. Am. Chem. Soc. 95, 5423-5425.

Büchi, G., Luk, K.C., Kobbe, B. and Townsend, J.M. (1977) Four new mycotoxins of *Aspergillus clavatus* related to tryptoquivaline. J. Org. Chem. 42, 244-246.

Burroughs, R. and Sauer, D.B. (1971) Growth of fungi in sorghum grain stored at high moisture contents. Phytopathol. 61, 767-772.

Capitaine, R. and Balouet, G. (1974) Etude histopathologique des lesions induites chez la souris par des injections intraperitoneales et intracerebrales de patuline. Mycopathol. Mycol. Appl. 54, 361-368.

Carter, E.P. and Young, G.Y. (1950) "Role of Fungi in the Heating of Moist Wheat". Circular 838. Washington, D.C., U.S. Department of Agriculture.

Channell, S., Blyth, W., Lloyd, M., Weir, D.M., Amos, W.M.G., Littlewood, A.P., Riddle, H.F.V. and Grant, I.W. B., (1969) Allergic alveolitis in maltworkers. Q. J. Med. 38, 351-376.

Christensen, C. M. (1975) "Molds, Mushrooms and Mycotoxins". Minneapolis, University of Minnesota Press.

Clardy, J., Springer, J.P., Büchi, G., Matsuo, K. and Wightman, R. (1975) Tryptoquivaline and tryptoquivalone, two tremorgenic metabolites of *Aspergillus clavatus*. J. Am. Chem. Soc. 97, 663-665.

Cohen, B.L. (1981) Regulation of protease production in *Aspergillus*. Trans. Br. Mycol. Soc. 76, 447-450.

Domsch, K.H., Gams, W. and Anderson, T. (1980) "Compendium of Soil Fungi", Vol. 1. Academic Press, London.

Douglas, P.E. and Flannigan, B. (1988) A microbiological evaluation of barley malt production. J. Inst. Brew. 94, 85-88.

Fassatiová, O., Kubátová, A., Prásil, K. and Vánová, M. (1991) Catalogue of Strains, Culture Collection of Fungi, Charles University, Prague, 3rd edn.

Flannigan, B. (1986) *Aspergillus clavatus* - an allergenic, toxigenic deteriogen of cereals and cereal products. Int. Biodeter. 22, 79-89.

Flannigan, B., Okagbue, R.N., Khalid, R. and Teoh, C.K. (1982) Mould flora of malt in production and storage. Brew. Dist. Int. 12, 31-33, 37.

Flannigan, B., Day, S.W., Douglas, P.E. and McFarlane, G.B. (1984) Growth of mycotoxin-producing fungi associated with malting of barley, in "Toxigenic Fungi" - Their Toxins and Health Hazard" (Kurata, H., and Ueno, Y., Eds.), pp. 52-60. Kodansha/Elsevier, Tokyo.

Flannigan, B., Healy, R.E. and Apta, R. (1986) Biodeterioration of dried barley malt, in "Biodeterioration 6" (Llewellyn, G.C., O'Rear, C.E., and Barry, S., Eds.), pp. 300-305. Commonwealth Agricultural Bureaux, Slough.

Forgacs, J., Carll, W.G., Herring, A.S. and Mahlandt, B.G. (1954) A toxic *Aspergillus clavatus* isolated from feed pellets. Am. J. Hyg. 60, 15-26.

Fritzsch, W. and Abadjieff, W. (1967) Mykologische Untersuchungen an Malzkeimproben. Arch. Tierernähr. 17, 463-474.

Gilmour, J.S., Inglis, D.M., Robb, J. and Maclean, M. (1989) A fodder mycotoxicosis of ruminants caused by contamination of a distillery by-product with *Aspergillus clavatus*. Vet. Record 124, 133-135.

Glinsukon, T. and Lekutai, S. (1979) Comparative toxicity in the rat of cytochalasin B and cytochalasin E. Toxicon 17, 137-144.

Glinsukon, T., Yuan, S.S., Wightman, R., Kitaura, Y., Büchi, G., Shank, R.C., Wogan, G.C. and Christensen, C.M. (1973) Isolation and purification of cytochalasin E and two tremorgens from *Aspergillus clavatus*. Plant Fds. Man 1, 113-119.

Glinsukon, T., Shank, R.C., Wogan, G.N. and Newberne, P.M. (1975*a*) Acute and subacute toxicity of cytochalasin E in the rat. Toxicol. Appl. Pharmacol. 32, 135-146.

Glinsukon, T., Shank, R.C. and Wogan, R.N. (1975*b*) Effects of cytochalasin E on fluid balance in the rat. Toxicol. Appl. Pharmacol. 32, 158-167.

Grant, I.W.B., Blackadder, E.S., Greenberg, M. and Blyth, W. (1976) Extrinsic allergic alveolitis in Scottish maltworkers. Br. Med. J. 1, 490-493.

Graves, R.R. and Hesseltine, C.W. (1966) Fungi in flour and refrigerated dough products. Mycopathol. Mycol. Appl. 29, 277-290.

Harwig, J., Blanchfield, B.J. and Jarvis, G. (1977) Effect of water activity on disappearance of patulin and citrinin from grains. J. Fd. Sci. 42, 1225-1228.

Hassan, M.N. and Selim, S.A. (1982) Some toxinogenic fungi associated with stored corn in Egypt. J. Egypt. Vet. Med. Assoc. 41, 5-12.

Jacquet, J., Villette, O. and Richou, R. (1958) Les phosphatases des filtrats microbiens. 2. Moisissures. Rev. Immunol. 2, 44-50.

Jacquet, J., Boutibonnes, P. and Cicile, J.P. (1963) Observations sur la toxicite d'*Aspergillus clavatus* pour les animaux. Bull. Acad. Vet. France 36, 199-208.

Jiang, C.S., Huang, S.S., Chen, M.Z., Yu, Z.Y., Chen, L.H. and Sang, Y.Z. (1982) Studies on mycotoxicosis in dairy cattle caused by *Aspergillus clavatus* from mouldy malt sprouts. Acta Vet. Zoo. Sin. 13, 247-254.

Kellerman, T.S., Pienaar, J.G., van der Westhuizen, G.C.A., Anderson, L.A.P. and Naude, T.W. (1976) A highly fatal tremorgenic mycotoxicosis of cattle caused by *Aspergillus clavatus*. Onderstepoort J. Vet. Res. 43, 147-154.

Kellerman, T.S., Newsholme, S.J., Coetzer, J.A. and van der Westhuizen, G.C.A. (1984) A tremorgenic mycotoxicosis of cattle caused by maize sprouts infected with *A. clavatus*. Onderstepoort J. Vet. Res. 51, 271-274.

Kurata, H. (1978) Current scope of mycotoxin research from the viewpoint of food mycology, in "Toxicology: Biochemistry and Pathology of Mycotoxins" (Uraguchi, K. and Yamazaki, Y., Eds.), pp. 223-268. Wiley, New York.

Lacey, J. (1971) The microbiology of moist barley storage in unsealed silos. Ann. Appl. Biol. 69, 187-212.

Lindenfelser, L.A., Ciegler, A. and Hesseltine, C.W. (1978) Wild rice as fermentation substrate for mycotoxin production. Appl. Environ. Microbiol. 35, 105-108.

Marasas, W.F.O. and Smalley, E.B. (1972) Mycoflora, toxicity and nutritive value of mouldy maize. Onderstepoort J. Vet. Res. 39, 1-10.

Marsh, P.B., Bollenbacher, K., Butler, M.L. and Raper, K.B. (1949) The fungi concerned in fiber deterioration. 2. Their ability to decompose cellulose. Text. Res. J. 19, 462-484.

Mehrotra, B.S. (1976) "Investigations of Selected Microorganisms Associated with Cereal Grains and Flours in India to Provide Basic Information Related to the Utilisation of Cereal Crains in Foods an Feed Stuffs". Final Report. University of Allahabad, Allahabad.

Mehrotra, B.S. and Dwivedi, P. K. (1977) Subepidermal fungi of wheat grains in India. Int. Biodeter. Bull. 13, 25-29.

Moreau, C. and Moreau, M. (1960*a*) Un danger pour le betail nourri de plantules fourrageres cultivees en germoirs; la pullulation d'une moisissure toxique, l'*Aspergillus clavatus*, cause des accidents mortels. Compt. Rend. Acad. Agric. France 66, 441-445.

Moreau, M. and Moreau, C. (1960*b*) Recherches sur la sporulation de l'*Aspergillus clavatus*. Compt. Rend. Acad. Sci., Series D, 251, 1556-1557.

Morozova, V.G. and Bezborodova, S.I. (1972) Extracellular acid phosphomonoesterase of *Aspergillus clavatus*. Mikrobiol. 41, 404-412.

Murtuza, M., Srivastava, L.L. and Ahmad, N. (1981) Lignoclastic activity of *Aspergillus clavatus*, *Penicillium martensii* and *Pythium proliferum*. Curr. Sci. 50, 722-724.

Nikov, S., Simov, J., Koroleva, V. and Zelev, V. (1965) Investigations of mass poisoning of cows on a farm by mouldy barley sprouts. Nauchn. Trud. Viss. Vet. Inst. 15, 169-180.

Nikov, P.S., Fadeeva, L.M., Imankulova, S.K. and Kolesnikova, G.A. (1977) Characteristics of the microflora in food products originating from the southwestern part of the Kazakhstan. Voprosy Pitan. Part 2, 59-63.

Northolt, M.D., van Egmond, H.P. and Paulsch, W.E. (1978) Patulin production by some fungal species in relation to water activity and temperature. J. Fd Prot. 41, 885-890.

Ogundero, V.W. and Osunlaja, S.O. (1986*a*) Glucoamylase production and activity by *Aspergillus clavatus* Des., a toxigenic fungus from malting barley. Mycopathol. 96, 153-156.

Ogundero, V.W. and Osunlaja, S.O. (1986*b*) The purification and activities of an alkaline protease of *Aspergillus clavatus* from Nigerian poultry feeds. J. Basic Microbiol. 26, 241-248.

Olutiola, P.O. (1977) Cellulolytic enzymes in culture filtrates of *Aspergillus clavatus*. J. Gen. Microbiol. 102, 27-32.

Panasenko, V.T. (1967) Ecology of microfungi. Bot. Rev. 33, 189-215.

Parrish, F.W., Wiley, B.J., Simmons, E.G. and Long, L. (1966) Production of aflatoxin and kojic acid by species of *Aspergillus* and *Penicillium*. Appl. Microbiol. 14, 139-144.

Pearce, A.R. (1990) Growth of *Aspergillus clavatus* and mycotoxin production in by-products of malting and brewing. M. Sc. Thesis, Heriot-Watt University.

Pepeljnjak, S. and Cvetnic, Z. (1984) Distribution of moulds on stored grains in households in an area affected by endemic nephropathy in Yugoslavia. Mycopathol. 86, 83-87.

Pohland, A.E. and Allen, R. (1970) Stability studies with patulin. J. Assoc. Offic. Anal. Chem. 53, 688-691.

Rabie, C.J. and Lübben, A. (1984) The mycoflora of sorghum malt. S. Afr. J. Bot. 3, 251-255.

Ragheb, H.S. and Fabian, F.W. (1955) Growth and pectolytic activity of some tomato molds at different pH levels. Fd Res. 20, 614-625.

Raper, K.B. and Fennell, D.I. (1965) "The Genus *Aspergillus*". Williams and Wilkins, Baltimore.

Reese, E.T. and Downing, M.H. (1951) Activity of the aspergilli on cellulose, cellulose derivatives, and wool. Mycologia 43, 16-28.

van Rensburg, S.J., Purchase, I.H.F. and van der Walt, J.J. (1971) Hepatic and renal pathology induced in mice by feeding fungal cultures, in "Symposium on Mycotoxins in Human Health" (Purchase, I.F.H., Ed.), pp. 153-161. London. Macmillan.

Riddle. H.F.V. (1974) Prevalence of respiratory symptoms and sensitisation by mould antigens among a group of maltworkers. Br. J. Ind. Med. 31, 31-35.

Riddle, H.F.V., Channell, S., Blyth, W., Weir, D.M., Lloyd, M., Amos, W.M.G. and Grant, I.W.B. (1968) Allergic alveolitis in a maltworker. Thorax 23, 271-280.

Robb, J. and Norval, M. (1985) The use of a cytotoxicity test for mycotoxins, in "Trichothecenes and Other Mycotoxins" (Lacey, J., Ed.), pp. 375-380. John Wiley, Chichester.

Saito, M., Enomoto, M., Umeda, M., Ohtsubo, K., Ishiko, T., Yamamoto, S. and Toyokawa, H. (1971) Field survey of mycotoxin-producing fungi contaminating human foodstuffs in Japan, with epidemiological background. 2. Biological effects of the mycotoxins produced by the fungi isolated from foodstuffs, in "Symposium on Mycotoxins in Human Health" (Purchase, I.F.H., Ed.), pp. 179-183. Macmillan, London.

Schultz, J. and Motz, R. (1973) Zum Nachweis des *Aspergillus-clavatus*-Toxins in Füttermitteln. Monatsh. Vet. 28, 790-791.

Seifert, R.M. and King, A.D. (1982) Identification of some volatile constituents of *Aspergillus clavatus*. J. Agric. Fd. Chem. 30, 786-790.

Shlosberg, A., Zadikov, I., Perl, S., Yakobson, B., Varod, Y., Elad, D., Rapoport, E. and Handji, V. (1991) *Aspergillus clavatus* as the probable cause of a lethal mass mycotoxicosis in sheep. Mycopathol. 114, 35-39.

Springer, J.P. (1979) The absolute configuration of nortryptoquivaline. Tetrahedron Lett. 4, 339-342.

Stankushev, K. (1969) Microscopical fungi on oat and barley grains used as winter forage. Mikol. Fitopatol. 3, 268-272.

Steyn, P.S., van Heerden, F.R. and Rabie, C.J. (1982) Cytochalasins E and K, toxic metabolites from *Aspergillus clavatus*, J. Chem. Soc. Perkin Trans. I 10, 541-544.

Stoloff, L., Trucksess, M., Anderson, P.W., Glabe, E.F. and Aldridge, J.G. (1978) Determination of the potential for mycotoxin contaminated pasta products. J. Fd. Sci. 43, 228-230.

Suzuki, T., Takeda, M. and Tanabe, H. (1971) A new mycotoxin produced by *Aspergillus clavatus*. Chem. Pharm. Bull. 19, 1786-1788.

Tomov, A. (1965) *Aspergillus clavatus* Desm. - a cause of poisoning of malt germ-fed cows. Vet. Nauk. 2, 997-1003.

Udagawa, S., Ichinoe, M. and Kurata, H. (1970) Occurrence and distribution of mycotoxin producers in Japanese foods, in "Proceedings of the First U.S. -Japan Conference on Toxic Micro-organisms", (Herzberg, M., Ed.), pp. 174-184. Washington, D. C., UJNR Joint Panels on Micro-organisms and U.S. Department of the Interior.

Varshney, J.L. (1981) Utilisation of different carbon sources by *Aspergillus clavatus* and some of its mutants. Geobios 8, 156-159.

Vesely, D., Vesela, D. and Jelinek, R. (1984) Use of chick embryo in screening for toxin-producing fungi. Mycopathol. 88, 135-140.

Vojtkova-Lepsikova, A. and Kockova-Kratochvilova, A. (1968) Amylolytic activity by some species of the genus *Candida*. Fermentation Type II species. Biologia Bratisl. 23, 422-430.

Waksman, S.A., Horning, E.S. and Spencer, E.L. (1942) The production of two antibacterial substances, fumigacin and clavacin, by *Aspergillus fumigatus* and *Aspergillus clavatus*. Science 96, 202-203.

Wallace, H.A.H. (1973) Fungi and other organisms associated with stored grain, in "Grain Storage: Part of a System" (Sinha, R.N. and Muir, W.E., Eds.), pp. 71-98. Westport, Connecticut, AVI.

Wallace, H.A.H. and Sinha, R.N. (1975) Microflora of stored grain in international trade. Mycopathol. 57, 171-176.

Yamazaki, M. (1971) Isolation of *Aspergillus ochraceus* producing ochratoxins from Japanese rice, in "Symposium on Mycotoxins in Human Health" (Purchase, I.F.H., Ed.), pp. 107-114. London, Macmillan, London.

# INDUSTRIAL FERMENTATION AND *ASPERGILLUS* CITRIC ACID

A. G. Brooke
Haarmann & Reimer
Bayer Plc. Denison Road
Selby, North Yorkshire
YO8 8EF UK

## INTRODUCTION

Citric acid (2-hydroxy-1, 2, 3-propanetricarboxylic acid) was first isolated from lemon juice by the Swedish chemist Scheele (1784), and first produced commercially by John and Edmund Sturge in England from 1823 onwards. Production was from Italian produced calcium citrate derived from lemon and lime juice from Martinique. Before the end of the century, this production method had also been established in France and Germany. However, by 1920 the Italians had also begun to produce citric acid themselves, establishing a monopoly with concomitant high prices. As a result alternative methods of citric acid production were sought.

As early as 1880 citric acid had been synthesised from glycerol (Grimoux and Adam, 1880), and since this time a number of syntheses from other raw materials by different routes have been described.

The first report of a fermentative route to citric acid was made by Wehmer (1893), following the discovery that a small number of *Penicillium* species were able to generate significant quantities of citric acid when grown on sugar solutions. Unfortunately, Wehmer was not successful in commercialising the process.

In 1917, Currie isolated a strain of *Aspergillus niger* capable of producing significantly better yields. Later Currie joined Pfizer and Co Inc in the US, and in 1923 a commercial citric acid plant opened using the new fermentation method. Over the next 25 years fermentation plants were established in England, Germany, Belgium and Czechoslovakia all employing the surface or tray process in which *A. niger* is grown on a static medium of sucrose and inorganic salts (Doelger and Prescott, 1934) in large, shallow, aluminium trays supported on racks in ventilated rooms. However, due to economical considerations, processes based upon the cheaper beet molasses as substrate were rapidly introduced.

Later came the development of submerged fermentation processes for citric acid production with *A. niger* using molasses (beet or cane) or purified glucose syrups. (Perlman, 1949)

By 1965 interest had arisen in the potential use of yeast-based fermentative routes using, initially carbohydrates, but latterly following the first patent in 1968, and due to

*The Genus Aspergillus*, Edited by Keith A. Powell *et al.*,
Plenum Press, New York, 1994

pricing considerations at that time, on n-alkanes (Stottmeister *et al.*, 1982). However, the economic advantage of petroleum-based feedstocks was not realised, thus favouring traditional carbohydrate feedstocks to this day.

## USES OF CITRIC ACID

Citric acid has gained universal acceptance as a safe food ingredient. The Food and Drug Administration in the USA lists citric acid and its sodium, potassium and calcium salts as multiple purpose generally recognised as safe (GRAS) food additives.

The joint FAO/WHO Expert Committee on Food Additives (JEFCA) has allocated to citric acid an Acceptable Daily Intake (ADI) of "not specified". This means that, on the basis of the available data (chemical, biochemical, toxicological and others), the total daily intake of the substance, arising from its use at the levels required to achieve the desired effect, does not represent a hazard to health.

As a consequence of the above, the occurrence of citric acid in a wide variety of fruits (lemon 2.5%, grapefruit 1.5%, orange 1% total weight), and its central role in human metabolism (1.5-2.0 $kgd^{-1}$ is produced and metabolised in the citric acid cycle), citric acid has been described as "natures acidulant".

Citric acid is a commodity chemical product. Along with its salts it finds numerous applications in food (20%), beverage (50%), pharmaceutical/cosmetic (15%), and industrial/technical processing areas (15%).

The main uses of citric acid and its citrates are summarised below:

### Food and Beverages

* as a flavour adjunct, to improve taste.
* as a pH control agent for gelation control, buffering and preservative enhancement.
* as a cheating agent - citric acid is able to complex metals such as iron and copper enabling its use as a stabiliser of oils and fats where it reduces oxidation catalysed by these metals, and prevents spoilage of foods such as seafood.

### Pharmaceutical

* citric acid is included in the formulation for many types of effervescent tablets, e.g. antacid and soluble aspirin preparations, where it produces the effervescent effect when combined with carbonates/bicarbonates.
* citric acid is used as the anion in a range of pharmaceutical preparations which employ a basic substance as the active agent.
* blood anticoagulant - trisodium citrate is routinely used to prevent the coagulation of stored blood by complexing calcium which is required for the clotting process.
* cosmetics industry - citric acid is included in various creams, ointments, and shampoos and zinc citrate is used as a plaque inhibitor in a range of toothpastes.

### Industrial

* citric acid and sodium citrate have cleansing and sequestering properties and can therefore be used instead of phosphates in detergents, industrial and domestic cleaners. Excess phosphates are environmental pollutants which create undesirable growth of algae in rivers and lakes, disrupting the normal biological balance. In some countries phosphates are subject to strict legislation, or are already forbidden.

* industrial cleaning - the low degree of attack on steels enables cleaning/pretreatment of special metals prior to coating. Citric acid is completely biodegradable and does not pollute the environment.

## PRODUCTION PROCESSES

### Fermentation Types

There are currently two organisms which are used for the commercial production of citric acid. These are:

a) *A. niger* - in the surface or submerged fermentation.
b) Yeast - in the submerged fermentation mode.

In the surface fermentation with *A. niger* beet molasses is used as the raw material, cane molasses is not practical in the surface mode because very low yields are obtained. The carbohydrate is suitably diluted, the pH is adjusted (5-7), and precise levels of additional nutrients and ferrocyanide are added. The medium is charged into shallow, high purity, aluminium trays and inoculated with spores which develop into a coherent mat of mould on the surface of the medium; into which the citric acid is released and from which it is subsequently recovered.

In the submerged *A. niger* fermentation beet molasses, cane molasses or pure sugars can be used as carbon source. In many cases a vegetative inoculum stage is used prior to transfer to the main fermentation stage in the larger "production" fermenter thereby maximising the potential of the production phase. The success of the fermentation is critically dependent upon the initial growth phase during which controlled development of the mould takes place. Key to success is the shape of the hyphae and the subsequent aggregation to form small spherical pellets. The hyphae should be abnormally short, stubby, forked and bulbous (Snell and Schweiger, 1951; Kisser *et al.*, 1980; Hustede and Rudy, 1976*a*), arising primarily from a deficiency of manganese (Kisser *et al.*, 1980). Factors resulting in the production of such pellets are the correct ferrocyanide level (Clark, 1964; Hustede and Rudy, 1976*b*), producing an iron concentration of 1ppm (Snell and Schweiger, 1951), and the correct concentration of manganese (Clark *et al.*, 1966). Since small variations in the levels of trace metals originating from the carbohydrate used, can have a major impact on the fermentation performance, strict feedstock quality checks are required.

Where glucose or glucose syrups are used as the carbon source, heavy metals are removed by ion-exchange treatment, however, pre-treatment of sucrose-based media by additions of ferrocyanide has also been proposed.

Yeast as a producer organism has a number of distinct advantages over *A. niger*, including the reported relative insensitivity to trace-metal effects, higher fermentation rates, and the potential for higher initial sugar concentrations by using osmophilic variants (Pfizer Inc., 1974).

The yield of citric acid in the fermentation is generally expressed as Kg citric acid monohydrate per 100 Kg carbohydrate supplied. Yields in the range 70-80% on this basis have been reported. It should be mentioned that the theoretical yield of citric acid monohydrate from sucrose, assuming no carbon is converted into biomass, carbon dioxide or other products is 123%.

Over the years a large number of reports have emerged describing a wide range of different possible feedstocks for citric acid fermentation processes. These investigations stem primarily from the need to keep to a minimum the cost of the fermentation stage, of which the cost of feedstock is a major component. Thus apart from the more commonly used molasses (beet and cane), glucose and sucrose a large number of other feedstocks have been shown to be acceptable in the process.

## METABOLIC PATHWAYS LEADING TO THE FORMATION OF CITRIC ACID IN *A. NIGER*

As described above citric acid is accumulated in very large quantities by *A. niger*, certain yeasts (and even some bacteria). This is not a normal phenomenon but results from development under certain key constraints, aided additionally by the selection of strains and mutants whose metabolism has been altered in some way which prediposes them to citrate accumulation. Much work has been published on the citric acid fermentation, mostly on *A. niger*. This work on the whole concentrates on nutritional requirements and the way nutritional constraints affect the amount and activity of critical enzymes at various stages throughout the fermentation. Some radio-labelled C-tracer investigations have been published, and there has been some work reported on the activity and proportion of enzymes isolated from mutants which differ in their ability to produce citric acid. Very little detailed genetic work on the process has been described in the literature.

Despite this and despite differences between different producer strains, it is generally accepted that a rapid flux of carbon through glycolysis to pyruvate, an anaplerotic carbon dioxide fixation (via pyruvate carboxylase) and a restriction in the Tricarboxylic Acid Cycle (TCA cycle) (preventing breakdown of citric acid once formed), bring about the rapid accumulation of citric acid (Figure 1). However, the exact mechanism of citric acid accumulation remains to be elucidated.

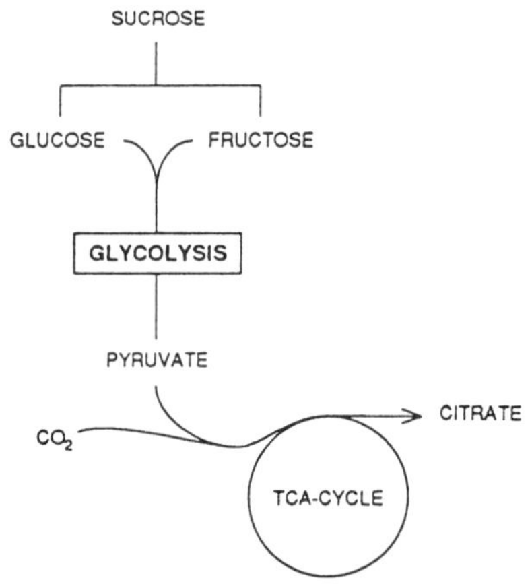


**Figure 1.** Pathways leading to citric acid in *Aspergillus niger*.

## PRODUCT RECOVERY

Once the producer organism has been removed from the final broth, two possible methods, dependent upon the nature of the fermentation feedstock used, can be employed to recover the citric acid.

### Classical Recovery Process

This method, which is particularly suitable for use with impure liquors derived from molasses-based fermentations (since it has the effect of separating from the citric acid most of the impurities originating from the substrate or arising during the fermentation), involves heating the fermented liquor and adding lime. This precipitates insoluble calcium citrate trihydrate, which is then subsequently washed and reacted with sulphuric acid to produce an aqueous solution of citric acid and calcium sulphate (gypsum) by-product.

### Solvent Extraction Method

Solvent extraction techniques are essentially restricted to use for extraction in processes employing cleaner feedstocks, e.g. glucose, since in this process impurities are extracted with the acid. Advantages, however, are that this process avoids the use of lime and sulphuric acid and the concomitant problem of gypsum disposal. The process involves extracting the citric acid from the fermented liquor by reacting with the solvent at a low temperature (<20°C), followed by subsequent stripping of the acid from the solvent at a higher temperature (80°C). Extractants used include butan-2-ol (Les Usines de Melle, 1961), and tributyl phosphate diluted with small quantity of kerosene (Les Usines de Melle, 1963).

## EFFLUENT DISPOSAL

Reference has already been made to the production of gypsum in the calcium citrate precipitation method of citric acid recovery, the disposal of which can cause a problem, although it can be incorporated into building materials. Disposal of the filtrate from the calcium citrate precipitation, especially when molasses is used as the starting material, creates an additional problem. Although non-toxic, this waste can have a very high oxygen demand making it unacceptable in rivers without further treatment. Braun *et al.*, (1979) have proposed the cultivation of yeasts on the effluent producing a material suitable for animal feed. Another possibility is to evaporate the effluent to produce a concentrated molasses-like material usually called condensed molasses solubles (CMS) which can be used in animal feedstuff formulations. Another method of treatment is anaerobic digestion which has the advantage of producing a fuel gas as a by-product (De Zeeue and Lettinga, 1980; Frostell, 1981; Binot *et al.*, 1982). Mycelium can be dried and used as an additive to fertilisers or as an animal feedstock (Kubicek and Rohr, 1986), and recently it has been suggested that waste mycelium can also be used as a source of chitin for use as a biosorbant (Muzzarelli *et al.*, 1980).

## SUMMARY

For the past 60 plus years citric acid has been produced by fermentation of carbohydrates, initially employing the surface process using *A. niger* and eventually with beet molasses replacing pure sugar as carbon source. Latterly submerged fermentations of beet or cane molasses, or glucose syrups by *A. niger* have been introduced. More recently still *A. niger* has been replaced in part by strains of yeast which exhibit the potential for higher productivity and less sensitivity to variations in the crude carbohydrate media.

Work continues to further improve the overall productivity of the production process for citric acid and to identify novel applications for by-products/co-products of the process in an effort to still further improve the economics of the commercial production of citric acid.

## REFERENCES


Binot, R., Bol, T., Naveau, H-P. and Nyns, E-J. (1982) Biomethanation by immobilised fluidised cells. IAW PR Specialised Seminar Copenhagen, Denmark.

Braun, R., Meyrath, J., Stuparek, W. and Zerlauth, G. (1979) Feed yeast production from citric acid waste. Pr ocess Biochem, 14 (1) 16-20.

Clark, D. S. (1964) (National Research Council of Canada, Citric acid production, US Pat. 3 118 821. Chem Abstr., 61 4926d.

Clark, D. S., Ito, K. and Horitsu, H. (1966). Effect of manganese and other heavy metals on submerged fermentation of molasses. Biotechnol Bioeng., 8, 455-471.

Currie, J. N. (1917) The citric acid fermentation of *A. niger*. J Biol. Chem., 31, 15-17

De Zeeuw, W. J. and Lettinga, G. (1980) Use of anaerobic digestion for waste water treatment. Antonie van Leeuwenhoek, 46, 110-112.

Doulger., W. P. and Prescot, S. C. (1939) Citric acid fermentation. Ind. Eng. Chem., 261142-1149.

Grimoux, E and Adam, P. (1880) Synthase de l'acide citrique. C. R. Hebd Seance Acad Sci., 90, 1252-1255

Hustede, H. and Rudy, H. (1976*a*) Manufacture of citric acid by submerged fermentation. US Pat 3 941 656. Chem. Abstr., 84 178222w. (J. A. Benckiser ed.).

Hustede, H. and Rudy, H. (1976*b*) Manufacture of citric acid by submerged fermentation. US Pat. 3 940 315. Chem. Abstr., 85, 3869f. (J. A. Benckiser ed.).

Kisser, M., Kubicek, C. P. and Rohr, M. (1980) Influence of manganese on morphologyand cell wall composition of *A. niger* during citric acid fermentation. Arch. Microbiol., 128, 26-33.

Kubicek, P. and Rohr. (1980) Citric acid fermentation. CRC Critical Review of Biotech., 3, 331-373.

Les Usines de Melle (1961) Process for extracting citric acid from aqueous solutions thereof. Br. Pat. 874030. Chem. Abstr., 55 18010c.

Les Usine de Melle (1963) Process for extracting citric carboxylic acids produced by fermentation. Br. Pat. 936 339 Chem. Abstr. 57, 15620a.

Perlman, D. (1949) Mycological production of citric acid - the submerged culture method. Econ. Bot., 3, 360-374.

Pfizer Inc. (1974) Fermentation process for the production of citric acid. Br. Pat. 1 364 094. Chem. Abstr. 78, 70192y.

Snell, R. L. and Schweig, B. (1951) Citric acid by fermentation, Br. Pat. 653 808. Chem. Abstrc. 45 8719a

Stottmeister, U., Behrens, U., Weissbrodt, E., Barth, G., Franke-Rinker, D. and Schulze, E. (1982) Nutzumg von Paraffinen und anderen Nichtkohlenhydrate - Kohlenstoffquellan zur mikrobiellen Citronensaure synthese. Z. Allg. Mikrobiol., 22, 399-424.

Wehmer, C. (1893) Note sur fermentation citrique. Bull Soc. Chim. Fr., 9, 728-730.

# REGULATION OF ORGANIC ACID PRODUCTION BY ASPERGILLI

C. P. Kubicek[1], C. F. B. Witteveen[2] and J. Visser[2]

[1]Abteilung für Mikrobielle Biochemie
Institut für Biochemische Technologie und Mikrobiologie
TU Wien, Getreidemarkt 9/1725
A1060 Wien, Austria

[2]Section Genetics
Department of Genetics
Agricultural University
Dreijenlaan 2, 6703 HA
Wageningen, The Netherlands

## INTRODUCTION

Within the fungal kingdom, the accumulation of organic acids in considerable amounts is one of the major domains of aspergilli. These acids may be summarised as falling into two groups, i.e. such derived from sugars by simple oxidation (i.e. gluconic acid, kojic acid) and those, which are related to tricarboxylic acid intermediates (citric, cis-itaconic, malic, oxalic and epoxy-succinic acid). The mechanisms by which aspergilli accumulate these organic acids have attracted the interest of numerous researchers through several decades (for review see Roehr *et al.*, 1983*a*). The basic metabolic routes involved in the formation of these acids have been established, Figure 1: aspergilli can utilise two different pathways for glucose catabolism, i.e. the hexose-bisphosphate (Embden-Meyerhoff-Parnas) pathway and the hexosemonophosphate (pentose phosphate) shunt. It has been shown that the hexose-bisphosphate pathway prevails under conditions of high acid accumulation. In addition to these pathways, *Aspergillus niger* and some other species (as well as several other fungi) form a glucose oxidase, which catalyses the formation of glucono-δ-lactone from glucose, hence connecting glucose utilisation with gluconate breakdown (Roehr *et al.*, 1983*b*). From Figure 1, it is evident that the accumulation of sugar acids as gluconic acid is mainly a consequence of the synthesis of the corresponding oxidase, i.e. glucose oxidase, whereas the metabolic steps involved in the accumulation of tricarboxylic acid cycle related acids may require the coordinate operation of at least three subsequent steps, i.e. (a) glycolytic flow to pyruvate; (b) formation of oxaloacetate from pyruvate; and (c) possible steps within the tricarboxylic acid cycle. All of these steps are present in most eucaryotic organisms and serve catabolic and anabolic functions. Overproduction of various organic acids will therefore only occur under very delicate imbalances of metabolism, which by-pass metabolic regulation. This review will attempt to describe the present knowledge in this context.

*The Genus Aspergillus*, Edited by Keith A. Powell *et al.*,
Plenum Press, New York, 1994

## REGULATION OF GLUCOSE OXIDASE BIOSYNTHESIS AND GLUCONATE ACCUMULATION

Because of the high $K_m$ for oxygen of glucose oxidase ($K_m$ = 0.48mM at 27°C; Gibson *et al.*, 1964) the dissolved oxygen level is a key parameter in the process kinetics of gluconate formation (Reuss *et al.*, 1986). Since process optimisation using dextrose up to 240 $gl^{-1}$ has already resulted in high yields (95%) in less than 24 h, strain improvement has not received much attention (Zidwick, 1992). Higher amounts of glucose oxidase only lead to increased product formation when high air pressure is applied simultaneously. Physiological aspects of gluconate biosynthesis and regulation of the enzyme levels operating in this pathway have also been neglected. The biochemistry involved is rather straightforward, as shown in Figure 2: the initial step of the pathway is the oxidation of glucose to glucono-δ-lactone which is catalysed by the flavoprotein glucose oxidase (EC

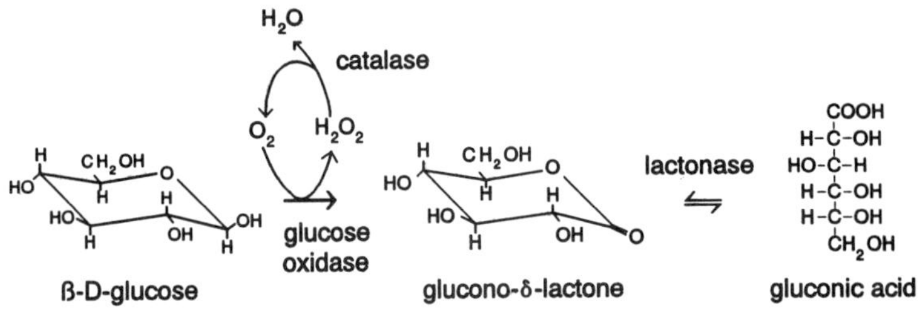


**Figure 1.** Involvement of glucose oxidase, catalase and lactonase in the formation of gluconic acid by *Aspergillus niger*.

1.1.3.4) with concomitant reduction of oxygen to hydrogen peroxide. The glucono-δ-lactone can be hydrolysed by a lactonase (EC 3.1.1.17) or it hydrolyses spontaneously. Hydrogen peroxide is detoxified by catalases (EC 1.11.1.6). The physiological function of glucose oxidase is not quite clear. The most likely function is the contribution to the competitiveness of the organism by removal of glucose, by concomitant acidification of the environment and/or by the formation of hydrogen peroxide to which *A. niger* itself is very resistant. Biocontrol of the phytopathogenic fungus *Verticillium dahliae* by *Talaromyces flavus* has been ascribed to hydrogen peroxide (Kim *et al.*, 1988). In some white-rot fungi like *Phanerochaete chrysosporium* it has been suggested that glucose oxidase provides the hydrogen peroxide required for lignin degradation by lignin peroxidases (Ramasamy *et al.*, 1985; Kelley and Reddy, 1986) although it has been pointed out that glyoxal oxidase is a more likely candidate (Kersten and Kirk, 1987; Kersten, 1990).

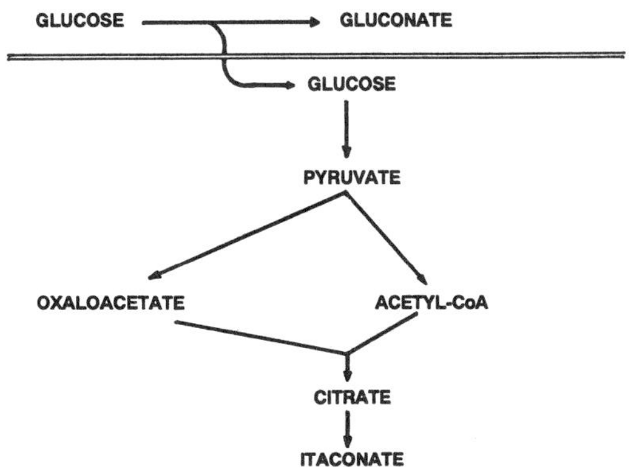


**Figure 2.** General metabolic relationships between organic acids produced by *Aspergillus niger* and*Aspergillus terreus*.

## Localisation of the enzymes involved in gluconate formation

For a long time it was assumed that in *A. niger* glucose oxidase was located intracellularly (Pazur, 1966). Cytochemical and ultrastructural studies by Van Dijken and Veenhuis (1980) seemed to support this view. The observed positive staining of microbodies caused by catalase as a result of $H_2O_2$ generation was thought to indicate that the glucose oxidase generating $H_2O_2$ was also localised in the peroxisomes. Mischak *et al.* (1985) showed that under manganese deficient growth conditions glucose oxidase activity was found almost quantitatively in the medium. Their explanation for this was that a cell-wall localised glucose oxidase enters the culture fluid because of an altered cell wall composition resulting from manganese deficiency. Three arguments are in favour of an extracellular localisation of glucose oxidase: (1) the enzyme is quite strongly glycosylated which has never been observed for any peroxisomal protein. (2) the $K_m$ value for glucose is high (15 mM), whereas the intracellular glucose concentration is relatively low. This is not compatible with the high fermentation efficiencies of gluconate production. (3) The gene for glucose oxidase from *A. niger* has recently been cloned and sequenced (Kriechbaum *et al.*, 1989; Frederick *et al*,., 1990; Whittington *et al.*, 1990). The mature enzyme consists of 583 amino acids, contains eight potential sites for N-linked glycosylation and is preceded by a 22 amino acid N-terminal preprosequence which has a putative signal peptidase processing site and a monobasic processing site (Frederick *et al.*, 1990), indicative for a secretory protein. Using immuno-cytochemical methods Witteveen *et al.*, (1992) recently definitely demonstrated that glucose oxidase is indeed localised in the fungal cell wall whereas cytochemical staining of catalase activity identifies this activity to be present both in the cell wall and in the peroxisomes. These authors also demonstrated the presence of 4 different catalases in *A. niger* using native gel electrophoresis followed by a negative staining procedure to identify activity. Two of these catalases are more or less constitutive, one (CAT I) is localised intracellularly whereas the other (CAT II) occurs in the cell wall. Under glucose oxidase inducing conditions two other catalases appear, one intracellular form (CAT III) and one extracellular form (CAT IV), thus protecting the fungal cell against hydrogen peroxide by a double barrier. The lactonase activity is also secreted and found to be distributed in equal amounts over culture fluid and cell wall (Witteveen *et al.*, 1992). The relative amounts of glucose oxidase (80%), catalase (>95%) and lactonase (50%) retained by the cell wall seem to correlate with differences in size of the proteins. These enzyme localisation studies explain why mycelium could be successfully reused for subsequent fermentations (cf. Reuss *et al.*, 1986).

## Glucose oxidase overproducing and negative mutants

Glucose oxidase overproducing mutants are interesting a) for gluconate production purposes under high pressure fermentation conditions, and b) to improve the enzyme yield as such since glucose oxidase itself is also of commercial interest. Several groups reported the isolation of glucose oxidase overproducing strains (Kundu and Das, 1985; Fiedurek *et al.*, 1986; Markwell *et al.*, 1989) using plate screening methods. These are usually based on acidification of the medium around a colony. Witteveen *et al.*, (1990) developed a more sensitive colony screening method which is based on the localisation of glucose oxidase in the cell wall. Selection of mutants with an altered induction was achieved by plating out UV-mutagenised spores under conditions [fructose, low glucose (<2mM), low oxygen levels (<2% oxygen) ] where the parental strain hardly produces glucose oxidase. After 40 h a buffered solution containing glucose and horseradish peroxidase is added. Glucose oxidase containing colonies can be easily monitored since these turn brownish red in colour. Besides a variety of glucose oxidase overproducing mutants one glucose oxidase negative mutant was obtained. A phenotypical analysis of the overproducing mutants indicated that oxygen and carbon source dependent induction is due to different mechanisms. The *gox* mutations have been genetically localised and belong to nine different complementation groups (Swart *et al.*, 1990). Three *gox* loci (*gox*B, *gox*F, *gox*C) belong to linkage group II, one (*gox*I) to linkage group III, two (*gox*D, *gox*G) to linkage group V, two (*gox*A, *gox*E) to linkage group VII, whereas *gox*H could not be assigned. All *gox* mutations were found to be recessive. Swart *et al.*, (1990) have outlined a strategy to construct improved strains which carry up to four different *gox* overproducing mutations. Glucose oxidase overexpression has also been achieved by transforming *A. niger* with the glucose oxidase gene (Whittington *et al.*, 1990; Witteveen *et al.*, 1993). In multicopy transformants only a three-to fourfold increase in glucose oxidase activity was observed when using optimal induction conditions (Witteveen *et al.*, 1993). Whittington *et al.*, (1990) report a higher increase relative to the wild type but used different culture conditions. The glucose oxidase gene was introduced and expressed in yeast under control of the regulated ADH2-GAPDH hybrid promoter (Frederick *et al.*, 1990) and in *Aspergillus nidulans* under its own promoter (Whittington *et al.*, 1990; Witteveen *et al.*, 1993). In yeast the enzyme is secreted and becomes overglycosylated (MW350-400 kDa) whereas the MW's (150-180 kDa) in *A. niger* and *A. nidulans* are indistinguishable. However, Witteveen *et al.*, (1993) observed host dependant differences in the isoelectric focusing patterns of the two glucose oxidases which are probably the result of slight differences in glycosylation.

## Induction of the glucose oxidase system

The availability of the *gox* structural gene as well as the regulatory mutants have led Witteveen *et al.*, (1993) to investigate the induction characteristics of glucose oxidase, catalase and lactonase activities. The *gox*C17 mutant in which glucose oxidase activity is absent can be complemented by the glucose oxidase structural gene. Most important in these studies is the observation that none of the three enzyme activities is induced in this mutant so induction requires an active glucose oxidase. Subsequent experiments in wild type showed that all three activities are induced by hydrogen peroxide even in the absence of glucose. Amongst the glucose oxidase overproducing mutants the most pronounced phenotypes are in the *gox*B and the *gox*E loci which are most likely regulatory genes. In the *gox*B mutants all three enzymes are induced regardless the carbon source or the level of oxygenation. The *gox*B gene product is hypothesised to mediate hydrogen peroxide induction and regulates the expression of glucose oxidase, catalase III and IV and lactonase in a coordinate way. Phenotypically the effects of mutations in *gox*E are very similar to those in *gox*B except that these are limited to a carbon independent induction of glucose oxidase and lactonase whereas induction still depends on high levels of aeration. This indicates that in gluconate biosynthesis at least two different control systems operate which have partially overlapping targets.

## REGULATION OF ACCUMULATION OF TRICARBOXYLIC ACID CYCLE RELATED ACIDS

### Type and concentration of sugar used trigger tricarboxylic acid cycle related acid accumulation via bypassing glycolytic regulation

In contract to gluconic acid formation, the biosynthesis of tricarboxylic acid cycle related acids requires catabolism of glucose via several enzymic steps, and accumulation in notable amounts is only observed under various conditions of nutrient imbalance (Table 1).

**Table 1.** Optimal nutrient parameters for citric acid accumulation

| Parameter | Optimal Condition |
|---|---|
| Sugar type | Easy metabolizable |
| Sugar Concentration | High (>100 g/l) |
| Nitrogen source and conc. | Low, $NH_4$ > $NO_3$ |
| Phosphate Conc. | Low; to be balanced with trace metals |
| Trace metal ions | $Fe^{3+}$, $Zn^{2+}$ low; $Mn^{2+}$ very low (to be balanced with phosphate conc.) |
| pH | <3 |

Among the various parameters influencing citric acid accumulation, the type and concentration of the sugar used most severely determines the extent of acid production: while the optimal concentrations of trace metals, phosphate and nitrogen are interrelated, the concentration and type of the sugar used are the only parameters which cannot be antagonised by appropriate manipulation of the others (Xu *et al.*, 1989). A comprehensive report on the types of sugars suited or not suited for citric acid production by *A. niger* has been given Xu *et al.*, (1989) and is briefly summarised in Table 2: only sugars which are rapidly taken up by the fungus provide both high yield and high rate of acid accumulation. The biochemical basis of this finding appears to be to bypass glycolytic regulation: the regulation of the hexosebisphosphate pathway in aspergilli involves at least three key regulatory enzymes, i.e. hexokinase (EC 2.7.1.1), 6-phosphofructokinase (EC 2.7.1.11; PFK 1), and pyruvate kinase (EC 2.7.1.40) (Smith and Ng, 1972; Kubicek, 1988*b*). PFK 1 is susceptible to various fine controls, i.e. activation by fructose-2,6-diphosphate (Fru-2,6-$P_{2)}$, AMP and $NH_4$+ ions, and inhibition by phosphoenol-pyruvate, citrate and ATP (Habison *et al.*, 1983) and therefore a potential major point of flux regulation.

Activation by Fru-2,6-$P_2$ links PFK 1 activity to activity and regulation of 6-phosphofructo-2-kinase (EC 2.7.1.105; PFK 2). The induction of citric acid accumulation by high sugar concentrations is accompanied by a rise in the intracellular concentration of Fru-2,6-$P_2$ (Kubicek-Pranz *et al.*, 1990). Higher concentrations of Fru-2,6-$P_2$ are also observed upon cultivation on carbon sources which allow higher yields of citric acid. Recently, we partially purified and characterised PFK 2 (the enzyme forming Fru-2,6-$P_2$ from *A. niger* to identify the regulatory signals which reflect the elevated carbohydrate concentration (Harmsen *et al.*, 1992): our data suggest that in contrast to PFK2 from mammalian tissues and yeast its activity is not regulated by phosphorylation, and is not regulated by the metabolites citrate, phosphoenol-pyruvate or glycerol-3-phosphate. Hence the *in vivo* activity of *A. niger* PFK 2 is then modulated mainly by the cellular fructose-6-phosphate (Fru-6-P) level. Fru-2,6-$P_2$ concentrations under various cultivation conditions were measured and higher levels were found with high external sugar concentrations. The result of this would be that the *A. niger* PFK 1 activity strongly increases due to this higher activator concentration.

**Table 2.** Carbon sources giving high (H), intermediate (I) and low (L) yields of citric acid accumulation [1]

| H | I | L |
|---|---|---|
| Sucrose | Fructose | Galactose |
| Mannose | Sorbitol | Starch |
| Glucose | | Cellobiose |
| Maltose | | Xylose |
| | | Mannitol |

[1] yields (g/g) for H, I and L are >70, 20 - 70, and <20 %, respectively

Another important control point for citric acid accumulation is hexose phosphorylation. We have therefore recently purified and characterized the hexose phosphorylation enzymes from *A. niger* (Steinböck *et al.*, 1993): there is a single major hexokinase present, which has only low specificity for fructose and other hexoses, and may therefore rather be called a glucokinase (EC 2.7.1.). Several attempts to identify other glucose phosphorylating enzymes under various culture conditions have failed (F. Steinböck, M. Röhr, C. P. Kubicek, unpublished data). Increased enzyme amounts are formed on high glucose or sucrose concentrations. Its kinetic characteristics are close to the mammalian enzyme, yet it is not inhibited by glucose-6-phosphate. In order to prove that the activity of hexokinase determines the rate of Glu-6-P and of Fru-6-P (because of the high activities of glucose phosphate isomerase) formation, we isolated 2-desoxyglucose-resistant mutants of *A. niger*, and investigated their citric acid production (Steinböck *et al.*, 1993). In accordance with other authors (Fiedurek *et al.*, 1988; Kitimura *et al.*, 1992), a class of mutants was found, whose citric acid production had dropped down considerably. A closer inspection of these mutants revealed that they exhibited significantly reduced glucokinase activity. It is difficult to explain the decrease in citrate production in these mutants solely on the basis of their reduced glucokinase activity, however. Electrophoresis of cell-free extracts and zymostaining revealed that the molecular properties of the glucokinase formed by the mutants were the same as those in the parent strain. These findings suggest that the reduced glucokinase activity is rather due to less efficient biosynthesis of the enzyme. Glucose transport by these mutants was studied to find out, whether regulation of glucose transport and glucokinase activity is linked in *A. niger.*

Based on these findings, a mechanism by which increased sugar concentrations may "trigger" high acid accumulation is proposed in Figure 3. According to this model, high sugar concentrations bypass glycolytic "fine-regulation" by the synthesis of increased glucokinase activities, which finally results in high cellular concentrations of the positive activator Fru-2,6-$P_2$. Increased activities of other glycolytic enzymes probably do not lead to increased citric acid production. An example, gene amplification of *pki*A (encoding pyruvate kinase) did not lead to improved producer strains.

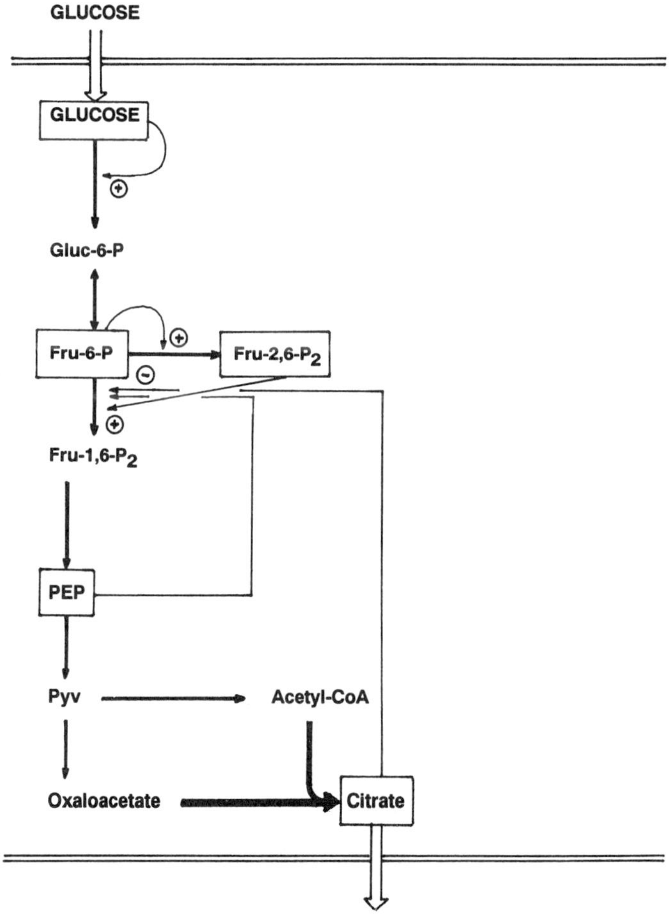


**Figure 3.** A summary of regulatory events which are involved in the stimulation of citric acid production by elevated sucrose concentrations in *Aspergillus niger.*

## Pyruvate Carboxylase and the Role of Metabolite Compartmentation

Independent of its regulation, the end product of aerobic glycolysis is pyruvate. In "normal" cells, the further catabolic fate of pyruvate is to enter the mitochondrion and to be used for the formation of acetyl-CoA. Aspergilli and also several other fungi, however, possess a pyruvate carboxylase (EC 6.4.1.1), which forms oxaloacetate from pyruvate and carbon dioxide (Osmani and Scrutton, 1983; Bercovitz *et al.*, 1990; Jaklitsch *et al.*, 1990; 1991). This enzyme is also induced in the presence of high carbohydrate concentrations (Feir and Suzuki, 1969; Hossain *et al.*, 1984). The involvement of pyruvate carboxylase in the formation of citric, *cis*-itaconic, oxalic as well as malic acid has been proven by tracer experiments (Cleland and Johnson, 1954; Bentley and Thiessen, 1957; Winskill, 1983; Kubicek *et al.*, 1988; Peleg *et al.*, 1988), and exit gas measurements from fermentations (Kubicek *et al.*, 1979).

An essential difference of *A. niger* and *A. terreus* pyruvate carboxylase to that of most other eukaryotic organisms, exists in the location of pyruvate carboxylase; whereas most eucaryotic organisms contain this enzyme in the mitochondria (Osmani and Scrutton, 1985; Purohit and Ratledge, 1988), its localisation in these two *Aspergillus* sp. appears to be exclusively in the cytosolic fraction (Jaklitsch *et al.*, 1990). This may be an essential advantage to acid accumulation; oxaloacetate cannot be transported by the mitochondria, and needs first to be reduced to malate. Since the tricarboxylate carrier uses dicarboxylic acids as partner substrate, this may ensure a high rate of citrate efflux from the mitochondria.

**Does the tricarboxylic acid cycle play a role in organic acid accumulation?**

Most of the previous work on the accumulation of di- and tricarboxylic acids has been dedicated to the role of the tricarboxylic acid cycle in their biosynthesis, but conflicting reports had been obtained: whereas numerous workers claimed the necessity of an inhibition of a step in the cycle for the accumulation of organic acids, particularly citrate (for review see Berry *et al.*, 1977; Roehr *et al.*, 1983*a*; Kubicek and Roehr, 1986), other workers, using isolated mitochondria; provided evidence for the presence of an intact cycle (Ahmed *et al.*, 1972; Winskill 1983). The finding of a cytosolic location of pyruvate carboxylase may help to clarify this situation to some extent: oxalic and malic acid are formed without any involvement of mitochondrial metabolism, which was proven by applying $^{13}C$-NMR spectroscopy (Peleg *et al.*, 1989). Therefore, malic acid biosynthesis is totally located in the cytosol and self-contained in ATP and NADH generation. Similar findings have been obtained with respect to the biosynthesis of oxalic acid: $^{14}C$-labelling has shown that this metabolite originates from oxaloacetate (Kubicek *et al.*, 1988); the corresponding enzyme, oxaloacetate hydrolase,has been purified from *A. niger* (Lenz *et al.*, 1976), and recently demonstrated to be located solely in the cytoplasm (Kubicek *et al.*, 1988). High molar yields of both acids from glucose (128%) have been reported to occur only in the presence of extracellular $CaCO_3$ (Peleg *et. al.*, 1988), which suggests that *Aspergilli* can take up -$HCO_3$ from the extracellular medium and can use it for oxaloacetate formation.

Citrate and itaconate on the other hand clearly require at least one mitochondrial step, i.e. citrate synthase which is located exclusively in the mitochondria (Osmani and Scrutton 1983; Kubicek *et al.*, 1988; Jaklitsch *et al.*, 1990). As has been suggested above, it may be possible that the export of citrate from the mitochondria is solely triggered by the increase in its counter-ion malate (Kubicek, 1988*a*). Whether or not a rise in the mitochondrial citrate concentration is necessary for citric or itaconic acid accumulation is unknown. Such a rise may occur, however, either as a consequence of the regulatory properties of some tricarboxylic acid cycle enzymes, i.e. inhibition of NADP-specific isocitrate dehydrogenase by citrate (Mattey, 1977), inhibition of NAD-specific isocitrate dehydrogenase by the "catabolic reduction charge" [NADH/(NAD+NADH] ratio (Kubicek, 1988*a*) and inhibition of α-ketoglutarate dehydrogenase by oxaloacete, *cis*-asconitate and NADH (Meixner-Monori *et al.*, 1986).

The chemical relationship of itaconic acid to *cis*-aconitate suggests an obvious linkage of its biosynthesis to the reactions of the citric acid cycle. This hypothesis was strengthened by the demonstration of a *cis*-aconitate decarboxylase in *A. terreus* (Bentley and Thiessen, 1957). The enzyme is specifically induced under conditions of itaconic acid production (Jaklitsch *et al.*, 1990), and its involvement in the formation of itaconic acid was established by $^{14}C$-tracer studies (Bentley and Thiessen, 1957; Winskill, 1983). Interestingly, *cis*-aconitate decarboxylase is localised in the cytoplasm, whereas aconitate hydratase is exclusively mitochondrial (Jaklitsch *et al.*, 1990). This implies that *cis*-aconitate and not citrate is exported from the mitochondrion in this species of *Aspergillus*. It is not know whether this occurs by the general tricarboxylate transporter, or whether a specific permease is involved in this process.

## ACKNOWLEDGEMENTS

Experimental work carried out at Wageningen Agricultural University on gluconate metabolism was supported by the Netherlands Technology Foundation (STW), Utrecht grant no WBI 47,0637.

## REFERENCES

Ahmed, S.A., Smith, J.E. and Anderson, J.A. (1972) Mitochondrial activity during citric acid production by *Aspergillus niger*. Trans. Br. Mycol. Soc. *59, 51-61.*

Bentley, R. and Thiessen, C.P. (1957) Biosynthesis of itaconic acid in *Aspergillus terreus*. I. Tracer studies with $^{14}C$-labelled substrates. J. Biol. Chem. 226, 673-687.

Bercovitz, A., Peleg, Y., Battat, E., Rokem, J.S. and Goldberg, I. (1990) Localisation of pyruvate carboxylase in organic acid producing *Aspergillus* strains. Appl. Envir. Microbiol. 56, 1594-1597.

Berry, D.R., Chmiel, A. and Al Obaidy, Z. (1977) Citric acid accumulation by *Aspergillus niger*. In: Genetics and Physiology of Aspergillus (Smith, J.E. and Pateman,J.A., eds.)pp. 405-423. Academic Press, London.

Cleland, W.W. and Johnson, M.J. (1954) Tracer experiments on the mechanism of citric acid formation by *Aspergillus niger*. J. Biol. Chem. 208, 679-692.

Feir, H.A. and Suzuki, I. (1969) Pyruvate carboxylase from *Aspergillus niger*: kinetic study of a biotin-containing carboxylase. Can. J. Biochem. 47, 697-710.

Fiedurek, J., Szczodrak, J. and Ilczuk, Z. (1988) Citric acid synthesis by *Aspergillus niger* mutants resistant to 2-desoxyglucose. Acta Microbiol. Polon. 36, 303-307.

Fiedurek, J., Rogalski, J., Ikczuk, Z. and Leonowicz (1986) Screening and mutagenesis of moulds for the improvement of glucose oxidase production. Enzyme Microb. Technol. 8, 734-736.

Frederick, K.R., Tung, J., Emerick, R.S., Masiarz, F.R., Chamberlain, S.H., Vasavada, A., Rosenberg, S., Chakraborty, S., Schopfer, L.M. and Massey, V. (1990) Glucose oxidase from *Aspergillus niger*. Cloning, gene sequence, secretion from *Saccharomyces cerevisiae* and kinetic analysis of a yeast-derived enzyme. J. Biol. Chem. 265, 3793-3802.

Gibson, O.H., Swoboda, B.E.P.and Massey, B. (1985) Kinetics and mechanism of action of glucose oxidase. J. Biol. Chem. 239, 3927-3924.

Habison, A., Kubicek, C.P. and Roehr, M. (1983) Partial purification and regulatory properties of phosphofructokinase from *Aspergillus niger*. Biochem. J. 209, 669-676.

Harmsen, H.J.M., Kubicek-Pranz, E.M., Roehr, M., Visser, J. and Kubicek, C.P. (1992) Regulation of 6-phosphofructo-2-kinase from the citric acid accumulating fungus *Aspergillus niger*. Appl. Microbiol. Biotechnol. 37, 784-788.

Hossain, M., Brooks, J.D. and Maddox, I.S. (1984) The effect of sugar source on citric acid production by *Aspergillus niger*. Appl. Microbiol. Biotechnol. 19, 383-391.

Jaklitsch, W.M., Kubicek, C.P. and Scrutton, M.C. (1990) The subcellular localisation of itaconate biosynthesis in *Aspergillus terreus*. J. Gen. Microbiol. 137, 533-539.

Jaklitsch, W.M., Kubicek, C.P. and Scrutton, M.C. (1991) Intracellular location of enzymes involved in citrate accumulation in *Aspergillus niger*. Can. J. Microbiol. 37, 823-827.

Kelley, R.L. and Reddy, C.A. (1986) Purification and characterisation of glucose oxidase from ligninolytic cultures of *Phanerochaete chrysosporium*. J. Bacteriol. 166, 269-274.

Kersten, P.J. and Kirk, T.K. (1987) Involvement of a new enzyme, glyoxal oxidase, in extracellular $H_2O_2$ production by *Phanerochate chrysosporium*. J. Bacteriol. 169, 2195-2201.

Kersten, P.J. (1990) Glyoxal oxidase of *Phanerochaete chrysosporium*: its characterisation and activation by lignin peroxidase. Proc. Natl. Acad. Sci. USA. 87, 2937-2940.

Kim, K.K., Fravel, D.R. and Papavizas, G.C. (1988) Identification of a metabolite produced by *Talaromyces flavus* as glucose oxidase and its role in the biocontrol of *Verticillium dahliae*, Phytopathol. 78. 488-492.

Kitmura, K., Sarangbin, S., Rugsaseel, S. and Usami, S. (1992) Citric acid production by 2-desoxyglucose resistant mutant strains of *Aspergillus niger*. Appl. Microbiol. Biotechnol. 36, 573-577.

Kriechbaum, M., Heilman, H.J., Wientjes, F.D., Hahn, M., Jany, K.D. and Gassen, H.G. (1989) Cloning and DNA sequence analysis of the glucose oxidase gene from *Aspergillus niger* NRRL-3. FEBS Letts. 255, 63-66.

Kubicek, C.P. and Roehr, M. (1986) Citric acid fermentation. CRC Crit. Rev. Biotechnol. 3, 331-373.

Kubicek, C.P. (1988*a*) Regulatory aspects of the tricarboxylic acid cycle in filamentous fungi - a review. Trans. Br. Mycol. Soc. 90, 339-349.

Kubicek, C.P. (1988*b*) The role of the citric acid cycle in fungal organic acid fermentations Biochem. Soc. Symp. 54, 113-126.

Kubicek, C.P. Zehentgruber, O. and Roehr, M. (1979) An indirect method for studying the fine control of citric acid formation by *Aspergillus niger*. Biotechnol. Letts. 1, 57-62.

Kubicek, C.P., Schreferl-Kunar, G., Wöhrer, W. and Roehr, M. (1988) Evidence for a cytoplasmic pathway of oxalate biosynthesis in *Aspergillus niger*. Appl. Envir. Microbiol. 54, 633-637.

Kubicek-Pranz, E.M., Mozelt, M., Roehr, M. and Kubicek, C.P. (1990) Changes in the concentration of fructose-2,6-bisphosphate in *Aspergillus niger* during stimulation of acidogensis by elevated sucrose concentrations. Biochem. Biophys. Acta 1033, 250-255.

Kundu, P.N. and Das, A. (1985) A note on crossing experiments with *Aspergillus niger* for the production of sodium gluconate. J. Appl. Bacteriol. 59, 1-5.

Lenz, H., Wunderwald, P. and Eggerer, H. (1976) Partial purification and some properties of oxalacetase from *Aspergillus niger*. Eur. J. Biochem. 65, 225-233.

Markwell, J., Frakes, L.G., Brott, E.C., Osterman, J. and Wagner, F.W. (1989) *Aspergillus niger* mutants with increased glucose oxidase production. Appl. Microbiol. Biotechnol. 30, 166-169.

Mattey, M. (1977) Citrate regulation of citric acid accumulation by *Aspergillus niger*. FEMS Microbiol. Letts. 2, 71-74.

Meixner-Monori, B., Kubicek, C.P., Harrer, W., Schreferl-Kunar, G. and Roehr, M. (1986) NADP-specific isocitrate dehydrogenase from the citric acid accumulating fungus *Aspergillus niger*. Biochem. J. 236, 549-557.

Mischak, H., Kubicek, C.P. and Roehr, M. (1985) Formation and location of glucose oxidase in citric acid producing mycelia of *Aspergillus niger*. Appl. Microbiol. Biotechnol. 21, 27-31.

Osmani, S.A. and Scrutton, M.C. (1983) The subcellular location of pyruvate carboxylase and some other enzymes in *Aspergillus nidulans*. Eur. J. Biochem. 133, 551-560.

Osmani, S.A. and Scrutton, M.C. (1985) The subcellular localisation of pyruvate carboxylase from *Rhizopus arrhizus*. Eur. J. Biochem. 147, 119-128.

Pazur, J.H. (1966) Glucose oxidase from *Aspergillus niger*. Meth. Enzymol. 9, 82-86.

Peleg, Y., Stieglitz, B. and Goldberg, I. (1988) Malic acid accumulation in *Aspergillus flavus*. I. Biochemical aspects of acid biosynthesis. Appl. Microbiol. Biotechnol. 28, 69-75.

Peleg, Y., Barak, A., Scrutton, M.C. and Goldberg, I. (1989) Malic acid accumulation by *Aspergillus flavus*. III. $^{13}$CNMR and isoenzyme analysis. Appl. Microbiol. Biotechnol. 30, 176-183.

Purohit, H.J. and Ratledge, C. (1988) Mitochondrial location of pyruvate carboxylase in *Aspergillus niger*. FEMS Microbiol. Letts. 55, 129-132.

Ramasamy, K., Kelley, R.L. and Reddy, C.A. (1985) Lack of lignin degradation by glucose oxidase negative mutants of *Phanerochaete chrysosporium*. Biochem. Biophys. Res. Comm. 131, 436-441.

Reuss, M., Fröhlich, S., Kramer, B., Messerschmidt, K. and Pommerening, G. (1986) Coupling of microbial kinetics and oxygen transfer for analysis and optimisation of gluconic acid production by *Aspergillus niger*. Bioproc. Eng. 1, 79-91.

Roehr, M. Kubicek, C.P. and Kominek, J. (1983*a*) Citric acid In: Biotechnology, Vol.3, (Rehm,H.J. and Reed,G., eds.) pp. 331-373. Verlag Chemie Weinheim.

Roehr, M., Kubicek, C.P. and Kominek, J. (1983*b*) Gluconic acid. In: Biotechnology, Vol. 3, (Rehm H.J. and Reed,G., eds.) pp. 455-465. Verlag Chemie Weinheim.

Smith, J.E. and Ng, W.S. (1972) Fluorometric determination of glycolytic intermediates and adenylates during sequential changes in replacement culture of *Aspergillus niger*. Can. J. Microbiol. 18, 1657-1664.

Steinböck, F., Choojun, S., Held, I., Roehr, M. and Kubicek, C.P. (1993) manuscript in preparation.

Swart, K., van den Vondervoort, P.J.I., Witteveen, C.F.B. and Visser, J. (1990) Genetic localisation of a series of genes affecting glucose oxidase levels in *Aspergillus niger*. Curr. Genet. 18, 435-439.

Van Dijken, J.P. and Veenhuis, M. (1980) Cytochemical localisation of glucose oxidase in peroxisomes of *Aspergillus niger*. Eur. J. Appl. Microbiol. Biotechnol. 9, 275-283. Verlag chemie Weinheim

Whittington, H., Kerry-Williams, S., Bidgood, K., Dodsworth, N., Peberdy, J.F., Dobson, M., Hinchliffe, E. and Ballance, D.J. (1990) Expression of the *Aspergillus niger* glucose oxidase gene in *A. niger, A. nidulans* and *Sachharomyces cerevisiae.* Curr. Genet. 18, 531-536.
Winskill, N. (1983) Tricarboxylic acid cycle activity in relation to itaconic acid biosynthesis in *Aspergillus terreus.* J. Gen. Microbiol. 129, 2877-2883.
Witteveen, C.F.B., van den Vondervoort, P., Swart, K. and Visser, J. (1990) Glucose oxidase overproducing and negative mutants of *Aspergillus niger.* Appl. Microbiol. Biotechnol. 33, 683-686.
Witteveen, C.F.B., Veenhuis, M. and Visser, J. (1992) Location of glucose oxidase and catalase activities in *Aspergillus niger.* Appl. Envir. Microbiol. 58, 1190-1194.
Witteveen, C.F.B., van den Vondervoort, P.J.I., van den Broeck, H.C., van Engelenburg, F.A.C., de Graaff, L.H., Hillebrand, M.H.B.C., Schaap, P.J. and Visser, J. (1993) The induction of glucose oxidase, catalase and lactonase in *Aspergillus niger.* Curr. Genet. in press.
Xu, D.-B., Madrid, C.P., Roehr, M. and Kubicek, C.P. (1989) The influence of type and concentration of the carbon source on production of citric acid by *Aspergillus niger.* Appl. Microbiol. Biotechnol. 30, 553-558.
Zidwick, M.J. (1992) Organic acids. In "Biotechnology of Filamentous Fungi. Technology and Production" (Finkelstein, D.B. and Ball, C., eds.) pp. 304-334. Butterworth-Heinemann, Boston.

# *ASPERGILLUS* ENZYMES AND INDUSTRIAL USES

Karen Oxenbøll

Industrial Biotechnology
Novo Nordisk A/S
Novo Allé
DK-2880 Bagsvaerd
Denmark

## INTRODUCTION

Industrial enzymes have recently been reviewed by Berka *et al.* (1992) and Bigelis (1992). These reviews offer excellent and detailed information on most industrial applications of *Aspergillus* enzymes.

In this paper another approach has been taken. The view will be that of a microbiologist - looking upon *Aspergillus* as a source for new enzymes of commercial interest. The present applications of the enzymes will be summarised only broadly, whereas a personal view of future possibilities will be the main theme.

## THE DOMINANT POSITION OF *ASPERGILLUS*

The industrial exploitation of microbial enzymes in the western world started 100 years ago, in 1894, with the patent of Takamine on production of alpha-amylase from *Aspergillus oryzae.* Ever since *Aspergillus* has held a dominant position. As illustrated in Table 1 *Aspergillus* is the major source of enzymes for use in food and beverage manufacturing. Thus *Aspergillus* enzymes are used for such diverse applications as starch processing, cheese manufacturing, juice clarification, brewing, dough conditioning, wine modification, food preservation and instant tea production. However there are areas where *Aspergillus* enzymes apparently cannot work. For example, wherever pH requirements are alkaline, as in detergents, or temperature requirements are well above 60°C, as in starch liquefaction. Moreover *Aspergillus* seems not to be a good source of true cellulolytic activity, where the enzymes from *Trichoderma* have taken the lead, or for the production of oxidoreductases which are active on lignin and similar aromatic compounds, where the basidiomycetes, especially the white rot fungi are the richest source.

## WHY IS *ASPERGILLUS* SO DOMINANT ?

The biodiversity of microorganisms is immense. Not less than 70,000 species of fungi exist, and Hawksworth (1991) has estimated that the total number of species may be close to 1.5 million. So why is it, that *Aspergillus* has become so dominating? What are the

*The Genus Aspergillus*, Edited by Keith A. Powell *et al.*,
Plenum Press, New York, 1994

**Table 1.** Major industrial applications of microbial enzymes

| Application | Enzyme | Source |
|---|---|---|
| Detergents | Protease | *Bacillus* |
| | Amylase | *Bacillus* |
| | Lipase | *Humicola, Pseudomonas* |
| | Cellulase | *Bacillus, Humicola* |
| Starch industry | Amylase | *Bacillus* |
| | Glucoamylase | *Aspergillus* |
| | Glucose isomerase | *Bacillus, Streptomyces* |
| Dairy | Protease | *Rhizomucor* |
| | Lipase | *Aspergillus* |
| | Lactase | *Klyveromyces, Aspergillus* |
| | Sulfhydryloxidase | *Aspergillus* |
| Wine & Juice | Pectinase | *Aspergillus* |
| | Cellulase | *Aspergillus, Trichoderma* |
| | Cellobiase | *Aspergillus* |
| | Glucose oxidase | *Aspergillus* |
| | Polyphenol oxidase | *Trametes* |
| Distilling industry | Amylase | *Aspergillus* |
| | Glucoamylase | *Aspergillus* |
| Brewery | ß-glucanase | *Aspergillus, Bacillus* |
| | Acetolactatedecarboxylase | *Bacillus* |
| Bakery | Amylase | *Aspergillus, Bacillus* |
| | Protease | *Aspergillus* |
| | Glucose oxidase | *Aspergillus* |
| Textiles | Amylase | *Bacillus* |
| | Cellulase | *Trichoderma, Humicola* |
| | Catalase | *Aspergillus* |
| Animal Feed | Phytase | *Aspergillus* |
| | Cellulase | *Trichoderma, Humicola, Aspergillus* |
| | Plant cell wall degrading enzyme | *Aspergillus* |
| Pulp and paper | Xylanase | *Trichoderma, Bacillus* |
| Leather | Protease | *Aspergillus* |
| Tea | Tannase | *Aspergillus* |

**Table 2.** Examples of extracellular enzymes from *Aspergillus niger*

| Number | Commonly used name | Systematic name |
|---|---|---|
| 1.1.3.4 | Glucose oxidase | ß-D-Glucose:oxygen 1-oxidoreductase |
| 1.11.1.6 | Catalase | Hydrogen-peroxide:hydrogen-peroxide oxidoreductase |
| 3.1.1.3 | Lipase | Triacylglycerol acylhydrolase |
| 3.1.1.11 | Pectinesterase | Pectin, pectylhydrolase |
| 3.1.1.20 | Tannase | Tannin acylhydrolase |
| 3.1.3.8 | Phytase | myo-Inositol-hexakisphosphate-3-phosphohydrolase |
| 3.2.1.1 | α-Amylase | 1,4-α-D-Glucan glucanohydrolase |
| 3.2.1.3 | Glucoamylase | 1,4-α-D-Glucan glucohydrolase |
| 3.2.1.4 | Cellulase | 1,4-(1,3;1,4)-ß-D-Glucan-4-glucanohydrolase |
| 3.2.1.6 | ß-glucanase | 1,3-(1,3;1,4)-ß-D-Glucan 3(4)-glucanohydrolase |
| 3.2.1.7 | Inulinase | 2,1-ß-D-Fructan fructanohydrolase |
| 3.2.1.8 | Xylanase | 1,4-ß-D Xylan xylanohydrolase |
| 3.2.1.15 | Polygalacturonase | Poly(1,4-α-D-galacturonide) glucanohydrolase |
| 3.2.1.21 | Cellobiase | ß-D-Glucoside glucohydrolase |
| 3.2.1.22 | α-Galactosidase | α-D-Galactoside galactohydrolase |
| 3.2.1.23 | ß-Galactosidase | ß-D-Galactoside galactohydrolase |
| 3.2.1.24 | α-Mannosidase | α-D-Mannoside mannohydrolase |
| 3.2.1.26 | Invertase | ß-D-Fructofuranoside fructohydrolase |
| 3.2.1.55 | α-L-Arabinofuranosidase | α-L-Arabinofuranoside arabinofuranohydrolase |
| 3.2.1.57 | Isopullulanase | Pullulan-4-glucanohydrolase |
| 3.2.1.67 | Poly(galacturonate) hydrolase | Poly(1,4-α-D-galacturonide) galacturonohydrolase |
| 3.2.1.78 | Mannanase | 1,4-ß-D-Mannan mannanohydrolase |
| 3.2.1.89 | Galactanase | Arabinogalactan 4-ß-D-galactanohydrolase |
| 3.4.16.1 | Carboxypeptidase | Peptidyl-L-amino-acid hydrolase |
| 4.2.2.2 | Pectate lyase | Poly(1,4-α-D-galacturonide) lyase |
| 4.2.2.10 | Pectin lyase | Poly(methoxygalacturonide) lyase |

special attributes of this genus and of the species *Aspergillus niger* and *Aspergillus oryzae* in particular?

First of all *Aspergillus* species are major agents of decomposition and decay and thus possess the capability to produce a broad range of enzymes. It is possible that no- one has ever counted the number of enzymes secreted for the decomposition of plant cell material. However taking into consideration the heterogeneous nature and the multiple linkages in this substrate an estimate of at least 100 different enzymes from one of the more versatile species is realistic. Furthermore *Aspergillus* cultures seems to be flexible and adaptive and thus capable of utilising a wide range of substrates, as recently demonstrated by Lusta *et al.* (1991), who discovered the capability of *Aspergillus terreus* to grow on methanol and even demonstrated a great increase in the occurrence of microbodies with crystalloids (as known from methanotrophic yeasts) as an adaptive change to growth on this substrate. Table 2 shows some of the enzymes produced by *A. niger*. As new enzymes are discovered regularly the list is not complete, which illustrates the richness of *A. niger* as a source.

Another reason for the success of *A. niger* in the industrial environment is of course the ease of handling cultures. *Aspergillus* cultures typically grow rapidly on cheap substrates and secrete their enzymes into the medium, which makes them easy to recover. Furthermore the abundance of conidia ease preservation and strain development work. This also means that they are frequently isolated from environmental samples.

Thirdly *A. oryzae* and *A. niger* have been used for a long time, meaning that a number of enzyme types have obtained generally recognised as safe (GRAS) status from the FDA

That is for instance the case for carbohydrases, meaning that any new carbohydrase from any of these species will not require a pre-market approval.

## FUTURE PERSPECTIVES FOR *ASPERGILLUS* ENZYMES

Will *Aspergillus* continue to be the dominant source for fungal enzymes ? Yes and no. *Aspergillus* is likely to be less dominating, however still an important source to industrial enzymes. This is illustrated in Table 3 by the number and the variety of patent applications published in 1992 on enzymes from *Aspergillus*. The list is not complete, but demonstrates a high level of activity regarding new processes which apply enzymes from *Aspergillus*. The majority deals with new applications for "old" enzymes. Only about 10% of the applications claim totally new enzymes which is not so impressive but still significant when taking into consideration that the traditional narrow range of *Aspergillus* species still seems to be the main source. The patent literature of these years also contains an increasing number of applications on DNA sequences on known enzymes. Such applications have not been included in Table 3.

**Table 3.** *Aspergillus* patent applications/enzymes 1992

| Enzyme | Use | Patent application number |
|---|---|---|
| lipase | preparation of interesterified glyceride(s) | EP 170431 |
| lipase | production of optically active 1 phenyl 1,3 propendiol | JP 4258297 |
| lipase | production of benzenedicarboxylacid monoester | JP 4158789 |
| lipase | enzymatic resolution of mercaptoalkanoic acids | US 5106736 |
| lipase | preparation of acrylic acid alkylamino alkyl ester | JP 407988 |
| lipase | production of optically active glycolderivs. | EP 197484 |
| lipase | production of fatty acid esters(s) | JP 62195292 |
| lipase | production of percarboxylic acid | WO 9104333 |
| lipase | hydrolysis of resin in wood pulp | WO 9107542 |
| lipase | salad oil having good flavour | JP 4066052 |
| lipase | hair tonic for preventing dandruff | EP 117087 |
| endohemicellulase | production of partially decomposed mannan | JP 4131089 |
| cellulase/ hemicellulase | removal of inert plant material from seeds | WO 9207456 |
| cellulase | decoration of textile articles by degrading carrier materials | EP 290027 |
| cellulase | preparation of isomaltooligosaccharide from glucose or maltose | JP 4112797 |

**Table 3.** *Continued*

| Enzyme | Use | Patent application number |
|---|---|---|
| cellulase | detergents | JP 4027385 |
| ß1,4 galactanase | degradation/modification of plant cell walls | EP 498137 |
| pentosanase | fine grained starch separation | EP 378522 |
| l-fucosidase | preparation of oligosaccharide containing l-fucose (for analysis) | JP 4099492 |
| inulinase | product of inulooligosaccharide | JP 4190789 |
| glucuronidase | preparing ethyl-α-glucoside, smell enhancer | JP 4112798 |
| α-amylase | removal of residues of cyclodextrin from fats and oils | DE 4041386 |
| α-amylase | preparation of starch oligosaccharides | JP 4066094 |
| glucoamylase | enzyme electrode for measuring maltooligo-saccharides | EP 335167 |
| phytase | reducing phytate content of soy protein | EP 380343 |
| phytase | steeping of cereals | EP 321004 |
| protease | degrading ß-lacto globulin in cows milk serum protein | EP 335399 |
| protease | production of white laver (seaweed) | WO 911767 |
| glucoseoxidase | preparation of glucono-delta-lactone without going via gluconic acid | JP 60047693 |
| urateoxidase | diagnosis of hyperuricaemia, acute leukaemia | FR 2664286 |
| mixtures | maceration during work up of flax | DE 4012351 |
| mixtures | pet food with improved palatability and elasticity | EP 223484 |
| mixtures | polysaccharide products for beverage by extraction of yeast cells | JP 4058893 |

**Table 4.** *Aspergillus* patent applications/cultures 1992

| Use | Patent application number |
|---|---|
| High yield vanillin production from ferulic acid by microbial fermentation in presence of sulphydryl compound | US 5128253 |
| Conversion of ß-damascone to 4-hydroxy-ß-damascone-10-ol a modifier of tobacco flavour | JP 4273854 |
| Production of α-hydroxy-carboxylic acid to hydrolysis of α-hydroxy-nitrile | JP 63222696 |
| Processing seaweed to remove bad smell | JP 4258273 |
| Preparation of malted rice food | JP 4281756 |
| Production of cheese-like fermented food | JP 63269946 |
| Treating malodorous gases | JP 4007017 |
| Durable microorganism-based deodorant containing alcohol | JP 4009158 |

Table 4 illustrates that although enzyme technology dominates the biotechnological exploitation of microorganisms, new processes applying microbial cultures in for instance biochemical conversion or fermentation of food products are continuously developed. There are three factors giving rise to new possibilities for *Aspergillus* enzymes:

1) New applications
2) New technologies
3) New enzymes

**New Applications**

Developments in food technology (improvements in flavour, colour, texture, diet products) and environmental problems require new technological solutions. Enzyme technology is an attractive approach, which is today expanding into areas which were traditionally more chemically oriented.

New applications can be used more extensively which do not necessarily require new enzymes. The patent survey has highlighted the application of lipases in stereo-specific hydrolysis, also various oligosaccharides have been used within the food industry for dietary products or as agents to improve texture or nutritional value.

In addition to these proteases from *Aspergillus* are being increasingly used for production and flavour modification of meat and vegetable hydrolysates and in 'enzymic' dehairing of hides.

**New Technologies.**

It is expected that the progress of gene technology offers new possibilities for *Aspergillus* enzymes. Monocomponent enzyme products will be produced in bulk quantities from recombinant cultures and be available for industrial applications. No doubt this development will widen the range of applications for *Aspergillus* enzymes. Within fruit processing a number of new possibilities are being pursued, for example the use of accelerated juice clarification processes in cider production based on pectin esterase preparations and polygalacturonase mediated puree production, when compared with conventional boiling processes, the newer processes are superior regarding preservation of vitamins, flavours and natural colour. Furthermore the advent of monocomponent products allow new and well-defined mixed products.

Another advance offered by gene technology is the concept of "boosting" the enzyme activity, where an enzyme product is specifically enriched with one of the key activities by introduction into the production strain of extra gene copies coding for the enzyme in question. For classical *Aspergillus* pectinase products, which contain a wide range of plant cell-wall degrading enzymes, this is the key to improved performance. This has been demonstrated by Dörreich *et al.* (1992) in apple liquefaction where boosting with low quantities of ß1,4-galactanase has given rise to a synergistic increase in the total liquefaction power of the enzyme product.

A third perspective of gene technology is protein engineering offering the possibility of changing the properties of the enzymes by altering the amino acid composition. So far only very few *Aspergillus* enzymes have been the subject of protein engineering. However, the engineering of other microbial enzymes has clearly demonstrated the potential of this technology and there is no doubt that modified *Aspergillus* enzymes will be introduced to the industrial market in the future. *Aspergillus* enzymes may not be produced which are able to work at pH 10, but changes in substrate specificities has been demonstrated by Sierks *et al.*, (1989) for glucoamylase from *A. niger*, and enhancement of thermo stability was demonstrated by Ikegaya *et al.*, (1992) for alkaline *A. oryzae* protease, developments which are likely to be introduced.

### New Enzymes

Finally there is still a chance that it may be possible to find completely new enzymes in *Aspergillus* spp. As touched upon earlier and demonstrated by the patent search new *Aspergillus* enzymes continue to be found even in the "traditional" species.

Why is it always the same few species which are being used from a total of almost 200 species of *Aspergillus*? Are the others of no use? It is difficult to tell why only a few species are being utilised. Many of course are myxotoxin producers, but that is not the whole explanation. Very little is known about the ecology of the various species of *Aspergillus*. It is difficult to know the habitat for cultures isolated from soil samples but isn't it likely that the various species represent different habitats and various capabilities for production of enzymes? It is probable that the potential of *Aspergillus* as an enzyme source has not yet been fully exploited.

## CONCLUSION

The advent of gene technology offers a dramatic expansion in the range of microbial sources for new industrial enzymes. Therefore microbiologists who carry out screening programmes need no longer limit themselves to microorganisms that are easy to handle in large scale culture. Although in the long term this means the importance of *Aspergillus* as a source for future industrial enzymes will diminish. There is, however, every reason to believe that *Aspergillus* will remain the single most dominant genus on the arena of industrial enzymes from fungi for many years to come.

## REFERENCES


Berka, R.M., Dunn-Coleman,N. and Ward M. (1992) Industrial Enzymes from *Aspergillus*, in "*Aspergillus*, Biology and Industrial Applications" (Bennett, J.W. and Klich, M.A. ed.), Butterworth-Heinemann, USA.

Bigelis, R. (1992) Food Enzymes, in "Biotechnology of Filamentous Fungi" ( Finkelstein,D.B. and Ball, C. ed.), Butterworth-Heinemann, USA.

Dörreich, K. , Dahlboge,H.,Mikklesen,J.M. and Christensen, F.M. (1992) Sg(b)-1,4-galactanase and a DNA sequence, WO 92/13945.

Hawksworth, D.L. (1991) The fungal dimension of biodiversity: magnitude, significance and conservation, Mycol. Res. 95(6), 641-655.

Ikegaya, K., Ishida, Y., Murakami K., Masaki, A., Sugio,N., Takechi, K., Murakami, S., Tatsumi, H., Ogawa, Y., Nakano, E., Motai H. and Kawabe H. (1992) Enhancement of the thermostability of the alkaline protease from *Aspergillus oryaze* by introduction of a disulfide bond. Biosci. Biotech Biochem. 56 (2), 326-327.

Lusta, K.A., Sysoev O.V. and Sharyshev A.A. (1991) Cytobiochemical characterization of *Aspergillus terreus* 17p utilizing various carbon substrates. J. Basic Microbiol. 31(4), 265-277.

Sierks, M.R., Reilly, P.J., Ford, C. and Svensson, B. (1989) Site-directed mutagenesis at the active site Trp120 of *Aspergillus awamori* glucoamylase, Protein Engineering 2(8), 621-625.

# INDUSTRIAL ASPECTS OF SOY SAUCE FERMENTATIONS USING *ASPERGILLUS*

Kofi E. Aidoo[1], John E. Smith[2], Brian J.B. Wood[2]

[1]Division of Human Nutrition and Dietetics
Glasgow Caledonian University, Jordanhill Campus
Glasgow G13 1PP

[2]Department of Bioscience and Biotechnology
University of Strathclyde, 204 George Street
Glasgow G1 1XW

## INTRODUCTION

Soy sauce is a brown, salty liquid used in cooking and as a table condiment in China, Japan and South East Asia. It is of very ancient lineage, and its production today includes every level of sophistication from primitive, village level or domestic scale, to advanced and highly controlled facilities producing enormous amounts of a very high quality product. It has achieved significant market penetration outside its homelands, with the U.S.A. leading the non-Oriental world both in production and in consumption per capita.

*Miso,* although less familiar to the Western consumer than soy sauce, is becoming more readily available. It comes from the same geographical areas as soy sauce, and is used as a condiment in cooking, for example of soups and casseroles. Miso is a Japanese word for a range of products which could be described as fermented soy pastes, but in the West the term has become generally adopted for the whole class of fermented soy pastes, irrespective of the country of origin. On the other hand, a Japanese name for soy sauce, *Shoyu,* is generally restricted to denoting products of Japanese origin in the West.

Both groups of products are made from the soy bean, *Glycine max,* generally with the addition of a cereal - wheat for soy sauce, rice or barley for miso; it should be noted however that both classes of products include representatives which are made from soybeans alone. Fermentation with *Aspergillus* spp. is central to production in both cases. The mould's principal function is the elaboration and release of a range of hydrolytic enzymes, including amylases, proteases, cellulases, invertase, etc. These enzymes hydrolyse materials present in the beans and cereals, and the mould-growth (or koji) stage has thus been likened to the malting stage in the production of beverages from barley. The product from this stage is then subjected to a further fermentation in the presence of salt and added water, and involving the participation of lactic acid bacteria and yeasts; this is called the salt mash or moromi fermentation. After further processing, the product from the moromi fermentation is ready for sale and use.

*The Genus Aspergillus,* Edited by Keith A. Powell *et al.*,
Plenum Press, New York, 1994

## HISTORICAL BACKGROUND

The history of soy sauce is necessarily a matter for speculation and inference at least as much as it is one for hard fact. This arises from its great antiquity and the way in which it is interwoven into the lives, beliefs and value-systems of the peoples among whom it originated. In preparing this historical introduction to our presentation we have therefore opted to attempt a synthesis of the views held by various authorities, as represented in their writings on the subject, rather than essay a spurious scholasticism with an excessive number of references to original sources. Among these sources we include Yong and Wood (1974); Abiose *et al.* (1982); Djien (1982); Wood (1982); Shurtleff and Aoyagi (1983); Steinkraus (1985); Yokotsuka (1985); Fukushima (1986); Campbell-Platt (1987) and Ebine (1989).

Fermentation as an aid to preservation of foodstuffs must be practically as old as humankind, such is the need to ensure a steady supply despite the vagaries of supply and demand. It is the tradition among coastal and lacustrine peoples of South-East Asia to ferment surplus and trash fish and fish offals. Fermentation will not only preserve foods (if properly conducted), but will also change it. The fish, particularly if small material such as comminuted offal was being preserved, would yield a liquid extract after pounding with salt, then holding at ambient temperatures often in excess of 30°C; Vietnamese *Nuoc-Mam* is an example from the many such fish sauces of S.E. Asia. Such preparations are important sources of flavour, both directly and through flavour enhancement, in diets which can be rather bland without them. Some workers suggest that the rise of Buddhist thought may have made such animal-based products less acceptable and stimulated a search for alternatives, so leading to the development of products based on the soy bean. On the other hand, the soy bean is so important in the traditional agriculture of S.E. Asia that it is quite reasonable to suggest the independent rise of technology for its fermentation.

Fermentation of the beans as a paste, yielding a form of miso, has been suggested as the earliest process for soy bean utilisation. During the fermentation a liquid layer sometimes separates, and can be decanted off; this, it has been suggested, is the origin of both the term *tamari* and the concept of a liquid fermentation yielding a "sauce". There is some evidence for these ideas, but they must remain essentially speculative in nature. It is certain that by the time of Confucius soy sauce was important enough to have a place in religious rites associated with the Chinese Emperor.

Similarly, the idea that the Japanese imported the production of soy sauce from China around the end of the 1st Millenium A.D. is speculative at best, although there is some evidence to support it. The Japanese effected considerable developments in the production technology, long before the development of modern microbiology, and the products from a Company such as the Kikkoman Corporation are distinctive and easily recognisable.

Soy sauce is, in reality, a single name for a spectrum of products, although variations are perhaps more subtle, and may tend to be between rather than within regions. The Japanese varieties which are best known in the West tend to be light (for soy sauce) in colour, with a distinctive aroma, and made from fermentation products with a minimum of additives. On the other hand, British and Dutch expectations are for a rather dark, more viscous, sweetish material containing such additions as caramel and toddy (sugar from the sap of certain palm trees). These qualities are typical of those favoured in Indonesia (Kecap) and Hong Kong, although the lighter, saltier varieties are also esteemed. It is an European error to suppose that the range of types, and of uses for, soy sauce is fairly limited, and it can be fascinating to witness gourmet Southeast Asians discriminate between a selection of products which appear virtually indistinguishable to a the Western palate. For convenience we have tended to refer to the light, unsweetened types as "Japanese" and the heavier, sweetened ones as "Chinese", although this does scant justice to either culture, and to save frequent qualification this practice will again be followed here. The large-scale industrial manufacture of soy sauce involves five main unit operations, viz. the preparation of the raw materials, the koji process, mash or moromi production, pressing and, finally, refining (Figure 1).

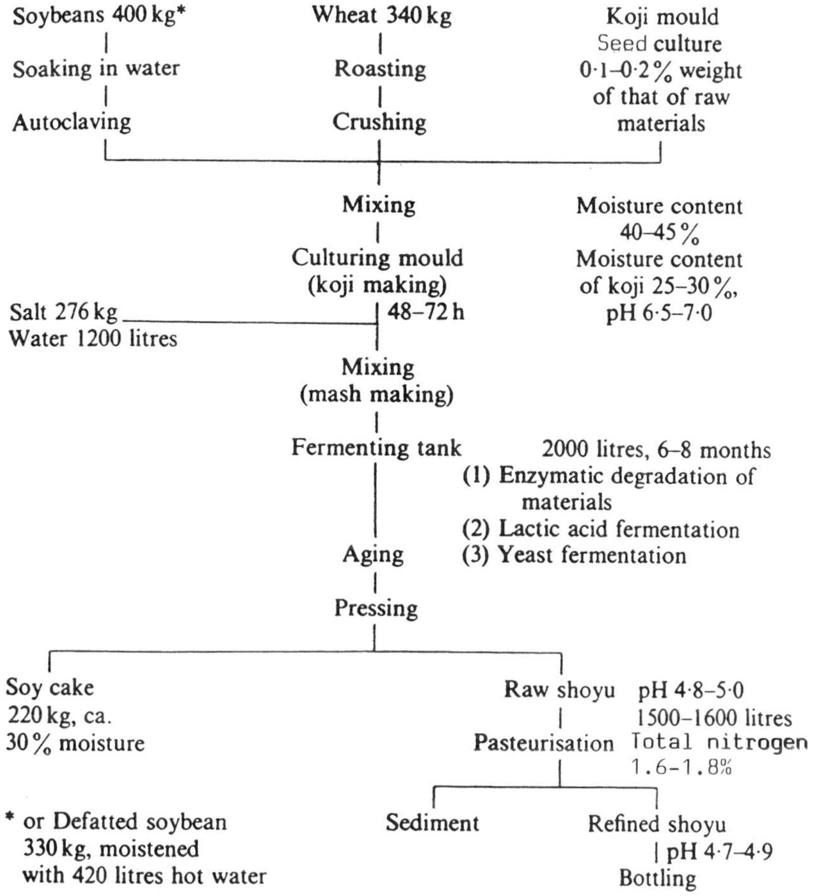


**Figure 1.** Koikuchi shoyu fermentation.
Reproduced with permission from Yokotsuka (1985).

Thus, there are two specific fermentation procedures involved in soy sauce production, viz. an aerobic koji (mould growth) fermentation and an anaerobic moromi fermentation or salt mash.

## Raw Material Preparation

Before cooking, the soy beans normally need to be soaked for several hours or overnight in order to allow full water saturation and swelling. Under tropical and subtropical conditions with ambient temperatures in excess of 30°C, considerable bacterial growth will take place (Mulyowidarso *et al.*, 1989, 1991) with production of organic acids and other metabolites having an effect on the subsequent fermentations which has not yet been fully quantified. This unintentional fermentation is largely anaerobic, leading to the metabolism of sugars diffusing from the beans, with the production of malic and lactic acids, which diffuse back into the beans. No in-depth investigations into the effects (if any) of these acids on subsequent stages of soy sauce production have been reported. Japanese work does however recognise that ethanoic acid may be beneficial in subsequent stages of the process, this having been suggested as another possible product of these bacterial activities, and patents exist covering adding controlled amounts of the acid to beans about to undergo the next stage of fermentation.

The fully hydrated beans are drained of excess water, and are then cooked. Traditionally this is by prolonged boiling at ambient pressure, but this inefficient process has been replaced in many cases by cooking in autoclaves, which is quicker, more economical on fuel, and gives a more complete sterilisation of the beans. Further benefit is gained if, at the end of cooking the pressure is reduced to atmospheric as rapidly as possible, as this tends to swell the beans so aiding penetration of the fungal hyphae in the koji fermentation, and it also helps to dry off surplus moisture from the surface of the beans.

In the most modern plants, the use of whole beans has been superseded by that of defatted soy bean meal or grits. Some authorities claim that the product using whole beans has a superior flavour, but the defatted meal is more suited to continuous processing of the raw materials. The comparison between whole and defatted soy beans or grits is based on cost, enzymatic digestibility of proteins, fermentation time, koji growth characteristics and quality of soy sauce in terms of chemical components, organoleptic evaluation and stability (Yokotsuka, 1983).

The wheat is in the form of whole meal flour or white flour in much of S.E. Asia, but Japanese practice favours using whole grains of wheat, which are roasted in a fluidised sand bed, and when cooled down are then crushed. The Japanese practice gives colour and extra flavour to the finished product, and breakdown products from the lignin present in the outer cell walls contribute aromatic alcohols which are important in ameliorating the salty "bite" of the soy sauce.

## THE KOJI PROCESS

### Basic Considerations

The koji process is an excellent and well-studied example of solid substrate fermentation (SSF). Such processes involve the growth of microorganisms (the koji moulds) on and within solid organic materials (soya beans/wheat) in the absence or near absence of free water. The koji process is probably one of the longest practised forms of food-related SSF (Moo-Young *et al.*, 1983).

The types of organisms that grow well in SSFs is determined largely by the water activity factor, $a_w$ which is defined as the relative humidity of the gaseous atmosphere in equilibrium with the substrate. Certain filamentous fungi and a few yeasts can grow at $a_w$ values between 0.6 and 0.7. The $a_w$ of the substrate quantitatively expresses the water requirements for microbial activity and is the most critical condition that determines successful growth of the microorganisms. The $a_w$ of the medium is a fundamental parameter for mass transfer of water and solutes across the cell membrane. Continued regulation of the $a_w$ of the solid substrate can be controlled by the relative humidity (RH) of the air. Values of $a_w$ will vary depending on substrate formulation and type of fermentation. When the moisture level is too high, there will be a decrease in porosity or intracellular spaces, lower oxygen diffusion, decreased gas exchange, decreased substrate degradation and increased risk of contamination. Conditions of low moisture will cause decreased substrate swelling and decreased microbial growth. Furthermore, during a fermentation, additional water loss can occur by evaporation and microbial activity. This can be ameliorated by adjusting the humidity of air passing through the system ( Smith and Aidoo, 1988; Pandey, 1992).

Together with substrate moisture level and humidity, temperature, aeration and agitation are the other important factors which will determine the degree of monitoring and control of SSF (Gervais and Bensoussan, 1993).

Temperature regulation of SSF is directly related to water potential and aeration, and will strongly influence organism growth. Furthermore, during aerobic microbial growth, large quantities of heat will be generated in the fermenting mass. A major limitation of SSF is the difficulty in removing excess heat, due primarily to the low thermal conductivity of the solid substrate matrix. In the closed mass of a koji fermentation such heat build up can adversely affect the temperature distribution and level within the fermentation mass.

In SSFs, oxygen is essential and is usually supplied by forcing air through the fermenting mass or by intermittent turning. The level of aeration will be dependent on the type of microorganism and the specific requirements of the process. The degree of aeration is also strongly related to the need to dissipate heat generated in the fermentation, and to the removal of metabolic $CO_2$ and other volatile byproducts of the fermentation which could in some cases become inhibitory. Such mass transfer steps are critically important to the koji process. Aeration rate will also depend on the thickness of the fermenting layer and the porosity of the medium. Particle size will largely determine the amount of space (the void space) within the mass that can be occupied by air. Aeration also has important effects on hydration properties and heat regulation in SSF.

Interparticle mass transfer is essential for the transfer of oxygen from the void space within the mass to the growing microorganism. The dew point of the used air must be considered as a function of the air inlet temperature in order to control the relative humidity of the solid substrate. Changes in the air status occur when the air is passing through the mass of the SSF. The relative humidity of the air will generally increase through the mass due to evaporation. However, this effect can be localised to lead to heterogeneity in the medium if the flow rate is too slow (Gervais and Bensousson, 1993). In practice, only air movement is used for controlling the temperature within SSF. The heat capacity of the air at maximal water saturation is smaller than the heat capacity of water for cooling in submerged fermentations. Consequently, SSF needs large quantities of air exceeding the amount necessary for microbial respiration.

The complex physico-chemical requirements of solid substrate fermentations have recently been discussed by Gervais and Bensousson (1993) and their conclusions are of particular significance to the overall koji process, in particular to commercial optimisation, viz:

1. To ensure optimum $O_2$ requirements for the microorganism, without limiting the liquid diffusion of nutrients, it is necessary for the water content to remain just behind the water holding capacity of the solid substrate.
2. The relative humidity of the forced aeration should be controlled to optimise removal of $CO_2$ and other volatiles and to satisfy the $O_2$ requirements of the microorganism.
3. It is essential to appreciate the effects of the depth of the fermentation bed, air temperature and air flow rate to prevent saturation phenomena and condensation in the upper part of the bed. Residence time of air in the fermentation mass must be sufficiently short to prevent saturation at the exit.
4. Changes in water sorption properties of the polymers in the SSF by extracellular enzyme activity could cause major decreases in water activity for the same water content; thus, adjustment of the forced aeration must be practised during the fermentation.

## The Koji Mould

The yellow-green aspergilli (koji moulds) used in Asian soybean fermentations have long been a subject generating serious debate amongst fungal taxonomists (Flegel, 1988). In *Aspergillus* classification (Table 1) the koji moulds belong to the Section Flavi Gains *et al.* (= *Aspergillus flavus* group Roper and Fennell) (Samson, 1993). Historically, the use only of morphological criteria have shown that these isolates can be difficult to separate conclusively. Christensen (1981) considered that the true koji moulds comprised *Aspergillus oryzae* (Raper and Fennell, 1965), *Aspergillus* sojae (Murakami *et al.*, 1982) and *Aspergillus tamarii* Kita

(Raper and Fennell, 1965), the distinction between them being based on conidial head colour, growth at 37°C and dimensions of conidiophores, vesicles and conidia. The absence of pink conidia on Czapck agar slants containing 0.05% anisaldehyde was a further diagnostic determinant (Murakami, 1971; Murakami *et al.*, 1982).

**Table 1** Accepted *Aspergillus* species Section Flavi (Samson, 1993)

| |
|---|
| *A. avenaceus* Smith |
| *A. clavato-flavus* Raper and Fennell |
| *A. flavus* Link |
| *A. leporis* |
| *A. nomius* Kurtzman *et al.* |
| *A. oryzae* (Ahl6) Cohn |
| *A. parasiticus* Speare |
| *A. sojae* Sakaguchi and Yamada |
| *A. subolivaceus* Raper and Fennell |
| *A. tamarii* Kita (= *A. flavo-furcatis* Batista and Maia) |
| *A. zonatus* (Kwon and Fennell) Raper and Fennell |

The relationship between the koji moulds and the toxigenic *A. flavus* and *A. parasiticus* has been a source of much investigation. While there are clear morphological features separating all the species, the existence of intermediate forms (Flegel, 1988) does indicate some relationship between *A. oryzae* and *A. flavus* and between A. *sojae* and *A. parasiticus.* Wicklow (1984*a*) concluded that *A. oryzae* and *A. sojae* had been differentiated from *A. flavus* and *A. parasiticus* respectively. Later he was to further confirm these conclusions by demonstrating that the domesticated koji moulds had significantly larger, more variable and less roughened conidia than the *A. flavus* and *A. parasiticus* strains and that the fermentation rates were much faster (Wicklow, 1984*b*).

Studies are now showing that the relative amount and properties of various nucleic acids in organisms can be used to resolve taxonomic relatedness. DNA relatedness relies mainly on measuring the extent and stability of renatured DNA strands from the test pair, e.g. the fidelity of complementary base pairs. Kurtzman *et al.* (1986) have extensively examined *A. flavus, A. parasiticus, A. oryzae* and *A. sojae* and found significantly high complimentarity among the four groups to suggest that they were conspecific. Because there were some divergence between *A. flavus/A. oryzae* and *A. parasiticus/A. sojae* it was considered that the four taxa be divided into two subspecies and four varieties with the following designations: *A. flavus* Link:Fr. subsp. *flavus* var. *flavus; A. flavus* Link:Fr. subsp. *flavus* var. *oryzae* (Ahlburg) Kurtzman *et al.*, *A. flavus* Link:Fr. subsp. *parasiticus* (Speare) Kurtzman *et al.*, and *A. flavus* Link:Fr. subsp. *parasiticus* (Speare) Kurtzman *et al.* var. *sojae* (Sakaguchi and Yamada ex Murakami) Kurtzman *et al.*

By means of electrophoretic comparison of enzymes and ubiquinone systems, Yamatoya *et al.* (1990) considered that the four taxa could be grouped as *A. flavus* and *A. parasiticus.* Recently, Samson and Frisvad (1991) have shown that *A. flavus, A. parasiticus* and *A. sojae* consistently produced aspergillic acid while isolates of *A. oryzae* never produce this toxin.

In Japan strains of *Aspergillus* are purchased in the markets and used for food fermentations such as production of soy sauce. The selection of the strains used is based among others on

a) flavour and colour of the final product;
b) ability to sporulate for the preparation of the seed starter;
c) rapid growth and high enzyme production;
d) length of stalk;

e) genetic stability, and
f) inability to produce toxins (Yokotsuka, 1985).

In a study on koji mould starters purchased in a Japanese market, Terada *et al.* (1981) found that 69% of soy sauce koji mould starters and 79% for miso were composed of two types of mould. Sixty-five strains of the soy sauce koji mould were composed of 80% *A. oryzae* and 20% *A. sojae* while 63 strains for miso were composed of 89% *A. oryzae* and 11% *A. sojae*. The starter moulds for soy sauce production generally showed short stalk and high spore production whereas those for miso production had high amylolytic activity.

According to published information (Terada *et al.*, 1981; Hayashi *et al.*, 1981) the soy sauce koji cultured with *A. sojae* has the following characteristics compared to koji cultured with *A. oryzae*:

a) higher pH of koji but lower pH of raw shoyu;
b) lower starch utilisation during koji fermentation;
c) high endo-poly-galacturonase activity but lower activity of alpha-amylase, acid protease and acid carboxypeptidase activity during koji fermentation;
d) lower viscosity of mash or moromi;
e) lower enzyme activity and lower levels of reducing sugar, lactose, and ammonia in the raw shoyu.

After the discovery of the aflatoxins in the 1960s, there was extensive examination of the koji moulds for toxin production. While early studies by Japanese scientists demonstrated that during koji mould growth in soy sauce production there was the formation of compounds exhibiting $R_f$ values and fluorescences similar to the aflatoxins, further investigation clearly demonstrated that these compounds differed markedly in chemical structure from the aflatoxins (Yokotsuka *et al.*, 1966).

While no aflatoxins have ever been demonstrated in *A. oryzae, A. sojae* or *A. tamarii* strains, it is well accepted that all can produce other mycotoxins under specific environmental conditions, e.g. aspergillic acid, cyclopiazonic acid, kojic acid, maltoryzine and beta nitropropionic acid (Frisvad, 1986; Trucksess *et al.*, 1987) (Table 2). However, while these mycotoxins can be demonstrated under specific environmental conditions with individual moulds there is little evidence that they are being produced in commercial production processes.

**Table 2.** Toxins reported to be produced by koji moulds

| | |
|---|---|
| *Aspergillus oryzae* | Cyclopiazonic acid, kojic acid, maltoryzine, beta-nitropropionic acid |
| *Aspergillus sojae* | Aspergillic acid, kojic acid |
| *Aspergillus tamarii* | Kojic acid |

Why, then, should koji mould fermented products be traditionally free of mycotoxins? Certainly there are no documented cases of human toxicity that could be related to the consumption of such products. Undoubtedly, the most significant factor is the duration of the fermentation. Whereas under ideal conditions toxigenic fungi normally take several days to begin toxin formation, most koji fermentations seldom exceed 48-72 h (Yokotsuka, 1983). Under these conditions even unlikely toxin production by the koji moulds would be limited. Most evidence of toxin formation by some koji moulds was as a result of prolonged incubation (5-8 days).

A second, and equally important consideration with soy bean fermentation, is the well-documented fact that the soy bean is only weakly supportive of toxin formation by fungi, e.g. *A. flavus* and aflatoxin formation (Hesseltine *et al.*, 1966; Gupta and Venkitasubramanian, 1975). A final consideration must also be given to possible detoxification of any formed toxin by the moromi bacteria and yeasts. Microbial detoxification of aflatoxin and other toxins has been well-documented (Bol and Smith, 1989).

While it must be conceded that all koji mould fungi are potentially toxigenic, there is no evidence that toxin formation occurs in practice. However, strain selection programmes should be vigilant to any possible genetic changes that could increase any potential for toxin production (Kolayanaamitr *et al.*, 1987).

## The Koji Fermentation Process

In the koji fermentation of soy sauce production the following process features are of critical importance, viz:

1. to maximise protease, amylase, cellulase and other enzyme activity.
2. to prevent denaturation of these enzymes once they are produced.
3. to avoid contamination by other microorganisms.
4. to minimise the utilisation of the nutrients by the koji moulds (Yokotsuka, 1983).

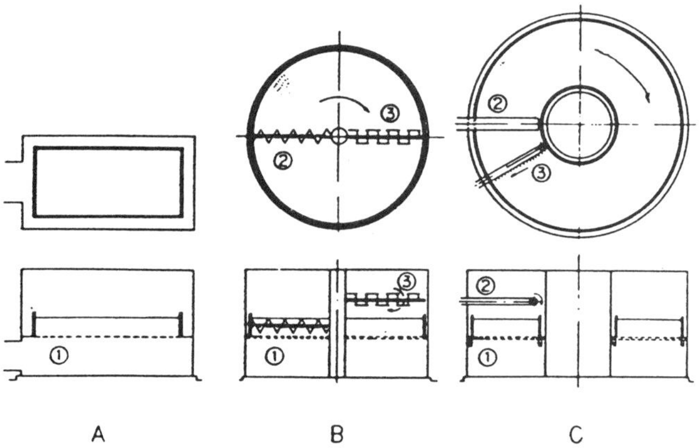


**Figure 2.** Koji culturing machines with through-flow system of aeration. A, rectangular type; 1, perforated plate. B, circular type (batch); 1, perforated plate; 2, feed and discharge screw; 3, mixer. C, circular type (continuous); 1, perforated plate; 2, feed conveyor; 3, discharge conveyor.
Reproduced with permission from Yokotsuka (1985).

The mechanical bioreactor types of koji manufacture have been almost exclusively developed in Japan and comprise batch and continuous systems (Figure 2) (Yokotsuka, 1983; Aidoo *et al.*, 1984).

Three major features which characterise these bioreactors can be summarised as follows:

1a. Batch-type with a rectangular perforated plate.
1b. Batch-type with a circular moving perforated plate.

1c. Continuous-type with a circular moving perforated plate.
2. Rotary drum.
3. Surface-flow system of aeration - the temperature and moisture controlled air flows over the fermenting mass.

These koji incubation chambers/bioreactors now involve automated inoculation of substrate, controlled mass transfer, automated heaping and turning of the solid fermentation mass and automated harvest of the finished koji.

In modern koji processes the *Aspergillus* spores (tane koji) are thoroughly mixed with the cooled soy beans and roasted wheat prior to charging the fermentation trays or deep bed systems now widely used. This gives a more uniform distribution of the mould propagules and consequently a more uniform development of the fermentation process. Traditional village level fermentations will be discussed elsewhere in this volume (Cook and Campbell-Platt) while this presentation will be restricted to mechanically ventilated koji bioreactors, in particular the rotary disc type koji method (Figs. 3 and 4). As an example of good industrial

**Figure 3.** Package type rotary Koji making equipment.
Reproduced with permission from Fugiwara, Japan

practice, these methods give excellent and prolonged control of temperature and humidity, regular ventilation with air and removal of inhibitory gases and volatiles and further ensure regular and complete mixing. In essence, this is the most advanced example of a solid substrate fermentation worldwide and is the endpoint of generations of study of this process. While this sophisticated technology is widely available in Japan and other Far East countries there are only a few examples in the West.

The mixed substrate and mould spores are dispersed into the insulated koji machine/room to a uniform thickness by means of the heaping machine (Figure 4.(4)). After heaping, air, controlled at the proper temperature and humidity, is delivered by blower into the koji room through the air supply duct (Figure 4.(5,6,7)). The air flows through the koji material upwards and exits by way of the air duct (Figure 4.(7)). The air is recycled and the correct temperature and humidity is constantly achieved. The temperature requirements of the koji process is regulated according to predetermined studies to ensure optimisation of mycelial growth and enzyme production but also to ensure that the formed extracellular enzymes are not subsequently inactivated by high temperature.

Mould growth is rapid and extensive, highly aerobic and strongly exothermic, and if the koji is not frequently mixed can create difficulties in ventilation and control of temperature. Thus, to ensure a homogeneous system, the fermentation mass is regularly turned mechanically (Figure 4.(3)).

The duration of the fermentation depends on the nature of the intended product but will normally be 48-72 hrs., seldom longer since the mould will then start to sporulate creating unwanted flavours and odour. At the end of the fermentation period the product, the finished koji, will be discharged (Figure 4(4)) through the chuting port (Figure 4.(ll)) and transferred to the next process by means of a conveyor belt or pneumatic conveyer.

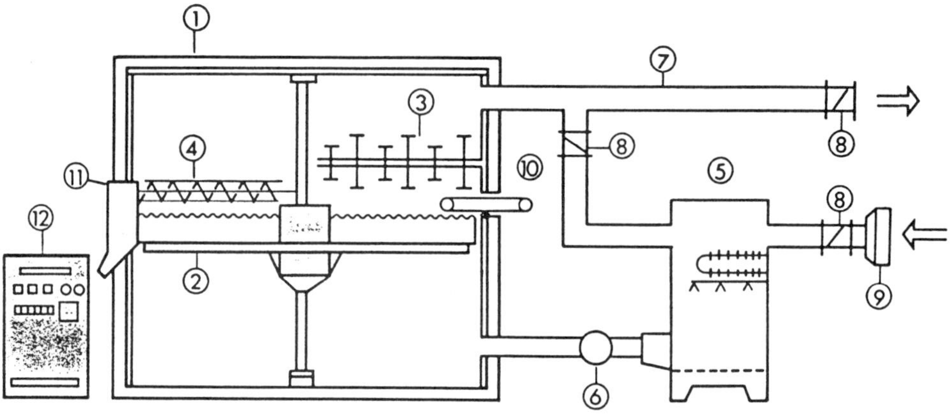

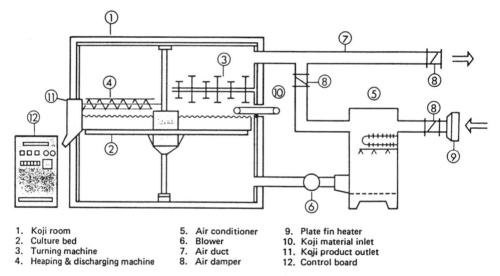

1. Koji room
2. Culture bed
3. Turning machine
4. Heaping & discharging machine
5. Air conditioner
6. Blower
7. Air duct
8. Air damper
9. Plate fin heater
10. Koji material inlet
11. Koji product outlet
12. Control board

**Figure 4.** Automated Koji making eqipment.
Reproduced with permission from Fugiwara, Japan.

The size of the koji making equipment depends on the bulk density and the amount of raw material to be used per batch. In continuously operated systems connected with a continuous soybean cooker and wheat roaster, one machine can deal with up to 50-100 tons of raw material per day (Yokotsuka, 1983).

Koji incubation chambers or bioreactors involving automated inoculation of the substrate, controlled mass transfer, automated heaping and turning of solid fermenting mass and the automated harvest of the finished koji are widely used in many commercial plants in Japan, Far East and the United States. The use of automated machinery for soy sauce production was initiated by the major Japanese soy sauce manufacturers such as Kikkoman, Yamasa and Nippon Maruten. The automatic koji equipment has wider application including production of miso (soy paste), sake, amazake, beer malt, vinegar, antibiotics and other enzyme preparations.

### The Major Enzymes Found During Koji Fermentation Process

According to Yokotsuka (1985) the quality of koji very much influences not only the degree and speed of enzymatic degradation of the raw material for the moromi stage but also the chemical and sensory attributes of the final product.

Carbohydrases (alpha-amylase, amyloglucosidase, maltase, sucrase, pectinase, beta-galactosidase, cellulase, hemi-cellulase and pentosan-degrading enzymes) and proteinases are the two main groups of enzymes produced by *A. oryzae* during koji fermentation although lipase activity has also been reported (Yong and Wood, 1976; 1977). These major enzymes hydrolyse carbohydrates and proteins to sugars and amino acids and low molecular weight peptides respectively and these soluble hydrolytic products are essential for the moromi fermentation. The combined action of the carbohydrases results in the production of fermentable sugars such as D-glucose, D-galactose, D-galacturonic acid, D-rhamnose, D-xylose and L-arabinose.

Although alkaline protease (optimum pH 7 to 10) is formed in the largest amounts (about 80%), neutral protease is most important for soy sauce manufacture. Three kinds of acid protease (optimum pH 3 to 4) are produced by *A. oryzae* in the koji fermentation and two neutral protease have also been isolated. Three kinds of aminopeptidase and four carboxypeptidases are produced by the koji mould and the former, in particular leucine-aminopeptidase, are associated with enzymatic formation of formolnitrogen and glutamic acid in the mash or moromi. The mould normally produces the highest protease activities at 20° to 30°C. The koji mould also produces glutaminase which converts glutamine to glutamic acid, an amino acid essential for a good quality soy sauce. Yamamoto and Hirooka (1974) reported on glutaminase hyperproductive mutants of *A sojae* and suggested that they be used in soy sauce fermentation, but recent reports (Hamada *et al.*, 1991) indicate that glutaminase from a yeast, *Candida versatilis*, may be of commercial importance.

## MOROMI FERMENTATION

The koji is mixed with 1.2 to 1.5 volumes of salt brine (23%) to make a mash or moromi which undergoes lactic acid bacteria and yeast fermentations for at least one year at ambient temperatures to produce quality soy sauce.

The strong salt brine creates a condition that limits the microbial activity to a few desirable organisms. At first *Pediococcus halophilus* grows and produces lactic acid which lowers the pH to 5.5 or less and then salt-tolerant, dominant yeast *Zygosaccharomyces rouxii* begins to grow and produces 2 to 3 percent alcohol and several compounds which add characteristic aromas to soy sauce. Other types of soy sauce yeasts such as *Candida versatilis* and *Candida ethellsii* produce phenolic compounds, 4-ethylguaiacol and 4-ethylphenol, which contribute to soy sauce aroma.

Under the moromi fermentation conditions the mould is rapidly killed, but its extracellular enzymes, possibly supplemented by enzymes released from the mycelium post mortem, continue to hydrolyse their various substrates, albeit slowly, in the salty (18% sodium chloride), acidic (down to pH 4.5 or less) brew, producing substrates for growth of the yeasts and lactic acid bacteria. Experimental studies indicate that most of the hydrolysis occurs during this stage, with relatively little (beyond that needed to supply the mould's needs) in the koji stage. The very high enzyme yields from the highly selected *Aspergillus* strains help to ensure remarkably high levels of solubilisation of the proteins and starches in the substrate.

This is further aided by the advanced technology for cooking the beans and wheat grains.

The submerged moromi fermentation is usually carried out in a wood, concrete or steel tank with a capacity of 10 to 30 $m^3$. In Taiwan the traditional ceramic pots and wooden tanks are now replaced with epoxy resin-coated cement tanks with a capacity of 27 to 36 kilolitres.

The most important function of the koji fungi in soy bean fermentations is the production of extracellular and exocellular enzymes, in particular, protease and carbohydrase complexes. These complexes must be produced in suitable quantities to ensure good maceration and flavour development.

## PRESSING, PASTEURISATION, FILTRATION AND BOTTLING

The moromi is ultimately separated by a pressing machine into raw soy sauce and cake. Pressing is carried out for 2 to 3 days. Although such mechanised systems add extra cost, the lower pressures and longer filtration times result in loss of flavour. The cooled soy sauce is then blended with added sweeteners and other additives such as benzoic acid or propyl- or butyl-p-hydroxybenzoate, according to ultimate user demand and then pasteurised at 70°-80°C and finally bottled.

## RECENT DEVELOPMENTS IN SOY SAUCE FERMENTATION

There appears to be little information on the use of mutant strains of *Aspergillus* for koji making. Nasuno and Ono (1972) reported an increase of 2-6 % in protein digestibility in the production of soy sauce by using an induced mutant of *A. sojae* of which alkaline protease activity was increased by six times compared to that of the parent strain.

One of the major developments in soy bean fermentation technology over the past two decades is the design of bioreactors, particularly for koji making, because of their wide applications mentioned earlier. However, application of new biotechnology to soy sauce fermentation was reported recently by the Kikkoman Corporation of Japan, (Hamada *et al.*, 1989, 1991). They reported on continuous production of good quality soy sauce by immobilised cells in a bioreactor. The new technique of koji making, fermentation and ageing of moromi and product refinement takes about two weeks compared with at least six months for conventional methods. The bioreactor method differs from the conventional method in the following ways:

a) proteases from continuous submerged culture are used.
b) fermentation is carried out in liquid state,
c) the fermentation period is about 2 days.

The process involved enzymatic hydrolysis of liquidised cooked soy bean (108 kg) and roasted wheat (61 kg) with a broth culture of *A. oryzae* (400 l) in the presence of 8.5% NaCl solution at 45°C for 3 days in a stirred tank. The hydrolysed mash was then filtered in a filter press with cloth. The liquid fraction in a feed tank, after adjusting the NaCl, total nitrogen contents, and pH to 13%, 2% and 6.0 respectively, was then passed through a series of immobilised bioreactors which took 2 days to complete the fermentation processes (Fig. 5). The first reactor contained immobilised glutaminase from *Candida famata* to increase glutamic acid content of the fraction while the second reactor contained immobilised cells of *Pediococcus halophilus* to carry out lactic acid fermentation; both were plug-flow reactors with a total working volume of 1.8 and 7.5 l respectively. The residence time for the immobilised glutaminase bioreactor was 0.7 h maintained at 40°C and 6.1 h for the *P. halophilus* bioreactor at 27°C.

The third and fourth bioreactors were airlift types with immobilised cells of

*Zygosaccharomyces rouxii* and *C. versatilis* respectively both maintained at 27°C and set in parallel because *Z. rouxii* inhibits the growth of *C. versatilis* in a mixed state (Hamada *et al.*, 1991). The *Zygosaccharomyces rouxii* bioreactor, with a residence time of 25.5 hr and total volume of 27 l, produced alcohol constantly and the *Candida versatilis* bioreactor, residence time 10.7 h and total volume of 1 l, produced phenolic compounds such as 4-ethyl-guaiacol. A schematic diagram of the bioreactor system for continuous production of soy sauce is shown in Fig. 6.

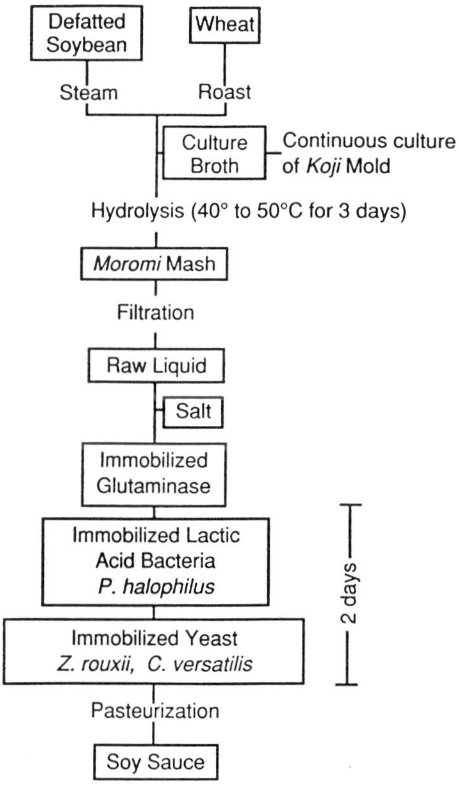


**Figure 5.** Bioreactor method of soy sauce production.
(Reproduced with permission from Hamada *et al.*, 1991)
A: Feed tank; B: Glutiminase reactor; C: *Pediococcus halophilus* reactor; D: Buffer tank; E: *Zygosaccharomyces rouxii* reactor; F: *Candida versatilis* reactor; G: Product tank; P: Feed pump; Air P: Air pump.

The system was run for more than 100 days without problems of microbial contamination. Although the organoleptic properties as well as chemical composition of the bioreactor soy sauce compared favourably to those of conventional soy sauce, the authors conceeded that more detailed studies on conditions of fermentation would be needed to improve the quality of soy sauce produced by this new technology.

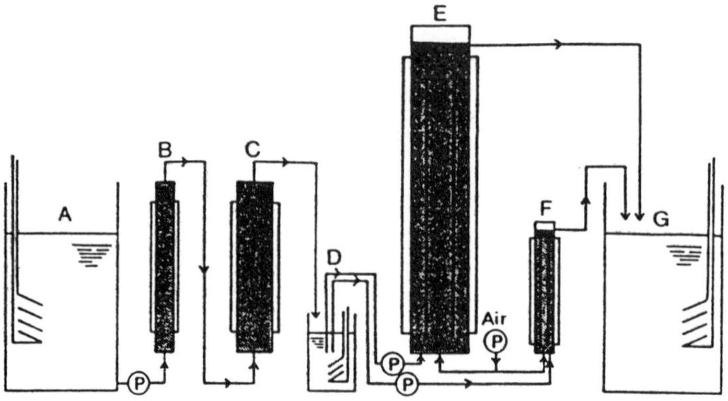


**Figure 6.** Schematic diagram of a bioreactor system for soy sauce production. Reproduced with permission from Hamada *et al.* (1991).

## REFERENCES


Abiose, S.H., Allen, M.C. and Wood, B.J.B. (1982) Microbiology and biochemistry of miso (soy paste) fermentation. Adv. Applied Microb. 28 , 201-237.

Aidoo, K.E., Hendry, R. and Wood, B.J.B. (1984) Mechanised fermentation systems for the production of experimental soy sauce koji. J. Food Technol. 19, 389-398.

Bol, J. and Smith, J.E. (1990) Biotransformation of aflatoxin. Food Biotech. 3, 127-144.

Campbell-Platt, G. (1987) "Fermented Foods of the World". pp. 130-131. 197-198. Butterworths Press, London.

Christensen, M. (1981) A synoptic key and evaluation of species in the *Aspergillus flavus* group. Mycologia 73, 1056-1084.

Djien, K.S. (1982) Indigenous fermented foods. in "Economic Microbiology", Vol 7, Fermented Foods (Rose, A.H., Ed.) pp. 15-38. Academic Press, London.

Ebine, H. (1989) Industrialisation of Japanese miso fermentation. in "Industrialisation of Indigenous Fermented foods" (Steinkraus, K.H., Ed). pp. 89-126. Marcel Dekker Inc., New York.

Flegel, T.W. (1988) Yellow-green *Aspergillus* strains used in Asian soybean fermentations. Asean Food J. 4, 14-30.

Frisvad, J.C. (1986) Taxonomic approaches to mycotoxin identification (taxonomic indication of mycotoxin content in foods), in "Modern Methods in the Analysis and Structural Fluidation of Mycotoxins" (Cole, R.J. Ed.), pp. 415- , Academic Press, New York.

Fukushima, D. (1986) Soy sauce and other fermented foods of Japan. in "Indigenous Fermented Food of Non-Western Origin" (Hesseltine, C.W. and Wang, Hwa L. Eds.). Mycologia Memoir No.11, pp. 121-150. New York Botanical Garden; J. Cramer, Berlin.

Gervais, P. and Bensoussan, H. (1993) Solid state fermentations of the genus *Aspergillus,* in "Biotechnology Handbooks" Vol. 7. (Smith, J.E. Ed.), Plenum Press, London.

Gupta, S.K. and Venkitasubramanian, T.A. (1975) Production of aflatoxin on soybeans. Appl. Microb. 29, 834-836.

Hamada, T., Ishiyama, T., and Motai, H. (1989) Continuous fermentation of soy sauce by immobilised cells of *Zygosaccharomyces rouxii* in an airlift reactor. Appl. Microb. Biotechnol. 31, 346-350.

Hamada, T., Sugishita, M., Fukushima, Y., Fukase, T. and Motai, H. (1991) Continuous production of soy sauce by a bioreactor system. Process Biochem. 26, 39-45.

Hayashi, K., Terada, M. and Mizunuma, T. (1981) J. Japan Soy Sauce Res. Inst. 7(4), 166.

Hesseltine, C.W., Shotwell, O.C., Ellis, J.J. and Stubblefield, R.D. (1966) Aflatoxin formation by *Aspergillus flavus.* Bacteriol. Rev. 30, 795-805.

Kolayanaamitr, A., Bhumiratana, A., Flegel, T.W., Glinsukm, T. and Shinmyo, A. (1987) Occurrence of

toxicity among protease, amylase and colour mutants of a non-toxic soy sauce koji mold. Appl. Environ. Microb. 53, 1980-1982.
Kurtzman, C.P., Smiley, M.J., Robnett, C.J. and Wicklow, D.T. (1986) DNA relatedness among wild and domestic species in the *Aspergillus flavus* group. Mycologia 78, 955-959.
Moo-Young, M.A., Moreira, R. and Tengerdy, R.P. (1983) Principles of solid substrate fermentations, in "The Filamentous Fungi", Vol. IV. (J.E. Smith, D.R. Berry and B. Kristiansen, eds). pp. 117-144. Edward Arnold, London.
Mulyowidarso, R.K., Fleet, G.H. and Buckle, K.A. (1989) The microbial ecology of soybean soaking for tempeh production. Intern. J. Food Microb. 8 , 35-46.
Mulyowidarso, R.K., Fleet, G.H. and Buckle, K.A. (1991) Changes in the concentration of carbohydrates during the soaking of soybeans for tempeh production. Intern. J. Food Sci. Technol. 26 , 595-614.
Murakami, H. (1971) Classification of the koji mould. J. Gen. Appl. Microb. 17, 281-309.
Murakami, H., Hayashi, K. and Ushijimi, S. (1982) Useful key characters separating three *Aspergillus* taxa - *A. sojae, A. parasiticus* and *A. toxicarius*. J. Gen. Appl. Microb. 28, 55-60.
Nasuno, S. and Ono, T. (1972) Seasoning Science 19 (10), 41.
Pandey, A. (1992) Recent process developments in solid-state fermentation. Proc. Biochem. 27, 109-117.
Raper, K.B. and Fennell, D.I. (1965) "The Genus *Aspergillus.* " Huntington, N.Y. R.E. Krieger Publishing Co.
Samson, R.A. (1993) Taxonomy, current concepts of Aspergillus systematics, in "Biotechnology Handbooks", Vol. 7. "*Aspergillus*". (Smith, J.E. Ed.) Plenum Press, London. in press.
Samson, R.A. and Frisvad, J.C. (1991) Taxonomic species concepts of Hyphomycetes related to mycotoxin production. Proc. Japan Assoc. Mycotox. 32, 3-10.
Shurtleff, W. and Aoyagi, A.(1983) "The Book of Miso". 278 pp. Tenspeed Press, Berkeley, CA. 278 pp.
Smith, J.E. and Aidoo, K.E. (1988) Growth of fungi on solid substrates. in "Physiology of Industrial Fungi". (Berry, D.R., Ed.), pp. 249-269. Blackwell Scientific Publications. Oxford.
Steinkraus, K.H. (1985) "Handbook of Indigenous Fermented Foods". pp. 671. Marcel Dekker Publishers, New York.
Terada, M., Hayashi, K. and Mizunuma, T. (1981) J. Jap. Soy Sauce Res. Inst. 7(4), 158.
Trucksess, M.W., Misliveck, P.B., Young, K., Bruce, V.E. and Page, S.W. (1987) Cyclopiazonic acid production by cultures of *Aspergillus* and *Penicillium* species isolated from dried beans, corn meal, macaroni and pecans. J. Assoc. Off. Anal. Chem. 70, 123-126.
Wicklow, D.T. (1984a) Adaptation in wild and domesticated yellow-green Aspergilli, in "Toxigenic Fungi - Their Toxins and Health Hazard". Vol. 7. "Developments in Food Science (Kurata, K. and Ueno, Y. Eds.), pp. 78-86. Elsevier, New York.
Wicklow, D.T. (1984b) Conidium germination rate in wild and domesticated yellow-green Aspergilli. Appl. Environ. Microb. 47, 299-300.
Wood, B.J.B. (1982) Soy sauce and miso, in "Economic Microbiology", Vol 7, (Rose, A.H. Ed.).pp. 39-87. Academic Press, London.
Yamamoto, K. and Hirooka, H. (1974) Production of glutaminase by *Aspergillussojae.* J. Ferment. Technol. 52, 564-576.
Yamatoya, T., Sugiyama, J. and Kuraishi, H. (1990) Electrophoretic comparison of enzymes as a chemotaxonomic aid among *Aspergillus* taxa (2). *Aspergillus* sect. *Flavi*, in "Modern Concepts of Penicillium and Aspergillus Classification "(Samson, R.A. and Pitt, J.I. Eds). pp. 295-406. New York and London, Plenum Press.
Yokotsuka, T., Sasaki, M., Kikuchi, T., Assao, Y. and Nabukara, A. (1966) Production of fluorescent compounds other than aflatoxin by Japanese industrial molds. in "Biochemistry of Some Foodborne Microbial Toxins" (Mateles, R.I. and Wogan, G.N. Eds.), pp. 111-152. M.I.T. Press, Cambridge, Mass.
Yokotsuka, T. (1983) Scale-up of traditional fermentation technology. Kor. J. Appl. Microb. Bioeng. 11, 353-371.
Yokotsuka, T. (1985) Fermented protein foods in the Orient, with emphasis on Shoyu and Miso in Japan, in "Microbiology of Fermented Foods", Vol. 1 (Wood, B.J.B. Ed.), pp. 197-248. Elsevier Applied Science Publishers, London.
Yong, F.-M. and Wood, B.J.B. (1974) Microbiology and biochemistry of soy sauce fermentation. Adv. Appl. Microb. 17, 157-194.
Yong, F.M. and Wood, B.J.B. (1976) Microbial succession in experimental soy sauce fermentation. J. Food Technol. 11, 525-536.
Yong, F.M. and Wood, B.J.B. (1977) Biochemical changes in experimental soy sauce koji. J. Food Technol.12, 163-175.

# *ASPERGILLUS* AND FERMENTED FOODS

Paul E Cook and Geoffrey Campbell-Platt

Department of Food Science and Technology
University of Reading
Whiteknights, P O Box 226
Reading, RG6 2AP

## INTRODUCTION

Fermentation is one of the oldest forms of food preservation as well as being a precursor of modern biotechnology. Food fermentations are used in many parts of the world and Campbell-Platt (1987) has documented some 250 categories of these foods and more than 3500 individual products. Preservation by fermentation offers several advantages over other food processing methods. Energy requirements are often lower and the fermentation process may result in desirable changes in texture and flavour. The nutritional value may be improved by the microbial synthesis of vitamins, the reduction in toxic and antinutritional components and the reduction or elimination of food-borne pathogenic bacteria.

The origin of food fermentation by moulds lies in the Orient where these fungi have been employed in food production for at least 3000 years. China is considered to be the centre of origin of most Oriental fermented foods and the wide variation in climate and types of cultivated crops have had an important influence on the development of food fermentations and the exploitations of different microfungi.

Aspergilli are probably the most important of the food fermentation moulds since they provide a source of enzymes for the production of a large number of savory foods, condiments, sweet desserts and alcoholic beverages (Leistner, 1986; Campbell-Platt, 1987; Campbell-Platt and Cook, 1989). In Western countries, aspergilli are less important for food production although Oriental foods such as soy sauce and miso are produced commercially in North America and Europe and aspergilli may have a minor role in the ripening and maturation of country-cured ham and salami.

## ORIGINS OF *ASPERGILLUS* FERMENTED FOODS

Various sources of enzymes can be used for hydrolysing plant and animal substrates for fermentation (Table 1). Malting of cereals is used in Europe and Africa for brewing and salivary amylases are used to hydrolyse maize and cassava for brewing beer in South

America. In Africa, amylase containing plant roots are used to hydrolyse maize and millet starch for the production of munkoyo sour maize beverage (Pauwels *et al.*, 1992). Malting of cereals was also practised in China (Yokotsuka, 1991*b*; Chang, 1993) but not to the same extent as in Europe or Africa. An alternative approach for food processing was to grow amylolytic and proteolytic moulds and yeasts on rice, wheat and millet to provide a source of enzymes termed chu, shi or qu in China and koji in Japan (Yokotsuka, 1991*a*).

Aspergilli and mucoraceous moulds grow well on cereals but aspergilli require more aerobic conditions and grow better when cereal grains are in a granular form. Balls or cakes made from cereal flour provide less aerobic conditions and favour the growth of *Rhizopus* and *Mucor* spp. many of which are capable of growth under anaerobic conditions (Hesseltine *et al.*, 1985). Balls or cakes of cereal flour containing microorganisms were probably first produced in China and dissemination of this technology *via* trade routes gave rise to the amylolytic starter cultures or ragi which are today, found in most Southeast Asian countries, Northern India and Nepal. These balls or cakes are used for the production of alcoholic beverages and foods but differ from *Aspergillus* koji since the fermentative organisms are in a dried, inactive state and the balls or cakes provide an inoculum rather than a direct source of enzymes or substrate for the fermentation. Surveys of the mycoflora of these cakes have revealed that although aspergilli occur in these starters the dominant moulds are *Amylomyces, Mucor* and *Rhizopus* spp. (Hesseltine *et al.*, 1988).

## Domestication of Aspergilli

Relatively few filamentous fungi are actively involved in food fermentations with perhaps some 20-30 species amongst the genera *Actinomucor, Amylomyces, Aspergillus, Eurotium, Monascus, Mucor, Neurospora, Penicillium and Rhizopus.* This paucity of food fermentative moulds is somewhat surprising considering the large number of filamentous fungi associated with foods and food spoilage. Within the aspergilli the important species are *Aspergillus tamarii,* the yellow-green aspergilli *Aspergillus oryzae* and *Aspergillus sojae* and the black aspergilli *Aspergillus awamori, Aspergillus usami* and *Aspergillus kawachii.* Of the telemorphic genera, only *Eurotium* appears to be significant in fermented food production with the important species being *E. repens* and *E. rubrum.*

Many of the aspergilli used in food fermentations have been domesticated over a period of many centuries although little is known about the domestication process since most studies have focused on the fermentations. Aspergilli and other fungi isolated from fermented foods frequently differ from their wild counterparts in morphological and physiological characteristics. Molecular techniques have begun to unravel the relationship between fermented food domesticates and their wild counterparts. *Aspergillus oryzae* and *A. sojae* are important moulds in koji production for soy sauce and miso manufacture in the Orient and *A. oryzae* has GRAS status and is an important source of industrial enzymes (Campbell-Platt and Cook, 1989). Using DNA complementarity, Kurtzman *et al.* (1986) showed that these moulds were closely related to the morphologically similar but aflatoxin producing species *A. flavus* and *A. parasiticus.* Kusters-van Someren *et al.* (1991) examined the relationship between the black aspergilli using restriction fragment length polymorphisms. Fermentative strains such as *Aspergillus awamori* and *Aspergillus usami* showed a close relationship with the non-domesticated *Aspergillus niger.*

Over many centuries a continuing cycle of starter culture and fermented food production has resulted in the selection of particular strains of microorganisms. This selection is likely to have been on the basis of morphological and physiological features which favour the production of biomass, enzymes and fermentation products. The size and number of asexual propagules produced by microfungi are thought to influence colonising ability.

**Table 1.** Examples of fermented foods produced using plant, animal and microbial enzymes. (Source: Steinkraus (1983); Campbell-Platt (1987); Mowat (1989); Steinkraus (1989); Yokotsuka (1991*a, b*); Pauwels *et al.* (1992)).

| Source (type) | Substrate(s) Fermented | Typical products | Production Region |
|---|---|---|---|
| Malt (plant) | Barley *(Hordeum vulgare)*, Sorghum (*Sorghum bicolor*), Millet (*Pennisetum* spp., *Eleusine coracana*) | Alcoholic beverages: Beer, Whisky, Bantu beer, Bouza | Worldwide |
| Saliva (mammalian) | Maize (*Zea mays*), Cassava (*Manihot esculenta*), Rice (*Oryza sativa*) | Alcoholic beverages: Chicha, Manioc beer, Saké (pre koji) | South America, Japan |
| Plant roots (*Eminia, Rhynchosia,* & *Vigna* spp.) | Maize meal (*Z. mays*) or Millet (*Pennisetum* spp.) | Sour beverage: Munkoyo | Zaire, Zambia |
| Ragi (Microbial) | Rice (*O. sativa*), Cassava (*M. esculenta*), Millet (*Pennisetum* spp.) | Alcoholic foods and beverages: Tapé, Lao chao Brem, Rice beer, Arak | East and Southeast Asia, Indian sub-continent |
| Koji (microbial) | Rice (*O, sativa*), Wheat (*Triticum* spp.), other cereals, Soybeans (*Glycine max*) | Alcoholic beverage, sweet and savory condiments: Saké, Mirin, Miso Shoyu, Vinegar | East and Southeast Asia |

Larger spores provide more endogenous resources to support germ tube growth and rapid establishment of mycelium in situations where there may be competitors. Larger spores which germinate faster and establish mycelium more rapidly may be favoured in food fermentations. This appears to be the case in koji fermentations using *A. oryzae* and *A. sojae*. Wicklow (1984) has demonstrated that conidia of *A. oryzae* are larger and germinate faster than those of the non-domesticated *A. flavus*.

Janzen (1977) has suggested that microorganisms are likely to be under strong selection pressure to protect themselves and their food resources from other microbes. Some strains of *A. flavus* and *A. parasiticus* produce the potent aflatoxins but the domesticated strains *A. oryzae* and *A. sojae* used in koji appear to have lost the ability to produce these toxins. Aflatoxins probably offer no selective advantage in a koji environment where enzyme production, rapid spore germination and growth rate are likely to be more important than aflatoxins, production of which might impose a drain on metabolic resources (Wicklow, 1983). However, domestication has not eliminated the ability of some moulds to produce the mycotoxin cyclopiazonic acid. Cyclopiazonic acid is produced by both the domesticated *A. oryzae* and wild type *A. flavus* and by *Penicillium camembertii* the domesticate of *P. commune* used to make brie cheese (Pitt *et al.*, 1986; Engel and Teuber, 1990; Geisen *et al.*, 1990*a*). Presumably the metabolic cost of producing this compound is relatively small or it has a key role in metabolism.

Domesticated plants often survive poorly in competition with wild relatives yet this aspect has not been examined in detail with food fermentative fungi Fungal fermented foods pose a number of intriguing questions about the domestication process. For instance, what characteristics make *A. oryzae* and *A. sojae* better koji organisms than *A. flavus* and *A. parasiticus*?, why do the domesticated strains lose the ability to produce aflatoxins? and do the domesticated strains lose their competitive ability? With the development of serological and molecular techniques for aspergilli it should be possible to conduct experiments to answer some of these questions. For example, reciprocal transplant experiments have been used in plant ecology to examine the performance of plants in different habitats. A similar approach could be used to examine growth and toxin production by *A. flavus* and *A. parasiticus* in koji and growth and survival of *A. oryzae* and *A. sojae* in the soil or on plant litter.

## KOJI FERMENTATION

The majority of western fermentation technology utilises aseptic, liquid fermentations whereas solid substrate koji systems are widely used in the Orient for fermented food production. Koji is a good source of macerating enzymes. It can be dried and still retain high levels of enzyme activity. Solid substrate fermentations have a number of advantages over liquid fermentations. These include a lower capital investment in fermentation equipment, lower energy and water requirements, aseptic conditions are not usually required and there are fewer problems with the disposal of liquid waste. However, there are a number of problems associated with these fermentations and in particular, a) predicting and controlling heat and mass transfer, b) quantifying microbial biomass, and c) fermenter design and the scale-up of production.

Although manual koji production is still undertaken, the development of koji making machines have enabled better control of temperature, humidity and aeration in large-scale industrial production. The temperature and length of koji fermentation depends on the type of food product being produced but is usually between 30-45°C for up to 72 hours. More than 50 different types of enzyme have been identified in koji with the amylases, proteases and peptidases being the most important. Proteolytic and amylolytic enzymes have different temperature optima and consequently the temperature for rice koji will be different from soybean/cereal koji. In wheat/soybean koji for soy sauce production the temperature is allowed to rise initially to just under 40°C. Periodic mixing then reduces the temperature to between 30-35°C to obtain maximum proteolytic activity. With rice koji the temperature is allowed to increase steadily to 40-45°C to achieve a high amylolytic activity required for saké, vinegar or mirin production. In rice koji fermentation using *A. awamori* or *A. usami* the temperature is initially kept at 43°C and is reduced to around 30°C after 30 hours. It is not clear why a high initial temperature is required in koji using black aspergilli. It has been suggested that this is to control the growth of bacteria which reduce the quality of the final product (Yokotsuka, 1991*b*).

### Fermented Cereal Products

Table 2 gives examples of fermented cereal foods and beverages which involve aspergilli in their production. Further examples, together with details of the processes involved are given in Steinkraus (1983, 1989), Saono *et al.* (1986), Campbell-Platt (1987) and Yokotsuka (1991*b*). The use of aspergilli to produce staple foods from cereals appears to be rare and most products are either alcoholic beverages or condiments.

In the Andean sierra region of Ecuador, rice with a water content of 17-22% is often fermented for 3-10 days by spreading on cement or cane floors and covering with tarpaulin (Herzfeld, 1957). During fermentation, the rice may reach temperatures of 50-80°C and develops a brownish yellow colour. At 3000m. water boils at 87°C and processing the rice by fermentation prior to cooking probably reduces cooking time for this food which is eaten as a staple (Van Veen *et al.*, 1968). Van Veen *et al.* (1968) isolated a mixture of fungi and bacteria from the fermented product including *Aspergillus candidus, A. flavus, A. fumigatus, Absidia corymbifera, Rhizopus rhizopodiformis, Bacillus subtilis* and an actinomycete. Aflatoxins were not detected in any of the fermented rice samples and two isolated strains of *A. flavus* failed to produce detectable aflatoxins on a whole wheat medium.

Alcoholic beverages based on koji fermentations are important in parts of East and Southeast Asia. Glutinous and non-glutinous varieties of rice are the major substrates in these fermentations although sweet potato may sometimes be used. In China, a wider range of cereals are used to produce alcoholic beverages than in Japan. Many of the fermentations in China involve the use of mucoraceous moulds rather than aspergilli. The types of moulds used for these fermentations may relate to climatic and dietary differences between China and Japan or the preference in China for the use of cereal dough starter cultures which in China and Southeast Asia appear to favour the growth of a mucoraceous mould flora (Hesseltine *et al.*, 1988; Yokotsuka, 1991*b*).

The traditional Japanese alcoholic beverages are the rice wine known as saké and a distilled saké known as shochu. Saké brewing in Japan was first reported in the 8th century although a process similar to the current technology was not established until the 16th century (Yokotsuka, 1991*b*). Ankei (1986) has reported that a koji process involving moulds is used to produce a saké-like beverage in central Africa.

Yoshizawa and Ishikawa (1989) reported that some 74 million hectolitres of alcoholic beverages were consumed in Japan in 1985 of which saké, shochu and beer accounted for 18.3, 8.4 and 65.5% of the market respectively.

Yoshizawa and Ishikawa (1989) and Yokotsuka (1991*b*) have provided detailed reviews of the production of saké and related alcoholic beverages. Most fermented cereal products use koji prepared with *A. oryzae* although members of the *A. niger* group are used in southern Japan and China to produce koji for shochu spirit and rice vinegar production (Yokotsuka, 1991*b*).

Yeasts and lactic acid bacteria play an important role in the production of alcoholic beverages such as saké and shochu but aspergilli are the only microorganisms used in mirin production in Japan. The process uses an *A. oryzae*, non-glutinous rice koji which is mixed with steamed glutinous rice and ethanol or shochu and fermented for 40-60 days at 20°C. After filtering followed by 1-3 months clarification the mirin, which is used as a seasoning, has 12-14% ethanol, 35-48% sugar and a pH value of 5.3-5.5 (Saono *et al.*, 1986; Yokotsuka 1991*b*). Although *A. oryzae* is used to make koji for mirin production, *A. usamii* mut. *shiro-usamii* and *Rhizopus oligosporus* may also be suitable for this process (Oyashiki *et al.*, 1989).

## Fermented Legume Products

Aspergilli are important fungi in the production of a number of legume or legume/cereal fermented foods. Historically, these foods have their origins in China where soybean, wheat, sorghum and millet are grown in the North and rice and sweet potato grown in the South (Simoons, 1991). Soybean cultivation became widespread in Northern China by the 4th and 3rd centuries BC and was introduced into Japan sometime after the

**Table 2.** Cereal fermented foods which involve *Aspergillus* spp. in their production. (Source: Steinkraus (1983); Saono *et al.* (1986); Campbell-Platt (1987))

| Fermented food | Use[1] | Substrate(s) fermented | Aspergilli involved | Country |
|---|---|---|---|---|
| Arroz fermentado (Sierra rice) | St | Rice (*Oryza sativa)* | *Aspergillus candidus*, *A. flavus* | Ecuador |
| Takju | Wi | Rice (*O. sativa*), Barley (*Hordeum vulgare*), Maize (*Z. mays*), Wheat (*Triticum* sp.) or Millet (*Panicum* or *Setaria* spp.) | *A. oryzae*, *A. usami* | Korea |
| Saké | Wi | Rice (*O. sativa*) | *A. oryzae* | Japan |
| Huang-chiu | Wi | Glutinous Sorghum (*Sorghum* sp.) or Millet (*Panicum* or *Setaria* spp.) | *A. oryzae* or *A. niger* group | China |
| Kao-liang chiu liquor | Sp | Sorghum (*Sorghum* sp.), Maize (*Z. mays*), Sweet Potato (*Ipomoea batatas*) | *A. niger* group | China |
| Shochu | Sp | Rice (*O. sativa*), Sweet potato (*I. batatas*), Barley (*Hordeum* sp.), Buckwheat (*Fagopyrum esculentum*) | *A. oryzae*, *A. awamori*, *A. kawachii*, *A. usami* | Japan |
| Bupja | Wi | Glutinous or non-glutinous Rice (*O. sativa*) | *A. oryzae* | Korea |
| Mirin | Se | Glutinous and non-glutinous Rice (*O. sativa*) | *A. oryzae*, *A. sojae* | Japan |
| Vinegar | Co | Glutinous and non-glutinous Rice (*O. sativa*), Wheat, (*Triticum* sp.), Barley (*Hordeum* sp.), Sweet potato (*I.batatas*) | *A. oryzae*, *A. niger* group | China Japan Thailand |

[1]Co = Condiment; Se = Seasoning; Sp = Distilled spirit; St = Staple; Wi = Wine

first century AD (Simoons, 1991). Soaking, sprouting, grinding, heating and fermentation were developed in ancient China to overcome the bitterness, trypsin inhibitors and flatulent producing oligosaccharides present in unprocessed soybeans (Simoons, 1991). Soybean fermentation in China is though to date from late in the Chou dynasty (1200BC-221BC) or early in the Han dynasty (206BC-220AD) (Anderson, 1988). Soy sauce and miso fermentations appear to have been introduced into Japan before the 7th century but large scale production of soy sauce only began in the 17th century. Fukushima (1989) and Yokotsuka (1991*a*) have provided a detailed account of the origins of soy sauce and the development of the fermentation industry in Japan. Products produced from soybean or other legumes can broadly be divided into three types based on the final product being a liquid, a paste or solid beans. These foods are produced in many countries in East and Southeast Asia and a selected list of local names are presented in Table 3. These foods may be sweet or salty and some contain spices. Their production, microbiology and chemistry have been described by Steinkraus (1983), Saono *et al.* (1986), Campbell-Platt (1987) and Yokotsuka (1991*a*).

The principal factors which influence the nature of the final product are a) the proportion of different substrates used, b) the koji fermentation conditions, c) length of the moromi or brine stage and d) the processing treatments used. For the production of most types of soy sauce the koji is usually prepared with a mixture of soybean and cereal substrate. In some of the paste-like products, typified by miso fermentation, a cereal koji is only mixed with the soybeans at the brine stage. Under these conditions, the hydrolysis of soybean proteins occurs in the brine at a reduced water activity and the paste like consistency is retained because enzymic hydrolysis is less complete than in soy sauce fermentation. With hama-natto, the final product is a soft bean which retains its structure because enzymic activity is arrested by the sun-drying process.

As with the production of alcoholic beverages, mucoraceous moulds are also used to produced fermented legume products similar to those produced by aspergilli. Tauco is a sweet and savory paste-like soybean condiment produced in Indonesia. Some producers use *A. oryzae* for the koji whereas others use *Rhizopus* spp. originating from tempe, a

**Table 3.** Fermented legume[1] foods which involve *Aspergillus* spp. in their production. The foods are arranged according to the physical nature of the product. (Source: Steinkraus (1983); Saono *et al.* (1986); Campbell-Platt (1987); Yokotsuka (1991*b*)

| Whole Beans | Paste | Liquid |
|---|---|---|
| Haka-natto<br>Itohhiki-natto (Japan) | Miso (Japan, North America, Brazil) | Shoyu (Japan, North America,. Brazil, Europe) |
| Tou-shih (China) | Wei-chiang<br>Doubanjiang<br>Tou-chiang (China) | Inyu<br>Chijhi<br>Chiang-yu (China) |
| Taotjo (Indonesia) | Tauco (Indonesia) | Ketjap<br>Kecap (Indonesia) |
| Tao-si (Philippines) | Hishio (Japan)<br>Tao-chiew (Thailand)<br>Doenjang (Korea)<br>Tao-cheo (Malaysia)<br>Tuong (Vietnam) | Toyo (Philippines)<br>Ce-ieu (Thailand)<br>Kanjang (Korea)<br>Kicap (Malaysia) |

[1]Usually soybean (*Glycine max*) but broad bean (*Vicia faba*) sometimes used in China. Chickpea (*Cicer arietinum*) and winged bean (*Psophocarpus tetragonolobus*) have also been used for miso production.

moulded soybean cake. This fermentation would provide a good opportunity to compare the fermentative abilities of aspergilli and mucoraceous moulds, particularly as these organisms show differences in protein hydrolysis (Ebine, 1968; Oyashiki *et al.*, 1989).

## GENETIC MANIPULATION OF KOJI MOULDS

Although large-scale, efficient solid-substrate koji processes have been developed, less attention has been paid to improving the physiological properties of the *Aspergillus* strains. Aspergilli used in fermented foods often show differences in their physiological properties and these differences relate in part to the nature of the substrates used for fermentation. Takahashi and Yamamoto (1913) found differences in the amylolytic and proteolytic activity of yellow-green aspergilli isolated from koji used to produce saké, soy sauce and tamari. Amylolytic activity was highest in the sake´ koji strains and proteolytic activity highest in those strains isolated from soy sauce and tamari koji. At the molecular level, *A. oryzae* which is used to produce saké koji has three alpha amylase genes whereas *A. sojae* used predominantly for soy sauce koji has only one (Hara *et al.*, 1992).

Genetic manipulation is being used to select and 'improve' koji moulds and this has involved interspecific and intraspecific protoplast fusion and more recently DNA transformation (Table 4). Because of the lack of a sexual cycle in koji moulds the application of protoplast fusion has been an important step in their genetic manipulation. Use of mutation alone has met with limited success since yields of enzymes are generally poor and mutated strains often show reduced growth and conidiation. Protoplast fusion has been used to develop high protease and glutaminase producing strains of *A. sojae*, high alpha amylase and glucoamylase producing strains of *A. oryzae* and strains of *A. awamori* with increased amylase activity and citric acid production. Many of the heterozygous diploids derived from protoplast fusions have characteristics which are intermediate between the parent strains and additional treatments such as mutagenesis or haploidisation are often used to improve enzyme yields. Interspecific and intrageneric protoplast fusions have also been undertaken in an attempt to confer novel characteristics on selected koji moulds. This has included citric acid production by *A. oryzae* and red pigment production by *A. oryzae* following fusion with *Monascus anka* (Hara *et al.*, 1992).

Recombinant DNA technology has yet to make a major impact on *Aspergillus* koji fermentations although some 10 genes have been cloned and sequenced from the koji moulds *A. oryzae* and *A. awamori* (Hara *et al.*, 1992). The technology should eventually result in significant strain improvements in koji production and major benefits for the fermentation industry could be derived by increasing the copy number of genes encoding for the production of enzymes such as alpha amylase, glucoamylase, alkaline protease and glutaminase. Tada *et al.* (1989) reported the transformation of *A. oryzae* using cloned genes for alpha amylase with the transformant having alpha amylase activity 2-5 times that of the parent strains. A similar approach was adopted by Hata *et al.* (1991) who observed that a transformant with a cloned glucoamylase gene produced enzyme yields up to eight times that of the parent strain. DNA transformation has also been used to improve production of alkaline proteases which are particularly important in soy sauce production (Cheedvadhanarak *et al.*, 1991).

### Raw starch hydrolysis

Another objective for improving koji might be to enhance the raw starch affinity of *A. oryzae* glucoamylase. Starch containing substrates are usually cooked before they are suitable for hydrolysis by amylolytic enzymes. Cooking starch may require considerable amounts of energy and in the case of ethanol production from sweet potato some 30-40% of the energy required comes from cooking the starch prior to enzymatic

**Table 4.** Examples where mutation and genetic manipulation have been used to improve *Aspergillus* spp. used in fermented foods.

| Species | Objective | Approach used | Reference |
|---|---|---|---|
| *Aspergillus oryzae* | Increase proteases for fish hydrolysis | Mutation using $^{60}Co$ irradiation | Anglo and Orillo (1977) |
| *A. oryzae* | Construct plasmid vector system and markers | Protoplast fusion and transformation | Iimura *et al.*, (1987) |
| *A. sojae* | Increase glutaminase and protease production | Protoplast fusion | Ushijima and Nakadai (1987) |
| *A. oryzae* | Increase alpha amylase production | DNA transformation | Tada *et al.* (1989) |
| *A. sojae* and *A. oryzae* | Modify morphology and increase alkaline proteases | Protoplast fusion, hybridisation and haploidisation | Ushijima *et al.* (1990) |
| *A. awamori* var. *kawachi* and *A. usamii* mut. *shiro-usamii* | Increase amylase and citric acid production | Protoplast fusion, hybridisation and haploidisation | Ogawa *et al.* (1990) |
| *A. oryzae* | Increase alkaline protease | DNA transformation | Cheevadhanarak *et al.* (1991) |
| *A. oryzae* | Increase glucoamylase production | DNA transformation | Hata *et al.* (1991) |

hydrolysis (Matsuoka *et al.*, 1982). The ability to hydrolyse raw starch at acid pH has been reported from relatively few organisms but includes moulds of the genera *Rhizopus* (Yamazaki and Ueda, 1951) and *Aspergillus* (Ueda *et al.*, 1984) and the yeast *Saccharomycopsis fibuligera* (Ueda and Saha, 1983). If koji fermentation processes could be developed which utilise raw starch then there could be a significant saving in energy costs for saké, shochu and mirin production. Strains of *A. oryzae* have a lower affinity for raw starch than the black aspergilli *A. awamori* and *A. awamori* var. *kawachi* (Mitsue *et al.*, 1979). Genetic manipulation might offer a route for the production of *A. oryzae* glucoamylase with a high affinity for raw starch.

## ASPERGILLI AND FERMENTED MEAT AND FISH

### Salami and Ham

Many fermented meat products are characterised by the presence of a surface mould flora which, in some cases, may entirely cover the surface of the meat. As with many cheese fermentations, the processes for sausage and ham production were developed locally as 'cottage industries' which has resulted in a diverse range of products. Although there are similarities between some of these meat fermentations, local preferences influence the nature of the final product. Whereas a surface mould growth is encouraged in parts of Eastern Europe and Mediterranean countries in other regions it may be avoided or removed (Leistner, 1986).

Leistner and Ayres (1968) conducted an extensive survey of the mould flora of fermented meats including salami type sausages and country-cured hams. Although *Penicillium* spp. formed the bulk of the moulds isolated, *Aspergillus* and *Eurotium* spp. also formed a significant proportion of the mould flora, particularly in the case of country-cured hams. With mould fermented sausages the mycoflora tends to be dominated by *Penicillium* spp. but country-cured hams differ from fermented sausages in being aged for 1-2 years and attain a lower water activity. *Aspergillus* and *Eurotium* spp. develop more extensively on country-cured hams and the presence of particular species may provide some indication of the degree of ripening (Monte *et al.*, 1986; Huerta *et al.*, 1987*a*). In sausage production the presence of *Aspergillus* and *Eurotium* spp. is regarded as undesirable since *Penicillium* spp. are the most important moulds during ripening.

Fermented meats may undergo a number of chemical changes during fermentation due to the presence of meat enzymes and the production of enzymes and organic acids by bacteria, yeasts and moulds. During the initial bacterial fermentation the pH value decreases rapidly from an initial value of between pH 5.9-6.2 to around 5.2-5.4 after 5-10 days following casing. In those meat products which involve mould ripening, the pH often rises again due to the utilisation of lactate and acetate and the production of ammonia from protein hydrolysis. Grazia *et al.* (1986) examined selected physical and chemical characteristics of salami inoculated with nine *Penicillium* strains and one *Aspergillus* strain isolated from sausage factories. After 30 days fermentation with selected moulds the pH values were between 6.02 and 6.50 whereas control salami, which had been treated with sorbate to prevent mould growth, had a pH value of 5.34. Ammonia was detected in all samples but in only four of the mould cultures was it higher than the controls treated with sorbate.

Although the presence of moulds contributes to flavour production, colour and chemical changes in some sausage fermentations the significance of moulds in the curing of raw hams is less clear. Huerta *et al.* (1987*b*) screened enzyme activities of 33 *Aspergillus* and 41 *Penicillium* strains isolated from Spanish dry-cured hams. Proteolytic activity was shown by 31% of *Aspergillus* and 20% of *Penicillium* strains. All *Aspergillus* strains, and 81% *Penicillium* strains showed lipolytic activity. Toldrá and Etherington (1988) demonstrated that pork muscle proteinases (cathepsins B, D, H and L) were still active after eight months of curing for serrano dry-cured hams. Plate counts of microorganisms suggested that numbers were insufficient to account for the enzyme activity in the interior of the ham. Since most fungal growth on hams occurs on the surface, the contribution of *Aspergillus* or *Eurotium* to the flavour of ham is likely to be minimal.

Many more traditional fermentations which involve fungal ripening rely on passive inoculation by the indigenous "house flora" associated with the buildings, rooms and equipment used for meat fermentation and storage. Examples of passive inoculation by a "house flora" of moulds occurs in some cheese fermentations involving *Penicillium camembertii* and *P. roquefortii* and koji fermentations using *Aspergillus* spp. Many producers still rely on the "house flora" to provide a natural inoculum for mould coating and ripening of salami and raw ham. Since many of the fungi isolated from ripened hams and salami are known mycotoxin producers there has been increasing concern about the continued use of undefined strains of moulds for meat fermentation (Leistner *et al.*, 1989). Some producers discourage mould growth by dipping the meat in potassium sorbate or pimaricin (Holley, 1986).

## Fermented Fish and Shellfish

In many tropical countries in East and Southeast Asia, fish are frequently sun-dried with or without salting. Like meat, salted and dried fish products will support the growth of a wide range of filamentous fungi and particularly *Aspergillus, Eurotium, Polypaecilum* and *Penicillium* spp. (Hitokoto *et al.*, 1976; Ichinoe *et al.*, 1977; Phillips and Wallbridge, 1977; Tanimura *et al.*, 1979; Wheeler *et al.*, 1986). Mycelial growth may be visible on the surface of the fish; species such as *Polypaecilum pise* may dominate the growth in some cases. Japan is one of the largest producers of dried and salted fish products. Although the presence of moulds on fish is usually considered undesirable, fungi are used intentionally in katsuobushi production to assist drying and flavour production. Katsuobushi may be produced using a number of fish and shellfish species including Bonito (*Katsuwonus pelamis*), Mackerel (*Scomber* sp.), Frigate mackerel *Auxis tapeinosoma*) and the mollusc Abalone (*Haliotis* sp.). Doi *et al.* (1991) found that boiled and dried anchovy could also be processed using katsuobushi moulds and Kinosita *et al.* (1968) reported that at katsuobushi plants, washed bonito intestines are fermented by fungi and mixed with salt to form an appetizer called shuto or shiokara. Bonito is the most popular substrate for katsuobushi. The product is a bone-hard, dried fish which is shaved with special knives to produce a seasoning used in Japanese cooking. Kastsuobushi is produced on the warm southern Pacific coast of Japan with some 150 producers and a total production in 1990 of 33,000 tonnes (Doi., personal communication). Katsuobushi is also produced in Taiwan, Palau, Papua New Guinea, Irian Jaya and Maldives (Tanimura *et al.*, 1979) although it is not known if these products involve an intentional moulding stage.

Figure 1 shows the production of katsuobushi using Bonito. The method involves both smoking, drying and moulding stages. After boiling and cooling, the bones are removed to avoid twisting during drying and difficulty in shaving the dried product. Bonito flesh is firm and thick and it is difficult to remove all the water efficiently since drying too rapidly cracks the flesh. If the process is too slow, bacteria may grow and spoil the fish. The drying and smoking involves a series of heating and cooling stages with the objective of maintaining the surface dryness whilst removing the internal water. The initial heating raises the internal temperature of the bushi to 70°C. Hardwood and sawdust are burnt to provide the heat and the smoke contributes phenolic flavour components, inhibits the growth of spoilage organisms, provides colour and reduces oxidation of fats. The drying is a critical stage since if it is too little, the product will spoil rapidly. If it is too excessive then the product will be difficult to mould. The moulding process is conducted in a wooden container for 40-60 days at 25-28°C and at a relative humidity of 80-85%. The bushi is either inoculated directly with a mould culture or it becomes inoculated by the "house flora" present in the box and the factory environment. The surface mould growth is removed every 10-14 days and the moulding process repeated up to four times. The periodic removal of the surface mould growth appears to assist the drying process by removing water from within the hard bushi tissue. The water content of the smoked and dried bushi is about 35% but this decreases to a level of 15-16% after moulding. A detailed review of the production of katsuobushi in Japan has been provided by Ota (1983*a, b, c*).

Grey Ormers or Abalone (*Haliotis* sp.) are also dehydrated and moulded. Pickled Ormers are boiled and dehydrated by sun drying or by heat from a fire. The partially dried product is then allowed to become covered in mould growth which is removed and the product dried in the sun.

Tanikawa (1971) listed *Aspergillus glaucus*, *Penicillium glaucum* and *Aspergillus melleus* as desirable species of moulds on dried katsuobushi. Doi *et al.* (1989*a, b*) isolated four strains of *Aspergillus glaucus*, five strains of *Aspergillus repens* and two strains of *Aspergillus candidus* from katsuobushi. Doi *et al.* (1992) reported that in practice a mixture of *Aspergillus* strains are used for the moulding process at katsuobushi factories in Japan. In addition to lowering the moisture content, the moulds are thought to reduce oxidation, hydrolyse the fats and proteins, modify the phenolic smoke components and contribute colour to the final product (Ota, 1983*a, b, c*). A decrease in the fat content of bushi is considered to be one of the most important consequences of the moulding stage in katsuobushi production (Kanazawa and Kakimoto, 1958). Following inoculation with moulds, the acid value of the katsuobushi tissue increases and the peroxide value decreases. These changes are more pronounced in the outer 5-10mm of the tissue. There is a corresponding increase in free fatty acids, a decrease in polar lipid and little change in neutral lipids (Ota, 1983*b*).

The smoky aroma and flavour of katsuobushi is known to be derived from the smoke generated by the hardwood fire and the pungent smoke components are moderated by the moulding process. Doi *et al.* (1989*a, b*) compared the levels of volatile phenols before and after moulding. They found that *Eurotium repens* (= *Aspergillus repens)* could convert a number of phenolic components found in katsuobushi to *o*-methylated derivatives which could result in the milder, less pungent, smoky flavour characteristic of katsuobushi. Similarly, Doi *et al.* (1992) found that *Eurotium repens* reduced acetophenone stereo-selectively to the R(+) isomer of alpha phenylethyl alcohol. This isomer was reported to have a milder aroma than the S(-) isomer and yields were highest at 25°C a temperature close to that used in the commercial moulding process. Inosinic acid is also thought to contribute to the flavour of katsuobushi and constitutes 0.3-0.9% of the dried product (Ota 1983*b*).

Many fungi have the ability to modify lipids (Koritala *et al.*, 1987) and aspergilli have been used to process fish waste. Jeffreys and Krell (1965) and Jeffreys (1970) developed a process for producing powdered, odourless fish protein concentrates by enzymically digesting cooked, comminuted fish using an *Aspergillus* wheat bran koji. Yeast (*Saccharomyces cerevisiae*) was used to ferment the proteolytically liquified cooked fish. *A. oryzae* is able to degrade lipids in sardine meal koji (Kunimoto *et al.*, 1989) and *A. terreus* has been used to produce a fish feed by treatment of fish waste (Hossain *et al.*, 1989). There would appear to be considerable potential for combining enzyme preparations such as *Aspergillus* koji with waste meat or fish protein.

## MYCOTOXINS IN FERMENTED FOODS

Considerable attention has focused on whether mycotoxins, and in particular aflatoxins, are present in fermented foods produced using aspergilli. In many cases koji and moromi fermentations appear to have little effect on any aflatoxin present in the substrate prior to fermentation or produced following contamination by *A. flavus* or *A. parasiticus*. Maing *et al.*, (1973) examined the production of aflatoxin in soy sauce koji prepared using *A. parasiticus* (NRRL 2999) and *A. oryzae* (NRRL 1988). No aflatoxin was detected in the *A. oryzae* koji but large quantities of aflatoxin were produced in a mixed culture koji with *A. parasiticus* and *A. oryzae*. There was little reduction in aflatoxin produced by *A. parasiticus* during a subsequent 6 week moromi brine stage. However, Sardjono *et al.* (1992) found that in glucose ammonium nitrate broth, production of aflatoxin by *A. flavus* was significantly reduced in a mixed culture

SLICE AND BOIL BONITO FLESH FOR 45-90 MINUTES
(Water content 70%)

REMOVE BONES, SKIN AND FAT

SMOKE/DRY FOR 1 HOUR OVER FIRE MADE OF HARDWOODS
(Air temperature 110-114°C)

REPAIR THE DAMAGED FLESH
(mend cracks with additional fish tissue)

REPEAT SMOKING/DRYING UP TO 10 TIMES
DRY EVERY 2 DAYS AFTER 4-5 CYCLES
EVERY 3 DAYS AFTER 7-8 CYCLES
(Water content reduced to 35%)

PLACE ROUGH BUSHI IN WOODEN CONTAINER FOR 3 DAYS

PLANE TO SHAPE THE FISH AND REMOVE FAT AND TAR

INOCULATE WITH MOULDS IN A BOX AT 25-28°C, RH 80-85%

INCUBATE FOR 10-14 DAYS THEN BRUSH OFF MOULD

REPEAT MOULDING PROCESS UP TO 4 TIMES

KATSUOBUSHI
(Water content 15-16%, Water activity $<0.64$)

**Figure 1.** Production of katsuobushi using Bonito (adapted from Ota, 1983*a, b, c*).

with *A. oryzae*. Manabe and Matsuura (1972) added aflatoxins at the beginning of miso fermentation and found that after one month of fermentation there was no degradation of aflatoxin $B_2$ and $G_2$ and only a 50% reduction in aflatoxins $B_1$ and $G_1$. Park *et al.* (1988) found that aflatoxin production in Korean soy paste and soy sauce was affected by the fermentation and ripening conditions with degradation of aflatoxins being more rapid during ripening in water rather than brine.

A number of surveys have been made to screen koji moulds and *Aspergillus* fermented foods for the presence of aflatoxins. Kinosita *et al.* (1968) examined 24 samples of commercial or home-made fermented foods and fermentation starters from Japan including miso, soy sauce and katsuobushi. From 24 samples, 37 fungal strains were isolated and of these 21 were found to produce toxic culture filtrates and extracts. Many of the strains produced kojic and 3-nitropropionic acids but aflatoxins were not detected.

Aflatoxins have been detected in some miso and soy sauce products but the majority of these were produced in homes or small-scale factories (Shank *et al.*, 1972; Purchio, 1976; Kim *et al.*, 1977; Tank and Long, 1977; Sardjono *et al.*, 1992). In some cases, defined starter cultures are not used by small-scale producers and traditional methods such as sun-drying the koji or moromi may provide additional opportunities for contamination by toxigenic strains of *A. flavus* and *A. parasiticus*.

Industrially produced fermented soybean products and defined starter cultures used for koji have been found to be free of detectable aflatoxins (Hesseltine *et al.*, 1966; Matsuura, 1970). Cyclopiazonic acid, aspergillic acid, kojic acid and 3-nitopropionic acid are produced by some koji strains but aspergillic acid is not usually formed under the conditions of koji fermentation and cyclopiazonic acid and kojic acid are degraded by yeasts during the moromi a brine fermentation (Shinshi *et al.*, 1984).

Although the conditions used for ripening country-cured ham and salami can render these foods susceptible to the growth of toxigenic fungi, the incidence of *A. flavus* and *A. parasiticus* on these foods is low (Leistner and Ayres, 1968; Sutic *et al.*, 1972). It has been demonstrated, however, that aflatoxins can be produced when fermented meats are inoculated with *A. flavus* or *A. parasiticus* prior to ripening (Alvarez-Barrera *et al.*, 1983). Aflatoxins have been detected in hams after 6-9 months but not in hams after 12 months (Bullerman et *al.*, 1969*a, b*). There is less information on fermented fish products although aflatoxin producing strains have been isolated from katsuobushi (Tanimura *et al.*, 1979).

*A. ochraceus* and *A. versicolor* are sometimes isolated from fermented meats. Halls and Ayres (1973) demonstrated that *A. versicolor* was capable of producing sterigmatocystin in cured hams and Escher *et al.* (1973) found that *A. ochraceus* isolated from country-cured hams could produce ochratoxin A and B in culture as well as in hams.

As with all food fermentations it is desirable to use starter cultures which have been selected for their fermentation characteristics and the absence of toxicity or pathogenicity. In the case of koji moulds, commercial starter cultures are available but less is known about the safety of *Aspergillus* and *Eurotium* species used for katsuobushi production. Commercial starter cultures of *Penicillium* spp. are available for meat and cheese fermentations (Leistner, 1986; Lücke and Hechelmann, 1987; Leistner *et al.*, 1989) and genetic manipulation is being investigated as a means of strain improvement for these fungi (Geisen *et al.*, 1988, 1990*b*). Since the mould growth on country-cured ham is of minor importance to the fermentation it would be desirable to prevent growth by chemical treatments (Kemp *et al.*, 1981, 1983; Holley, 1986) or, where a moulded appearance is required, inoculate with a defined starter culture such as the *Penicillium* strains used for salami ripening (Lücke and Hechelmann, 1987).

## FUTURE PROSPECTS

Fermented foods involving aspergilli are likely to remain an important part of our food supply with products ranging from meaty peptide sauces and pastes to dried fish seasoning, sweet condiments and alcoholic beverages. Although some progress has been made in strain development, molecular genetics will have an increasingly important role in the development of safe starter cultures and more efficient koji fermentations through the enhancement of enzyme production and fermentation products. Strain improvement of *A. oryzae* for koji fermentation may come indirectly as a result of its GRAS status and value as a source of specific enzymes and potential host for the production of heterologous proteins.

Although soy sauce and saké production are now large, efficiently run processes, many related products are still produced by small-scale industries in East and Southeast Asia. These processes are often labour intensive and, in some cases, appear to have changed little over many centuries. Although large industries are using well defined, non-toxigenic starter cultures to produce koji, some small-scale manufacturers or home producers still rely on the naturally contaminating "house flora" for inoculum. Well defined, non-toxigenic starter cultures should be used in preference to a "house flora" since with natural inoculation there will always be the possible risk of contamination by aflatoxin producing strains of *A. flavus* and *A. parasiticus.*

Strain improvement will be the key to further development of fermented foods which involve aspergilli. However, the diverse range of strains isolated from fermented foods represents a unique genetic resource which should be maintained for the future. These strains can provide much valuable information about the process of microbial domestication and loss of toxicity as well as providing an important resource for food technology and biotechnology.

## REFERENCES


Alvarez-Barrea, V., Pearson, A.M., Price, J.F. , Gray, J.I. and Aust, S.D. (1983) Some factors influencing aflatoxin production in fermented sausages. J. Food Sci. 47, 1773-1775.

Anderson, E.N., Jr. (1988) "The Food of China", Yale University Press, New Haven.

Anglo, P.G. and Orillo, C.A. (1977) Production of proteolytic enzyme. II. Study of some factors influencing the activity of the enzyme produced by irradiated strains of *Aspergillus oryzae* (Ahlburg) Colin, Philippine J. Sci. 106, 1-10.

Ankei, T. (1986) Discovery of Saké in Central Africa: Mould fermented liquor of the Songola Journ d'Agric Trad. et de Bota. Appl. 33, 29-47.

Bullerman, L.B., Hartman, P.A. and Ayres, J.C. (1969*a*) Aflatoxin production in meats. I Stored meats. Appl. Microbiol. 18, 714-717.

Bullerman, L.B., Hartman, P.A. and Ayres, J.C. (1969*b*) Aflatoxin production in meats. II. Aged dry salamis and aged country cured hams. Appl. Microbiol. 18, 718-722.

Campbell-Platt, G. (1987) "Fermented Foods of the World: A Dictionary and Guide", Butterworths, London.

Campbell-Platt, G. and Cook, P.E. (1989) Fungi in the production of foods and food ingredients. J. Appl. Bact. Symp. Suppl. 117S-131S.

Chang, L-T. (1993) Maltose was made in old China. Starch 44, 117.

Cheevadhanarak, S., Renno, D.V., Saunders, G. and Holt, G. (1991) Cloning and selective over expression of an alkaline protease-encoding gene from *Aspergillus oryzae*. Gene 108, 151-155.

Doi, M., Matsui, M., Shuto, Y. and Kinoshita, Y. (1989*a*) O-Methylation of phenols by *Aspergillus repens.* Agric. Biol. Chem. 53, 3031-3032.

Doi, M., Ninomiya, M., Matsui, M., Shuto, Y. and Kinoshita, Y. (1989*b*) Degradation and O-methylation of phenols among volatile flavour components of dried bonito (katsuobushi) by *Aspergillus* species. Agric. Biol. Chem. 53, 1051-1055.

Doi, M., Yamauchi, H., Matsui, M., Shuto, Y. and Kinoshita, Y. (1991) Microbial Degradation of BHA in *Niboshi* (Boiled and Dried Anchovy) by *Aspergillus* species. Agric. Biol. Chem. 55, 1095-1098.

Doi. M., Matsui, M., Kanayama, T., Shuto, Y. and Kinoshita, Y. (1992) Asymmetric reduction of acetophenone by *Aspergillus* species and their possible contribution to katsuobushi flavour. Biosci. Biotech. Biochem. 56, 958-960.

Ebine, H. (1968) Application of *Rhizopus* for shoyu manufacturing. Seasoning Sci. 15, 10-18.

Engel, G. and Teuber, M. (1990) Toxic metabolites from fungal cheese starter cultures (*Penicillium camemberti* and *Penicillium roqueforti*), in "Mycotoxins in dairy products" (Van Egmond, H.P., Ed.), pp. 163-192. Elsevier Science Publishers, London.

Escher, F.E., Koehler, P.E. and Ayres, J.C. (1973) Production of ochratoxins A and B on country cured ham. Appl. Microbiol. 26, 27-30.

Fukushima, D. (1989) Industrialisation of fermented soy sauce production centering around Japanese shoyu, in "Industrialisation of indigenous fermented foods" (Steinkraus, K.H., Ed.), pp. 1-88. Marcel Dekker, New York.

Geisen, R., Glenn, E. and Leistner, L. (1988) Development of a transformation system for *Penicillium nalgiovense*. Proc. Jap. Assoc. Mycotox., Suppl. 1, 47-48.
Geisen, R., Glenn, E. and Leistner, L. (1990*a*) Two *Penicillium camembertii* mutants affected in the production of cyclopiazonic acid. Appl. Environ. Microbiol. 56, 3587-3590.
Geisen, R., Standner, L. and Leistner, L. (1990*b*) New mould starter cultures by genetic modification. Food Biotechnol. 4, 497-503.
Grazia, L., Romano, P., Bagni, A., Roggiani, D. and Guglielmi, G. (1986) The role of moulds in the ripening process of salami. Food Microbiol. 3, 19-25.
Halls, N.A. and Ayres, J.C. (1973) Potential production of sterigmatocystin on country-cured ham. Appl. Microbiol. 26, 636-637.
Hara, S., Kitamoto, K. and Gomi, K. (1992) New developments in fermented beverages and foods with *Aspergillus*, in "*Aspergillus*: Biology and Industrial Applications" (Bennett, J.W. and Klich, M.A., Ed.), pp. 133-153. Butterworth-Heinemann, London.
Hata, Y., Tsuchiya, K., Kitamoto, K., Gomi, K., Kumagai, C., Tamura, G. and Hara, S. (1991) Nucleotide sequence and expression of the glucoamylase-encoding gene (glaA). Gene 108, 145-150.
Herzfield, H.C. (1957) Rice fermentation in Ecuador. Econ. Bot. 11, 267-270.
Hesseltine, C.W., Shotwell, O.L., Ellis, J.J. and Stubblefield, R.P. (1966) Aflatoxin formation by *Aspergillus flavus*. Bacteriol. Rev. 30, 795-805.
Hesseltine, C.W., Featherston, C.L., Lombard, G.L. and Dowell, Jr. V.R. (1985) Anaerobic growth of moulds isolated from fermentation starters used for foods in Asian countries. Mycologia 77, 390-400.
Hesseltine, C.W., Rogers, R. and Winarno, F.G. (1988) Microbiological studies on amylolytic oriental starters. Mycopath. 101, 141-155.
Hitokoto, H., Morozumi, S., Wauke, T., Sakai, S., Zen-Yoji, H. and Benoki, M. (1976) Studies on fungal contamination of foodstuffs in Japan: Fungal flora on dried small sardines marketing in Tokyo. Ann. Rep. Tokyo Metr. Res. Lab. P.H. 27, 36-40.
Holley, R.A. (1986) Effect of sorbate and pimaricin on surface mould and ripening of Italian dry salami. Alia.-Wiss. u.-Technol. 19, 59-65.
Hossain, M.A., Furuichi, M. and Yone, Y. (1989) Growth and feed efficiency in red sea bream fed on (fish) waste lipid treated with mould (*Aspergillus terreus*). Bull. Jp. Soc. Sci. Fish. 55, 657-660.
Huerta, T., Sanchis, V., Hernandez-Haba, J. and Hernandez, E. (1987*a*) Mycoflora of dry salted Spanish ham. Microbiologie - Aliments - Nutrition 5, 247-252.
Huerta, T., Sanchis, V., Hernandez-Haba, J. and Hernandez, E. (1987*b*) Enzymic activities and antimicrobial effects of *Aspergillus* and *Penicillium* strains isolated from Spanish dry cured hams. Microbiologie -Aliments - Nutrition 5, 289-294.
Ichinoe, M., Suzuki, M. and Kurata, H. (1977) Microflora of commercial slice dried fishes including bonito. Bull. Nat. Inst. Hyg. Sci. 95, 96-99.
Iimura, Y., Gomi, K., Uzu, H. and Hara, S. (1987) Transformation of *Aspergillus oryzae* through plasmid mediated complementation of the methionine-auxotrophic mutation. Agric. Biol. Chem. 51, 323-328.
Janzen, D.H. (1977) Why fruits rot, seeds mould, and meat spoils. Am. Nat. 111, 691-713.
Jeffreys, G.A. (1970) US Patent 3,547,652.
Jeffreys, G.A. and Krell, A.J. (1965) US Patent 3, 170, 794.
Kanazawa, A. and Kakimoto, D. (1958) Change of the chemical constituent of "Katsuwobushi" by *Aspergillus ruber* and *Aspergillus repens*. Mem. Fish. Kagoshima Univ. 6, 144-147.
Kemp, J.D. Langlois, B.E. and Fox, J.D. (1981) Effect of vacuum packaging and potassium sorbate on yield yeast and mould growth, and quality of dry-cured hams. J. Food Sci. 46, 1015-1017, 1024.
Kemp, J.D., Langlois, B.E. and Fox, J.D. (1983) Effect of potassium sorbate vacuum packaging on the quality and microflora of dry-cured, intact and boneless hams. J. Food Sci. 48, 1709-1714.
Kim, Y.H., Hwangbo, J.S. and Lee, S.R. (1977) Detection of aflatoxins in some Korean foodstuffs, Korean J. Food Sci. Technol. 9, 73-80.
Kinosita, R., Ishiko, T., Sugiyama, S., Seto, T., Igarasi, S. and Goetz, I.E. (1968) Mycotoxins in fermented food. Cancer Res. 28, 2296-2311.
Koritala, S., Hesseltine, C.W., Pryde, E.H. and Mounts, T.L. (1987) Biochemical modification of fats by microorganism. J. Am. Oil. Chem. Soc. 64, 509-513.
Kunimoto, M., Hoshino, T. and Nakano, M. (1989) Decomposition of lipid and occurrence of lipase in mould-inoculated sardine meal koji. Bull. Jap. Soc. Sci. Fish 55, 1097-1102.
Kurtzman, C.P., Smiley, M.J., Robnett, C.J. and Wicklow, D.T. (1986) DNA relatedness among wild and domesticated species in the *Aspergillus flavus* group. Mycologia 78, 955-959.
Kusters-van Someren, M.A., Samson, R.A. and Visser, J. (1991) The use of RFLP analysis in classification of the black aspergilli: reinterpretation of the *Aspergillus niger* aggregate. Curr. Genet. 19, 21-26.

Leistner, L. (1986) Mould-ripened foods. Fleischwirtsch. 66, 1385-1388.
Leistner, L. and Ayres, J.C. (1968) Moulds and meat. Fleischwirtsch. 1, 62-65.
Leistner, L., Geisen, R. and Fink-Gremmels, J. (1989) Mould-fermented foods of Europe: Hazards and developments, in "Mycotoxins and Phycotoxins 88" (Natori, S., Hashimoto, K. and Ueno, Y., Ed.) pp. 145-154. Elsevier Science Publishers, Amsterdam.
Lücke, F.-K. and Hechelmann, H. (1987) Starter cultures for dry sausages and raw ham. Composition and effect. Fleischwirtsch. 67, 307-314.
Maing, I.Y., Ayres, J.C. and Koehler, P.E (1973) Persistence of aflatoxins during the fermentation of soy sauce. Appl. Microbiol. 25, 1015-1017.
Manabe, M. and Matsuura, S. (1972) Studies on the fluorescent compounds in fermented foods. Part IV. Degradation of added aflatoxin during miso fermentation. J. Food Sci. Technol. 19, 275-279.
Matsuoka, H., Koba, Y. and Ueda, S. (1982) Alcoholic fermentation of sweet potato without cooking. J. Ferment. Technol. 60, 599-602.
Matsuura, S. (1970) Aflatoxins and fermented foods in Japan. Japan Agr. Res. Q. 5, 46-51.
Mitsue, T., Saha, B.C. and Ueda, S. (1979) Glucoamylase of *Aspergillus oryzae* cultured on steamed rice. J Appl. Biochem. 1, 410-421.
Monte, E., Villanueva, J. and Dominguez, A. (1986) Fungal profiles of Spanish country-cured hams. Int. J. Food Sci. Technol. 3, 355-359.
Mowat, L. (1989) "Cassava and Chicha". Shire Ethnography No.11. Shire Publications Ltd., Aylesbury.
Ogawa, K., Ito, M., Tazaki, S. and Nakatsu, S. (1990) Breeding by interspecific protoplast fusion of *koji* moulds, *Aspergillus awamori* var. *kawachi* and *Aspergillus usamii* mut. *shirousamii*. J. Ferment. Bioeng. 70, 63-65.
Ota, S. (1983*a*) Katsuobushi I. New Food Industry 25, (5) 35-41.
Ota, S. (1983*b*) Katsuobushi II. New Food Industry 25, (6) 17-25.
Ota, S. (1983*c*) Katsuobushi III. New Food Industry 25, 54-60.
Oyashiki, H., Uchida, M., Obayashi, A. and Oka, S. (1989) Evaluation of Koji prepared with various moulds for Mirin-making. J. Ferment. Bioeng. 67, 163-168.
Park, K.Y., Lee, K.B. and Bullerman, L.B. (1988) Aflatoxin production by *Aspergillus parasiticus* and its stability during the manufacture of Korean soy paste (Doenjang) and soy sauce (kanjang) by traditional method. J. Food Prot. 51, 938-945.
Pauwels, L., Mulkay, P., Ngoy, K. and Delaude, C. (1992) Eminia, Rhynchosia et Vigna (Fabacées) à comlexes amylolytques employés dans la région Zambésienne pour la fabrication de la bière "Munkoyo". Belg. Journ. Bot. 125, 41-60.
Phillips, S. and Wallbridge, A. (1977) The mycoflora associated with dry salted tropical fish. Proceedings of the Conference on the handling, processing and marketing of tropical fish, London 5-9th July 1976, pp. 353-356. Tropical Products Institute, London.
Pitt, J.I., Cruickshank, R.H. and Leistner, L. (1986) *Penicillium commune*, *P. camembertii*, the origin of white cheese moulds, and the production of cyclopiazonic acid. Food Microbiol. 3, 363-371
Purchio, A. (1976) Search for $B_1$ aflatoxin and similar fluorescent compounds in miso. Mycopath. 58, 13-17.
Saono, S., Hull, R.R. and Dhamcharee, B. (1986) "A Concise Handbook of Indigenous Fermented Foods in the Asca Countries, The Indonesian Institute of Sciences (LIPI), Jakarta.
Sardjono, Kapti, R. and Sudarmadji, S. (1992) Growth and aflatoxin production by *Aspergillus flavus* in mixed culture with *Aspergillus oryzae*. ASEAN Food J. 7, 30-33.
Shank, R.C., Wogan, G.N., Gibson, J.B. and Nondasuta, A. (1972) Dietary aflatoxins and human liver cancer. II, Aflatoxins in market foodstuffs of Thailand and Hong Kong. Food Cosmet. Toxicol. 10, 61-69.
Shinshi, E., Manabe, M., Goto, T., Misawa, S., Tanaka, K. and Matsuura, S. (1984) Studies on the fluorescent compound in fermented foods 7. J. Japan Soy Sauce Res. Inst. 10, 151-155.
Simoons, F.J. (1991) "Food in China, A Cultural and Historical Inquiry". CRC Press, Boca Raton, Florida.
Steinkraus, K.H. (1983) "Handbook of Indigenous Fermented Foods". Marcel Dekker, New York.
Steinkraus, K.H. (1989) "Industrialisation of Indigenous Fermented Foods". Marcel Dekker, New York.
Sutic, M.J., Ayres, J.C. and Koehler, P.E. (1972) Identification and aflatoxin production of moulds isolated from country cured hams. Appl. Microbiol. 23, 656-658.
Tada, S., Iimura, Y., Gomi, K., Takahashi, K., Hara, S. and Yoshizawa, K. (1989) Cloning and nucleotide sequence of the genomic Taka amylase A gene of *Aspergillus oryzae*. Agric. Biol. Chem. 53, 593-599.
Takahashi, T. and Yamamoto, T. (1913) On the physiological difference of the varieties of *Aspergillus oryzae* employed in the three main industries in Japan, namely saké, shoyu and tamari manufacture. J. Coll. Agric. Tokyo Imp. Univ. 5, 153-161.
Tang, S.Y. and Long, K.H. (1977) Mycotoxins in soy sauce and fermented soybean paste of Taiwan. J. Formosan Med. Assoc. 76, 482-487.

Tanikawa, E. (1971) Marine products in Japan. Koseisha-Koseikaku Company, Tokyo.
Tanimura, W., Iitoi, Y., Mura, K., Aizawa, T. and Tanaka, R. (1979) Chemical compositions and hygienic inspection of foods in Southeast Asia. J. Agric. Sci. Tokyo, Suppl. 3, 3-15.
Toldrá, F. and Etherington, D.J. (1988) Examination of cathepsins B, D,H and L activities in dry-cured hams Meat Sci. 23, 1-7.
Ueda, S. and Saha, B.C. (1983) Behaviour of *Endomycopsis fibuligera* glucoamylase towards raw starch. Enz. Microb. Technol. 5, 196-198.
Ueda, S., Saha, B.C. and Koba, Y. (1984) Direct hydrolysis of raw starch. Microbiol. Sci. 1, 21-24.
Ushijima, S. and Nakadai, T. (1987) Breeding by protoplast fusion of koji mould *Aspergillus sojae*. Agric. Biol. Chem. 51, 1051-1057.
Ushijima, S., Nakadai, T. and Uchida, K. (1990) Breeding of new Koji-moulds through interspecific hybridisation between *Aspergillus oryzae* and *Aspergillus sojae* by protoplast fusion. Agric. Biol. Chem. 54, 1667-1676.
Van Veen, A.G., Graham, D.C.W. and Steinkraus, K.H. (1968) Fermented rice, a food from Ecuador. Arch Latinamer. Nut. 18, 363-373.
Wheeler, K.A., Hocking, A.D., Pitt, J.I. and Anggawati, A.M. (1986) Fungi associated with Indonesian dried fish. Food Microbiol. 3, 351-357.
Wicklow, D.T. (1983) Taxonomic features and ecological significance of sclerotia, in "Aflatoxin and *Aspergillus* in corn" (Diener, U.L., Asquith, R.L. and Dickens, J.W., Ed.), pp. 6-12. Southern Cooperative Series Bull. 279, Alabama Agricultural Experimental Station, Auburn University, Alabama.
Wicklow, D.T. (1984) Conidium germination rate in wild and domesticated yellow-green aspergilli,. Appl. Environ. Microbiol. 47, 299-300.
Yamazaki, I. and Ueda, S. (1951) Action of black-koji amylase on raw starch. Nippon Nogeikagaku Kaishi 24, 181-185.
Yokotsuka, T. (1991*a*) Proteinaceous fermented foods and condiments prepared with koji moulds, in "Handbook of Applied Mycology - Volume 3 - Foods and Feeds" (Arora, D.K., Mukerji, K.G. and Marth, E.H., Ed.), pp. 329-373. Marcel Dekker, New York.
Yokotsuka, T. ((1991*b*) Nonproteinaceous fermented foods and beverages produced with koji mould, condiments prepared with koji moulds, in "Handbook of Applied Mycology - Volume 3 - Foods and Feeds" (Arora, D.K., Mukerji, K.G. and Marth, E.H., Ed.), pp. 293-328. Marcel Dekker, New York.
Yoshizawa, K. and Ishikawa, T. (1989) Industrialisation of Sake Manufacture, in "Industrialisation of Indigenous Fermented Foods" (Steinkraus, K.H., Ed.), pp. 127-168. Marcel Dekker, New York.

# THE ARp1 *ASPERGILLUS* REPLICATING PLASMID

John Clutterbuck, David Gems and Scott Robertson

Department of Genetics
Glasgow University
Glasgow G11 5JS
Scotland, UK

## INTRODUCTION

Most plasmids used in the transformation of filamentous fungi depend for their maintenance on integration into the chromosome (Ballance 1991). Previous attempts to isolate plasmids capable of autonomous replication were unsuccessful. However, Johnstone (1985) had used a gene bank consisting of genomic inserts in the *argB*$^+$ plasmid pILJ16 (Figure 1) primarily to clone developmental genes, and reasoned that some members of this gene bank might contain inserts able to promote autonomous replication. Such a plasmid might replicate or be partitioned less efficiently than the chromosome and would therefore be identifiable by the instability of its transformants. On this basis three unstable transformants were selected and from one of these a plasmid was recovered which retransformed *argB* strains much more efficiently than the original vector. The properties of the *Aspergillus* replicating plasmid (ARp1) have been described by Gems *et al* (1991) and are summarised here.

## PROPERTIES OF ARp1

ARp1 transforms *A. nidulans* at frequencies up to 2000 times higher than those obtained with the integrating parent plasmid pILJ16 (Figure 1). The resulting transformants showed vegetative loss of the plasmid: approximately 55% of conidia (asexual spores) formed on selective medium were Arg$^-$. A proportion of ascospores from selfed cleistothecia of a transformant also carried the plasmid. Southern hybridisation of transformant DNA revealed free plasmids in both supercoiled and open circle forms, and these could be recovered by transformation of *Escherichia coli* much more readily than would be the case for a chromosomally integrated plasmid.

The restriction map of the recovered plasmid suggests that the insert, which is designated *AMA1*, consists of a short unique central region, flanked by inverted repeats of 2.7 kb each. The central *Sal*I fragment has been sequenced, confirming that it consists of 376 bp of unique sequence, flanked by the start of the inverted repeats.

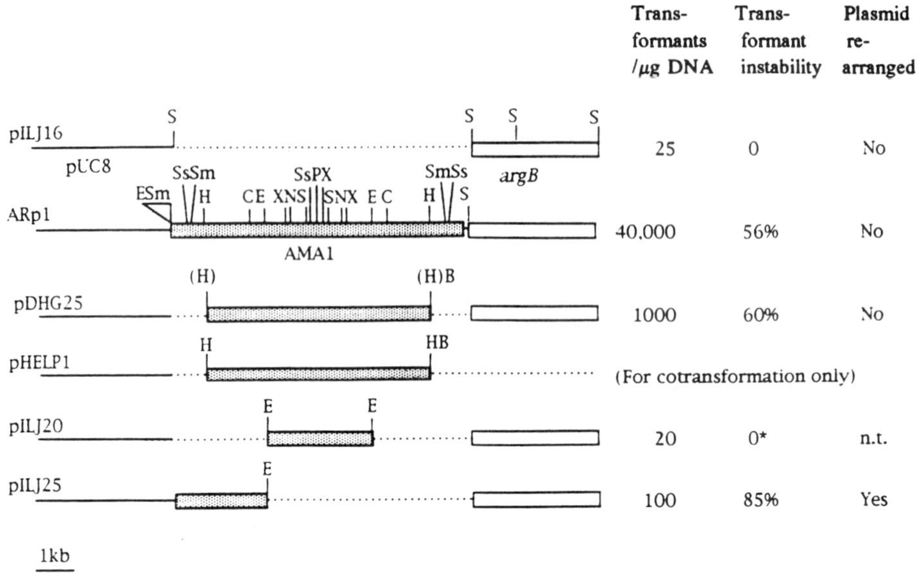


**Figure 1.** Linearised diagrams of circular plasmids employed in *Aspergillus* transformation
Single line: bacterial vector (pUC8 or derivatives), open box: the *Aspergillus argB* gene, shaded box: the AMA1 insert, dotted line: segments absent from this plasmid. Restriction sites: C; *Cla*I, E: *Eco*RI, H: *Hin*dIII, N:*Nru*I, P: *Pst*I, S: *Sal*I, Sm: *Sma*I, Ss: *Sst*I, X: *Xho*I. Only sites relevant to their construction are shown for plasmids other than pILJ16 and ARp1. The *Hin*dIII sites shown in parenthesis in pDHG25 were used in construction of the plasmid, but have been deleted by blunt-ending.
Transformant stability denotes the percentage of conidial (asexual) progeny of transformants which have lost the $argB^+$ plasmid.
"n.t." = not tested for plasmid rearrangement.
*Some of these transformants were initially very unstable, but produced fully stable sectors.

Plasmid DNA could also be detected as a distinct band, separate from the chromosomes in CsCl/ethidium bromide gradients, and this material, on examination with an electron microscope, was seen to contain plasmid circles of sizes appropriate for monomers, dimers, trimers and tetramers, as well as probable higher multimers. This material, when denatured and allowed to renature before examination, showed monomers and dimers with stem-loop structures, such as would be predicted from pairing of the inverted repeats.

The function of the inverted repeats is a matter of speculation. One possibility is that it has a role in replication analagous to that of a similar inverted repeat in the *Saccharomyces cerevisiae* 2 μm plasmid. If replication is initiated in one arm of the 2 μm circle, giving bidirectional replication forks, recombination between the inverted repeats is believed to convert this configuration to a double rolling circle (Futcher, 1986). The result of this switch is the ability to reel off a large number of copies in one replication cycle; the copies being formed as a large multimer which could be broken down to monomers by further recombination. In yeast both recombination steps are achieved by means of a site-specific recombination mechanism operating on sites within the inverted repeats and using an enzyme (FLP) encoded by the plasmid. Such a system is unlikely for ARp1, which if it does use a similar procedure, probably has to rely on mitotic recombination. This would fit with two features of ARp1: firstly the instability of transformants, despite a calculated copy number of 5-10 per nucleus, can be explained if many of the plasmid copies are present as multimers. Secondly, the apparent relationship between transformation efficiency and the size of the AMA1 insert (see below) may relate to its function as a recombination substrate, for which purpose the larger it is the better. There is also evidence that flipping of the two arms of ARp1 does occur; a plasmid containing a further insert into one arm of AMA1 was recovered in two forms from the same transformant, and restriction digestion of these indicated that they differed in the AMA1 arm which contained the insert.

## ANALYSIS OF SUBCLONES

Analysis of AMA1 by transformation with deleted plasmids failed to reveal specific functional elements. Virtually all derivatives tested transformed at frequencies higher than the pILJ16 integrating control, but substantially lower than the complete ARp1. Plasmids were scored for transformation frequency, instability of transformants, and rearrangement of the plasmid in the fungus (see Figure 1).

ARp1 is nearly always recovered unrearranged from *Aspergillus* transformants. pDHG25 is a slightly smaller plasmid which behaves similarly. It contains all but the outer 0.5 kb of AMA1 on a *HindIII* fragment, (Figure 1) and it transformed with approximately 1/40 of the frequency of ARp1 although 50 times better than pILJ16. Phenotypically unstable transformants were obtained, indicative of autonomous replication, and from these unaltered pDHG25 was again recovered.

In contrast to this pILJ20, a plasmid retaining only the central *Eco*R1 fragment of the AMA1 insert (Figure 1), transformed no better than the pILJ16 control, and the transformants, although in some cases initially very unstable, became mitotically stable, implying that the plasmid had probably integrated into a chromosome. A variety of other subclones, consisting of ARp1 with the central *Nru*I, *Xho*I or *Cla*I fragments deleted or contained varying sized fragments of one arm were also tested (data not shown). All gave at least slightly elevated transformation frequencies and produced unstable transformants. However, transformants with these plasmids differed from those containing ARp1 or pDHG25 in that plasmids were rarely recoverable in *E. coli* and Southern hybridisation analysis of transformant DNA revealed extensive rearrangement. For instance, five independent transformants with pILJ25, which contains only part of one arm of AMA1 (Figure 1), were passaged through 4 conidial subcultures on selective medium. DNA was then extracted from each subculture, restricted with *Sal*I and probed with the *argB* gene. Hybridisation analysis of pILJ25 transformants, using the *argB* gene as probe, revealed a pattern of large, rearranged fragments, most of which were identical in all the transformants, and which added up to a figure approaching 100 kb. This implies that pILJ25 had suffered a series of programmed rearrangements, similar in all transformants, to give either one very large replicating plasmid, or possibly a variety of smaller ones, none of which could be transferred to *E. coli*.

To test the role of the fungal DNA component of AMA1, the insert was hybridised to chromosomal DNA of *A. nidulans*, *P. chrysogenum* and *Acremonium chrysogenum*. No hybridising bands were seen in the *Acremonium* DNA, but a number of bands hybridised in DNA from the other two species. Two lambda clones of *P. chrysogenum* DNA (λPEN1 and λPEN2) that hybridised with AMA1 were then isolated. They were tested by cotransformation (see below) for promotion of plasmid replication in both hosts, where λPEN1 gave positive results. Similarly, DNA restriction fragments from an *A. nidulans* cosmid that hybridised to AMA1 promoted plasmid replication in the parent species.

It was concluded that the AMA1 insert gives a stable plasmid structure promoting autonomous replication. Fragments of AMA1 can also be rearranged in the fungus to give replicating plasmids, but the rearrangements required often destroy sequences required for plasmid recovery in bacteria.

## COTRANSFORMATION, PLASMID RECOMBINATION AND LIGATION

ARp1 is very effective in cotransformation. The efficiency of transformation of a *trpC*⁻ strain with the normally integrating *trpC*⁺ plasmid pTA11 was increased about 100-fold by addition of ARp1 to the transformation mixture. Addition of the markerless ARp1 derivative pHELP1 (Figure 1) to transformations with pILJ16, increased efficiency by a similar figure. In both cases the majority of the resulting transformants were unstable for the selective marker, implying that the two plasmids had recombined to form a replicating dimer. In some instances, it has been possible to recover such a dimer in *E. coli*. Addition of the selective marker on a linear molecule instead of a plasmid was almost as efficient, and linearisation of the replicating plasmid as well (e.g. at the unique *Bam*H1 site in pHELP1) did not reduce its effectiveness. It was concluded that in the fungus a replicating plasmid recombines readily with either linear or circular cotransforming DNA, and that ligation of two linear DNAs is equally efficient.

It has also been demonstrated that a replicating plasmid already in the cell can recombine with a second plasmid introduced by subsequent retransformation. Some of the resulting transformants were unstable for markers carried by both plasmids, again suggesting formation of replicating heterodimers. Other transformants were stable for both markers, implying that both DNAs had in these instances recombined with a chromosome.

## THE "INSTANT GENE BANK"

The ability of *Aspergillus* to ligate linear DNAs has suggested the use of the plasmid in gene cloning by means of an "instant gene bank". This procedure obviates *in vitro* gene bank construction by presenting the fungus with a replicating plasmid plus chromosomal DNA fragments which it will ligate together of its own accord. With this method, genes can be selected from any donor DNA that will complement available *A. nidulans* mutants (or mutants of other recipients in which ARp1 derivatives can function). Sonicated or restriction endonuclease partially digested donor DNA has been used. The most suitable plasmids are pHELP1 or pDHG25 which can be linearised at a unique *Bam*HI site thus directing insertion of the chromosomal DNA at this position. The efficacy of the method with a variety of genes, using donor DNA from either *A. nidulans* or *P. chrysogenum* has been demonstrated, and other workers have extended these experiments to *A. niger* and *Gaeumannomyces gramini* (see Table 1). It is a regular observation that recovery of the complementing plasmids into *E. coli* is not possible for a proportion of the transformants (probably due to plasmid rearrangement), therefore a number of transformants should be isolated for each gene. It also appears that re-isolation may be more efficient if the DNAs in the initial transformation step are present at very high concentrations: in one successful instance A.J. Aleksenko, L.M. Belenky and Y.P. Vinetski (personal communication) used 200$\mu$g of pHELP1 and 500$\mu$g of genomic DNA per $10^6$-$10^7$ protoplasts to isolate the *trp*C gene from *P. canescens* (see Table 1).

In addition to *A nidulans, A. oryzae* and *A. niger argB* mutants can be transformed by ARp1 at frequencies substantially above those of the pILJ16 control, and instability of the resulting transformants implies that the plasmid replicates autonomously in these fungi (Table 2). Transformation of *P. chrysogenum* with an *oliC* (oligomycin resistance)

**Table 1.** Genes cloned by the instant gene bank method.

| Donor DNA | Gene | Reference |
|---|---|---|
| *Aspergillus nidulans/Eco*RI | *argB* | This work |
| *Penicillium canescens/Sau*3A1 | *trpC* | 1 |
| *Gaemuannomyces graminis/Sau*3AI | *argB* | 2 |
| *Gaeumannomyces graminis/Sau*3AI | *pyrG* | 2 |
| *Aspergillus niger* cosmids | *nicB* | 3 |

1. A. Aleksenko, Ł. Belenky and Y. Vinetski (personal communication)
2. P. Bowyer, A.E. Osborn and M.J. Daniels (personal communication)
3. J.C. Verdoes, P. van den Berg, P. J. Punt and C.A.M.J.J. van den Hondel (personal communication): using an ARp1 derivative (pAB4-arp1) carrying the *A. niger pyrG* gene and employing an *A. niger pyrG⁻* strain as recipient for transformation.

derivative of ARp1 also gave unstable transformants at a frequency that was low but significantly above that of the integrating control plasmid. ARp1 is therefore usable directly in other *Aspergillus* species, and as described above, it may be possible to isolate hybridising sequences from other fungi which will provide replicative function in these species.

## CONCLUSIONS AND FUTURE DEVELOPMENT

The transformation efficiency of ARp1 in comparison with integrating plasmids suggests that, given an effective transformation protocol, the main barrier to transformation for standard plasmids is at the chromosomal integration step. This conclusion was supported by an experiment demonstrating that ARp1 is unable to increase transformation rates where a requirement for chromosomal recombination is reimposed. This experiment employed a plasmid containing a truncated *argB* gene which could only regenerate a complete copy by recombination with the chromosome; as predicted, cotransformation with a replicating plasmid had no effect on transformation frequencies in this system.

In contrast to the rarity of plasmid-chromosome recombination, plasmid-plasmid recombination appears to be a frequent event, leading to formation of replicating co-integrates if one of the components is capable of autonomous replication. This observation also favours an explanation for the effectiveness of cotransformation for non-replicating plasmids in terms of recombination of constituent plasmids with each other, followed by recombination of the resulting co-integrate with the chromosome.

The main value of ARp1 and derivatives is due to their effectiveness in enhancing transformation frequencies in *A. nidulans* and related species. They are also particularly suitable as cotransformation vehicles facilitating the introduction of any DNA species, linear of circular, into the fungus, e.g. for qualitative tests of the activity of manipulated gene constructs.

The instability of ARp1 transformants makes them unsuitable where stable gene expression is required. Indeed we have relied on instability as an indicator of autonomous replication and more stably replicating plasmids would have been ignored. A deliberate search for derivatives of ARp1 with more efficient replication or partitioning, and/or increased copy number is therefore a serious prospect.

**Table 2.** Fungi transformed with ARp1 or derivatives.

| SPECIES | SELECTIVE MARKER | TRANSFORMATION/$\mu g$ | (INTEGRATING CONTROL PLASMID) | TRANS-FORMANTS UNSTABLE? | REFERENCE |
|---|---|---|---|---|---|
| *Aspergillus nidulans* | *argB* | 40,000 | (25) | yes | (1) |
| *Aspergillus niger* | *argB* | 1,000 | (50) | yes | (1) |
| *Aspergillus oryzae* | *argB* | 6,000 | (200) | yes | (1) |
| *Aspergillus parasiticus* | *argB* | 327 | (0) | yes | (2) |
| *Penicillium chrysogenum* | *oliC* | 12 | (5) | yes | (3) |
| *Gibberella fujikuroi* | *argB* | 10 | (4) | yes | (4) |

1. Gems *et al.* 1991
2. M.A. Moreno, C. Pascal, A. Gibello, S. Ferrar, S.J. Bos, A.J.M. Debets and G. Suárez, 1993 (Abstracts of FEMS/BMS meeting Canterbury, Kent 5-8 April)
3. Scott Robertson unpublished
4. Brückner *et al.* 1992

The instant gene bank method also offers an alternative method of gene cloning which may be particularly suitable where a constructed gene bank is not available for the donor strain, or if passage of the DNA through *E. coli* before cloning is undesirable.

## REFERENCES


Ballance, D.J. (1991) Transformation systems for filamentous fungi, in "Molecular Industrial Mycology", (Leong, S.A. and Berka, R.M., Eds) pp. 1-29. Marcel Dekker, New York.

Brückner, B., Unkles, S.E., Weltring, K. and Kinghorn, J.R. (1992) Transformation of *Gibberella fujikuroi*: effect of the *Aspergillus nidulans* AMA1 sequence on frequency and integration. Curr. Genet. 22, 313-316.

Futcher, A.B. (1986) Copy number amplification of the 2$\mu$m circle plasmid of *Saccharomyces cerevisiae*. J. Theor. Biol. 119,197-204.

Gems, D.H. and Clutterbuck, A.J. (1993) Cotransformation with autonomously replicating plasmids facilitates gene cloning from an *Aspergillus nidulans* gene library. Curr. Genet. (in press).

Gems, D.H., Johnstone, I.L. and Clutterbuck, A.J. (1991) An autonomously replicating plasmid transforms *Aspergillus nidulans* at high frequency. Gene 98, 61-67.

Johnstone, I.L. (1985) Transformation of *Aspergillus nidulans*. Microbiol. Sci. 2, 307-311.

# GENETICS OF PENICILLIN BIOSYNTHESIS IN *ASPERGILLUS NIDULANS*

G. Turner

Department of Molecular Biology and Biotechnology
Krebs Institute for Biomolecular Research
University of Sheffield
P.O. Box 594, Western Bank
Sheffield, S10 2UH

## INTRODUCTION

Ever since the development of penicillin as an antibiotic in the 1940s, there has been an interest in understanding the chemistry, biochemistry and genetics of the biosynthesis of penicillin and other β-lactam antibiotics. Of course, the main interest was with *Penicillium chrysogenum* and *Acremonium chrysogenum* (formerly named *Cephalosporium acremonium*), which have always been the commercial sources of these antibiotics, but in the late 1940s, screening of a wide range of fungi for new antibiotics led to the detection of penicillin production by *Aspergillus nidulans* (Dulaney, 1947). Although this has never been of any commercial significance, the development of *A. nidulans* as a model organism for genetic studies (Pontecorvo *et al.*, 1953) led to its occasional use in projects to investigate the genetics of penicillin production (Ball, 1983).

Holt and Macdonald (1968) studied penicillin yield improvement in the progeny of sexual crosses between wild-type isolates and laboratory (Glasgow) strains derived from NRRL 194. In later studies, specific mutations (*pen*) giving rise to penicillin yield improvement were characterised and mapped to the *penA*, *B* and *C* loci (Ditchburn *et al.*, 1976). In addition, mutations leading to deficiencies in penicillin production (Npe) were isolated and mapped to four loci, *npeA*, *B*, *C* and *D*, although most mapped to the *npeA* locus (Edwards *et al.*, 1974; Holt *et al.*, 1976; Makins *et al.*, 1983).

A biometric approach was adopted (Merrick and Caten 1975; Simpson and Caten 1979*a*) to study variation in penicillin titre in naturally-occurring strains. As a result of mutation and recombination studies it was concluded that penicillin yield in *A. nidulans* was a polygenic trait.

The development of transformation systems for *A. nidulans* (Ballance *et al.*, 1983; Tilburn *et al.*, 1983; Yelton *et al.*, 1984) led to renewed interest in the application of molecular genetics to identify the genes responsible for penicillin biosynthesis.

*The Genus Aspergillus*, Edited by Keith A. Powell *et al.*,
Plenum Press, New York, 1994

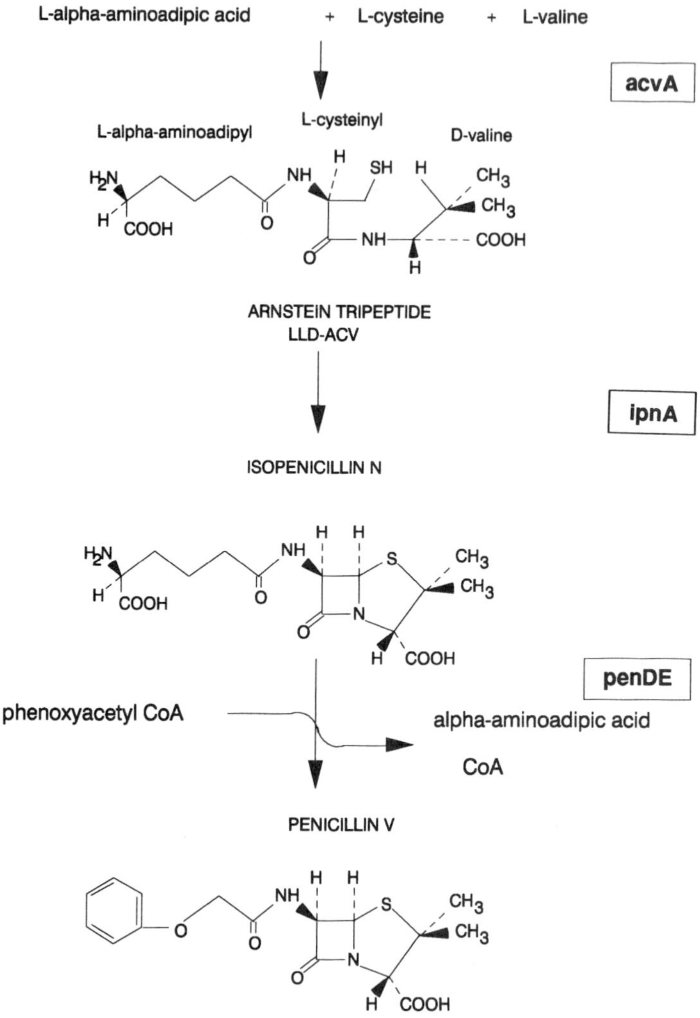


**Figure 1.** The penicillin biosynthetic pathway in *Aspergillus nidulans*. The genes responsible for each step are shown: *acvA*, δ-(L-α-aminoadipyl) cysteinyl valine synthetase; *ipnA*, isopenicillin N synthetase, *penDE*, acyl coenzyme A:6-amino penicillinanic acid acyltransferase. Gene nomenclature is based on the first reported description of each gene, following the guidelines for naming of *A. nidulans* genes (Clutterbuck, 1973).

## ISOLATION OF THE BIOSYNTHETIC GENES

The biosynthetic pathway for penicillin from amino acid precursors is shown in Figure 1. This pathway is common to both *P. chrysogenum* and *A. nidulans.*

Research on the genetics of antibiotic production has been largely stimulated by commercial considerations, therefore gene isolation and characterisation focussed initially on *P. chrysogenum* and *A. chrysogenum* . Isolation of genes from *A. nidulans* was often a secondary aim but the potential of this species as a model organism was always acknowledged particularly since the former, commercially-used species have no sexual cycle and very little background genetic information had accumulated for them.

The earlier history of fungal genetics did lead to an incorrect supposition about the organisation of the genes. At the time when the search for the biosynthesis genes began, it had already become clear that genes for antibiotic biosynthesis in the streptomycetes were clustered (Malpartida and Hopwood, 1984) but it was believed that the fungal genes would not be clustered. This belief was partly based on the generalisation that prokaryotic genes for pathways tended to be clustered, while those in eukaryotes were not, and the earlier work on *A. nidulans npeA*, *B*, *C* and *D* also reinforced this view, since these loci were reported to be unlinked (Makins *et al.*, 1983).

The first β-lactam biosynthetic gene to be cloned was the isopenicillin synthetase (IPNS) gene of *A. chrysogenum*, using oligonucleotides based on partial sequence of purified enzyme (Samson *et al.*, 1985). Although *A. chrysogenum* produces cephalosporin C and not penicillin, the first two steps in the pathway (Figure 1) were common to both penicillin and cephalosporin producers. Availability of this gene quickly led to the isolation of IPNS genes from *P. chrysogenum* (Carr *et al.*, 1986) and *A. nidulans* (Ramon *et al.*, 1987) by heterologous hybridisation.

As the molecular biology of *A. nidulans* developed, attempts were made to isolate the biosynthetic genes by transformation and complementation of available Npe mutants using gene libraries constructed from wild-type *A. nidulans*, but without success (D.J. Smith, J. Bull, and G. Turner, unpublished; J.R. Kinghorn, personal communication). However, the discovery that a cephamycin biosynthetic gene cluster from the Gram-negative bacterium, *Flavobacterium* spp. SC12,154, hybridised to the IPNS gene of *P. chrysogenum*, enabled the identification of the region encoding the δ-(α-aminoadipyl)-L-cysteinyl D-valine synthetase (ACVS) and IPNS in four species, including *A. nidulans* (Burnham *et al.*, 1989; Smith *et al.*, 1990*b*). Transformation of an *A. nidulans npeA*0022 mutant with a cosmid vector containing the gene cluster from *P. chrysogenum* restored penicillin production, but a smaller plasmid containing only the IPNS gene did not. The same cosmid was inserted into *Aspergillus niger* and *Neurospora crassa* by transformation, resulting in penicillin production by these species, which do not naturally contain any detectable homologous DNA sequence, nor do they produce penicillin (Smith et al. 1990*a*). This confirmed that a contiguous stretch of *P. chrysogenum* DNA of about 35kb contained all the essential genetic information for penicillin production from primary metabolic intermediates. Three transcribed regions detectable in this region of the *P. chrysogenum* genome were also present in the same arrangement in *A. nidulans* (see Figure 2), and it was concluded that these represented the genes encoding the three steps shown in Figure 1 (Smith *et al.* 1990*b*). The remarkable conservation of sequence of the IPNS encoding genes between distantly related microorganisms has also led to a hypothesis of horizontal gene transfer between bacteria and fungi (see below).

In parallel with this work, mutants mapping to the *npeA* locus were also studied (MacCabe *et al.* 1990), since they were the most most frequent category, and had complete loss of penicillin production. In addition to the *npeA* mutants obtained by mutagenesis, wild-type isolates of *A. nidulans* were also found which lacked the ability to produce penicillin and

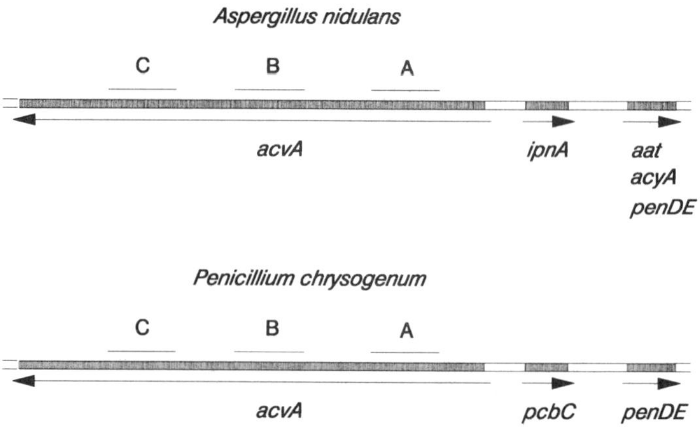


**Figure 2.** Organisation of the biosynthetic genes in *P. chrysogenum* and *A. nidulans*. Gene symbols as in Figure 1. A, B and C denote the regions encoding conserved domains. Some alternative names have been given. Direction of transcription is indicated by arrows.

the mutations responsible were mapped to the *npeA* locus (Cole *et al.*, 1976). These naturally occuring Npe strains belonged to the heterokaryon compatibility (h-c) groups F and G. The mutation *npeA*0022, isolated following mutagenesis of the laboratory strain (Edwards *et al.*, 1974), appeared to result from a gross rearrangement of the DNA sequence which included the IPNS coding region as detected by hybridisation experiments. The rearrangement could be explained by insertion or translocation affecting the *npeA* locus. Mutation *npeA*0049/1, found in a natural isolate belonging to the h-c group F, resulted from a deletion of at least 15kb, including the IPNS coding region. Complementation of *npeA*0022 was not possible by IPNS alone, although larger fragments did restore penicillin biosynthesis, suggesting the presence of further biosynthesis genes in this region. Transcript analysis of the *npeA* region in $Npe^+$ strains suggested the presence of three genes. The *npeA*0049 deletion resulted in the loss of all three genes.

While enzyme purification and partial sequencing had been a route to the isolation of the *pcbC* gene (encoding IPNS) of *A. chrysogenum*, purification of the ACVS from this species was not easily achieved, probably due to the instability of the enzyme. It was not clear whether one or two enzymes were required to synthesise ACV from the precursor amino acids, or even what type of enzyme was involved. The ACVS of *A. nidulans* was the first to be purified sufficiently for some detailed enzymology, and sufficient amino acid sequence was obtained to make oligonucleotide probes, which also showed that the gene encoding ACVS was adjacent to the gene encoding IPNS at the *npeA* locus (MacCabe *et al.*, 1990).

The isolation of the *penDE* gene, encoding the final step in penicillin biosynthesis (acyl coenzyme A:6-aminopenicillanic acid acyltransferase) was initially from *P. chrysogenum* (Barredo *et al.*, 1989), using oligonucleotides designed from the N-terminal amino acid sequence of the purified enzyme, and it was found to be adjacent to the *pcbC* (IPNS) gene. The equivalent gene of *A. nidulans* was isolated using the *P. chrysogenum penDE* gene as a probe (Montenegro *et al.*, 1990), and it was clear that the gene organisation in *A. nidulans* was very similar to that in *P. chrysogenum* (see Figure 2).

Smith *et al.* (1990*c*) first reported the sequence of the ACVS of *P. chrysogenum*, and MacCabe *et al*, (1991) sequenced the equivalent gene *A. nidulans*. The most striking feature was the presence of three domains, each of some 600 amino acids (Figure 2), which showed homology to each other, and to domains of peptide synthetases from other microorganisms.

The three major structural genes for the biosynthetic pathway have now been isolated and sequenced, but there may be other genes specific for penicillin biosynthesis which have not yet been found. For example, the substrates for the acyltransferase are isopenicillin N and a CoA derivative of the side chain precursor, such as phenylacetyl CoA (see Figure 1). It is not yet clear how the CoA derivative is formed, there is enzymological evidence that acetyl CoA synthetase is able to carry out this step (Martinez-Blanco *et al.*, 1992). However, strains of *A. nidulans* carrying mutations at the *facA* locus, which encodes acetyl CoA synthetase (Connerton *et al.*, 1990), are unaffected in penicillin biosynthesis as determined by a bioassay (Holt and Macdonald, 1968) (H.J.R. Rolfe and G. Turner, unpublished). Other fatty acyl CoA synthetases may also be able to carry out this reaction yet it is not clear whether there exists a specific precursor-CoA ligase in penicillin-producing fungi.

## REGULATION OF GENE EXPRESSION

Much attention has been given to achieving the optimum fermentation conditions for the commercial production of β-lactam antibiotics by *P. chrysogenum* and *A. chrysogenum* (Martin and Aharonowitz, 1983), but these organisms are not amenable to genetic analysis, and little work has been done until recently on *A. nidulans*. The availability of the structural genes has now begun to alter this situation. High glucose and lysine concentrations with oxygen deficiency are all factors which decrease penicillin yield from *P. chrysogenum*. Repression of both penicillin production and expression of structural genes has been examined in *A. nidulans*. The effect on gene expression has been studied using both reporter genes (Brakhage *et al.*, 1992*a,b*) and steady state mRNA levels (Espeso and Penalva, 1992). Since the *ipnA* and *acvA* genes are divergently transcribed (see Figure 2), and separated by only 872 bp, it was possible to construct a double reporter gene vector by translational fusion of the two reporter genes derived from *E. coli*, *lac*Z (β-galactosidase) and *uidA* (β-glucuronidase), to the 5' ends of the *acvA* and *ipnA* genes (see Figure 3). This vector carries an *argB* gene as a selectable marker for transformation, and permits the insertion of the vector, in single copy, at the *argB* locus of a recipient strain of *A. nidulans*. Strains carrying this vector were used to study expression of genes and measure penicillin biosynthesis under a range of fermentation conditions. Defined media are used for most biochemical and genetical studies, but penicillin production is generally very poor on such media, so complex medium containing corn steep liquor and lactose, close to the commercial medium used for *P. chrysogenum*, was used. Glucose and lysine were both found to repress penicillin production, though only to about 50% of the derepressed level. Glucose repression seemed to act at the level of transcription, lowering the steady state level of IPNS mRNA (Espeso and Penalva, 1992).

Expression of both *acvA* and *ipnA* was also repressed to a similar extent by lysine, but glucose appeared to exert its effect only on *ipnA*, and not on *acvA* (Brakhage *et al.*, 1992*a,b*). These results indicate that expression of the penicillin biosynthesis genes is regulated by media and growth conditions with the identification of the genes involved in this regulation being of current interest.

Glucose is a preferred carbon source for *A. nidulans*, and represses the expression of a number of genes involved in utilisation of alternative carbon sources, such as ethanol, acetate, acetamide, and proline. In some cases, this repression is mediated via *creA*, *creB*, and *creC* genes (Bailey and Arst, 1975; Hynes and Kelly, 1977), since mutations at these loci result in varying degrees of loss of the glucose repression. The *creA* gene has been cloned (Dowzer and Kelly, 1989), and its action on the alcohol utilisation genes has been studied in some detail (Kulmburg *et al.*, 1993). However, glucose repression of utilisation of other carbon sources, such as quinic acid, is unaffected by mutations at the *cre* loci, indicating the existence of other mechanisms of catabolite repression (Kelly and Hynes, 1977). The allele *creA*$^{d}$*1*, which exerts a strong derepression effect on *alcA*,and *creB304*, with a lesser effect, do not relieve glucose repression of penicillin biosynthesis, suggesting the involvement of other regulatory gene(s).

The repression of penicillin biosynthesis by lysine in *P. chrysogenum* has been explained in terms of its feedback effect on lysine biosynthesis, thus decreasing the level of α-aminoadipate substrate for the ACV synthetase. Lysine inhibition of homocitrate

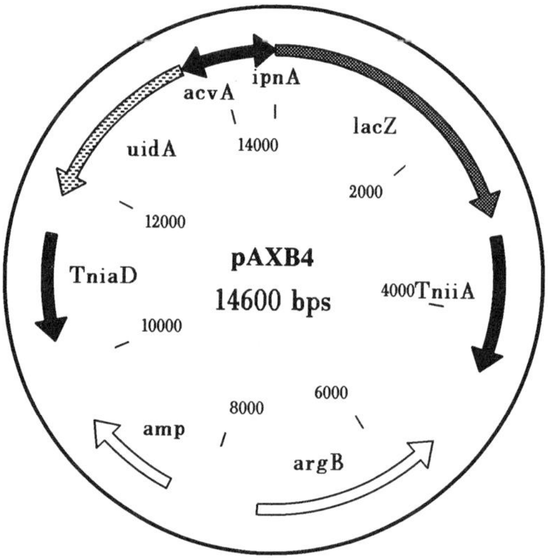


**Figure 3.** Twin reporter gene vector pAXB4 used in studies of gene regulation (Brakhage *et al.*, 1992*a*). *uidA* and *lacZ* are reporter genes; *acvA* and *ipnA* denote the intergenic region carrying the 5' (promoter) regions of these genes (Figure. 2), TniaD and TniiA are termination sequences from nitrate and nitrite reductase genes respectively, *argB* is the selectable marker for transformation of *A. nidulans*, and *amp* the ampicillin resistance gene for transformation of *E. coli*.

synthetase, an early step in the pathway, has been reported (Somerson *et al.*, 1961; Demain and Masurekar, 1974; Luengo *et al.*, 1980).

The decrease in penicillin biosynthesis could be explained in terms of insufficient precursor (Honlinger and Kubicek, 1989), but it also seems that repression of the ACV synthetase gene expression occurs (Brakhage and Turner, 1992*b*). While relatively high levels of lysine (0.1M) are required for repression, an equivalent concentration of arginine in the growth medium had no significant effect on penicillin biosynthesis.

Expression of the reporter genes by individual fungal colonies can be monitored directly on agar plates with the appropriate indicators, the artificial substrates X-Glu and X-Gal, and this offers a way of selecting for mutants altered in expression of the biosynthesis genes. Such an approach has now been initiated, in an effort to identify trans-acting regulatory genes (Brakhage, 1993).

The term "secondary metabolism" is sometimes associated with the idiophase of microbial growth, when active, exponential growth has slowed down (Martin and Demain, 1980). In the case of some prokaryotes, antibiotic biosynthesis begins at the onset of the idiophase (Marahiel *et al.*, 1987; Guthrie and Chater, 1990). Penicillin biosynthesis does not seem to fit into such a clear picture as no strong temporal regulation of the biosynthesis genes, or of penicillin biosynthesis itself, is seen in *A. nidulans* (Brakhage *et al.*, 1992*a*).

Finally, disruption of the *acvA* gene by transformation completely blocks penicillin biosynthesis, but does not greatly affect expression of the other two genes, suggesting that the genes are not subject to sequential induction (A. A. Brakhage, P. Browne and G. Turner, unpublished).

The earlier studies on penicillin yield in natural isolates and mutants of *A. nidulans* do suggest the involvement of many other genes in the control of penicillin biosynthesis. Some of these may be involved in primary metabolism, and affect penicillin yield indirectly. The *npeC* and *npeD* loci, which result in poor penicillin yield, have been recently re-examined, and appear to be synonymous with the *palA* and *palC* loci, respectively (Shah *et al.*, 1991), which are involved in gene regulation in response to external pH, along with *palB*, *E* and *F*. Penicillin yield is better at alkaline than acid pH. Mutations at *pacC*, which affect pH regulation, and mimic growth at alkaline pH, result in improved penicillin yield, while mutations at *palA,B,C,E* and *F* loci mimic acid pH, and result in lower penicillin yield. The earlier mapping data (Makins *et al.*, 1983) has also been revised, placing *npeC* on linkage group III (previously IV) and *npeD* on linkage group VII (previously II).

## ACV SYNTHETASE

While no close relatives of the genes encoding IPNS and ACT have been found to help explain their evolutionary derivation, the ACVS genes are members of the much wider family of peptide synthetases (Turgay *et al.*, 1992; Zhang and Demain, 1992). Gramicidin synthetases 1 and 2, responsible for the synthesis of the cyclic peptide gramicidin S by *Bacillus brevis*, were studied many years ago by Lipmann (1971) because of their ability to form peptide bonds by non-ribosomal peptide synthesis. Amino acids are activated by adenylation with ATP, and become covalently bound to the enzymes. The amino acids are then translocated to form a growing peptide, still bound to Gramicidin synthetase 2, from which they are finally released. One of the amino acids (L-phenylalanine) is racemised to the D-isomer (as is L-valine by ACVS), and a cyclised, head to tail dimer of the pentapeptide is formed.

Van Liempt *et al.*, (1989) demonstrated that the ACV synthetase of *A. nidulans* could carry out an ATP/PPi exchange reaction (diagnostic for the amino acid adenylation), and could form LLD-ACV from $^{14}$C-L-valine. Sequencing of the *P. chrysogenum* (Smith *et al.*, 1990*c*; Diez *et al.*, 1990) and *A. nidulans* (MacCabe *et al*, 1991) *acvA* genes and translation of the 12kb open reading frame, revealed amino acid motifs characteristic of enzymes capable of ATP/PPi exchange, including firefly luciferase, and many other motifs, of unknown function, which are present in all three domains, and in all peptide synthetases analysed to date (Turgay *et al.*, 1992). Sequencing of the gramicidin synthetase genes revealed one domain in GS1, and four domains in GS2. All the circumstantial evidence now points to one domain per amino acid, which presumably recognises, activates, and binds each amino acid. The availability of peptide synthetase genes from a wide range of microorganisms has now given the potential for genetic manipulation of these genes to learn more about how the enzymes function, and *A. nidulans*, with its well-developed genetics, is amenable to structural and functional analysis of the ACVS.

## DISTRIBUTION AND EVOLUTION OF Β-LACTAM BIOSYNTHESIS GENES

The hypothesis of horizontal transfer, probably from prokaryotes to eukaryotes, has been put forward to explain the high sequence identity of the genes encoding ACVS and IPNS in bacteria and fungi (Weigel *et al.*, 1988; Cohen *et al.*, 1990; Landan *et al.*, 1990). They are much more closely related than are the 16s rRNA genes, commonly used as standards for molecular phylogeny. Of course, it can also be argued that the β-lactam genes are not evolving as fast as the 16s rRNA genes, perhaps constrained by functional considerations, but the divergence of the IPNS genes within the three fungal species which were examined did not appear support this idea (Landan *et al.*, 1990).

β-lactam biosynthesis is not widely distributed amongst the fungi, and to date, genes from only three species have been characterised. It would be interesting to learn more about the distribution of the β-lactam biosynthesis pathway in relation to fungal phylogeny. The limited distribution of the pathway amongst fungal species noted to date has also been used as an argument for horizontal transfer events (Penalva *et al.*, 1990). The pathway has been detected in a few members of the Plectomycetes, Pyrenomycetes and related Deuteromycetes (Imperfect Fungi, lacking a sexual stage). Horizontal transfer to a common ancestor of these Series, followed by differential loss, would be one explanation. As mentioned previously, strains of *A. nidulans* which have lost the penicillin biosynthetic pathway can be isolated from the natural environment. Although persistence of the β-lactam antibiotic pathways in a number of microorganisms for $10^9$ years or so would suggest some selective advantage, however slight, it is clear that penicillin production is by no means essential for survival of *A. nidulans* in its natural environment. An alternative proposal is that independent horizontal transfer events have occurred for Plectomycetes and Pyrenomycetes (Penalva *et al.*, 1990).

Earlier studies on the pathway distribution was by screening for the products, but these are often made in very small quantities, and availability of gene probes should permit the detection of biosynthesis genes even where no products are detectable by bioassays.

There have been claims for the presence of penicillin biosynthesis by other species of *Aspergillus*, including *A. niger*, *A. oryzae*, and *A. flavipes* (Foster and Karow, 1945). *A. niger* was used as a recipient strain to test the function of the penicillin biosynthetic gene cluster (Smith *et al.*, 1990*a)*, however neither penicillin production or the biosynthetic genes were detected in the recipient strain that was used.

## FUTURE DEVELOPMENTS

*A. nidulans* is now an excellent model organism for genetic analysis of secondary metabolism, a widespread phenomenon in fungi, paralleling work with *Streptomyces* in prokaryotes. While the β-lactam pathways are now very well-characterised, there have been some recent exciting developments in the genetics of fungal polyketide biosynthesis. Aflatoxin biosynthesis by *A. flavus* and *A. parasiticus* has been the centre of attention, although these asexual species are not amenable to genetic analysis. However, *A. nidulans* is able to synthesise sterigmatocystin, a late intermediate in aflatoxin biosynthesis, and heterologous probing using gene sequences from the aflatoxin producers is being used to identify equivalent genes in *A. nidulans* (Keller and Adams, 1993). It is likely that *A. nidulans* will thus also provide a model genetic system for study of fungal polyketide biosynthesis.

A further interesting development is the identification of another peptide synthetase gene in *A. nidulans* var. *echinulatus,* a member of the *A. nidulans* species group (Xuei and Skatrud, 1993). Echinocandin is a complex peptide produced by this species which has been shown to have anti-yeast activity and may have therapeutic potential (Benz *et al.,* 1974).

## REFERENCES


Bailey, C. and Arst, H.N.Jr. (1975) Carbon catabolite repression in *Aspergillus nidulans.* Eur.J.Biochem. 51, 573-577.

Ball, C. (1983) Genetics of ß-lactam-producing fungi, in "Antibiotics Containing the Beta-Lactam Structure I" Demain, A.L. and Solomon, N.A., Eds.), pp. 147-162. Springer-Verlag, Berlin

Ballance, D.J., Buxton, F.P. and Turner, G. (1983) Transformation of *Aspergillus nidulans* by the orotidine-5'-phosphate decarboxylase gene of *Neurospora crassa.* Biochem.Biophys.Res.Commun. 112, 284-289.

Barredo, J.L., Van Solingen, P., Diez, B., Alvarez, E., Cantoral, J.M., Kattevilder, A., Smaal, E.B., Groenen, M.A.M., Veenstra, A.E. and Martin, J.F. (1989) Cloning and characterization of the acyl-coenzyme A:6-aminopenicillanic acid-acyltransferase gene of *Penicillium chrysogenum.* Gene 83, 291-300.

Benz, F., Knusel, F., Nuesch, J., Treichler, H., Voser, W., Nyfeler, R. and Keller-Schlierlein, W. (1974) Stoffwechselprodukte von Mikroorganismen. Echinocandin B, ein neuartiges polypeptid-antibioticum aus *Aspergillus nidulans* var. *echinulatus*: Isolierung und Bausteine. Helv. Chim. Acta . 57, 2459-2477.

Brakhage, A.A., Browne, P. and Turner, G. (1992a) Regulation of *Aspergillus nidulans* pencillin biosynthesis and penicillin biosynthesis genes *acvA* and *ipnA* by glucose. J. Bacteriol. 174, 3789-3799.

Brakhage, A.A. and Turner, G. (1992b) L-Lysine repression of penicillin biosynthesis and the expression of penicillin biosynthesis genes *acvA* and *ipnA* in *Aspergillus nidulans.* FEMS Microbiol. Lett. 98, 123-128.

Brakhage, A.A. (1993) Identification of trans-acting mutations affecting the regulation of the expression of *Aspergillus nidulans* penicillin biosynthetic genes. Seventeenth Fungal Genetics Conference, Asilomar. Poster abstract.

Burnham, M.K.R., Earl, A.J., Bull, J.H., Smith, D.J. and Turner, G. (1989) DNA encoding ACV synthetase. European Patent Application 88311655.0.

Carr, L.G., Skatrud, P.L., Scheetz, M.E., Queener, S.W. and Ingolia, T.D. (1986) Cloning and expression of the isopenicillin N synthetase gene from *Penicillium chrysogenum.* Gene 48, 257-266.

Clutterbuck, A.J. (1973) Gene symbols in *Aspergillus nidulans.* Genet. Res. 21, 291-296.

Cohen, G., Shiffman, D., Mevarech, M. and Aharonowitz, Y. (1990) Microbial isopenicillin N synthase genes: Structure, function, diversity and evolution. TIBTECH 8, 105-111.

Cole, D.S., Holt, G. and Macdonald, K.D. (1976) Relationship of the genetic determination of impaired penicillin production in naturally occurring strains to that in induced mutants of *Aspergillus nidulans.* J. gen. Microbiol. 96, 423-426.

Connerton, I.F., Fincham, J.R.S., Sandeman, R.A. and Hynes, M.J. (1990) Comparison and cross-species expression of the acetyl-CoA synthetase genes of the ascomycete fungi, *Aspergillus nidulans* and *Neurospora crassa.* Mol. Microbiol. 4, 451-460.

Demain, A.L. and Masurekar, P. (1974) Lysine inhibition of *in vivo* homocitrate synthesis in *Penicillium chrysogenum.* J. gen. Microbiol. 82, 143-148.

Diez, B., Gutierrez, S., Barredo, J.L., Van Solingen, P., Van der Voort, L.H.M. and Martin, J.F. (1990) The cluster of penicillin biosynthetic genes:identification and characterization of the *pcbAB* gene encoding α-aminoadipyl-cysteinyl-valine synthetase and linkage to the *pcbC* and *penDE* genes. J.Biol.Chem. 265, 16358-16365.

Ditchburn, P., Holt, G. and Macdonald, K.D. (1976) The genetic location of mutations increasing pencillin yield in *Aspergillus nidulans*, in "Second International Symposium on the Genetics of Industrial Microorganisms (Macdonald, K.D., Ed.), pp. 213. Academic Press, London.

Dowzer, C.E.A. and Kelly, J.M. (1989) Cloning of the *creA* gene from *Aspergillus nidulans*: a gene involved in carbon catabolite repression. Curr.Genet. 15, 457-459.

Dulaney, E.L. (1947) Some aspects of penicillin production by *Aspergillus nidulans*. Mycologia 39, 570-582.

Edwards, G.F.S.T.L., Holt, G. and Macdonald, K.D. (1974) Mutants of *Aspergillus nidulans* impaired in penicillin biosynthesis. J. gen. Microbiol. 84, 420-422.

Espeso, E.A. and Penalva, M.A. (1992) Carbon catabolite repression can account for the temporal pattern of expression of a penicillin biosynthetic gene in *Aspergillus nidulans*. Mol. Microbiol. 6, 1457-1465.

Foster, J.W. and Karow, E.O. (1945) Microbial aspects of penicillin VIII. Penicillin from different fungi. J. Bacteriol. 49, 19.

Guthrie, E.P. and Chater, K.F. (1990) The level of a transcript required for production of a *Streptomyces coelicolor* antibiotic is conditionally dependent on a tRNA gene. J.Bacteriol. 172, 6189-6193.

Holt, G. and Macdonald, K.D. (1968) Isolation of strains with increased penicillin yield after hybridization in *Aspergillus nidulans*. Nature 219, 636-637.

Holt, G., Edwards, G.F.S.T.L. and Macdonald, K.D. (1976) The genetics of mutants impaired in the biosynthesis of penicillin, in "Second International Symposium on the Genetics of Industrial Microorganisms" (MacDonald, K.D., Ed.), pp. 199-211. Academic Press, London.

Honlinger, C. and Kubicek, C.P. (1989) Regulation of δ-(L-α-aminoadipyl)-L-cysteinyl-D-valine and isopenicillin N biosynthesis in *Penicillium chrysogenum* by the α-aminoadipate pool size. FEMS Microbiol. Lett. 65, 71-76.

Hynes, M.J. and Kelly, J.M. (1977) Pleiotropic mutants of *Aspergillus nidulans* altered in carbon metabolism. Mol.Gen.Genet. 150, 193-204.

Keller, N.P. and Adams, T.H. (1993) Isolation of an *Aspergillus* gene, *verA*, encoding a putative reductase activity in the sterigmatocystin/aflatoxin biosynthetic pathway. Seventeenth Fungal Genetics Conference, Asiloma Poster abstract.

Kelly, J.M. and Hynes, M.J. (1977) Increased and decreased sensitivity to carbon catabolite repression of enzymes of acetate metabolism in mutants of *Aspergillus nidulans*. Molec. gen. Genet. 156, 87-92.

Kulmburg, P., Mathieu, M., Dowzer, C., Kelly, J. and Felenbok, B. (1993) Specific binding sites in the *alcR* and *alcA* promoters of the ethanol regulon for the CREA repressor mediating carbon catabolite repression in *Aspergillus nidulans*. Mol.Microbiol. 7, 847-857.

Landan, G., Cohen, G., Aharonowitz, Y., Shuali, Y., Graur, D. and Shiffman, D. (1990) Evolution of isopenicillin N synthase genes may have involved horizontal gene transfer. Mol.Biol.Evol. 7, 399-406.

Lipmann, F. (1971) Attempts to map a process evolution of peptide biosynthesis. Science 173, 875-884.

Luengo, J.M., Revilla, G., Lopez-Nieto, M.J., Villanueva, J.R. and Martin, J.F. (1980) Inhibition and repression of homocitrate synthetase by lysine in *Penicillium chrysogenum*. J. Bacteriol. 144, 869-876.

MacCabe, A.P., Riach, M.B.R., Unkles, S.E. and Kinghorn, J.R. (1990) *The Aspergillus nidulans npeA* locus consists of three contiguous genes required for penicillin biosynthesis. EMBO J. 9, 279-287.

MacCabe, A.P., Van Liempt, H., Palissa, H., Unkles, S.E., Riach, M.B.R., Pfeifer, E., Von Doehren, H. and Kinghorn, J.R. (1991) δ-(L-α-aminoadipyl)-L-cysteinyl-D-valine synthetase from *Aspergillus nidulans* - Molecular characterization of the *acvA* gene encoding the first enzyme of the penicillin biosynthetic pathway. J. Biol. Chem. 266, 12646-12654.

Makins, J.F., Holt, G. and Macdonald, K.D. (1983) The genetic location of three mutations impairing penicillin production in *Aspergillus nidulans*. J. gen. Microbiol. 129, 3027-3033.

Malpartida, F. and Hopwood, D.A. (1984) Molecular cloning of the whole biosynthetic pathway of a *Streptomyces* antibiotic and its expression in a heterologous host. Nature 309, 462-464.

Marahiel, M.A., Zuber, P., Czekay, G. and Losick, R. (1987) Identification of the promoter for a peptide antibiotic biosynthesis gene from *Bacillus brevis* and its regulation in *Bacillus subtilis*. J. Bacteriol. 169, 2215-2222.

Martin, J.F. and Aharonowitz, Y. (1983) Regulation of biosynthesis of ß-lactam antibiotics, in "Antibiotics Containing the Beta-Lactam structure I" (Demain, A.L. and Solomon, N.A., Eds.), pp. 229-254. Springer Verlag, Berlin.

Martin, J.F. and Demain, A.L. (1980) Control of antibiotic biosynthesis. Microbiol.Rev. 44, 230-251.

Martinez-Blanco, H., Reglero, A., Fernandez-Valverde, M., Ferrero, M.A., Moreno, M.A., Penalva, M.A. and Luengo, J.M. (1992) Isolation and characterization of the acetyl-CoA synthetase from *Penicillium chrysogenum* - involvement of this enzyme in the biosynthesis of penicillins. J. Biol. Chem. 267, 5474-5481.

Merrick, M.J. and Caten, C.E. (1975) The inheritance of penicillin titre in wild-type isolates of *Aspergillus nidulans*. J. gen. Microbiol. 86, 283-293.

Montenegro, E., Barredo, J.L., Gutierrez, S., Diez, B., Alvarez, E. and Martin, J.F. (1990) Cloning, characterization of the acyl-CoA:6-amino penicillanic acid acyltransferase gene of *Aspergillus nidulans* and linkage to the isopenicillin N synthase gene. Mol. Gen. Genet. 221, 322-330.

Penalva, M.A., Moya, A., Dopazo, J. and Ramon, D. (1990) Sequences of isopenicillin N synthetase genes suggest horizontal gene transfer from prokaryotes to eukaryotes. Proc. R. Soc. Lond. B 241, 164-169.

Pontecorvo, G., Roper, J.A., Hemmons, L.M., Macdonald, K.D. and Butron, A.W.J. (1953) The genetics of *Aspergillus nidulans*. Adv. Genet. 5, 141-238.

Ramon, D., Carramolino, L., Patino, C., Sanchez, F. and Penalva, M.A. (1987) Cloning and characterization of the isopenicillin N synthetase gene mediating the formation of the betalactam ring in *Aspergillus nidulans*. Gene 57, 171-181.

Samson, S.M., Belagaje, R., Blankenship, D.T., Chapman, J.L., Perry, D., Skatrud, P.L., Van Frank, R.M., Abraham, E.P., Baldwin, J.E., Queener, S.W. and Ingolia, T.D. (1985) Isolation, sequence determination and expression in *Escherichia coli* of the isopenicillin N synthetase gene from *Cephalosporium acremonium*. Nature 318, 191-194.

Shah, A.J., Tilburn, J., Adlard, M.W. and Arst, H.N., Jr. (1991) pH regulation of penicillin production in *Aspergillus nidulans*. FEMS Microbiol.Lett. 77, 209-212.

Simpson, I. and Caten, C.E. (1979) Induced quantitative variation for penicillin titre in clonal populations of *Aspergillus nidulans*. J. gen. Microbiol. 110, 1-12.

Simpson, I.N. and Caten, C.E. (1979) Recurrent mutation and selection for increased penicillin titre in *Aspergillus nidulans*. J. gen. Microbiol. 113, 209-217.

Smith, D.J., Burnham, M.K.R., Edwards, J., Earl, A.J. and Turner, G. (1990*a*) Cloning and heterologous expression of the penicillin biosynthetic gene cluster from *Penicillium chrysogenum*. Bio/Technology 8, 39-41.

Smith, D.J., Burnham, M.K.R., Bull, J.H., Hodgson, J.E., Ward, J.M., Browne, P., Brown, J., Barton, B., Earl, A.J. and Turner, G. (1990*b*) ß-Lactam antibiotic biosynthetic genes have been conserved in clusters in prokaryotes and eukaryotes. EMBO J. 9, 741-747.

Smith, D.J., Earl, A.J. and Turner, G. (1990*c*) The multifunctional peptide synthetase performing the first step of penicillin biosynthesis in *Penicillium chrysogenum* is a 421,073 dalton protein homologous to *Bacillus brevis* peptide synthetases. EMBO J. 9, 2743-2750.

Somerson, N.L., Demain, A.L. and Nunheimer, T.D. (1961) .Reversal of lysine inhibition of penicillin production by α-aminoadipic acid or adipic acid. Arch.Biochem.Biophys. 93, 238-241.

Tilburn, J., Scazzocchio, C., Taylor, G.G., Zabicky-Zissman, J.H., Lockington, R.A. and Davies, R.W. (1983) Transformation by integration in *Aspergillus nidulans*. Gene 26, 205-221.

Turgay, K., Krause, M. and Marahiel, M.A. (1992) Four homologous domains in the primary structure of GrsB are related to domains in a superfamily of adenylate-forming enzymes. Mol. Microbiol. 6, 529-546.

Van Liempt, H., Von Döhren, H. and Kleinkauf, H. (1989) δ-(L--aminoadipyl)-L-cysteinyl-D-valine synthetase from *Aspergillus nidulans*. The first enzyme in penicillin biosynthesis is a multifunctional peptide synthetase J. Biol.Chem. 264, 3680-3684.

Weigel, B.J., Burgett, S.G., Chen, V.J., Skatrud, P.L., Frolik, C.A., Queener, S.W. and Ingolia, T.D. (1988) Cloning and expression in *Escherichia coli* of isopenicillin N synthetase genes from *Streptomyces lipmanii* and *Aspergillus nidulans*. J. Bacteriol. 170, 3817-3826.

Xuei, X. and Skatrud, P. (1993) Electrophoretic molecular karyotypes of echinocandin B-producing strains of *Aspergillus nidulans*. Seventeenth Fungal Genetics Conference, Asilomar. Poster abstract.

Yelton, M.M., Hamer, J.E. and Timberlake, W.E. (1984) Transformation of *Aspergillus nidulans* by using a *trpC* plasmid. Proc. Natl. Acad. Sci. USA 81, 1470-1474.

Zhang, J. and Demain, A.L. (1992) ACV synthetase. Critical Reviews in Biotechnology 12, 245-260.

# MOLECULAR GENETICS OF THE *bimB* AND *bimD* GENES OF *ASPERGILLUS NIDULANS*, TWO GENES REQUIRED FOR MITOSIS

Gregory S. May, Steven H. Denison, Cydne L. Holt, Carol A. McGoldrick and Paul Anaya

Department of Cell Biology
Baylor College of Medicine
1 Baylor Plasa
Houston, TX 77030

## INTRODUCTION

Advances in understanding how cells duplicate and segregate their constituents through the cell cycle have been greatly facilitated by the genetic analysis of these processes in microbial eukaryotic organisms. The most notable of these organisms are the yeasts *Saccaharomyces cerevisiae* and *Schizosaccharomyces pombe* and the filamentous fungus *Aspergillus nidulans*. Much of this work has been directed towards understanding the regulators of major transition points in the cell cycle such as the transitions from G1 to S-phase and G2 to mitosis. For reviews of this literature the reader is referred to the following references: Doonan; 1992; Hartwell and Weinert, 1989; Murray, 1992. The advances made in *A. nidulans* were made possible by the isolation of a collection of heat sensitive (hs$^-$) mutants (Morris, 1976). This collection of hs$^-$ mutants contained a class of mutations that altered the cell's ability to traverse the cell cycle at restrictive temperature and in some cases resulted in a block in mitosis. Those mutations that resulted in a block in mitosis were designated *bim* and it is the *bim* genes that will be discussed here in some detail.

There have been seven *bim* genes defined by mutation. What is known about each of these genes and their polypeptide products is given in Table 1. The *bim* mutations have identified genes whose products might be expected to find to be involved in mitosis. Two such examples are the *bimC* and *bimG* genes. The former gene encodes a polypeptide with extensive homology to the kinesin heavy chain polypeptide, a protein that transports vesicles along microtubules and an activity that had been suggested might also be found to function in mitosis (Enos and Morris, 1990). *bimG* is the structural gene for a type 1 phosphoprotein phosphatase (Doonan and Morris, 1989). Phosphatases would have been predicted since entry into mitosis is regulated by protein phosphorylation, one would

*The Genus Aspergillus*, Edited by Keith A. Powell *et al.*,
Plenum Press, New York, 1994

expect to find protein phosphatases that are required to exit mitosis. Other genes encode polypeptides which are novel products of unknown function. These genes are the *bimA*, *bimB*, *bimD* and *bimE* genes. The *bimA* gene product is apparently a component of the spindle pole (Mirabito and Morris, 1993). Finally there is the *bimF* gene for which nothing is currently known (O'Donnell *et al.*, 1991; May *et al.*, 1992; Denison and May, 1993 Submitted; Engle *et al.*, 1990).

**Table 1.** The *bim* Genes of *Aspergillus nidulans*.

| Gene | Properties and Identity or Probable Function |
|---|---|
| *bimA* | Similar to *nuc2* of *S. pombe* and other genes of *S. cerevisiae*. Localises to the spindle pole body. |
| *bimB* | Mutations in this gene uncouple DNA replication from the completion of mitosis, leading to large polyploid nuclei with multiple spindle pole bodies. The c-terminus has similarity to the c-termini of the fission yeast *cut1*$^{+}$ gene and the budding yeast *ESP1* gene. |
| *bimC* | Gene encodes a kinesin-like heavy chain polypeptide. The *bimC3* mutation is defective for spindle pole separation. |
| *bimD* | Mutations in this gene result in spindles lacking some microtubules, including the spindle pole to chromosome microtubules and sensitivity to mutagens. Gene encodes a polypeptide with sequence motifs found in the leucine zipper class of transcription factors. |
| *bimE* | Negative regulator of entry into mitosis, encoding a large polypeptide with membrane spanning domains. |
| *bimF* | Unknown. |
| *bimG* | Encodes a type 1 phosphoprotein phosphatase required to complete mitosis. |

Studies have been made on the *bimB* and the *bimD* genes from *A. nidulans*. Mutations in either of these genes result in a temporary delay in progression through mitosis, the previously termed, blocked in mitosis phenotype. It has been shown that for the *bimB3* and *bimD6* mutations that the mitotic block at restrictive temperature is temporary. Upon reentry into the cell cycle these two mutations produce strikingly different terminal phenotypes. Some of our work on the mutant phenotypes for each of these mutations, details of the cloning of these genes and models for the function of each of these genes, is described in this chapter.

## PHENOTYPES OF *bimB3* AND *bimD6* MUTANT STRAINS

### The *bimB3* Mutant Phenotype

As had been previously reported both the *bimB3* and the *bimD6* mutations result in an increased chromosomal and spindle mitotic index upon a shift to restrictive temperature (Morris, 1976). For both of these mutants though the mitotic index soon declines and the nuclei take on an appearance very different from that of a nucleus blocked in mitosis. In a *bimB3* mutant strain grown for 14 hours at the restrictive temperature the cells are uninucleate and the nuclei are highly enlarged (Figure 1). The appearance of these nuclei in comparison to those of a wild type strain grown at the same temperature, or the mutant strain grown at permissive temperature is suggestive of an increased DNA content or ploidy. In addition the chromosome mitotic cycle appeared to be continuing because nuclei were present that had chromatin in a condensed state similar to that found at mitosis. Examination of the DNA content of wild type nuclei and those of a *bimB3* mutant strain to determine if there was a true increase in DNA content in the nuclei of the mutant at restrictive temperature revealed that the nuclei of the mutant had approximately 6 times more DNA when grown for 14 hours at restrictive temperature as compared to the nuclei of cells grown at permissive temperature (Table 2).

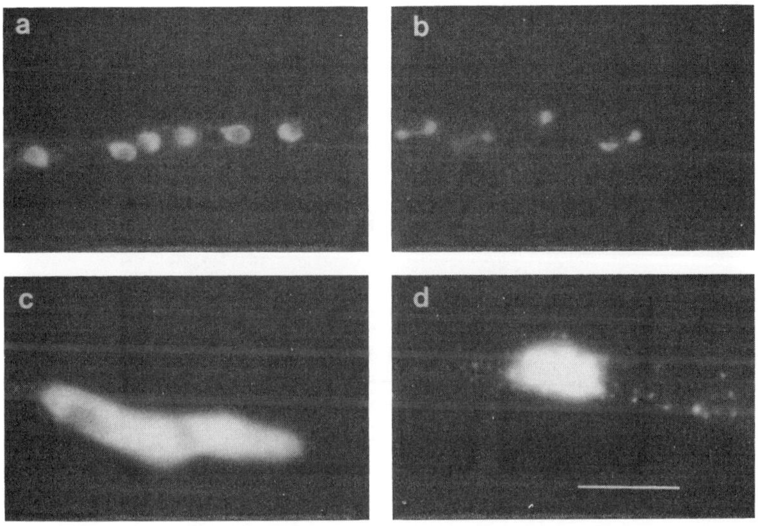


**Figure 1.** DAPI fluorescence micrographs of a wild type strain (a and b) and a *bimB3* mutant strain (c and d) grown for 14 hours at 42°C. The wild type strain shows normal nuclear morphology during interphase (a) and mitosis (b) while the *bimB3* mutant strain shows the single large nucleus during interphase (c) and mitosis (d). The scale bar in panel d is 10$\mu$m.

**Table 2.** Nuclear DNA content of a *bimB3* mutant strain at permissive and restrictive temperatures.

| Gowth temperature | Average fluorescence intensity | Average nuclear area | Total average fluorescence intensity |
|---|---|---|---|
| °C | | | |
| 32 | 1076[a] ($n$=6)[b] | 13114 | 13,3142,103 (1X)[c] |
| 42 | 1185 ($n$=6) | 73558 | 87,0467,748 (6.1X) |

[a] Values are averages S.E.M.
[b] $n$ is the number of individual nuclei analysed. The nuclei were in different cells.
[c] This is the -fold increase in total average fluorescence intensity measure relative to cells grown at permissive temperature. Units of fluorescence intensity are arbitrary.

Microtubule arrays were examined in a *bimB3* mutant strain by immunofluorescence microscopy. Cells grown at permissive temperature and shifted to restrictive temperature for four hours exhibited abnormal spindle morphology when compared to spindles of cells maintained at permissive temperature (Figure 2). Normal bipolar spindles stained with antitubulin antibodies appear as discrete bars with the chromatin residing on the microtubules (Figure 2, panels a and b). In contrast the spindle of mitotic cells kept at restrictive temperature for 4 hours take on an appearance that is suggestive of multiple spindle poles in that the microtubule staining appears as square- or chevron-shapes (Figure 2, panels c and d). An attempt was made to examine the microtubule arrays of the mutant grown at restrictive temperature as described for Figure 1, however, microtubules could not be found in these cells but were discovered in wild type cells. Using the nuclear morphology as a guide to the genotype, it was shown that microtubules could be seen in the wild type but not the mutant cells. These data would suggest that microtubule assembly in the *bimB3* mutant strain after extended periods at restrictive temperature has generally failed. Results from the mutant phenotype analysis suggest that the cells enter the next round of DNA replication following a failure of the spindle to segregate the

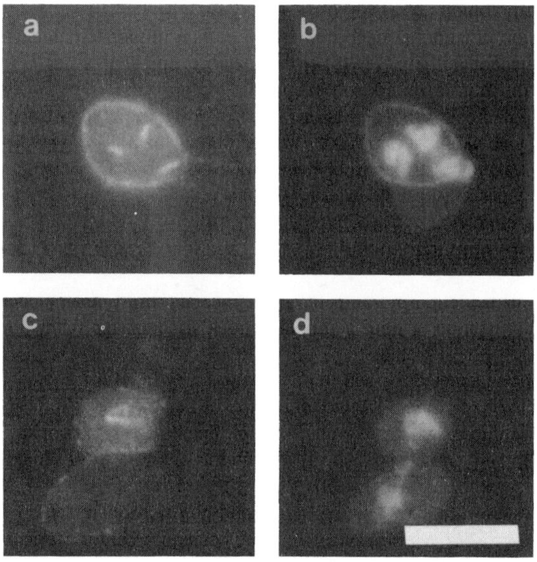


**Figure 2.** Immunofluorescence localisation of tubulin in a *bimB3* mutant strain at permissive (a and b) and restrictive (c and d) temperatures. Microtubule staining with antitubulin antibodies (a and c) and DNA staining with DAPI (b and d). The scale bar in panel d is 10 $\mu$m.

chromosomes at mitosis. This would explain how cells grown at restrictive temperature from conidia for multiple nuclear division cycles remain uninucleate and yet have an increased DNA content. A similar phenotype can be observed if all microtubule functions in the cell are disrupted using the antimicrotubule compound benomyl (May *et al.*, 1992).

### The *bimD6* Mutant Phenotype

The *bimD6* mutation results in a dramatic phenotype at restrictive temperature that was first observed by Oakley (1981). At restrictive temperature *bimD* mutant strains have a deficiency of spindle microtubules and the chromosomes do not attach to the spindle. A comparison of the mitotic spindle and chromatin of *bimD6* mutant cells at permissive and restrictive temperatures illustrates this very clearly (Figure 3). During anaphase and telophase the chromatin of wild type cells, or a *bimD6* mutant strain grown at permissive temperature, resides as two discrete masses at the poles of the spindle. In contrast the chromatin of a *bimD6* mutant strain at restrictive temperature typically appears as three to four discrete masses, indicating the failure of the chromosomes to attach to the spindle. In addition the overall fluorescence intensity of the mutant spindle appears to be less, reflecting the reduced numbers of microtubules in the spindle. During the course of these studies it was also noted that nearly all the spindles observed for the mutant at restrictive temperature were of the same length, indicating early anaphase. The frequency of mitotic cells in the different stages of mitosis, metaphase, anaphase and telophase (Figure 4), was quantified for a *bimD6* mutant strain at permissive and restrictive temperatures (Table 3). These results suggest that at restrictive temperature, mitosis in a *bimD6* mutant strain arrests with an anaphase spindle configuration. Thus *bimD6* mutant spindles cannot progress into telophase, suggesting that migration of chromosomes to the poles is required for further spindle elongation at anaphase.

The *bimD6* mutation confers a second phenotype, sensitivity to mutagens at temperatures that are permissive for the mitotic defect. This phenotype was discovered during the course of our genetic mapping of the mutation. *bimD6* mutant strains are more sensitive to ultraviolet light (UV) and methylmethane sulfonate (MMS) than wild type

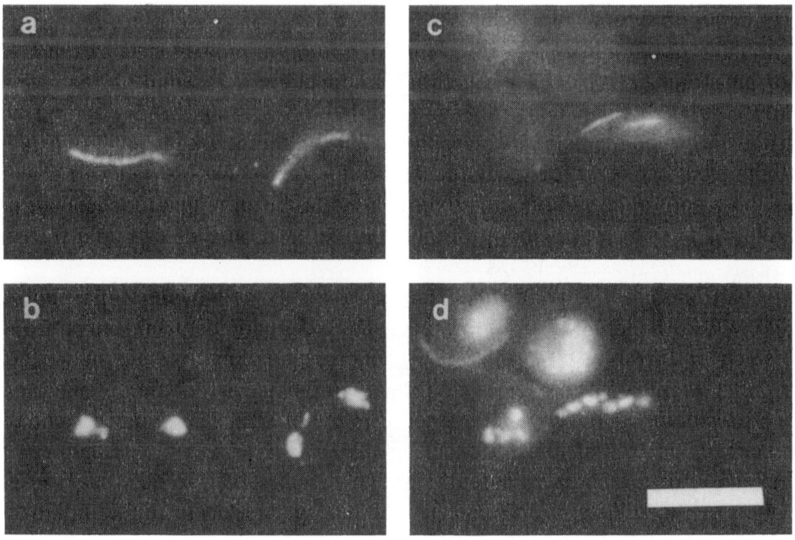


**Figure 3.** Fluorescence micrographs showing the localisation of tubulin (a and c) and DNA (b and d) during a normal mitosis (a and b) and that of a *bimD6* mutant cell at restrictive temperature (c and d). The scale bar in panel d is 10 $\mu$m.

strains (Figure 5). The sensitivity of the *bimD6* mutant strain to UV is less than that conferred upon a strain by the *uvsB110* mutation, but the two mutations have nearly identical sensitivity to MMS. The mechanism by which a single mutation confers the mitotic defect at restrictive temperature and the sensitivity to DNA damaging agents is unknown. Current evidence suggests that *bimD6* mutant strains exhibit a mitotic defect following MMS treatment that is similar to that seen at restrictive temperature. Thus, one hypothesis is that the mutagen sensitive phenotype is a hypomorphic one and that the hs⁻ mitotic defect is a loss of function phenotype. Future experiments and elucidation of *bimD* function should permit the determination of the reason for the two phenotypes.

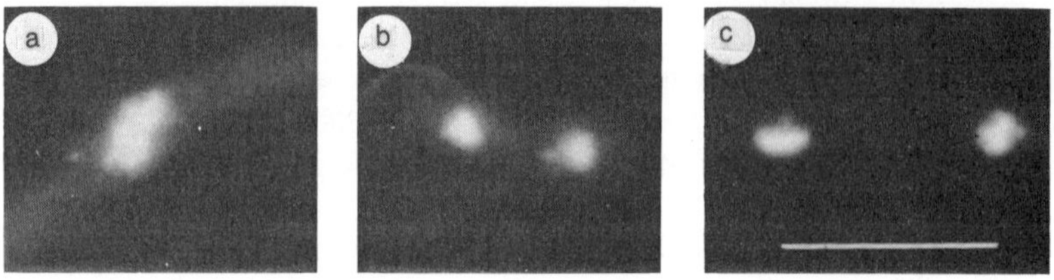


**Figure 4.** DAPI fluorescence micrographs showing metaphase (a), anaphase (b) and telophase (c) stages of a wild type mitosis. The scale bar in panel c is 10 $\mu$m.

**Table 3.** Frequency of mitotic stages of a *bimD6* mutant strain at permissive and restrictive temperatures.

| Growth temperature | Metaphase[1] | Anaphase | Telophase |
|---|---|---|---|
| Permissive | 55% | 3% | 42% |
| Restrictive | 6% | 94% | 0% |

[1]Values presented are based upon a sample size of 100 cells.

## CLONING THE *bimB* AND *bimD* GENES

### Cloning the *bimB* Gene

Early attempts to clone the *bimB* gene by complementation with existing plasmid libraries failed. A more directed approach was adopted to clone the *bimB* gene based upon genetic mapping information. Mapping of the *bimB3* mutation to locus placed it in the interval between *phenA* and *suBpro* on the right arm of linkage group III. The recombination frequencies between these two genes indicated that *bimB* lay very close to *amdS*, a previously cloned gene (Hynes *et al.*, 1983). Five cosmids that hybridised with *amdS* were co-transformed with the the *pyrG* gene into a *bimB3, pyrG89* double mutant strain to test for *bimB3* complementing activity. It was determined that two of the five cosmids tested complemented the *bimB3* mutation at high frequency. Restriction endonclease generated fragments from one of these cosmids were used to co-transform the *bimB3, pyrG89* double mutant and localised the complementing activity to an approximately 9 kilobase pair (kb) *Kpn*I, *Eco*RI fragment (Figure 6). Additional restriction endonuclease fragments were used to probe Northern transfers of polyadenylated messenger RNA and identified a transcriptional unit of approximately 6 kb in size that was flanked on either side by transcriptional units of approximately 2kb. These data combined with the complementation data indicated that the approximately 6kb transcriptional unit represented the *bimB* gene. Four overlapping cDNA clones (1,2,3 and 7) were identified based upon hybridisation with fragments of the genomic clone (Figure 6). These cDNA clones were completely sequenced, yielding a contiguous sequence of 6,316 base pairs.

This sequence contained an open reading frame of 2,068 amino acid residues in length encoding a polypeptide (BIMB) with a calculated molecular mass of 227,958 Da. When referring to the polypeptide product of the *bimB* or other genes all capital letters are used (to indicate the protein product is being referred to).

The derived amino acid sequence for the product of the *bimB* gene has no readily identifiable structural features based upon computer based structural prediction programs. A search of the protein data bases with BIMB identified two proteins that had a region of high similarity. These two proteins were CUT1 of *S. pombe* and ESP1 of *S. cerevisiae*. The region of similarity between these two proteins and BIMB encompasses approximately 400 amino acids and is located in the carboxyl terminus. This region had previously been shown to be similar between CUT1 and ESP1 (Uzawa *et al.*, 1990). Curiously, outside of this region of similarity there are no other regions of significant similarity between any

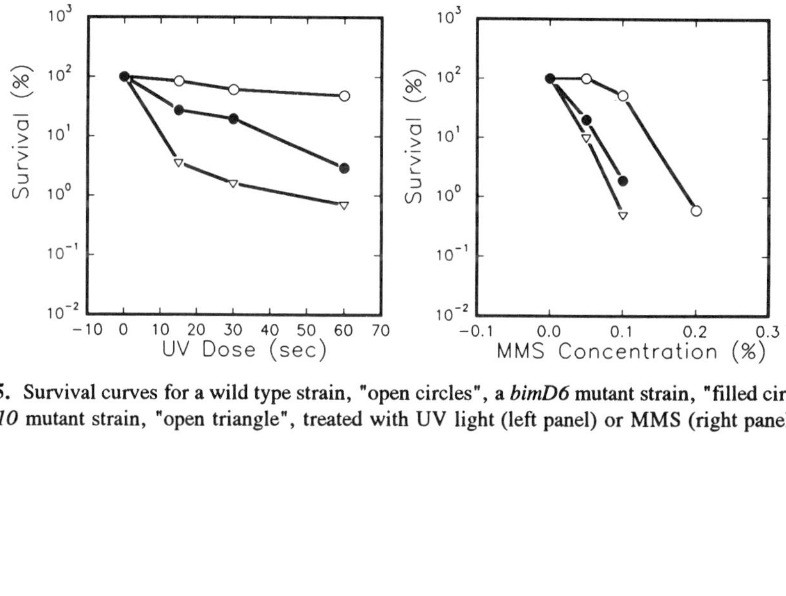


**Figure 5.** Survival curves for a wild type strain, "open circles", a *bimD6* mutant strain, "filled circles", and a *uvsB110* mutant strain, "open triangle", treated with UV light (left panel) or MMS (right panel).

of these polypeptides in pairwise comparisons. Thus, approximately 80% of each of the predicted polypeptides are not related. The structural relationship of the BIMB, CUT1 and ESP1 polypeptides to one another is shown schematically in Figure 7.

In addition to sharing a region of homology between these three proteins, mutations in these three genes give similar phenotypes at restrictive temperature (Baum *et al.*, 1988; McGrew *et al.* 1992; Uzawa *et al.*, 1990). The similarities would suggest that these three proteins have a common function in mitosis. Given this, it is somewhat surprising that the three gene products do not exhibit greater similarity. Alternatively it is possible that in the three different organisms the mutations define different members of a small gene family. We tested this hypothesis in *A. nidulans* by hybridising the conserved domain from the carboxyl terminus of *bimB* to a Southern transfer of total DNA digested with various restriction endonucleases (Figure 8). For each of the four restriction endonucleases used at least two bands of hybridisation were detected, where only one would be expected based on the DNA sequence. In contrast, a hybridisation probe generated from amino terminal domain that lacks homology with *cut1* or ESP1 hybridises

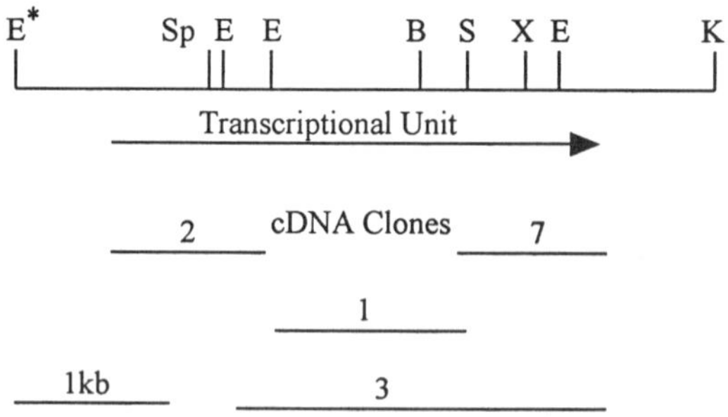


**Figure 6.** Partial restriction map for the *bimB* genomic clone. The direction of transcription and the limits of the *bimB* transcriptional unit are indicated by the arrow. The size and location of the overlapping cDNA clones used to determine the sequence of *bimB* are also shown. Sites for the restriction endonucleases *Bam*HI (B), *Eco*RI (E), *Kpn*I (K), *Sma*I (S), *Sph*I (Sp), and *Xba*I (X) are indicated on the genomic clone map. The *Eco*RI site indicated by E* is a site obtained from the cloning vector and is at the end of the genomic clone.

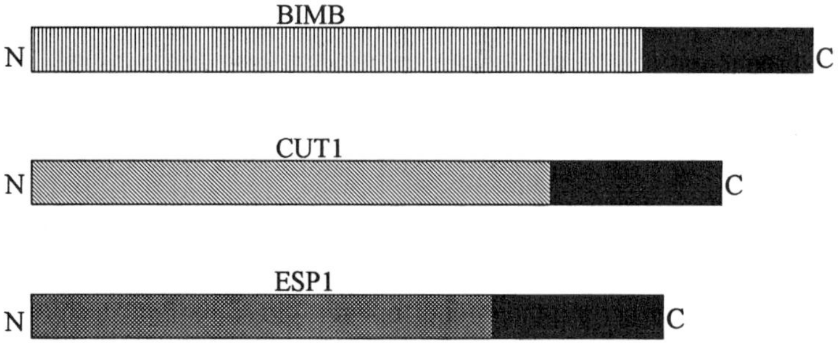


**Figure 7.** Schematic representation of the structure of the predicted BIMB, CUT1 and ESP1 polypeptides.The region of similarity at the carboxyl terminus is solid and the region of no similarity is hatched using a different pattern for each of the polypeptides to emphasise the differences.

with a single band as would be predicted for a single copy gene. Thus it would appear that more than one DNA sequence is present in the genome of *A. nidulans* that has homology to this conserved carboxyl terminal domain found in BIMB, CUT1 and ESP1. It will be interesting to determine the identity of the other DNA sequence in the genome of *A. nidulans* that is hybridising with this probe. One possiblity is that this sequence is from a gene that is more closely related to *cut1* or *ESP1* than to *bimB*.

## Cloning the *bimD* Gene

Compared to cloning the *bimB* gene, a somewhat less direct approach was used to clone the *bimD* gene. All the linkage group IV specific cosmids were first identified in two separate cosmid libraries (Brody *et al*, 1991). These sorted cosmids were grown

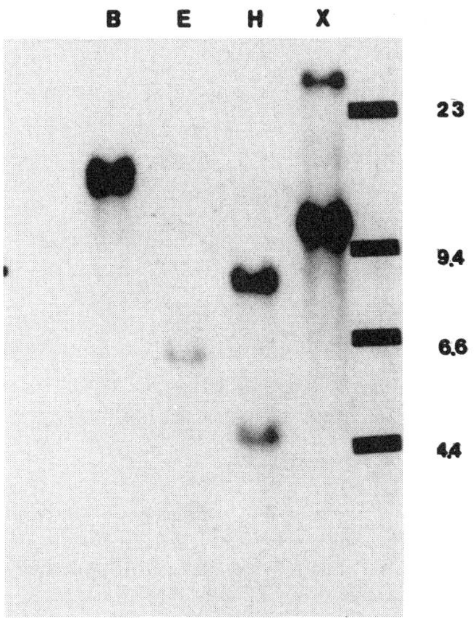


**Figure 8.** Southern transfer of total DNA from *A. nidulans* probed with the conserved carboxyl terminal domain from *bimB*. The conserved carboxyl terminal domain detects at least two bands of hybridisation. Total DNA was digested with *Bam*HI, *Eco*RI, *Hin*dIII or *Xba*I; lanes labelled B, E, H and X respectively.

individually and placed into pools of forty clones. The DNA from each of the pools of forty was co-transformed with the *pyrG* gene into a *bimD6, pyrG89* double mutant strain. A pool that had complementing activity was identified, the individual cosmids that constituted the pool were grown and DNA prepared from them. This cosmid DNA was co-transformed into the double mutant to identify the cosmid with complementing activity. As for the cloning of the *bimB* gene a large clone that needed to be subdivided was also obtained.

The purified cosmid that had the complementing activity was digested with several restriction endonucleases and *Kpn*I was found to produce a manageable number of reasonably sized fragments. Each of the *Kpn*I fragments was co-transformed into the *bimD6, pyrG89* double mutant strain. A *Kpn*I fragment of 12.7kb was identified that complemented the *bimD6* mutation at high frequency, indicating the presence of the *bimD* gene on this fragment. A detailed analysis of this *Kpn*I fragment was undertaken to identify the *bimD* transcriptional unit. Using restriction endonucleases various DNA fragments were generated, tested for their ability to complement the *bimD6* mutation and the messenger RNA species they hybridised with on Northern transfers. The *bimD* transcriptional unit was centrally located on the 12.7kb *Kpn*I fragment and was about 5kb in length (Figure 9).

Three cDNA clones for the *bimD* transcriptional unit were identified by hybridisation with genomic DNA fragments that spanned the transcriptional unit (Figure 9). Each of these clones was sequenced and one of these clones, #7, was found to be full length. The full length cDNA clone was 4,885 base pairs long and contained an open reading frame of 1506 amino acid residues, encoding a polypeptide of a calculated molecular mass of

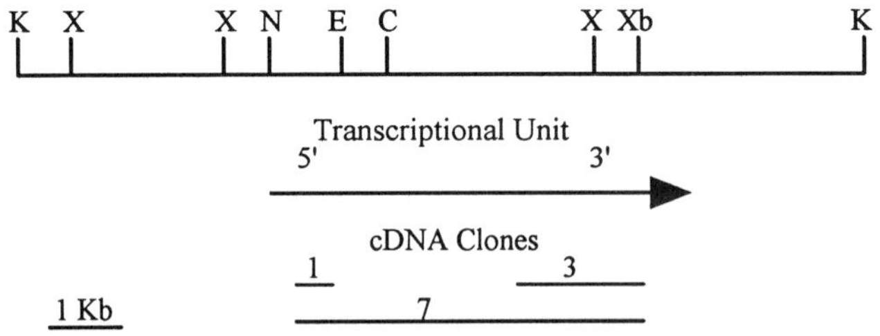


**Figure 9.** Partial restriction map for the *bimD* genomic clone. The direction of transcription and the limits of the *bimD* transcriptional unit are indicated by the arrow. The size and location of the overlapping cDNA clones used to determine the sequence of *bimD* are also shown. Sites for the restriction endonucleases, *Cla*I (C), *Eco*RI (E), *Kpn*I (K), *Nco*I (N), *Xba*I (Xb) and *Xho*I (X) are indicated on the genomic clone map.

166,295 Da. Three structural features of the predicted BIMD polypeptide are notable (Figure 10). The first, is the presence of three nuclear localisation signals. The second, is the presence of a leucine zipper motif preceded by a cluster of basic amino acids, features found in the bzip class of DNA binding transcription factors. The last feature is the presence of a region at the carboxyl terminus enriched in acidic amino acids, 28% over the last 140 amino acids as compared to 14% over the entire polypeptide. Taken together these features indicate that BIMD has the potential to be a DNA binding protein and possibly a transcription factor.

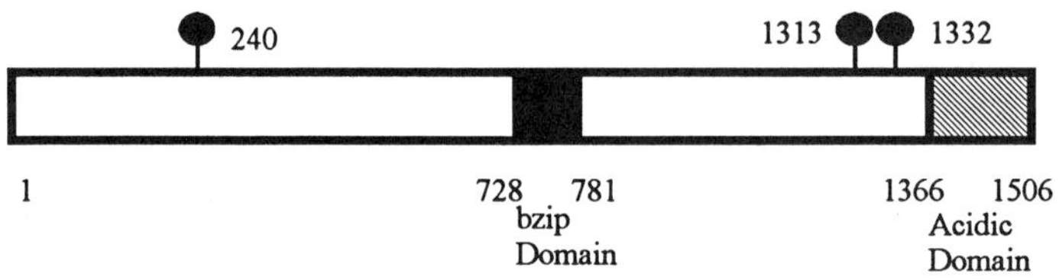


**Figure 10.** Schematic representation of the features identified in the BIMD polypeptide. The position and amino acid residues of the nuclear localisation signals are indicated by the lines with the circles at their ends. The amino acid residues encompassing the bzip domain, solid region, and the acidic domain, hatched region are also given.

## WHAT IS THE FUNCTION OF BIMD?

When studies of the *bimD* gene were started it was hypothesised that its product functioned in the formation or stabilisation of the spindle pole to kinetochore microtubules. This was based upon the fact that *bimD6* mutant strains lacked these microtubules and that the chromatin did not become attached to the spindle. The sequence of BIMD suggests two possible mechanisms by which it may achieve this. The first is that *bimD* utilizes the

DNA binding motifs of the bzip transcription factors to bind to centromeric DNA. The other possibility is that BIMD as a transcription factor may function to regulate genes whose products are required for the formation of these spindle structures. It was necessary to test each of these possiblities and to understand how these functions could explain the mutagen sensitivity of mutations in *bimD*.

To test whether BIMD was a transcription factor we placed the *bimD* cDNA was placed under the control of the regulatable *alcA* promoter. It was assumed that if BIMD activates genes that are required for the formation of mitotic structures we might see cells in mitotic-like states when BIMD is overexpressed. Primary transformants were tested for their ability to grow on media containing a repressing acetate, or inducing ethanol, carbon source. When placed on inducing media it was noted that there was a significant inhibition of growth in comparison to control transformed strains and that this inhibition of growth varied between strains. It was subsequently determined that the variation in growth correlated with the number of copies of the transforming plasmid, with increased copy number there was increased inhibition of growth. When a strain containing five copies of the transforming plasmid was examined for its nuclear morphology it was found to have a normal interphase appearance. The one unusual feature was that all the cells were uninucleate, suggesting an inhibition of the nuclear division cycle.

The block in the nuclear division cycle was unexpected and could have been because of a toxicity of high levels of BIMD and not represent a specific BIMD function. Therefore an examination was made to assess whether cells arrested in their cell cycle by BIMD overexpression could recover from the block and if so, where they were arrested in the cell cycle. Cells blocked by overexpression of BIMD were washed into repressing media having or lacking hydroxyurea, an inhibitor of DNA synthesis that will arrest cells in S phase (Bergen *et al.*, 1984). It is predicted that if the block in cell cycle progression was specific and reversible it would be possible to determine the point of the cell cycle arrest. The prediction is that cells blocked in G1 or S phase would remain uninucleate when washed into hydroxyurea, whereas cells blocked in G2 would become binucleate and then arrest at the start of the next S phase and that if the cells were blocked randomly throughout the cell cycle then would be expected a mixed result with some cells remaining uninucleate and others becoming binucleate.

It was found that the cells were reversibly blocked in G1 or S phase of the cell cycle (Figure 11). Cells germinated in the presence of ethanol were all uninucleate and when washed into repressing media were able to undergo nuclear division, thus the block was reversible. In contrast if the cells were placed into repressing media in the presence of hydroxyurea they remained uninucleate indicating that they were arrested in G1 or S phase of the cell cycle due to BIMD overexpression. This suggests that BIMD may act to negatively regulate cell cycle progression or alternatively may be titrating out a positively acting cellular factor that is required for cell cycle progression.

## DISCUSSION

### The *bimB* gene

Conditional mutations in *ESP1* and *cut1*$^+$, like the *bimB3* mutation, uncouple DNA replication from the completion of a successful mitosis. The shared homologous domain and terminal phenotypes for the conditional mutations in these three genes suggest similar functions for their gene products in mitosis. A question that remains to be addressed is if these genes are functionally equivalent or if they represent three individuals of a larger gene family. A larger family of genes is certainly suggested by the Southern analysis of *A. nidulans* in which at least two bands of hybridisation were detected with a probe from the conserved carboxyl terminal region. The possibility of functional interchangeability is particularly interesting given that the three predicted polypeptides exhibit sequence similarity over only approximately 400 amino acids, representing just 20-25% of each of these proteins. Does this mean that the remaining 75-80% of the encoded polypeptides have a unique function(s)? A lack of function for the region of nonhomology is highly unlikely given that the predicted *cut1*$^+$ polypeptide has sequence motifs for an ATP-binding site which are not found in *ESP1* or *bimB*. The lack of any similarity over 75-80% of the sequence of these proteins, including a putative ATP-binding site in *cut1*$^+$ that is absent in *ESP1* and *bimB*, makes it improbable that they are functionally interchangeable. *bimB, cut1*$^+$, and *ESP1* may be functionally related but not able to replace one another because of divergence in the components with which they interact. Alternatively, there are multiple functions that require proteins with a carboxyl terminal

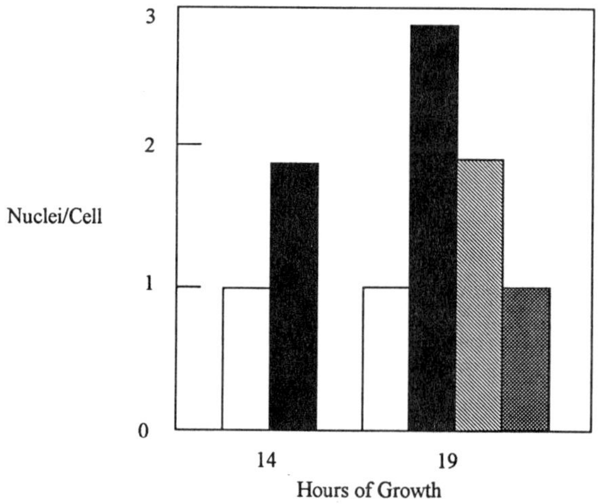


**Figure 11.** A *bimD* overexpressing strain (white bars, single hatched and double hatched bar) and a control strain (black bar) were germinated for fourteen hours at 32°C in ethanol (inducing) medium and samples were taken for nuclear number determination. At this time cells were either maintained in ethanol (white and black bars at 19 hours) or transferred to glucose (repressing) medium (single hatched bar) or glucose medium containing 90mM hydroxyurea (double hatched bar) and incubated a further five hours. Samples were then taken for determination of nuclear number. Values are the average for 100 cells for each condition.

domain similar to *bimB*. It has been shown that *ESP1* cannot complement a conditional *cut1* mutant. It will be of interest to determine whether *bimB* can complement conditional *cut1* mutants and if *cut1*$^+$ can complement the *bimB3* conditional mutation. It will also be important to determine the identity of the gene that corresponds to the second band of hybridisation in the *A. nidulans* genome. One possibility is that the second band of hybridisation is derived from an *ESP1* or *cut1*$^+$ homologue in *A. nidulans*.

It has been suggested that *cut1*$^+$ and *ESP1* function to regulate SPB duplication (Baum *et al.*, 1988; Uzawa *et al.*, 1990). An alternative explanation is that these gene products are not necessarily regulatory but instead may have a spindle function. There are precedents for this argument based on the effects of antimicrotubule agents in plant, animal, and fungal cells, as well as other conditional mitotic mutations and gene disruption experiments in *A. nidulans*. The production of polyploids in plants following treatment with the antimicrotubule drug colchicine is a long established phenomenon (Dustin, 1984). Similarly, it is possible to produce polyploid animal cells by interfering with microtubule function (Cabral, 1983). These observations are not unique to plants and animals and similar studies have been performed in fungi, including *A. nidulans* (Kunkel and Hadrich, 1977). Serial section electron microscopy was used in this study to demonstrate that the antimicrotubule compound methyl benzimidazol-2-yl carbamate, the active component of benomyl following hydrolysis in aqueous solution, produced cells containing multiple SPBs. It was suggested by this that the replication cycle of the SPB could take place in the absence of nuclear division. The *bimC4* mutation in *A. nidulans* produces multiple SPB and large nuclei that undergo rounds of replication in the absence of mitosis at restrictive temperature, similar to the phenotypes of conditional mutations in *bimB, cut1*$^+$, and *ESP1* genes (Enos and Morris, 1990). Yet, *bimC* does not encode a regulator of SPB duplication but is the structural gene for the heavy chain of a kinesin-like protein, a microtubule motor required for SPB separation. In addition, disruption of the $\gamma$-tubulin gene *mipA*, a component of the SPB, produces a similar phenotype (Oakley *et al.*, 1990). Like the previously cited studies it has been shown that the antimicrotubule drug benomyl results in polyploid nuclei. Therefore, it is proposed that the products of the *bimB, cut1*$^+$, and *ESP1* genes do not have to necessarily be regulators of SPB duplication but maybe genes whose products could function in the mitotic spindle. Our model is that the mutant gene products at restrictive temperature interfere with or block the normal functioning of the mitotic spindle resulting in a defective mitosis, which is then bypassed or exited from. The result is a cell containing a single nucleus with multiple SPB and an increased DNA content. It is the similarity of the effects of antimicrotubule agents in the production of polyploid cells, and in the case of *Aspergillus* multiple SPBs, that lead to the suggestion that these genes may function in the mitotic spindle and need not be regulators of SPB duplication.

### The *bimD* gene

The early models for BIMD function predicted that BIMD was a structural protein, possibly a kinetochore protein or a component of the mitotic spindle required to stabilise kinetochore microtubules. As determined by electron microscopy, the apparent absence of kinetochore microtubules from the mitotic spindle in the mutant supported this early conclusion (Oakley, 1981). Anti-tubulin immunofluorescence and nuclear staining with DAPI support this conclusion also, as in the mutant the chromatin is distributed randomly along the spindle rather than being separated into two discrete masses as in a wild type mitosis. The *bimD6* mitotic block occurs during anaphase and is characterised by incomplete chromatin separation and a mitotic spindle which does not elongate to the degree seen in a normal telophase. However, some spindle elongation does occur, as shorter, metaphase spindles are also seen at restrictive temperature. This suggests that kinetochore microtubule depolymerisation during anaphase A is not required for anaphase B spindle elongation, as the kinetochore microtubules are absent from the spindle in the mutant. Rather, a checkpoint may exist in the cell which prevents further spindle elongation, anaphase B, in the absence of complete anaphase A separation of chromatin, which normally occurs by kinetochore microtubule shortening.

The presence of a DNA binding motif could also indicate a role for BIMD as a kinetochore protein. This would explain the defect in chromatin separation of *bimD6* mutants. The centromere binding protein CBF1 contains a helix-loop-helix motif (Cai and Davis, 1990). Other centromere binding proteins may contain other DNA binding motifs. The presence in BIMD, however, of an acidic region found in some transcriptional activators in addition to a bZIP motif suggests a regulatory role in the cell (White *et al.*,

1992). The results from our BIMD overexpression experiments would seem to support this conclusion. Expression of a gene out of context has been used in a number of systems to examine the consequences to cellular proliferation and development. For example, overexpression of the *nimA* gene of *A. nidulans* causes premature entry into mitosis (Osmani *et al.*, 1988a) and in *S. pombe* overexpression of p53 blocks growth (Bischoff *et al.*, 1992). It was reasoned that if BIMD acts as a regulator of gene expression, the overexpression of the protein in a wild type cell may produce a specific effect on mitosis or cell cycle progression. If BIMD acts to positively regulate the cell cycle by regulating expression of genes required for mitosis, then overexpression may induce mitotic like events. The fact, however, that overexpression results in a cell cycle arrest in G1 or S suggests that BIMD may act as a negative regulator of cell cycle progression. BIMD may act to negatively regulate components required for mitosis, and overexpression may prevent the production of these components and therefore prevents mitosis from occurring. Alternatively, *bimD* may be a bifunctional protein, carrying out one function in S phase, regulatory for example and another, perhaps structural role in mitosis. For example, the CBF1 protein of *S. cerevisiae* acts as a centromere binding protein and may act as a transcriptional regulator required for methionine prototrophy (Cai and Davis, 1990).

The sensitivity of *bimD6* mutants to DNA damaging agents suggests that BIMD could regulate the transcription of genes required for DNA repair. One such gene has been identified in *S. cerevisiae* encoding the photolyase regulatory protein, which mediates the transcriptional induction of the DNA repair gene *PHR1* in response to DNA damage (Sebastian *et al.*, 1991). Alternatively, BIMD may itself play a more direct role in the repair of DNA damage. The increased sensitivity of *bimD6* strains to DNA-damaging agents is not seen in another of the *bim* mutants tested, *bimB3*, and is therefore not likely to be a general phenotype of *bim* mutants. Conversely, no mutagen sensitive *uvs* or *mus* alleles exhibit a conditional mitotic phenotype (Käfer and Mayor, 1986; Zhao and Käfer, 1992). It is likely that the sensitivity of *bimD6* mutant strains to DNA damaging agents at temperatures permissive for the mitotic defect is the result of partial loss of *bimD* function of the mutant gene product at permissive temperature and the mitotic defect seen at restrictive temperature the result of a complete loss of *bimD* function. Evidence for this is that MMS treatment of *bimD6* cells results in a catastrophic mitosis similar to that seen at restrictive temperature (data not shown). *bimD* function may therefore be required at S phase to carry out a role related to DNA metabolism, but the absence of its function may not be seen until M phase, when an abnormal mitosis occurs. The absence of kinetochore microtubules in this case could be due to a failure to assemble a kinetochore structure if DNA damage persists. The *bimD* gene product does not appear to have a checkpoint function, responding to DNA damage like the *RAD9* gene product of *S. cerevisiae*, as *bimD6* mutant cells, like wild type cells, show a cell cycle delay after DNA damage (data not shown).

The *bimD* gene product is a novel component of the cell involved in mitotic progression. In addition, its role seems to involve mitotic progression and repair of DNA damage. The relationships between these two events are not completely understood. Further understanding the role of *bimD* in the cell should enhance our understanding of these processes.

## ACKNOWLEDGMENTS

This work was supported by NIH grant GM41626 to G.S.M.

## REFERENCES


Baum, P., Yip, C., Goetsch, L. and Byers, B. (1988) A yeast gene essential for regulation of spindle pole duplication. Mol. Cell Biol. 8, 5386-5397.

Bergen, L. G., Upshall, A., and Morris, N. R. (1984) S-phase, G2, and nuclear division mutants of *Aspergillus nidulans*. J. Bact. 159, 114-119.

Bischoff, J. R., Casso, D., and Beach, D. (1992) Human p53 inhibits growth in *Schizosaccharomyces pombe*. Mol. Cell. Biol. 12, 1405-1411.

Brody, H., Griffith, J. Cuticchia, A. J., Arnold, J., and Timberlake, W. E. (1991) Chromosome-specific recombinant DNA libraries from the fungus *Aspergillus nidulans*. Nucl. Acids Res. 19, 3105-3109..

Cabral, F.R. (1983). Isolation of Chinese hamster ovary cell mutants requiring the continuous presence of taxol for cell division. J. Cell Biol. 97, 22-29.

Cai, M., and Davis, R. W. (1990) Yeast centromere binding protein CBF1, of the Helix-Loop-Helix protein family, is required for chromosome stability and methionine prototrophy. Cell 61, 437-446.

Doonan, J. H., and Morris, N. R. (1989) The *bimG* gene of *Aspergillus nidulans*, required for completion of anaphase, encodes a homolog of mammalian phosphoprotein phosphatase 1. Cell 57, 987-996.

Doonan, J.H., (1992) Cell division in *Aspergillus*. J. Cell Sci. 103, 599-611.

Dustin, P. (1984) in "Microtubules" pp.10-13, Springer-Verlag, New York.

Engle, D.B., Osmani, S. A., Osmani, A. H., Rosborough, S., Xiang, X., and Morris N. R. (1990) A negative regulator of mitosis in *Aspergillus* is putative membrane spanning protein. J. Biol. Chem. 265, 16132-16137.

Enos, A. P., and Morris, N. R. (1990) Mutation of a gene that encodes a kinesin-like protein blocks nuclear division in *A. nidulans*. Cell 60, 1019-1027.

Hartwell, L., and Weinert, T. A. (1989) Checkpoints: controls that ensure the order of cell cycle events. Science 246, 629-634.

Hynes, M. J., Corrick, C. M., and King, J. A. (1983) Isolation of genomic clones containing the *amdS* gene of *Aspergillus nidulans* and their use in the analysis of structural and regulatory mutations. Mol. Cell. Biol. 3, 1430-1439.

Käfer, E., and Mayor, O. (1986) Genetic analysis of DNA repair in *Aspergillus*: evidence for different types of MMS-sensitive hyperrec mutants. Mut. Res. 161, 119-134.

Kunkel, W. and Hadrich, H. (1977). Ultrastrukturelle Untersuchungen zur antimitotischen aktivitat von methylbenzimidazol-2-ylcarbamat (MBC) und seinen einfluss auf die replikation des kern-assoziierten organells ('centriolar plaque", "MTOC", "KCE") bei *Aspergillus nidulans*. Protoplasma 92, 311-323.

May, G. S., McGoldrick, C. A., Holt, C. L., and Denison, S. H. (1992) The *bimB3* mutation of *Aspergillus nidulans* uncouples DNA replication from the completion of mitosis. J. Biol. Chem. 267, 15737-15743.

McGrew, J. T., Goetsch, L., Byers, B., and Baum, P. (1992) Requirement for *ESP1* in the nuclear division of *Saccharomyces cerevisiae*. Mol. Biol. Cell 3, 1443-1454.

Mirabito. P. M. and Morris, N. R. (1993) BIMA, a TPR-containing protein required for mitosis, localizes to the spindle pole body in *Aspergillus nidulans*. J. Cell Biol. 120, 959-968.

Morris, N. R. (1976) Mitotic mutants of *Aspergillus nidulans*. Genet. Res. 26, 237-254.

Murray, A. W. (1992) Creative blocks: Cell cycle checkpoints and feed back controls. Nature 359, 599-604.

Oakley, B. R., 1981, "Mitosis and Cytokinesis". Academic Press Inc. NY, NY, 181-196.

Oakley, B.R., Oakley, C.E., Yoon, Y., and Jung, M.K. (1990). g-tubulin is a component of the spindle pole body that is essential for microtubule function in *Aspergillus nidulans*. Cell 61, 1289-1301.

O'Donnell, K. L., Osmani, A. H., Osmani, S. A., and Morris, N. R. (1991) *bimA* encodes a member of the tetratricopeptide repeat family of proteins and is required for the completion of mitosis in *Aspergillus nidulans*. J. Cell Sci. 99, 711-719.

Osmani, S. A., Pu, R. T., and Morris, R. N. (1988) Mitotic induction and maintenance by overexpression of a G2-specific gene that encodes a potential protein kinase. Cell 53, 237-244.

Sebastian, J., and Sancar, G. B. (1991) A damage-responsive DNA binding protein regulates transcription of the yeast DNA repair gene *PHR1*. Proc. Natl. Acad. Sci. U.S.A. 88, 11251-11255.

Uzawa, S., Samejima, I., Hirano, T., Tanaka, K. and Yanagida, M. (1990) The fission yeast *cut1*$^{+}$ gene regulates spindle pole body duplication and has homology to the budding yeast *ESP1* gene. Cell 62, 913-25.

White, J., Brou, C., Wu, J., Lutz, Y., Moncollin, V., and Chambon, P. (1992) The acidic transcriptional activator GAL-VP16 acts on preformed template-committed complexes. EMBO J. 11, 2229-2240.

Zhao, P., and Käfer, E. (1992) Effects of Mutagen-sensitive *mus* mutations on spontaneous mitotic recombination in *Aspergillus*. Genetics 130, 717-728.

# THE PROLINE UTILISATION GENE CLUSTER OF *ASPERGILLUS NIDULANS*

Victoria Gavrias*, Beatriz Cubero, Béatrice Cazelle,
Vicky Sophianopoulou** and Claudio Scazzocchio

Institut de Génétique et Microbiologie, Bâtiment 409, Unité Associée au CNRS, Université Paris-Sud, Centre d'Orsay, Orsay Cedex 91405 France

## THE PATHWAY

The proline utilisation pathway is one of the most appropriate examples of control of gene expression in *Aspergillus nidulans* and indeed in the fungi in general. The study of this pathway permits one to investigate the mechanism of specific induction, the interaction of carbon catabolite and nitrogen metabolite repression and provides interesting hints as to the function of gene clustering in eukaryotic microorganisms.

Proline is both a nitrogen and a carbon source for *A. nidulans*. Proline is taken up by the cell through a specific uptake system and through a lower-affinity, general amino acid permease (Arst *et al.*, 1980) The amino acid is converted to L-$\Delta^1$-pyrroline carboxylate, a reaction catalysed by L-proline oxidase, and this compound is then converted to L-glutamate, a reaction catalysed by the appropriate dehydrogenase.

L-$\Delta^1$-pyrroline carboxylate is also an intermediate in proline biosynthesis from glutamic acid. A futile cycle is prevented in *Saccharomyces cerevisiæ* because the biosynthetic enzymes are cytoplasmic and the catabolic enzymes mitochondrial (Brandriss and Magasanik, 1981). Older data have indicated that in *A. nidulans* proline oxidase is mitochondrial and $\Delta^1$-pyrroline dehydrogenase cytoplasmic (Arst *et al.*, 1981). This problem should be addressed again, to assess whether the apparent cytoplasmic localisation of the second enzyme of the pathway is an artefact or whether in *A. nidulans* the separation between the biosynthetic and the catabolic pathway is achieved by means other than a cytoplasmic/mitochondrial compartmentalisation.

## THE GENES AND THEIR SEQUENCE

The *prn* cluster has been cloned by a rather heterodox method (Green and Scazzocchio, 1985; Hull *et al.*, 1989) All the genes involved in this pathway are clustered in about 13 kb of DNA in chromosome VII. In *S. cerevisiæ*, on the other hand, the isofunctional genes are dispersed throughout the genome. A scheme of the cluster,

---

*Present address, Departement of Genetics, University of Georgia, Athens, Georgia, GA, 30602, USA.
** Present address, Institute of Molecular Biology and Biotechnology, Foundation for Research and Biotechnology, PO Box 1527, GR 71110, Heraklion, Crete, Grece.

including the direction of transcription of the genes is shown in figure 1. The *prnA* gene encodes a positive specific transcription factor (see below), *prnD* encodes proline oxidase, *prnC* L-$\Delta^1$-pyrroline carboxylate dehydrogenase, *prnB* the specific proline permease (Jones *et al.*, 1981). Mutants in all these genes, including null mutations, deletions, thermo- and cryosensitive alleles have been isolated and fine structure-mapped (Sharma and Arst, 1985). After cloning and sequencing of the cluster a new gene, coding for a protein of 40587 Daltons, *prnX*, was discovered. Inactivation of this gene does not prevent proline utilisation (Gavrias, 1992). Its function is at present unknown. All the *prn* cluster has been sequenced (Sophianopoulou and Scazzocchio, 1989; Hull, 1988; Gavrias, 1992). The two enzymes encoded by the *prnD* and *prnC* genes show clear similarities with isofunctional enzymes of other organisms, especially *S. cerevisiæ*. Gavrias (1992) has observed that the translated sequences of the genes coding for $\Delta^1$-pryrroline carboxylate dehydrogenase (*prnC* in *A. nidulans*, *PUT2* in *S. cerevisiæ*) show a typical consensus motif for the aldehyde dehydrogenases. This may indicate that the substrate of the enzyme is not the $\Delta^1$-pyrroline carboxylate but glutamic semialdehyde. These two compounds are non-enzymatically interconvertible in the cell. The permease coded by the *prnB* gene belongs, not surprisingly to the family of amino acid permeases, it shows strong similarities with the permeases for arginine and histidine of *S. cerevisiæ* (Sophianopoulou and Scazzocchio, 1989). The proline permease gene of the latter organism was subsequently sequenced and also belongs to this group (Vandenbol *et al.*, 1989). There are specific selective techniques to obtain mutations in many of the *prn* genes (Arst *et al.*, 1981). In the *prnB* gene in particular, the existence of a fine structure map, and of cryo- and thermo-sensitive mutations (Sharma and Arst, 1985) provides an excellent model for the dissection of the topogenesis and function of a whole class of amino acid transporters.

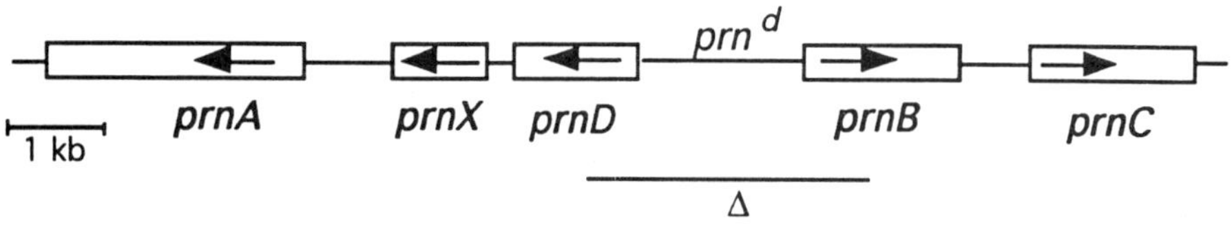


**Figure 1. Schematic representation of the *prn* cluster of *Aspergillus nidulans*.** Sizes of genes and intergenic regions to scale. The direction of transcription of the different genes is indicated according to Hull *et al.*, 1989; Sophianopoulou and Scazzocchio, 1989; Gavrias, 1992. The approximate position of the $prn^d$ mutations is indicated. Δ indicates any deletion with its termini in *prnB* and *prnD*, all these deletions have polar affects on *prnC* (see text).

## REGULATION- SPECIFIC INDUCTION- THE *prnA* GENE

Transcription of *prnB*, *C*, and *D*, only proceeds at a high rate in the presence of proline. Induction depends on the *prnA* gene. In contrast to what has been reported before (Hull *et al.*, 1989 ), this gene does not seem to regulate its own transcription, but it is subject, at least under some physiological conditions, to repression in the simultaneous presence of ammonia and glucose (see below). In strains carrying deletions of this gene the steady-state levels of the mRNAs for *prnB*, *C* and *D* are below the levels of detection in Northern blots. Transcription of the *prnX* gene seems also to be inducible, but its basal level is much higher. The PRNA protein belongs to the binuclear Zn cluster family of transcriptional activators. Interestingly this gene shows a clear correlation between functional domains and exons, at least at its 5' end. Exon 1 contains a putative nuclear entry signal, exon 2 the Zn binuclear cluster and exon 3 a putative activation signal. This contrasts with the ALCR and NIRA proteins of *A. nidulans*. These transcription factors, which belong to the same group of proteins as PRNA, have their Zn binuclear domains interrupted respectively by 1 and 2 introns (Kulmburg *et al.*, 1990;

Burger *et al.*, 1991). The large number of mutations extant in the *prnA* gene permits a thorough functional analysis of the protein. Among the mutants sequenced to date, some map in the putative nuclear entry signal and in the Zn finger, while other missense mutants map in regions of unknown function and thus could be used to define new functional domains of the PRNA transcription factor. A mutation mapping in the putative nuclear entry signal is cryosensitive and thus may affect the interaction of the amino-terminus of PRNA with proteins transporting transcription factors through the nuclear pore.

Gel retardation, footprint, and deletion studies are being used to establish the physiological binding sites of PRNA. It suffices to say here, that, like other transcription factors of this class PRNA recognises a GC rich sequence, and that most possibly the transcription of *prnB* and *prnD* is co-regulated from a common central region, and that *prnC* is regulated from an independent binding site upstream of this gene.

## REGULATION- CATABOLITE REPRESSION

The transcription of *prnD*, *prnB*, and *prnC* is affected by both nitrogen and carbon catabolite repression. In fact, these genes are only repressed efficiently when both a repressing carbon and a nitrogen source are present. This is shown in figure 2.

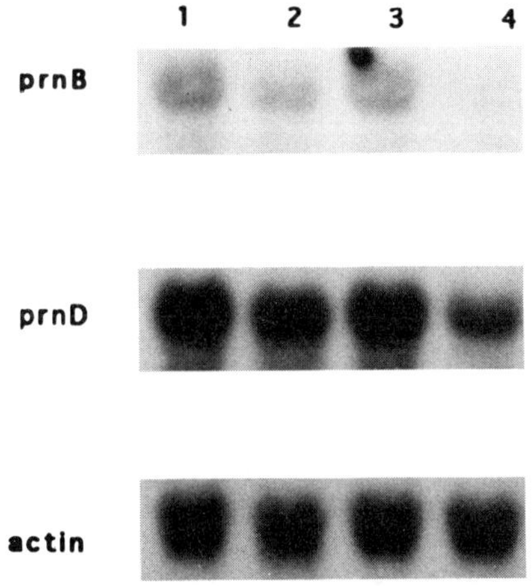


**Figure 2. Repression of the divergently transcribed *prnB* and *prnD* mRNAs.**
1, non-repressed; 2, repressed with glucose; 3, repressed with ammonia; 4, repressed with ammonia and glucose. The wild type strain was grown for 10 hours at 37° on 0.1% fructose and 5 mM urea and then total RNAs were extracted and analysed by Northern blotting. After 8 hours growth 20 mM L-proline was added to all cultures. In 2, 1% glucose was also added in 3, 20 mM ammonium D-tartrate, in 4 both 1% glucose and 20 mM ammonium D-tartrate. The three panels are probed respectively with *prnB* (upper panel) *prnD* (middle panel) and actin (lower panel) specific probes (as in Sophianopoulou *et al.*, 1993).

Carbon catabolite repression is mediated in *A. nidulans* by the negative-acting product of the *creA* gene (Bailey and Arst, 1975). The CREA protein belongs to a sub-family of Zn-finger proteins of which the best studied example is the mammalian protein Zif268 (Dowzer and Kelly, 1991; Pavletich and Pabo, 1991). In the presence of repressing carbon sources (glucose, sucrose) the CREA protein prevents the transcription of, for example the *alcA* and *aldA* genes, coding respectively for alcohol and aldehyde dehydrogenase, but also of the *alcR* gene, the positive transcription factor mediating ethanol induction (Lockington *et al.*, 1987). This effect is independent of the nitrogen source. The CREA binding sites upstream of the *alcA* and *alcR* genes have been determined and they respond to the consensus sequence 5'SYGGGG3' (Kulmburg *et al.*, 1993).

Nitrogen metabolite repression is mediated by the product of the *areA* gene. Ammonia and glutamine prevent, directly or indirectly the binding and/or activity of the AREA transcription factor. This factor belongs to a new family of Zn finger proteins. Its vertebrate counterparts, called GATA factors, are involved in the expression of different genes of the erythropoietic and some other cell lineages (Kudla *et al.*, 1990; Scazzocchio, 1990 for review). A large number of genes require an active AREA protein for effective transcription. (For early work see Arst and Cove, 1973, Arst and Scazzocchio, 1975, Hynes, 1975, Tollervey and Arst 1981 for recent work including Northerns or dot blots, see Suárez *et al.*, 1991, Stankovich *et al.*, 1993, Gorfinkiel *et al.*, 1993, Oestreicher and Scazzocchio, 1993). This effect is independent of the carbon source. The AREA protein binds, as other factors of this class, including the isofunctional protein (NIT2) from *Neurospora crassa* (Fu and Marzluf, 1990) to sites containing a core GATA sequence (A. Ravagnani, T. Langdon, and H. N. Arst Jr. unpublished).

**Table 1. Growth of a null *areA*⁻ mutant on different nitrogen and carbon sources.**

| | Nitrate Glucose | Nitrate Ethanol | Proline Glucose | Proline Ethanol | Acetamide Glucose | Acetamide Ethanol |
|---|---|---|---|---|---|---|
| *areA*$^+$ *creA*$^+$ | + | + | + | + | + | + |
| *areA*$^-$ *creA*$^+$ | - | - | - | + | - | + |
| *areA*$^-$ *creA*$^-$ | - | - | + | + | + | + |
| *areA*$^-$ *creA*$^+$ (*prn*$^d$) | - | - | + | + | - | + |

This table illustrates the behaviour of strains carrying *areA*$^-$ mutations and the physiological suppressor mutations of the latter. Growth on nitrate represents the growth of strains carrying *areA*$^-$ mutations on most nitrogen sources: they show a mutant phenotype irrespective of the carbon source. On the other hand, the same strains can utilise proline or acetamide as nitrogen sources when a repressing carbon source (glucose) is replaced by a non-repressing carbon source (ethanol). Loss-of-function mutations in the *creA* gene permit the growth of *areA*$^-$ strains on both proline and acetamide in the presence of glucose while the *cis*-acting *prn*$^d$ mutations are specific suppressors for growth on proline/glucose. "+" and "-"indicate whether a given strain utilises or not a given nitrogen source.

The transcription of *prnB*, *prnC*, and *prnD* also depends on an AREA active gene product. However this dependence is strongly influenced by the carbon source. On non-(or weakly) repressive carbon sources, the transcription of the structural genes of the *prn* cluster proceeds largely independently of the presence or absence of an active AREA protein. In other words, while some promoters are always dependent on AREA, others (like those of the structural genes of the *prn* cluster, but also *amdS*, coding for acetamidase and *gabA* coding for the γ-aminobutyric permease) require the transcription factor AREA only when repressed by CREA, in the presence of glucose or another repressing carbon source. It follows that a strain carrying a *areA*$^-$ null mutation cannot utilise proline as a nitrogen source when glucose is a carbon source, but it utilises proline as sole nitrogen and carbon source or when a non-repressing carbon source (ethanol, lactose) is present (Arst and Cove, 1973). Extragenic revertants of an *areA*$^-$ strain on proline-glucose belong to two classes. One class maps in the *creA* gene. These mutations result also in growth on some other nitrogen sources in the presence of glucose, like acetamide or γ-aminobutyric acid and on derepression of most and perhaps all the genes subject to carbon catabolite repression. Historically, this is the way the *creA* gene was discovered (Arst and Cove, 1973). The second class maps between the *prnB* and *prnD* genes. It is specific for proline, and results in derepression of the *prnB*, *prnD* and *prnC* (Arst and Cove, 1973; Arst and MacDonald, 1975; Arst *et al.*, 1980; Sophianopoulou *et al.*, 1993). These mutations, called $prn^d$ have been sequenced. One of them is a G/C to A/T transition at position -508 before the ATG of *prnB* and the second is a similar transition at position -512 (Sophianopoulou *et al.*, 1993).

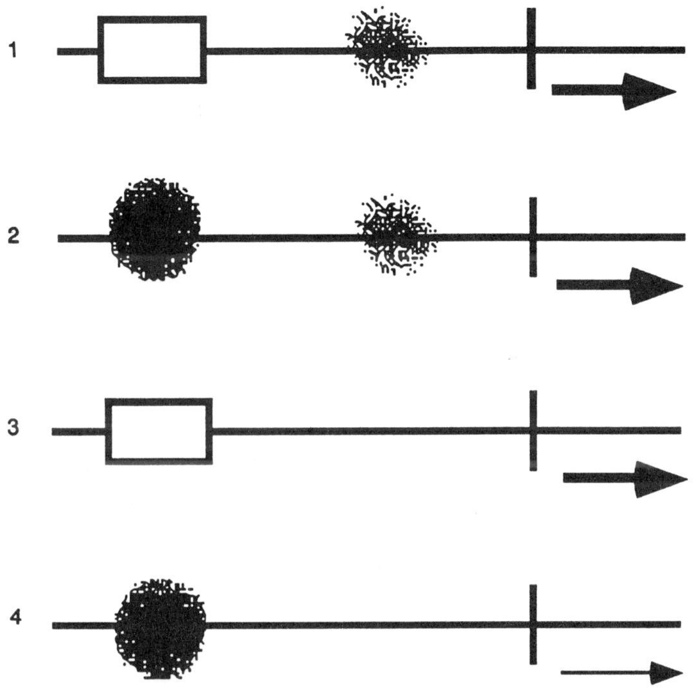


**Figure 3. Model of the repression of the *prnB* gene by carbon and nitrogen catabolite repression.**
1. Non-repressing conditions. Neither a repressing carbon (glucose) or nitrogen source (ammonium or glutamine) is present. An unknown transcription factor (rectangle) and the AREA protein (light cloud) bind to DNA. Transcription proceeds at full rate.
2. Glucose repression. CREA (dark cloud) binds to DNA and prevents either the binding or activity of the unknown transcription factor, but in the absence of nitrogen repressing sources AREA binds and permits full or almost full rate transcription.
3. Nitrogen repressing conditions. In the presence of ammonium or glutamine AREA does not bind to DNA, but in the absence of glucose the unknown transcription factor binds and permits full or almost full rate transcription.
4. Double repression. In the presence of ammonium or glutamine and glucose, the CREA protein prevents the binding or activity of the unknown transcription factor, the repressing nitrogen sources prevent the binding (or activity) of AREA, thus transcription proceeds at the low, basal rate.
This process may well act bi-directionally also on the expression of the *prnD* gene, and perhaps at a distance on the expression of *prnC*.
The Northern blot of figure 2 illustrates transcription of *prnB* and *prnD* under these four different situations.

Recent binding studies have shown that there are between *prnB* and *prnD* a number of CREA binding sequences. Two divergently orientated sequences are separated by only one base pair. Each of the $prn^d$ mutations eliminates completely *in vitro* binding to each one of the binding sites composing the divergent pair. The *in vivo* behaviour of *areA*⁻ mutations and their physiological suppressors is shown in table 1. Deletion studies of the sequences around and comprising the CREA binding pair have given surprising results. Being naive, one could expect these mutations to show a phenotype identical to strains carrying $prn^d$ mutations. In fact this is not the case. Strains carrying such deletions show an non-conditional dependence on the presence of an active AREA protein. This implies that together with the physiological CREA binding sites we have removed another element, which is positive-acting and allows the *prn* genes to be expressed efficiently in the absence of an AREA active protein.

Thus there must exist another transcription factor, which may be a rather general one, and CREA represses by preventing the binding or activity of this new, unknown (but not hypothetical) factor. The apparent paradox of the conditional dependence of the transcription of *prnB*, *D*, and *prnC* on AREA is solved by proposing that the transcription of this genes can be driven alternatively by AREA or a new, hitherto unknown factor, and that CREA prevents the action of the latter. The current model for the interaction of carbon and nitrogen metabolite repression in the *prn* gene cluster is shown in figure 3. The utilisation of some other metabolites (e. g., acetamide, γ-aminobutyric acid) which serve both as carbon and nitrogen is regulated in a similar fashion (Arst and Cove, 1973), and thus this model may well have a more general validity.

## THE ROLE OF GENE CLUSTERING

The role of gene clustering in eukaryotes, where mRNAs are monocistronic has been the object of some debate. Arst and co-workers have many years ago shown that deletions with termini in *prnB* and *prnD*, and thus deleting the *prnD*/*prnB* intergenic region have a polar effect on *prnC* (Arst and MacDonald, 1978), and qualitatively similar results were found at the level of the *prnC* transcript. Interestingly this effect is much stronger at 25° than at 37°.

In figure 1 we have indicated the notional span of any such deletion. At 25° deletions spanning *prnB* and *prnD* do not complement with mutations in the *prnC* gene.

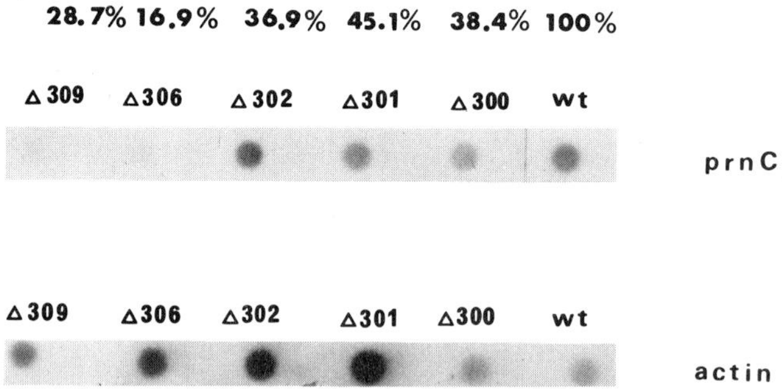


**Figure 4. Effects of *prnD*/*prnB* deletions on *prnC* transcription**
Upper panel, dot blot of polyA-mRNA probed with a *prnC* specific probe. Lower panel, monitoring of the loading by hybridising with an actin specific probe (Fidel *et al.*, 1988 ) Δ300, 301, 302, 306 and 309 are different deletion mutations; genetically mapped by Sharma and Arst, (1985) and physically mapped by Hull *et al.*(1989) wt, wild type. All polyA-mRNAs were prepared from mycelia grown at 25° on non-repressing, inducing conditions (minimal medium supplemented with 1% glucose, 5mM urea, 20mM L-proline). Above the dot blot the percent of *prnC* transcription, corrected for the differences in loading, relative to the wild type is indicated. There is no explanation presently available for the heterogeneity of the effect of the deletions, which all remove the *prnD*/*prnB* intergenic region. A heterogeneity was also found at the level of the enzyme assays (Arst and MacDonald, 1978; Arst *et al.*, 1981).

Arst and co-workers had proposed that this effect could be explained if a fraction of the *prnC* message was contained in a dicistronic mRNA initiated at the *prnB* promoter. This is not the case. Probes specific for the *prnC* gene reveal always only one monocistronic mRNA (Hull *et al.*, 1989; Sophianopoulou *et al.*, 1993). The steady state level of this mRNA is clearly affected by deletions encompassing the *prnD-prnB* intergenic region. This is shown in figure 4. These results imply that some element in the intergenic region affects at a distance the transcription of *prnC*.

The *prn*$^d$ mutations result in derepression of *prnB*, *prnD* and *prnC*. The question here is whether the derepression of *prnD* and *prnC* in strains carrying the *prn*$^d$ mutations is direct or indirect. Is the repression of these two genes an indirect effect of the repression of the permease (*prnB*) gene? The repression of the permease would result in inducer exclusion and would then prevent the efficient induction of *prnD* and *prnC*. On the other hand, the binding sites for CREA, AREA, and the new positive, unknown transcription factor, located in the *prnD/B* intergenic region could equally affect the divergent promoters of both genes. It is more difficult to invoke a direct effect of *prnC*. Upstream of the *prnC* gene there are possible AREA and CREA binding sites. As indicated above, *prn*$^d$ mutations result also in derepression for *prnC* transcription. This would indicate that the sites upstream of *prnC* are not physiological sites and that either the derepression of *prnC* in *prn*$^d$ mutants is a result of reversal of inducer exclusion or that indeed, the intergenic region between *prnB* and *prnD* can affect, at a distance, transcription of *prnC*. Other data indicate that *prnA*-mediated induction of *prnX* by proline may also be determined, at a distance, by sequences in the *prnB-prnD* intergenic region. If these roles of the *prnB-prnD* intergenic region are confirmed, the proline gene cluster would be a unique example of the function of gene clustering in gene regulation in microbial eukaryotes.

## AKNOWLEDGEMENTS

The recent work of our laboratory was supported by the Université Paris-Sud, the CNRS and grants of the European Community.

## REFERENCES

Arst, H. N. Jr. and Cove, D. J. (1973) Nitrogen metabolite repression in *Aspergillus nidulans*. Mol. Gen. Genet. 126,111-141.

Arst, H. N. Jr. and MacDonald, D. W. (1975). A gene cluster of*Aspergillus nidulans* with an internally located *cis*-actin regulatory region. Nature 254, 26-31.

Arst, H. N. Jr. and MacDonald, D. W. (1978) Reduced expression of a distal gene of the *prn* gene cluster in deletion mutants of *Aspergillus nidulans:* genetic evidence for a dicistronic messenger in a eukaryote. Mol. Gen. Genet. 163, 7-22.

Arst, H. N. Jr. and Scazzocchio, C. (1975) Initiator constitutive mutation with an "up promoter effect" in *Aspergillus nidulans*. Nature 254, 31-34.

Arst, H. N. Jr., MacDonald, D. W. and Jones, S. A. (1980) Regulation of proline transport in *Aspergillus nidulans*. J. Gen. Microbiol. 116, 285-294.

Arst, H. N. Jr., Jones, S. A. and Bailey, C. R. (1981) A method for selection of deletion mutations in the L-proline catabolism gene cluster of *Aspergillus nidulans*. Genet. Res. Camb. 38,171-195.

Bailey, C. R. and Arst, H. N. Jr. (1975) Carbon catabolite repression in *Aspergillus nidulans*. Eur. J. Biochem. 51, 573-577.

Brandriss, M. C. and Magasanik, B. (1981) Subcellular compartementation in the control of converging pathways for proline and arginine metabolism in *Saccharomyces cerevisiæ*. J. Bact. 145, 1359-1364.

Burger, G., Strauss, J., Scazzocchio, C. and Lang, F. B. (1991) *nirA*, the pathway-specific regulatory gene of nitrate assimilation in *Aspergillus nidulans*, encodes a putative GAL-4 type zinc finger protein and contains four introns in highly conserved regions. Mol. Cel. Biol. 11, 5746-5755.

Dowzer, C. E. A. and Kelly, J. (1991) Analysis of the *creA* gene, a regulator of carbon catabolite repression in *Aspergillus nidulans*. Mol. Cell. Biol. 11, 5701-5709.

Fidel, S., Doonan, J. H. and Morris, N. R. (1988) *Aspergillus nidulans* contains a single actin gene which has unique intron locations and encodes a g-actin. Gene 70, 283-293.

Fu, Y-H. and Marzluf, G. A. (1990) *nit-2*, the major positive-acting nitrogen regulatory gene of *Neurospora crassa*, encodes a sequence-specific DNA-binding protein. Proc. Natl. Acad. Sci. 87, 5331-5335.

Gavrias, V. (1992) Etudes moleculaires sur la régulation et la structure du groupe des gènes *prn* chez *Aspergillus nidulans*. Doctoral Thesis, Unversité Paris-Sud (XI) Orsay.

Green, P. M. and Scazzocchio C. (1985) A cloning strategy in filamentous fungi, in "Gene Manipulations in Fungi" pp. 345-353. Bennett, J. W. and Lasure, L. L. Eds. Academic Press, Orlando.

Gorfinkiel, L., Diallinas, G. and Scazzocchio C. (1993) Sequence and regulation of the *uapA* gene, encoding a uric acid-xanthine permease in the fungus *Aspergillus nidulans*. J. Biol. Chem. In press.

Hull, E. P. (1988) Molecular Analysis of the proline catabolism gene cluster of *Aspergillus nidulans* and sequencing of the regulatory gene. Ph. D. Thesis, Univesity of Essex, Cochester.

Hull, E. P., Green, P. M. , Arst, H. N. Jr and Scazzocchio, C. (1989) Cloning and physical characterization of the *L*-proline catabolism gene cluster of *Aspergillus nidulans*. Mol. Microbiol. 3, 553-559.

Hynes, M. J. (1975) Studies on the role of the *areA* gene in the regulation of nitrogen catabolism in *Aspergillus nidulans*. Aust. J. Biol. Sci. 28, 301-313.

Jones, S. A., Arst H. N. Jr. and MacDonald D. W. (1981) Genes roles in *prn* cluster of *Aspergillus nidulans*. Curr. Genet. 3, 49-56.

Kulmburg, P., Prangé, T., Mathieu, M., Sequeval, D., Scazzocchio, C. and Felenbok, B. (1990) Correct intron splicing generates a new type of a putative Zn-finger domain in a transcriptional activator of *Aspergillus nidulans*. FEBS Letter 280, 11-16.

Kulmburg, P., Mathieu, M., Dowzer, C., Kelly, J. and Felenbok, B. (1993) Specific binding sites in the *alcR* and *alcA* promoters of the ethanol regulon for the CREA repressor mediating carbon catabolite repression in *Aspergillus nidulans*. Mol. Micobiol. 7, 847-857.

Kudla, B., Caddick M. X., Langdon T., Martinez-Rossi N. C., Bennet C. F., Sibley S., Davies R. W. and Arst H. N. Jr. (1990) The regulatory gene *areA* mediating nitrogen metabolite repression in *Aspergillus nidulans*. Mutations affecting specifity of gene activation alter a loop residue of a putative zinc finger. EMBO J. 9, 1355-1364.

Lockington, R. A, Scazzocchio, C, Sequeval D, Mathieu, M, and Felenbok,.B. (1987) Regulation of *alcR*, the positive regulatory gene of the ethanol utilization regulon of *Aspergillus nidulans*. Mol. Microbiol. 1, 275-281.

Oestreicher, N. and Scazzocchio, C. (1993) Sequence, regulation and mutational analysis of the gene encoding urate oxidase in *Aspergillus nidulans*. J. Biol. Chem. In press.

Pavletich, N. and Pabo, C. O. (1991) Zinc finger-DNA recognition: Crystal tructure of the Zif268-DNA complex at 2.1 Å. Science 252, 809-817.

Scazzocchio, C. (1990) Of moulds and men, or two fingers are not better than one. Trends in Genet. 6, 311-313.

Sharma, K. K. and Arst, H. N. Jr. (1985) The product of the regulatory gene of the proline catabolism gene cluster in *Aspergillus nidulans* is a positive acting protein. Curr. Genet. 9, 299-304.

Sophianopoulou, V. and Scazzocchio, C. (1989) The proline transport protein of *Aspergillus nidulans* is very similar to amino acid transporters of *Saccharomyces cerevisiae*. Mol. Microbiol. 3, 705-714.

Sophianopoulou, V., Suárez, T., Diallinas, G. and Scazzocchio, C. (1993) Operator derepressed mutations in the proline utilisation gene cluster of *Aspergillus nidulans*. Mol. Gen. Genet. 236, 209-213.

Stankovich, M., Platt, A., Caddick, M. X., Langdon, T., Shaffer, P. M. and Arst, H. N. Jr. (1993) *C*-terminal truncation of the transcriptional activator encoded by *areA* in *Aspergillus nidulans* results in both loss-of-function and gain- of-function phenotypes. Mol. Microbiol. 7, 81-88.

Suárez, T., Oestreicher, N., Peñalva, M. A. and Scazzocchio C. (1991) Molecular cloning of the *uaY* regulatory gene of *Aspergillus nidulans* reveals a favoured region for DNA insertions. Mol. Gen. Genet. 230, 369-375.

Tollervey, D. W. and Arst, H. N. Jr. (1975) Mutations to constitutivity and derepression are separate and separable in a regulatory gene of *Aspergillus nidulans*. Curr. Genet. 6, 79-85.

Vandenbol, M., Jaumiaux, J. C. and Grenson, M. (1989) Nucleotide sequence of the *Saccharomyces cerevisiæ* PUT4 proline permease-encoding gene: similarities between CAN1, HIP1 and PUT4 permeases. Gene 83, 153-159.

# PHYSICAL KARYOTYPING : GENETIC AND TAXONOMIC APPLICATIONS IN ASPERGILLI

Klaas Swart, Alphons J.M. Debets, Edu F. Holub, Cees J. Bos, and Rolf F. Hoekstra

Department of Genetics
Agricultural University Wageningen
Dreijenlaan 2
NL-6703 HA Wageningen
The Netherlands

## INTRODUCTION

The karyotype is often used in genetics, either to characterise an individual as a member of a distinct species or to investigate chromosome recombination and segregation in crosses or in somatic cell division. A cytogenetic karyotype has been characterised in several fungi (see e.g. Fincham et al., 1979). Among those is *Aspergillus nidulans* in which 8 chromosomes have been recognised (Elliott, 1960), which compared quite well to the 8 linkage groups defined in genetic analyses. However, the morphology of the different chromosomes was not sufficiently different to study chromosome behaviour or anomalies and this approach was not continued. An additional drawback is that individual chromosomes could only be demonstrated in meiotic cells and therefore this method is not applicable to imperfect fungi.

The technique of pulsed field gel-electrophoresis (PFGE) was introduced a decade ago by Schwarts and coworkers (Schwarts *et al.*, 1982; Schwarts and Cantor, 1984) and by Carle and Olson (1984), enabling separation of DNA molecules larger than 50 kilobase pairs. Improvement of the apparatus and the technique enables the resolution chromosome-sized DNAs. At present the upper limit of resolution is somewhere between 10 and 15 megabase pairs, so complete chromosomes of the lower eukaryotes can be separated in discrete bands and physical karyotypes can be established. General aspects of molecular karyotype analysis in fungi have recently been described in a review by Skinner *et al.* (1991).

Physical karyotyping has been applied in aspergilli (i) to demonstrate the banding pattern of chromosomes, (ii) to allocate genes to chromosomes when no mutant phenotype is available and a probe can be made, (iii) to map transformed genes and to investigate copy numbers of the transformed sequences and their stability following subculturing or recombination with other strains, (iv) to prepare chromosome specific DNA libraries, and (v) to compare and to distinguish different isolates of the same species in order to investigate their evolutionary relationship.

## TECHNICAL ASPECTS

### Preparation of Embedded Chromosomal DNA

Protoplasts are easily prepared from aspergilli using commercially available enzymes like Novozym 234 (Novo BioLabs, Denmark); details have been described in several reports (Debets and Bos, 1986; Brody and Carbon, 1989; Keller *et al.*, 1992). Protoplasts are subsequently washed in isotonic medium containing EDTA (50 mM) and resuspended in 0.5% to 0.8% (final concentration) low gelling point agarose (e.g InCert agarose, FMC) in isotonic medium containing 500 mM EDTA and 1 - 2 mg $ml^{-1}$ proteinase K. The final concentration of protoplasts should be approximately $2 \times 10^8$ $ml^{-1}$. This suspension is then pipetted into a prechilled (on ice) mould to obtain plugs. Subsequently, the protoplasts have to be lysed *in situ*, which is done by incubation of the plugs in a mixture of 1% N-lauroylsarcosine, 500 mM EDTA and 1 mg $ml^{-1}$ proteinase K at 50° C for 48 h. Plugs are washed in 50 mM EDTA and stored in 50 mM EDTA at 4° C. Embedded chromosomal DNA remains intact for more than a year.

### Pulsed field gel electrophoresis

Several groups, including ours, have used contour-clamped homogeneous electric field (CHEF) gel electrophoresis to separate *Aspergillus* chromosomes. In the CHEF system the electric field alternates between two orientations at angles of 120° (Chu *et al.*, 1986). The main principle of separation is that DNA molecules that were electrophoresed in one direction have to reorient before they can migrate in a second direction. The time needed to achieve complete reorientation is inversely related to the size of the DNA molecule. After complete reorientation the molecule can move in the second direction. Hence, the net migration of large molecules is remarkably slower than that of relatively small molecules. The migration velocity of chromosome sized DNA is nonlinearly related to size. It is therefore important to use appropriate standards, e.g. chromosomal DNA of *Schisosaccharomyces pombe*. This yeast has three chromosomes of sizes 3.5, 4.7 and 5.7 megabasepairs (mb) which are commercially available to an electrophoresis standard (e.g. Bio-Rad, Promega).

To obtain maximum resolution it is often required to change pulse intervals during the electrophoretic run. Additionally, the electric field has to be homogeneous in both directions, so a controlled apparatus is required. The CHEF-DR II and its successor the CHEF-DR III systems of Bio-Rad meet these criteria and these systems have been used by several groups.

Gels are prepared of agarose (0.5% - 0.8%) of low electroendosmosis (EEO), like chromosomal-grade agarose (Bio-Rad), Seakem GTC (FMC), Megarose (Clontec). To get the plugs positioned and to have them correctly embedded in the gel, we placed the plugs next to the comb in an empty mould, sealed them with a little agarose and then poured gently the agarose into the mould. After cooling, the comb was removed and the same agarose was used to fill the holes. Gels are electrophoresed at constant temperature at 40 - 50 V (we use 9° C and 45 V) in e.g. 0.5 x TAE buffer. Run conditions will vary, a general scheme is: intervals of 55 min (72 h duration), 47 min (48 h) and 40 min (48 h). Then the total run-time is 7 days after which the gels are stained in 0.5 μg ml$^{-1}$ ethidium bromide for 1 h and destained in water for at least 1 h. Finally, gels are photographed under UV illumination.

### Identification of Chromosomes

Gels can be processed to denature the DNA and blotted to appropriate membranes using standard procedures. Probes of mapped genes have been used for hybridisation and after autoradiography the chromosomal band of hybridisation could be identified (e.g. Brody and Carbon, 1989, Debets *et al.*, 1990 *b*). In addition, probes of unknown genes can be used to map those genes to specific bands.

## PFGE APPLICATIONS IN ASPERGILLI

### Primary Karyotypes of Aspergilli

Among the aspergilli *A.nidulans* was the first species in which the physical karyotype has been determined (Brody and Carbon, 1989). Six chromosomal bands were separated, so two bands were supposed to contain two chromosomes. Hybridis-ation with linkage group (LG) specific probes was used to identify LGs I, III, VII and VIII. The other bands were characterised using translocation strains. In *A.nidulans* several well-characterised translocation strains are available. Breakpoints are known on the genetic map, hence it is feasible to estimate size increments and decrements of the chromosomes involved. Brody and Carbon (1989) analysed five different translocation strains on CHEF gels and thus were able to assign all linkage groups to the different bands. It was confirmed that four bands contained one chromosome only and that the two thicker bands comprised two chromosomes. The distribution of the LGs with decreasing size is: LG VIII (5.0 mb), VII (4.5 mb), II (4.2 mb), I and V (3.8 mb), III and VI (3.5 mb) and IV (2.9 mb) (see also Figure 1). Montenegro *et al.* (1992) mentioned resolution of LGs III and VI in *A.nidulans* ATCC 28901, but in the same gel no separation of these LGs was obtained in another wildtype strain (NRRL 194), nor in two members of the Glasgow collection. This can be due to chromosome length polymorphism in different isolates of the same species (*vide infra*) or is caused by a genome rearrangement (e.g. translocation).

The presence of eight linkage groups in *A.niger* has also been demonstrated (Debets *et al.*,1990 *a*, Debets *et al.*,1993). Analyses of chromosomes by cytological techniques were not done because of the lack of a sexual cycle, hence no meiotic chromosomes could be studied. A physical karyotype was obtained by CHEF electrophoresis and four chromosomal bands were separated (Debets *et al.*, 1990 *b*). In this species also, some bands appeared to contain doublets or even triplets. No clearly defined translocation strains were available, so further attempts to resolve doublets and triplets could not be undertaken. Probes of mapped *A.niger* genes were scarce and the distribution of the LGs among the chromosomal bands could not be demonstrated directly. However, taking advantage of the presence of several *amd*S transformed strains, obtained using the *A.nidulans amd*S gene, genetic mapping of the transformed genes was performed and it appeared that the *amd*S sequences in different transformants were integrated at different sites in the genome. Seven transformants with the integrated sequences on different chromosomes were selected and used for CHEF gel electrophoresis. They were all hybridised to a single probe, the *A.nidulans amd*S gene, and the linkage groups could be assigned to the different chromosomal bands. The overall physical karyotype of *A.niger* now is: LG IV (6.6 mb), LGs II and VI (5.0 mb), LG III and VIII (4.1 mb) and LGs V and I (3.5 mb). The intensity of the ethidium bromide stained bands suggests that the second band of 5.0 mb contains three chromosomes. Experiments are in progress to clone a LG VII specific gene which finally enables the identification of this linkage group in the physical karyotype.

Several members of *Aspergillus* section *flavi* have been analysed by Keller *et al.* (1992). These fungi also lack a sexual cycle and are genetically poorly defined. Further characterisation could be obtained by physical karyotyping. The chromosomal size range was of the same order as that of *A.niger*, so a few chromosomes are significantly larger than those of *A.nidulans*. It was apparent that only strains descending from the same progenitor gave the same physical karyotype. A characteristic species pattern (*A.flavus* and *A.parasiticus*) could not be obtained due to extensive chromosome length polymorphism. DNA hybridisation was done to map a few genes to chromosomes. Whereas little is known about the classical genetics of these species, this type of gene mapping can be useful, but the method is not always reliable because not all chromosomal DNAs resolve well.

## Gene Mapping

Pulsed field gel electrophoresis has been used to map genes for which probes were available, to specific chromosomal bands. Examples are summarised in Table 1. Genes can be allocated to a linkage group if the relationship of bands and linkage groups is known. We used the CHEF method to map, as an example, the rRNA genes (Debets *et al.*, 1990). Hybridisation was found to the second band in the *A.niger* physical karyotype, thus assigning this gene cluster to either LG II or VI. The rRNA genes of *A.nidulans* were found in the band comprising LG I and V. These two linkage groups differ significantly in genetic map length. Clutterbuck (1992) estimated the map length of chromosome I to be more than twice the length of chromosome V. An equal physical size can be due to a higher content of repetitive DNA in chromosome V. This can be explained, at least partly, by the presence of the rDNA repeats in chromosome V which does not extend the genetic map because meiotic recombination is extremely suppressed in the rDNA region (see e.g. Russell *et al.*, 1988). The rDNA cluster of *Neurospora crassa* is on LG V and here again this chromosome is genetically much smaller than LG I, which is the largest both genetically and physically, but LG V comigrates with LG I in CHEF electrophoresis (Orbach *et al.*, 1988). Indirectly, we now have evidence that the rDNA cluster of *A.nidulans* is on LG V, which agrees quite well with the suggestion of Brody *et al.* (1991) that chromosome V must contain multiple copies of a repeated element. For allocating the rDNA cluster of *A.niger* this reasoning cannot be applied because the map of *A.niger* is a mitotic recombination map and is therefore less saturated.

Ehinger *et al.* (1990) mapped core histone genes of *A.nidulans* by pulsed field gel electrophoresis. Mutant phenotypes of these genes are not known. The genes for H3, H4.1 (gene pair) and for H4.2 were all localised on linkage group VIII. The gene pair for H2A and H2B was assigned to the doublet band comprising LGs III and VI. Additional mapping is feasible, e.g. by using translocation strains that resolve these two LGs. However, this was not reported by these authors.

Aspergilli of the section *flavi* are genetically poorly characterised. Linkage groups have been defined in some strains, but very little is known about the precise nature of the marker genes. It is difficult then to compare the physical and the genetically defined chromosomes.

As mentioned earlier, probing with heterologous genes Keller *et al.* (1992) were able to map several of these genes to chromosomal bands which may enable further work on genome organisation. To some extent this approach might be an alternative to gene mapping by parasexual analysis in these fungi.

**Table 1.** Assignment of 'unknown' genes and transformed sequences to linkage groups by pulsed field gel electrophoresis.

| Organism | Gene (sequence) | LG | Reference |
|---|---|---|---|
| *A.niger* | rDNA | II or VI | Debets *et al.*,1990 |
| | *gla*A (transformed) | different[1] | Verdoes *et al.*,1993 |
| *A.nidulans* | rDNA | V | this paper |
| | penicillin cluster | VI | Montenegro *et al.*,1992 |
| | histone H3, H4 | VIII | Ehinger *et al.*,1990 |
| | histone H2, H2B | III or VI | ibid. |
| *A.flavus* and *A.parasiticus* | *arg*B, GPD1, *trp*C, *nia*D, *ben*A, *pyr*G, rDNA | not known[2] | Keller *et al.*,1992 |

[1] eight transformants were analysed and showed hybridisation to one band only. In several cases this was a shifted band due to multi-copy insertion of the transforming plasmid. In two cases the transformed genes were mapped definitely to one chromosome. In the other cases the location had to be confirmed either by hybridisation with LG specific probes or with classical genetic analysis.

[2] the relationship of LGs and chromosomal bands is not known in these *Aspergilli*.

Pulsed field gel electrophoresis can be applied as a powerful tool in mapping genes or sequences that were introduced by transformation. This principle was first shown by Debets *et al.* (1990) in analyses of *amd*S transformants of *A.niger*, although in that case the transformed genes were localised by classical methods as well. This study provided evidence for the shifting of a band of the chromosome on which multiple copies of the transforming plasmid were integrated. Accurate examination of the extent of band shifting then enabled the estimation of the number of plasmid sequences in the transformant. A similar approach has been applied by Verdoes *et al.* (1993) in an analysis of *gla*A transformants of *A.niger*. Out of eight transformants, four showed a clearly changed banding pattern. Upon hybridisation, the shifted bands were identified as containing the integrated *gla*A copies. Southern blot analysis with LG specific probes revealed the chromosomal location of the *gla*A genes in two transformants. The stability of the transformants with a shifted band can now be analysed upon CHEF gel electrophoresis as well. A partial or complete reversion of the band shift will be found if a substantial part of the inserted sequences become lost.

## Gene Cloning

Isolation and cloning of genes is highly facilitated by ordered gene libraries. Whilst this is more important for organisms with a large genome containing much repetitive DNA, it is convenient in lower eukaryotes as well. Brody *et al.* (1991) ordered DNA-clones from *A.nidulans* into chromosome specific libraries using hybridisation of CHEF gel electrophoresis separated chromosomal DNA to pre-existing random libraries. Clones that hybridised uniquely with DNA from a particular chromosome were gathered in a chromosome specific sub-collection. The eight chromosome specific sub-libraries from *A.nidulans* are now available at the Fungal Genetics Stock Centre at the University of Kansas.

## Taxonomy

Physical karyotyping is useful to identify individuals belonging to a particular species, just as cytological karyotyping in the characterisation of higher eukaryotes. We have recently studied several culture collection strains of the black aspergilli, often referred to as the *A.niger* aggregate (Megnegneau *et al.*, 1993). Variation in banding patterns was observed, although the estimated total genome sizes did not differ extensively (range from 36 to 42 mb, Figure 1). This may imply that most of the observed variation comes from chromosomal rearrangements during evolution of these sub-species. That extensive rearrangements occurred is not only reflected by the diversity in banding patterns but also by the allocation of the rDNA sequences to the chromosomal bands. The rRNA genes were mapped on chromosomes differing in size among strains (Figure 1.). In addition many attempts were made to make heterokaryons between different members of the *A.niger* group and it was never successful. This may either indicate that many heterokaryon incompatibility groups exist in the *A.niger* aggregate or that these isolates belong to genetically different subspecies. Further analysis of the *A.niger* aggregate is in progress (in cooperation with Drs Croft and Varga, Birmingham), using many isolates from nature.

Chromosome length polymorphisms were demonstrated also for the aspergilli section *flavi* by Keller *et al.* (1992). In both *A.flavus* and *A.parasiticus* more differences than similarities were found. These authors detected similarities in location of some conserved genes, e.g. hybridisation of *trp*C and *nia*D in one band and *pyr*G and rDNA in another. However, it is not possible to conclude genetic linkage from these results due to comigration of two or more chromosomes. In *A.nidulans* the genes *pyr*G and rDNA hybridise to a common band as well (Keller *et al.*, 1992) but apparently they are in different linkage groups, LG I and LG V, respectively.

Recently, the physical karyotypes of ten different isolates of *A.nidulans* which were classified in different heterokaryon compatibility groups have been examined (A. Coenen *et al.*, unpublished). Preliminary results show a high degree of similarity between different isolates, although a few significant differences were apparent as well. These differences may reflect a degree of inbreeding.

## SUMMARY

Pulsed field gel electrophoresis is a useful method to separate chromosomes of aspergilli and to describe a physical karyotype of these fungi. Once a karyotype has been established it can be used to map genes to linkage groups if (i) probes of these genes are available and, (ii) the relationship of chromosomal bands and linkage groups is known. This approach is also advantageous to map transformed genes. The method is not fully adequate if two or more chromosomes comigrate in one band. However, this is occasionally solved by using translocation strains with altered banding patterns. In the case of transformed strains, a shift of the recipient chromosome may occur due to multiple copy insertion of the transformed gene. The latter can be used to estimate the copy number of the transforming sequences and the stability of these insertions. Chromosome separation has provided an elegant method to prepare chromosome specific gene libraries of *Aspergillus nidulans* DNA. Physical karyotyping is applied as well to analyse chromosome length polymorphisms between members of presumably the same species. Extensive variation was found in asexual 'species' (*A.niger*, *A.flavus*, and *A.parasiticus*), whereas the variation in the sexual species *A.nidulans* seems to be much less.

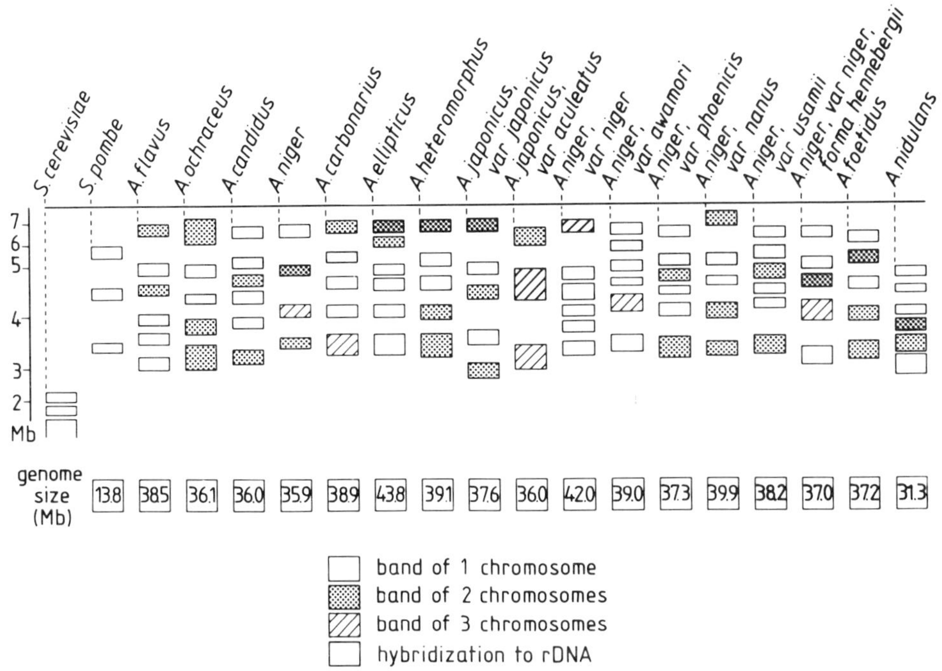


**Figure 1.** Compilation of the chromosomal banding patterns of culture collection strains of the *Aspergillus niger* aggregate. Chromosomal DNA was separated on CHEF gel at 45V in 0.5% agarose. Total electrophoresis time was 144 h, divided in three periods of 48 h with pulse intervals of 57, 47 and 35 mins., respectively.

## ACKNOWLEDGEMENTS

*Aspergillus niger* strains of the CBS culture collection were generously provided by Dr R.A. Samson, Centraal Bureau voor Schimmelcultures, Baarn, The Netherlands.

## REFERENCES

Brody, H. and Carbon, J. (1989) Electrophoretic karyotype of *Aspergillus nidulans*. Proc. Natl. Acad. Sci. USA. 86, 6260-6263.

Brody, H., Griffith, J., Cuticchia, A.J., Arnold, J. and Timberlake, W.E. (1991) Chromosome-specific recombinant DNA libraries from the fungus *Aspergillus nidulans*. Nucleic Acids Res. 19, 3105-3109.

Carle, G.F. and Olson, M.V. (1984) Separation of chromosomal DNA molecules from yeast by orthogonal-field-alternation gel electrophoresis. Nucl. Acids. Res. 12, 5647-5664.

Chu, G., Volrath, D. and Davis, R.W. (1986) Separation of large DNA molecules by contour-clamped homogeneous electric fields. Science 234, 1582-1585.

Debets, A.J.M. and Bos, C.J. (1986) Isolation of small protoplasts from *Aspergillus niger*. Fungal Genetics Newsletter 33, 24.

Debets, A.J.M., Swart, K. and Bos, C.J. (1990a) Genetic analysis of *Aspergillus niger*: Isolation of chlorate resistance mutants, their use in mitotic mapping and evidence for an eighth linkage group. Mol. Gen. Genet. 221, 453-458.

Debets, A.J.M., Holub, E.F., Swart, K., van den Broek, H.W.J. and Bos, C.J. (1990b) An electrophoretic karyotype of *Aspergillus niger*. Mol. Gen. Genet. 224, 264-268.

Debets, F., Swart, K., Hoekstra, R.F. and Bos, C.J. (1993) Genetic maps of eight linkage groups of *Aspergillus niger* based on mitotic mapping. Curr. Genet. 23, 47-53.
Ehinger, A., Denison, S.H. and May, G.S. (1990) Sequence, organisation and expression of the core histone genes of *Aspergillus nidulans*. Mol. Gen. Genet. 222, 416-424.
Elliott, C.G. (1960) The cytology of *Aspergillus nidulans*. Genet.Res., Camb. 1, 462-476.
Fincham, J.R.S., Day, P.R. and Radford, A. (1979) "Fungal Genetics", pp 34-47. Blackwell, Oxford.
Keller, N., Cleveland, T.E. and Bhatnagar, D. (1992) Variable electrophoretic karyotypes of members of *Aspergillus* section. *Flavi*. Curr Genet. 21, 371-375.
Megnegneau, B., Debets, F. and Hoekstra R.F. (1993) Genetic variability and relatedness in the complex group of black *Aspergilli* based on random amplification of polymorphic DNA. Curr. Genet. 23, 323-329.
Montenegro, E., Fierro, F., Fernandes, F.J., Gutierres, S. and Martin, J.F. (1992). Resolution of chromosomes III and VI of *Aspergillus nidulans* by pulsed-field gel electrophoresis shows that the penicillin biosynthetic pathway genes *pcbAB*, *pcbC*, and *penDE* are clustered on chromosome VI (3.0 megabases). J. Bacteriol. 174, 7063-7067.
Orbach, M.J., Vollrath, D., Davis, R.W. and Yanovski, C. (1988) An electrophoretic karyotype of *Neurospora crassa*. Mol. Cell. Biol. 8, 1469-1473.
Russel, P.J., Petersen, R.C. and Wagner, S. (1988) Ribosomal DNA inheritance and recombination in *Neurospora crassa*. Mol. Gen. Genet. 211, 541-544.
Schwarts, D.C. and Cantor, C.R. (1984) Separation of yeast chromosome-sized DNAs by pulsed field gradient gel electrophoresis. Cell 37, 67-75.
Schwartz, D.C., Saffran, W., Welsh, J., Haas, R., Goldenberg, M. and Cantor, C.R. (1982) New techniques for purifying large DNAs and studying their properties and packaging. Cold Spring Harbor Symp. Quant. Biol. 47, 189-195.
Skinner, D.Z., Budde, A.D. and Leong, S.A. (1991) Molecular Karyotype Analysis of Fungi, in "More Gene Manipulations in Fungi" (Bennett, J.W., and Lasure, L.L., Ed) pp 86-103. Academic Press, San Diego.
Verdoes, J.C., Punt, P.J., Schrickx, J.M., van Verseveld, H.W., Stouthamer, A.H. and van den Hondel, C.A.M.J.J. (1993) Glucoamylase overexpression in *Aspergillus niger*: molecular genetic analysis of strains containing multiple copies of the *gla*A gene. Transgenic Res. 2, 84-92.

# HETEROLOGOUS GENE EXPRESSION IN *ASPERGILLUS*

Robert F.M. van Gorcom, Peter J. Punt and
Cees A.M.J.J. van den Hondel

Medical Biological Laboratory TNO
P.O.Box 5815
2280 HV Rijswijk
The Netherlands

## INTRODUCTION

Filamentous fungi, including *Aspergillus* species, have a number of interesting properties which make them attractive to study biological processes at the molecular level. In particular the genetically and biochemically well-characterised species *Aspergillus nidulans* has been used extensively for molecular biological studies of eukaryotic gene organisation, regulation of gene expression and cellular differentiation.

In addition, the property of filamentous fungi, like *Aspergillus niger*, *Aspergillus oryzae* and *Trichoderma reesei* to secrete a large diversity of biopolymer degrading enzymes in significant quantities has stimulated the industrial application of these filamentous fungal strains. This property is the evolutionary consequence of the biological role these fungi have in degradation of (dead) complex biological material. Furthermore some fungal species like *Aspergillus niger* and *Penicillium chrysogenum* have been found to be excellent producers of primary (organic acids) and secondary metabolites (antibiotics, vitamins). Strain-improvement and fermentation technology is well developed for these species and at present industrial strains are available that secrete over 30 grams per litre of a specific protein (e.g. cellulases, (gluco-)amylases) in very large scale fermentation processes.

Another application of this class of microorganisms is the fermentation of food carried out by species like *Aspergillus oryzae* (soy sauce, tempeh), *Penicillium roquefortii* and *Penicillium camembertii* (cheese production). The last point of economic interest is the negative impact a large number of plant-pathogenic and mycotoxin-producing fungal species have in agriculture.

Because of the long-standing use of several fungal species for fermentation of food without any adversary effects on human health, products made by species like *A. niger* and *A. oryzae* easily obtain the FDA Generally Regarded As Safe approval.

In view of the scientific attractiveness and the biotechnological interest genetic

*The Genus Aspergillus*, Edited by Keith A. Powell *et al.*,
Plenum Press, New York, 1994

modification techniques have been developed for these organisms and several genes have been cloned in the past ten years (for recent reviews see Fincham (1989), Timberlake and Marshall (1989), Goosen *et al.*, (1992), van den Hondel and Punt (1991)). The availability of extensive genetic and biochemical knowledge together with the well-developed gene technology has also resulted in the application of *A. nidulans* as a host for cloning and analysing a number of heterologous fungal genes (e.g. Kos *et al.*, (1985), Gray *et al.*, (1986)).

All these points have been considered a good basis for a number of laboratories to investigate the possibilities of improving the production levels of current fungal products and to develop systems for the efficient production of new commercially interesting fungal and non-fungal proteins and metabolites. In these studies *Aspergillus niger* (and close relatives) and *Aspergillus oryzae* are most commonly used (for a review see van den Hondel *et al.*, 1991).

In this paper strategies used to obtain efficient production of proteins in *Aspergillus* will be illustrated and some examples will be given of the construction of *Aspergillus* strains which efficiently produce extracellular fungal and non-fungal proteins. Also some currently unsolved problems will be presented.

## PRODUCTION OF FUNGAL PROTEINS

A good example of the approach that has been used during the last few years to obtain *Aspergillus* strains which efficiently produce interesting fungal proteins, is the construction of *A. niger* strains that overproduce the enzyme glucoamylase (Verdoes *et al.*, 1993).

The *A. niger glaA* gene, which encodes glucoamylase, has been cloned and characterised (Boel *et al.*, 1984). Subsequent expression analysis studies showed that the *glaA* gene is preceded by a strong promoter (Fowler *et al.*, 1990). To obtain a glucoamylase overproducing *A. niger* strain, multiple copies of the *glaA* gene were introduced into the genome of the wild-type strain. Vectors were used which contain in addition to the *glaA* gene(s) the *amdS* gene of *A. nidulans* as selection marker (Kelly and Hynes, 1985). The latter marker facilitates selection of transformants containing multiple copies of the vector and, thus, of the *glaA* gene. Two vectors were constructed, pAB6-8 and pAB6-10, which contain one and four copies of the *glaA* gene, respectively, as well as one copy of the *amdS* gene (Verdoes *et al.*, 1993).

*A. niger* N402 (Bos, 1986) was transformed with these vectors. Transformants were first screened for their growth behaviour under different cultivation conditions. They could be divided into three groups: (1) growth on acetamide, (2) in addition growth on acrylamide and (3) in addition inhibition of growth on ω-amino acids, indicative of the presence of several copies of the *amdS* (Hynes *et al.*, 1989, Verdoes *et al.*, 1993).

Out of each group a number of transformants were chosen for analysis of the number of *glaA* gene copies and the production of glucoamylase. The number of copies was estimated by Southern blot analysis. As expected a clear relationship was found between the growth characteristics of the transformants (*amdS* copynumber) and the number of copies of the *glaA* gene. The amount of glucoamylase produced by these transformants in relation to the *glaA* copynumber is shown in Figure 1.

As shown in this figure there is no good correlation between the *glaA* copynumber and glucoamylase production, especially at high *glaA* copynumber. In the lower copynumber the glaA production of some transformants fits almost with the theoretical extrapolation from the wild-type (wt) one-copy production level (see insert Figure 1).

However, all transformants with more than 20 copies of the *glaA* gene produce (much) less than the theoretical level.

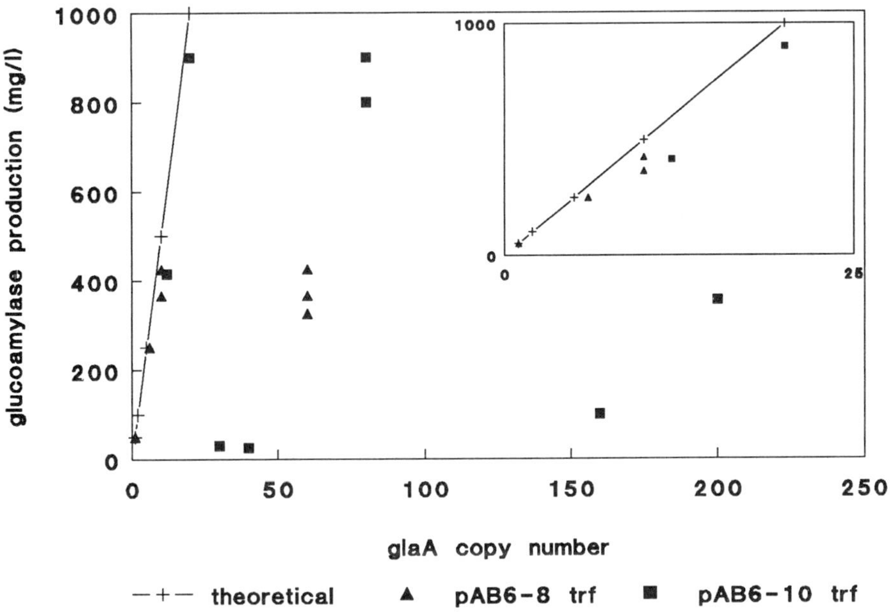


**Figure 1.** The relation *glaA* copynumber versus glucoamylase production found in different pAB6-8 (▲) and pAB6-10 (■) transformants. The line represents the theoretical extrapolation from the wildtype results (1 copy - 50 mg/l glucoamylase).

Factors, such as the site of integration, apparently influence expression levels strongly as has been shown before (van Gorcom *et al.*, 1985). The data presented in this figure (taken from Verdoes *et al.*, 1993) suggest that under the conditions used (shake flask cultures), there is a maximum glucoamylase production level of about 0.9 g/l (about eighteen times wildtype level).

This and other similar studies show that filamentous fungi can incorporate large stretches of new DNA into their genome. This DNA is stably maintained under non-selective conditions. A number of the transformants shown in Figure 1 have been analysed using CHEF electrophoresis and it was found that in most strains the vector copies were integrated in only one chromosome. The results of this analysis, combined with blotting data, indicate that strains, in which a high number of vector copies has been integrated, contain chromosomes which have increased in size up to 2 Mb (Verdoes *et al.*, 1993).

Several commercially interesting extracellular enzymes are secreted by *A. niger* and other filamentous fungi, however, often only in low amounts. An example is phytase, an enzyme that catalyses the conversion of the plant phosphate storage molecule phytic acid, into inositol and (six) phosphates. Addition of phytase to the feedstuffs of monogastric animals (like pigs and broilers, which are not able to convert phytic acid) increases the nutritional value (Simons *et al.*, 1990) and eliminates the need for additional phosphorus suppletion. The use of phytase decreases environmental problems concerning phosphorus in the manure, both by the removal of phytic acid and the absence of the need for phosphate addition to the feedstuffs.

The main reason for the low phytase production levels after induction of *A. niger* turned out to be the presence of a weak promoter upstream of the phytase gene. Replacement of the phytase promoter by the *glaA* promoter and introduction of multiple copies of this new expression cassette resulted in an increase of phytase production on laboratory scale from 0.1 u/ml by wildtype *A. niger* to 280 u/ml by a multicopy *A. niger* transformant (van Gorcom *et al.*, 1991; van Hartingsveldt *et al.*, 1993; Selten pers. comm.).

Using similar strategies as described above, efficient production of a number of other fungal proteins by *A. niger* and *A. oryzae* has been realised (for a review see van den Hondel *et al.*, 1991). Several of these proteins, produced by genetically engineering fungal strains, have been introduced on the market over the past few years.

## PRODUCTION OF MAMMALIAN PROTEINS

Because of their great secretion capacity, *A. niger* and *A. oryzae* are attractive organisms for the production of non-fungal (e.g. mammalian) proteins, especially for those of which the natural source is limited, such as the bovine milk-clotting enzyme (pro-)chymosin or human BRM's (biological response modifiers, e.g. interferons and interleukins). To investigate whether *A. niger* can be used for efficient production of this kind of proteins, genetic engineering techniques have been used to construct strains which produce some of these heterologous proteins. In our laboratory the production of bovine prochymosin and human interleukin 6 by *A. niger* have been studied as model systems (van Hartingsveldt et al., (1990*a*); Broekhuijsen et al., (1993)).

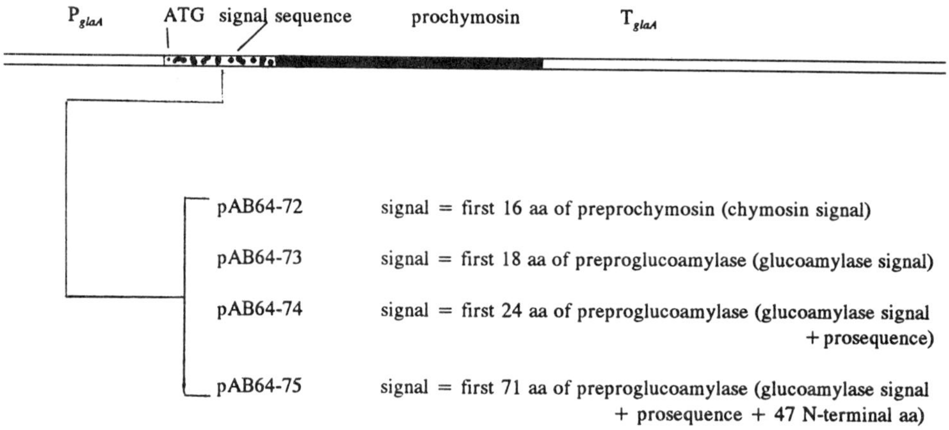


**Figure 2.** Schematic representation of prochymosin expression cassettes.

To study the expression of the gene encoding prochymosin, an enzyme traditionally used in cheese manufacturing, and the subsequent secretion of its gene product by *A. niger*, four almost identical expression cassettes were constructed. In each of these, the expression of the prochymosin gene was controlled by the expression signals of the glucoamylase gene of *A. niger* (*glaA*; Figure 2). To achieve secretion of prochymosin, in three of these cassettes the prochymosin encoding sequence was fused to sequences comprising the first 18 (signalpeptide), 24 (signal- and propeptide) or 71 (signalpeptide, propeptide and 47 mature N-terminal) amino acids of glucoamylase. In the fourth cassette the code for the signal peptide (16 amino acids) of prochymosin itself was present (Figure 2).

Proper comparison of prochymosin production levels by transformants containing these cassettes can only be accomplished if the transformants contain the same number of cassettes at identical genomic sites. Therefore, a strategy was worked out to generate *A. niger* transformants in which the glucoamylase gene is replaced by a single copy of one of the cassettes (Van Hartingsveldt *et al.*, 1990*b*). Such a gene replacement had occurred in 10-20% of the transformants obtained.

To analyse expression of the prochymosin gene single-copy transformants, each containing one of the four cassettes, and an untransformed control strain were grown under inducing culture conditions. Prochymosin mRNA levels were analysed by Northern blotting and levels of intra- and extracellular (pro)chymosin production were estimated by Western blot analysis of mycelial extracts and samples of growth medium. Chymosin activity in the medium was determined using a milk-clotting assay (MCA). The results are summarised in Table 1.

It can be concluded that the expression levels of prochymosin mRNA and prochymosin varied between transformants obtained with the different constructs. The highest chymosin level was found in growth medium of pAB64-73 transformants. In these transformants prochymosin was preceded by the glucoamylase signal peptide (first 18 amino acids).

The steady-state levels of prochymosin mRNA in the different transformants do not correlate with the levels of prochymosin found in the medium. The chymosin levels in the medium of pAB64-74 (24 aa) transformants is six times lower than in that of pAB64-73 transformants, although the levels of prochymosin mRNA are the same. A possible explanation for these results came from the analysis of mycelial extracts of the various transformants. This analysis revealed accumulation of glycosylated prochymosin in pAB64-74 transformants (Table 1). Although this was not tested further, it may be that the presence of the glucoamylase propeptide with a proteolytic processing site (Lys-Arg) at the junction of the glucoamylase and prochymosin sequences interferes with proper secretion. Introduction of several copies of plasmids similar to pAB64-72 and pAB64-73 (Figure 1) into *A. niger* var. *awamori* by Cullen *et al.*, (1987) resulted in almost identical production levels but, surprisingly, most of the chymosin accumulated intracellularly (Ward, 1989*a*). However, changing growth medium for these *A. niger* transformants resulted in improved secretion of prochymosin, indicating that also growth conditions can play an important role in efficient protein secretion.

In addition to intracellular accumulation mRNA stability and differential translation efficiency may also explain some of the results obtained. In transformants obtained with pAB64-72 (and to a lesser extent in pAB64-75 transformants) the steady-state level of prochymosin mRNA is clearly (up to 10 times) lower than that in pAB64-73 transformants, although identical transcription control sequences are present in each of these cassettes. However, the chymosin production in pAB64-72 transformants is only about two times lower. These results apparently indicate that prochymosin mRNA in pAB64-72 transformants is translated very efficiently but is relatively unstable.

In general, amounts of mammalian proteins produced by *Aspergillus* species are relatively low in comparison to the levels of fungal proteins obtained (0.2-25 mg/l vs. 1-5 g/l; for a review see Van den Hondel *et al.*, 1991). However, the proteins are biologically active and as far as examined resemble the authentic product (e.g. Upshall *et al.*, 1987; Archer *et al.*, 1990).

Recently Ward and co-workers (1989*b*, 1990) have shown that the secretion of (pro)chymosin can be increased 10-15 fold by the introduction of an additional N-glycosylation site in the protein or by in-frame fusion of the DNA sequence of prochymosin to the complete coding region of the *A. niger glaA* gene. In the latter case mature chymosin was obtained due to the autocatalytic removal of glucoamylase and prochymosin sequences at low pH.

The *glaA* fusion approach has also been successfully applied for the production of

**Table 1.** Analysis of (pro)chymosin production in *A.niger* 24 hours after induction.

| Strain/ transformant | 'Signal peptide' | Relative mRNA level | Chymosin (mg/l) from Western | MCA (U/ml) | Intracellular chymosin present |
|---|---|---|---|---|---|
| Wildtype | glucoamylase | 1 | 43* | 0 | - |
| AB64-72 | pre(prochymosin) | <0.1 | 6.2 | 8.6 | +/- |
| AB64-73 | first 18 aa of *glaA* | 1.0 | 11.3 | 19.5 | +/- |
| AB64-74 | first 24 aa of *glaA* | 1.0 | 4.1 | 3.2 | +++ |
| AB64-75 | first 71 aa of *glaA* | 0.4 | 10.2 | 19.1 | + |

* glucoamlyase production mg/l

human interleukin-6 (hIL-6) and porcine pancreatic phospholipase $A_2$ which, even if expressed under the control of efficient *Aspergillus* promoters, were secreted at almost undetectable levels into the culture medium (max. 25 µg/l), by *A. nidulans* or *A. niger* transformants (Carrez *et al.*, 1990; Broekhuijsen *et al.*, 1993; Roberts *et al.*, 1992). In the glucoamylase fusion approach used for the production of interleukin 6, a spacer peptide, containing a KEX-2 like (dibasic endo-protease) protein processing signal, was placed between the glucoamylase protein and the mature protein. The dibasic processing site is recognised and cleaved efficiently by an enzyme present both in *A. nidulans* and in *A. niger* resulting in biologically active, mature hIL-6 at levels up to 15 $mgl^{-1}$ with *A. niger* transformants (Contreras *et al.*, 1991; Broekhuijsen *et al.*, 1993).

## EFFECTS OF PROTEASES PRODUCED BY *ASPERGILLUS* ON PROTEIN SECRETION

A serious problem interfering with the efficient production of most mammalian proteins, but also with some fungal proteins, is the degradation by fungal proteases (Prt). An illustration of the negative effects of proteolysis is shown in Figure 3. An *A. niger* transformant carrying one copy of the chymosin expression cassette from pAB64-73 integrated at the *glaA* locus (see above) was cultivated in the presence and in the absence of protease inhibitors. Chymosin secretion into the medium was monitored at several timepoints by Western blotting. Clearly the addition of protease inhibitors increased the level of chymosin present in the medium. The same problem has been reported in several papers describing the expression of other mammalian proteins (e.g. Archer *et al.*, 1992, Broekhuijsen *et al.*, 1993).

To overcome the problem of proteolysis, strains deficient in extracellular protease activity have been generated either by deleting a major extracellular protease (aspergillopepsin A, an aspartyl protease: Berka *et al.*, 1990) or by mutagenesis and selection strategies. With the latter method several $Prt^-$ mutants have been isolated in the laboratory and these have been genetically and biochemically characterised (Mattern *et al.*, 1992). Some examples of mutants obtained in these studies are shown in Figure 4.

The use of protease deficient strains combined with the production strategy described above (fusion of the protein of interest to a carrier protein) is well accepted at the moment. Several mammalian proteins have been produced in *Aspergillus* species using these strategies at levels from 10 to 1000 mg/l without further optimisation (e.g. hen

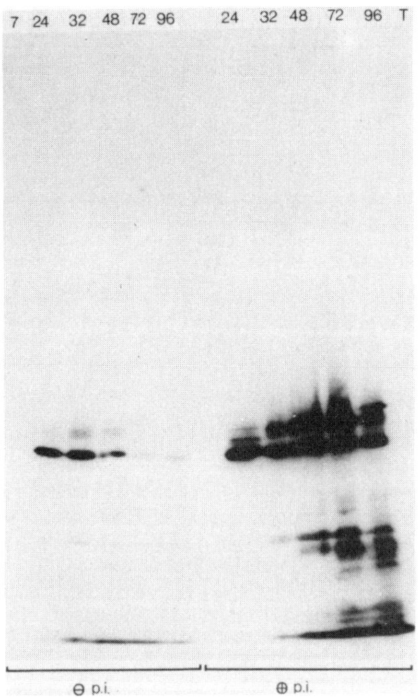


**Figure 3.** (Pro)chymosin production in time (T=hours) monitored by Western blotting of a AB64-73 *A. niger* transformant in the presence (⊕ p.i.) and absence (⊖ p.i.) of protease inhibitors.

egg-white lysozyme: 1 g/l with *A. niger*, Jeenes et al., 1993). A further improvement can be expected by inactivating the *Aspergillus* gene homologous to the *S. cerevisiae pep4* gene. This gene encodes a vacuolar protease that activates other (vacuolar) proteases (Ammerer *et al.*, 1986). Recently Dr. B.J. Bowman (UCSC; pers. comm.) reported the cloning of the *pep4* homologue of *Neurospora crassa.*

## CONCLUSIONS

The results obtained, by us and others, in studies on the production of extracellular fungal proteins show that, in general, these proteins are efficiently secreted by *Aspergillus* species. Concurrently, during the past few years the first products produced by recombinant fungal strains have been introduced on the market (e.g. Lipolase®, an *Humicola* lipase produced by Novo Nordisk using *A. oryzae*, and Natuphos®, an *Aspergillus* phytase produced by Gist-Brocades using *A. niger*).

The levels of extracellular production of mammalian proteins are in many cases much lower than those of fungal proteins, probably as a result of problems with mRNA stability, translation, secretion and/or degradation. However, the use of protease deficient strains in combination with expression cassettes in which the protein of interest is expressed as a fusion with a highly produced extracellular fungal protein has, however, improved these levels substantially. Although production levels of 10 to 100 mg per litre are "common practice" these days, these levels are still only a few percent of the levels obtained with fungal proteins. To identify and solve problems that are encountered at the level of gene expression (especially mRNA stability) and protein secretion a considerable research effort will be necessary regarding a better understanding of these processes.

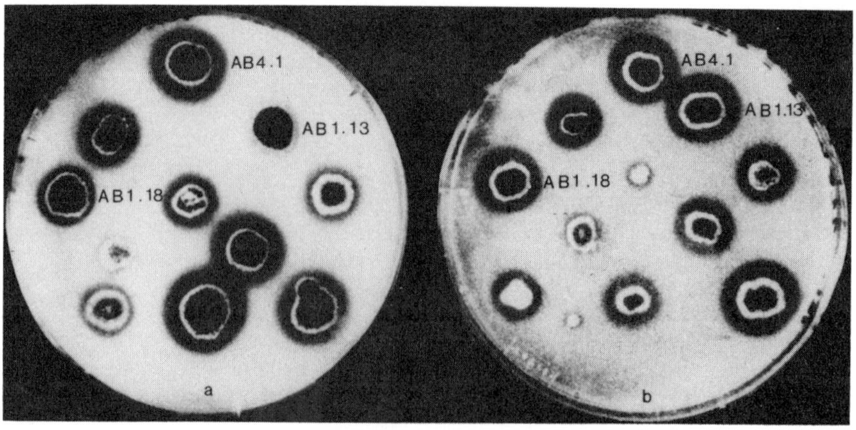


**Figure 4.** Analysis of *Aspergillus niger* mutants for reduced extracellular protease activity. The left plate contains skim milk to monitor protease activity. The right plate contains starch as control for general secretion mutants. AB4.1 is the wildtype strain, AB1.13 and AB1.18 are two strains with a strongly reduced extracellular protease activity level (resulting in a smaller halo on the milk plate but no effect on the size of the halo on the starch plate).

Since most of the (published) results on production levels are obtained in shake flasks, additional investigations are necessary to evaluate the production of these proteins under different fermentation conditions. It is expected that alterations of the growth characteristics (physiological) and/or the expression/secretion machinery (genetically) of the *Aspergillus* strains, used for the production of heterologous proteins, are necessary to reach the levels of production which are obtained with a number of enzymes made by the present "classical" *Aspergillus* production strains (e.g. >20 $gl^{-1}$).

It has, however, been shown by Ward and coworkers that a combination of modern molecular genetic techniques and classical genetics coupled to sophisticated screening-procedures can result in the isolation of *Aspergillus niger* strains that produce commercially interesting secreted levels of a mammalian protein (prochymosin; >1.5 $gl^{-1}$).

## ACKNOWLEDGEMENTS

We wish to thank our colleagues Martien Broekhuijsen, Wim van Hartingsveldt, Ineke Mattern, Jan Verdoes and Cora van Zeijl for their contribution to this paper. Part of the work described was supported by the Programme Commission Biotechnology or carried out as contract research for Gist-Brocades NV, Delft, The Netherlands.

## REFERENCES

Ammerer, G., Hunter, C.P., Rothman, J.H., Saari, G.C., Valls, L.A. and Stevens, T.H. (1986) PEP4 gene of *Saccharomyces cerevisiae* encodes proteinase A, a vacuolar enzyme required for processing of vacuolar precursors. Mol. Cell. Biol. 6, 2490-2499.

Archer, D.B., Jeenes, D.J., MacKenzie, D.A., Brightwell, G., Lambert, N., Lowe, G., Radford, S.E. and Dobson, C. (1990) Hen egg-white lysozyme expressed in, and secreted from, *Aspergillus niger* is correctly processed and folded. Bio/technology 8, 741-745.

Archer, D.B., MacKenzie, D.A., Jeenes, D.J. and Roberts, I.N. (1992) Proteolytic degradation of heterologous proteins expressed in *Aspergillus niger*. Biotechnol. Lett. 14, 357-362.

Broekhuijsen, M.P., Mattern, I.E., Contreras R., Kinghorn, J.R., Van den Hondel, C.A.M.J.J. (1993) Secretion of heterologous proteins by *Aspergillus niger*: Production of active human interleukin-6 in a protease-deficient mutant by KEX2-like processing of a glucoamylase-IL6 fusion protein, J.Biotech., in press.

Berka, R.M., Ward, M., Wilson, L.J., Hayenga, K.J., Kodama, K.H., Carlomagno, L.P. and Thompson, S.A. (1990) Molecular cloning and deletion of the aspergillopepsin A gene from *Aspergillus awamori*. Gene 86, 153-162.

Boel, E. Hansen, M.T., Hjort, I., Hoegh, I. and Fiil, N.P. (1984) Two different types of intervening sequences in the glucoamylase gene from *Aspergillus niger*. EMBO J. 3, 1581-1585.

Bos, C.J. (1986) Induced mutation and somatic recombination as tools for genetic analysis and breeding of imperfect fungi. PhD Thesis, Agricultural University Wageningen NL.

Carrez, D., Janssens, W., Degrave, P., Van den Hondel, C.A.M.J.J., Kinghorn, J.R., Fiers, W. and Contreras, R. (1990) Heterologous gene expression by filamentous fungi: secretion of human interleukin-6 by *Aspergillus nidulans*. Gene 94, 147-154.

Contreras, R., Carrez, D., Kinghorn, J.R., Van den Hondel, C.A.M.J.J. and Fiers, W. (1991) Efficient KEX2-like processing of a glucoamylase-interleukin-6 fusion protein by *Aspergillus nidulans* and secretion of mature interleukin-6, Bio/technology 9, 378-381.

Cullen, D., Gray, G.L., Wilson, L.J., Hayenga, K.J., Lamsa, M.H., Rey, M.W., Norton, S. and Berka, R.M. (1987) Controlled expression and secretion of bovine chymosin in *Aspergillus nidulans*, Bio/technology 5, 369-376.

Fincham, J.R. (1989) Transformation in fungi. Microbiol. Rev. 53, 148-170.

Fowler, T., Berka, R.M. and Ward, M. (1990) Regulation of the *glaA* gene of *Aspergillus niger*. Curr. Genet. 18, 537-545.

Goosen, T., Bos, C.J., and Van den Broek, H.W.J. (1992) Transformation and Gene Manipulation in Filamentous Fungi: An overview, in: "Handbook of Applied Mycology, Fungal Biotechnology", (Arora, D.K., Mukerji, K.G. and Elander, R.P. Eds.), pp.151, Vol. 4. Dekker, New York.

Gray, G.L., Hayenga, K., Cullen, D., Wilson, L.J. and Norton, S. (1986) Primary structure of *Mucor miehei* aspartyl protease: Evidence for a zymogen intermediate. Gene 48, 41-53.

Hynes, M.J., Andrianopoulos, A., Davis, M.A., Van Heeswijck, R., Katz, M.E., Littlejohn, T.G., Richardson, I.B. and Saleeba, J.A. (1989) Multiple circuits regulating the *amdS* gene of *Aspergillus nidulans*, *in*: "Proceedings of the EMBO-Alko Workshop on Molecular Biology of Filamentous Fungi" (Nevalainen, H. and Penttilä, M. Eds.), pp. 63-71. Foundation for Biotechnical and Industrial Fermentation Research 6. Helsinki.

Jeenes, D.J., Marczinke, B., MacKenzie, D.A. and Archer, D.B. (1993) A truncated glucoamylase gene fusion for heterologous protein secretion from *Aspergillus niger*, in press.

Kelly, J.M. and Hynes, M.J. (1985) Transformation of *Aspergillus niger* by the *amdS* gene of *Aspergillus nidulans*. EMBO J. 4, 475-479.

Kos, A., Kuijvenhoven, J., Wernars, K., Bos, C.J., Van den Broek, H.W.J., Pouwels, P.H. and Van den Hondel, C.A.M.J.J. (1985) Isolation and characterization of the *Aspergillus niger trpC* gene. Gene, 39, 231-238.

Mattern, I.E., Van Noort, J.M., Van den Berg, P., Archer, D.B., Roberts, I.A., and Van den Hondel, C.A.M.J.J. (1992) Isolation and characterization of mutants of *Aspergillus niger* deficient in extracellular proteases. Mol.Gen.Genet. 234, 332-336.

Roberts, I.N., Jeenes, D.J., MacKenzie, D.A., Wilkinson, A.P., Sumner, I.G. and Archer, D.B. (1992) Heterologous gene expression in *Aspergillus niger*: a glucoamylase-porcine pancreatic prophospholipase $A_2$ fusion protein is secreted and processed to yield mature enzyme. Gene 122, 155-161.

Simons, P.C.M., Versteegh, H.A.J., Jongbloed, A.W., Kemme, P.A., Slump, P., Bos, K.D., Wolters, G.E., Beudeker, R.F. and Verschoor, G.J. (1990) Improvement of phosphorus availability by microbial phytase in broilers and pigs. British Journal of Nutrition 64, 525-540.

Timberlake, W.E., and Marshall, M.A. (1989) Genetic engineering in filamentous fungi. Science, 244, 1313-1317.

Upshall, A., Kumar, A.A., Bailey, M.C., Parker, M.D., Favreau, M.A., Lewison, K.P., Joseph, M.L., Maraganore, J.M. and Mcknight, G.L. (1987) Secretion of active human tissue plasminogen activator from the filamentous fungus *Aspergillus nidulans*. Bio/technology 5, 1301-1304.

Van den Hondel, C.A.M.J.J. and Punt, P.J. (1991) Gene transfer systems and vector development for filamentous fungi. BMS Symposium Series Vol. 18: Applied Molecular Genetics of Fungi, (Peberdy, J.F., Caten, C.E., Ogden, J.E. and Bennett, J.W. Eds) British Mycological Society, Cambridge University Press, Cambridge, UK, 1-29.

Van den Hondel, C.A.M.J.J., P.J. Punt and Van Gorcom, R.F.M. (1991) Heterologous gene expression in filamentous fungi, in "More Gene Manipulation in Fungi", (Bennett, J.W. and Lasure, L.L., Eds.) pp. 396-428. Academic Press.

Van Gorcom, R.F.M., Pouwels, P.H., Goosen, T., Visser, J., Van den Broek, H.W.J., Hamer, J.E., Timberlake, W.E. and Van den Hondel, C.A.M.J.J. (1985) Expression of an *Escherichia coli* ß-galactosidase fusion gene in *Aspergillus nidulans*. Gene 40, 99-106.

Van Gorcom, R.F.M., Van Hartingsveldt, W., Luiten, R.G.M., Van Paridon, P.A., Selten, G.C.M. and Veenstra, A.E. (1991) Cloning and expression of microbial phytase, European patent application EP 0420358.

Van Hartingsveldt, W., Van Zeijl, C.M.J., Veenstra, A.E., Van den Berg, J.A., Pouwels, P.H., Van Gorcom, R.F.M. and Van den Hondel, C.A.M.J.J. (1990a) Heterologous gene expression in *Aspergillus*: analysis of chymosin production in single-copy transformants of *Aspergillus niger*, in "Proceedings of 6th International Symposium on Genetics of Industrial Microorganisms" (Heslot, H., Davies, J., Florent, J. Bibichon, L., Durant, G., Penasse, L. Eds) Société Française de Microbiologie, 107-116.

Van Hartingsveldt, W., Van den Hondel, C.A.M.J.J., Veenstra, A.E. and Van den Berg, J.A., 1990b, Gene replacement as a tool for the construction of *Aspergillus* strains, European Patent Application, EP 0357127.

Van Hartingsveldt, W., Van Zeijl, C.M.J., Harteveld, G.M., Gouka R.J., Suykerbuyk M.E.G., Luiten R.G.M., Van Paridon, P.A., Selten G.C.M., Veenstra A.E., Van Gorcom, R.F.M. and Van den Hondel C.A.M.J.J. (1993) Cloning, molecular characterization and overexpression of the phytase gene (*phyA*) of *Aspergillus niger*. Gene, in press.

Verdoes, J.C., Punt, P.J., Schrickx, J.M., Van Verseveld H.W., Stouthamer, A.H. and Van den Hondel, C.A.M.J.J. (1993) Glucoamylase overexpression in *Aspergillus niger*: molecular genetics analysis of strains containing multiple copies of the *glaA* gene. Transgenic Res. 2, 84-92.

Ward, M., 1989a, Production of calf chymosin by *Aspergillus awamori*, in: "Genetics and Molecular Biology of Industrial Microorganisms", (Hershberger, C.L., Queener, S.W. and Hegeman, G., Eds) American Society for Microbiology, Washington DC, 288-294.

Ward, M., 1989b, Heterologous gene expression in *Aspergillus*, in "Proceedings of the EMBO-Alko Workshop on Molecular Biology of Filamentous Fungi", (Nevalainen, H. and Penttilä, M., Eds) Foundation for Biotechnical and Industrial Fermentation Research 6, 119-128, Helsinki.

Ward, M., Wilson, L.J., Kodama, K.H., Rey, M.W. and Berka, R.M. (1990) Improved production of chymosin in *Aspergillus* by expression as a glucoamylase-chymosin fusion. Bio/technology 8, 435-440.

# APPLICATION : *ASPERGILLUS ORYZAE* AS A HOST FOR PRODUCTION OF INDUSTRIAL ENZYMES

Tove Christensen

Department of Fungal Genetics
Novo Nordisk A/S
Novo Allé
DK-2880 Bagsvaerd
Denmark

## INTRODUCTION

Several groups have described the use of *Aspergillus* species, predominantly *A. nidulans* and *A. niger*, as expression hosts for synthesis of heterologous proteins, as reviewed recently by Gwynne (1992). At Novo Nordisk an expression system has been developed in *A. oryzae* (Christensen *et al.*, 1988) and used for expression of several new industrial enzymes. Work in the Company has been focused on the production of secreted enzymes because purification of the product is simplified when the product can be isolated from the fermentation broth. Some of these enzymes are now available as commercial products, a situation which was made possible via high level expression in *A. oryzae*. In this paper some of the work carried out at Novo Nordisk in the development of *A. oryzae* into a host for production of industrial enzymes is described.

## CHOICE OF HOST STRAIN

The choice of a host strain for industrial production is influenced by several factors. It is mandatory that the strain can be manipulated by relatively simple techniques both in the laboratory and in the production process. With respect to the laboratory work, it is necessary that fundamental methods like DNA transformation can be performed without too much difficulty. It is also important that good large scale fermentation methods exist or can be developed. Naturally, the efficiency of the production must be considered as well. A commercially viable production of industrial enzymes requires product yields measured in grams per litre of culture medium. In addition, safety considerations makes it important that the chosen strain (as well as its products) is harmless to human beings.

After consideration of such points, *Aspergillus oryzae* was developed as a production host for industrial enzymes. *A. oryzae* was chosen in part because of its long history of use

*The Genus Aspergillus*, Edited by Keith A. Powell *et al.*,
Plenum Press, New York, 1994

in production of fermented foods in Asia, as well as its previous use in industrial production of for example amylases. *A. oryzae* and its enzymes are accepted by FAO/WHO as constituents of food. The safety aspects of *A. oryzae* has been reviewed by Barbesgaard *et al.* (1992).

In addition, since *A. oryzae* has been used as a classical production host for industrial enzymes for several decades, fermentation methods are well known. Also, molecular genetics methods have previously been used in *Aspergillus* species, predominantly *A. nidulans*, since the early eighties and the knowledge obtained has proven to be a good background for work with *A. oryzae*.

A further advantage of *A. oryzae* is that it seems to have a poor ability to establish itself in the environment. For example, we have not been able to isolate the production strain from the environment around a Novo Nordisk plant that had been used to produce *A. oryzae* derived amylase for more than twenty years.

## SELECTIVE MARKERS

DNA transformations are required for almost all molecular work, and it is one of the first methods that must be established when starting with a new organism. In filamentous fungi almost all established transformation systems are based on integration of the transforming DNA into the chromosome (see Swart *et al.*, this volume), and in all the experiments discussed below this has been the pattern of events. Table 1 shows the selective markers which have been used in *A. oryzae* and results obtained with the markers *benA*33, *amdS* and *niaD* will be described.

**Table 1.** Selective markers for transformation of *Aspergillus oryzae*.

| Gene | Donor organism | *A. oryzae* genotype | Reference |
|---|---|---|---|
| *amdS* | *A. nidulans* | wt | Christensen *et al.* (1988) |
| *benA*33 | *A. nidulans* | wt | this work |
| *niaD* | *A. oryzae* | *niaD* mutant | Unkles *et al.* (1989) |
| *sC* | *A. nidulans* | *sC* mutant | M. Egel-Mitani (unpublished) |
| *argB* | *A. nidulans* | *argB* mutant | Gomi *et al.* (1987) |
| *ble* | *E. coli* | wt | Cheevadhanarak *et al.* (1991) |
| *pyrG* | *A. niger* | *pyrG* mutant | Mattern *et al.* (1987) |

### *BenA*33

*BenA*33 is a benomyl resistant and temperature sensitive allele of the beta-tubulin gene from *A. nidulans* (May *et al.*, 1985). A selection method for this marker in *A. oryzae* was developed and some transformants were obtained. Two independent transformants have been analysed in detail. Transformant 1 grew slower than transformant 2 in the presence of 2 μg/ml benomyl at 28°C. On nonselective plates at 37°C transformant 1 grew well while transformant 2 grew weakly, but some faster growing colonies developed from the latter. Addition of benomyl to the growth medium stimulated growth of transformant 2 at 37°C. The phenotype of transformant 2 thus resembled the *benA*33 phenotype. *A. oryzae* has its own resident active β-tubulin gene; transformants will therefore contain a mixture of benomyl resistant, temperature sensitive and benomyl sensitive, temperature resistant $\beta$-tubulin proteins. Our results indicate that the ratio of *benA*33 derived ß-tubulin to endogenous *benA* derived $\beta$-tubulin is higher in transformant 2 than in transformant 1. Southern analysis of the two transformants shows that transformant 2 had more copies of the *benA*33 gene integrated in the genome than transformant 1.

A fast growing colony from transformant 2 on non-selective plates at 37°C was isolated. This phenotypic revertant is benomyl sensitive. Southern analysis was used to study possible rearrangements of *benA*33 DNA in the revertant. Equal amounts of chromosomal DNA from the untransformed strain A1560, from transformant 2 and from the revertant were digested with restriction enzymes *Eco*RI, *Pst*I or *Sac*I. A Southern blot was made and hybridised with a $^{32}P$ labelled 5.5kb fragment containing the *benA*33 gene. Figure 1 shows the result. The transforming *benA*33 plasmid carries the *benA*33 gene on a 5.5 kb PstI fragment cloned into a pUC19 vector. *Eco*RI cuts once in the plasmid, *Pst*I cuts out the insert from the vector and *Sac*I cuts twice in the *benA*33 insert. It can be seen from Figure 1 that transformant 2 has several copies of the *benA*33 gene in the chromosome and that the revertant has lost most but not all of the *benA*33 DNA. There is no internal *Sac*I fragment in the revertant and thus no functional *benA*33 gene is left.

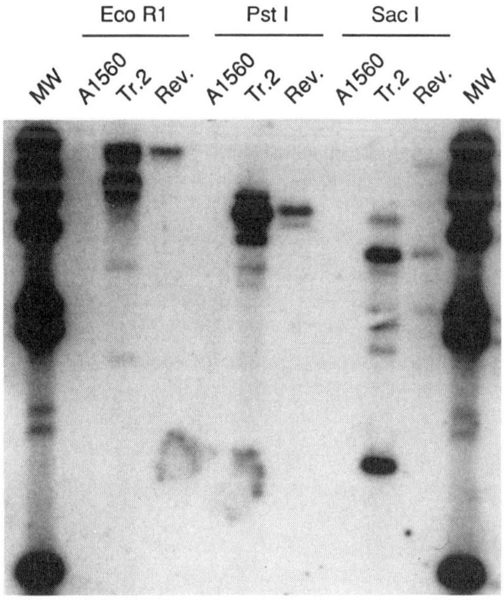


**Figure 1.** Southern analysis of transformant 2 and of a temperature resistant revertant. Equal amounts of DNA of A1560, transformant 2 and the revertant were digested with *Eco*RI, *Pst*I and *Sac*I. The DNA was separated on an agarose gel and blotted to a nylon membrane. The membrane was hybridised with a $^{32}P$ labelled 5.5 kb DNA fragment containing the *benA*33 gene. The MW lanes contain $^{32}P$ labelled lambda DNA digested with *Hind*III

## *amdS*

The use of the *amdS* gene from *A. nidulans* for transformation of *A. oryzae* has been described previously (Christensen *et al.*, 1988). The transformation frequency of the wildtype *amdS* allele in *A. oryzae* A1560 is 1-5 transformants per $\mu$g DNA. The same transformation frequencies were obtained with the original plasmid p3SR2 (Hynes *et al*, 1983), which has a 5.5 kb fragment carrying the *amdS* gene, and with a cutback plasmid containing a 2.7 kb *Xba*I fragment cloned into pUC19. The complete sequence of the 2.7 kb *Xba*I fragment has been published by Corrick *et al.* (1987).

*amdS* transformants normally had more copies of the marker gene integrated into the genome when compared to *argB* or *niaD* selected transformants. This could be the result of a high selective pressure in *amdS* selection, during which acetamide is used as the sole nitrogen source. It was decided to transform *A. oryzae* with the promoter-up mutant alleles of the *amdS* gene previously isolated by Hynes *et al.* (1988) and Katz *et al.* (1990). The rationale for doing this was that a stronger promoter might lead to a higher expression per copy of the gene. The requirement for a high mean copy number during the selection step might then not be as pronounced. This in turn might result in a higher apparent transformation frequency. This would of course only be true if mutants isolated in *A. nidulans* would have a similar phenotype when introduced into *A. oryzae*. M Hynes kindly provided the *amdS* alleles $I_9$, $I_9$ plus $I_{66}$ and $I_{666}$. $I_9$ is a cis acting promoter mutant which is positioned in the regulatory site for *fac*B, $I_{66}$ and $I_{666}$ are also cis acting mutants but within the regulatory site for *amdA* (Hynes *et al.*, 1988 and Katz *et al.*, 1990). A construct combining the $I_9$ and the $I_{666}$ mutations was made. As Table 2 shows, a higher apparent transformation efficiency was found when using DNA containing the stronger promoters. The number of transformants that can be isolated is at least ten times higher with $I_{666}$ and with $I_9$ plus $I_{666}$ than for the wild-type allele. An even stronger effect of $I_9$ plus $I_{666}$ than found in this particular experiment has often been observed. Transformants selected, with the mutant alleles, show faster growth compared to wild-type transformants on acetamide. The copy number of the transforming plasmid DNA in the different transformants was measured in a slot blot experiment using pUC19 as the hybridisation probe. The copy number was determined by comparing with the hybridisation signal obtained from a strain known to have one copy of pUC19 integrated in the genome. For all the *amdS* alleles some transformants with a low copy number were found, for the wild-type and Ig alleles some transformants with a very high copy number, were also seen. In general, there were more transformants with a high copy number among the wild-type and $I_9$ transformants than among the $I_9$ plus $I_{666}$ transformants. The co-transformation frequency of an unselected plasmid expressing the *E. coli uidA* gene was also determined. The β-glucuronidase expressed from this gene can easily be measured on plates. Strain A1560 was transformed with equal amounts of selected and unselected plasmid, the result is shown in Table 2. The 80% co-transformation observed with the wild-type allele is consistent with previous experience, where a co-transformation frequency of 80-100% for the *amdS* wild-type allele is usually obtained. The co-transformation frequency can be raised for the mutant *amdS* alleles by using a higher ratio of unselected to selected plasmid.

The fact that the mutant *amdS* alleles which were selected in *A. nidulans* have a similar phenotype in *A. oryzae* makes it plausible that *A. oryzae* contains regulatory proteins closely related to regulatory proteins in *A. nidulans*.

### *amdS* versus *niaD* selection

In *A. oryzae* A1560, *amdS* selection usually leads to a high copy number of the transformed DNA, while *niaD* selection gives a low copy number in most of the transformants; often only one copy of the plasmid is integrated at the *niaD* locus (Unkles *et al.*, 1989 and T. Christensen, data not shown). The co-transformation frequency is also higher for *amdS* than for *niaD*. It was necessary to establish whether these differences have any influence on the yield obtained from an expression plasmid introduced via co-transformation. In this analysis an expression plasmid encoding a preprolipase, which is secreted to the culture medium, was used. It is not clear whether processing of the lipase

**Table 2.** Effect of *amdS* allele on transformation.

| *amdS* allele | Transformants per $\mu$g DNA | Days from plating to sporulation | Copy no. | Co-transformation frequency |
|---|---|---|---|---|
| wt | 1 - 5 | 5 - 6 | 2 - >100 | 80% |
| $I_9$ | 1 - 5 | N.D. | 4 - >100 | 80% |
| $I_{66} + I_9$ | 20 | N.D. | N.D. | 60% |
| $I_{666}$ | 50 - 100 | 3 - 4 | N.D. | 50% |
| $I_{666} + I_9$ | 50 - 100 | 3 - 4 | 3 - 20 | 60% |

Approximately $2x10^7$ protoplasts of A1560 in 100 $\mu$l were transformed with 1 $\mu$g of plasmid DNA. The various plasmids contained the different *amdS* alleles on a 2.7 kb *Xba*I fragment cloned into pUC19. The number of transformants was determined on acetamide plates. The number of days from plating of the protoplasts to sporulating colonies could be seen was noted. The copy number of integrated DNA was determined in a slot blot experiment. Six three fold dilutions of chromosomal DNA from 10 independent transformants of each type were loaded onto a nitrocellulose membrane. Two identical sets of membranes were made. One set was hybridised with $^{32}$P labelled pUC19 DNA and the other was hybridised with a $^{32}$P labelled fragment of the *A. oryzae tpi* gene as a control of the amounts of DNA loaded. The copy number was determined by comparing transformants to a strain known to have one copy of pUC19 integrated in the genome of A1560. The co-transformation frequency was determined by co-transforming $2x10^7$ protoplasts with 2.8 $\mu$g of a $\beta$-glucuronidase expression plasmid and 2.8 $\mu$g of *amdS* plasmid. 10 transformants of each *amdS* plasmid were reisolated on acetamide plates and assayed for $\beta$-glucuronidase activity towards 4-methylumbelliferyl-ß-D-glucoronide.

occurs in the culture medium or during its secretion. The lipase expression plasmid was co-transformed into A1560 or A1560*niaD* with either *amdS* or *niaD* as the selective marker. The following strains were isolated: 17 lipase transformants in A1560 selected with *amdS*, 17 lipase transformants in A1560*niaD* selected with *amdS*, 20 lipase transformants in A1560*niaD* selected with *niaD*. All the strains were fermented in shake flasks in duplicates and the lipase yield was measured as enzyme activity. Figure 2 shows

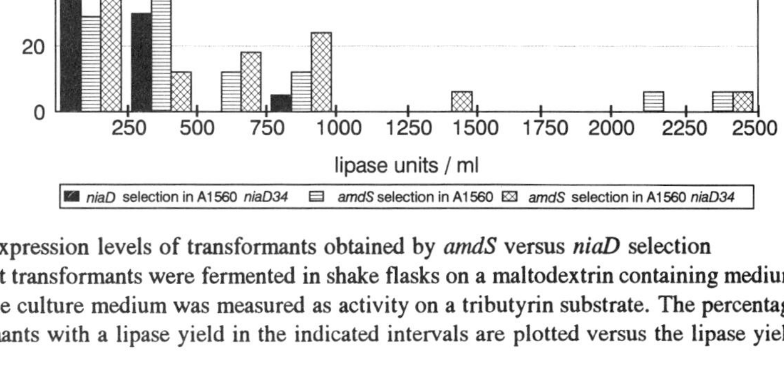

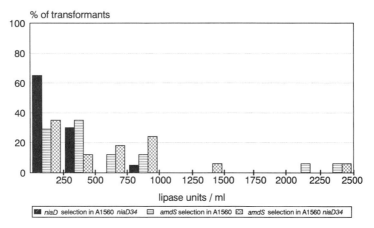


**Figure 2.** Expression levels of transformants obtained by *amdS* versus *niaD* selection
The different transformants were fermented in shake flasks on a maltodextrin containing medium. The lipase content in the culture medium was measured as activity on a tributyrin substrate. The percentage of each set of transformants with a lipase yield in the indicated intervals are plotted versus the lipase yield.

the result of this experiment. Most of the *niaD* selected transformants produced relatively low amounts of lipase, only one gave a yield above 500 lipase units per ml. The lipase yields in the *amdS* selected transformants are distributed over a larger range, and when compared to the *niaD* results, a significantly higher percentage of *amdS* selected transformants has high expression of the co-transformed gene.

## EXPRESSION SYSTEM

Obviously, an efficient expression and secretion system is necessary to produce commercially interesting amounts of industrial enzymes. In many cases, efficient secretion has been obtained using the signal sequences from the relevant enzyme encoding genes themselves, but signal sequences have also been successfully exchanged between some secreted fungal proteins (data not shown).

To ensure high levels of transcription of genes to be expressed the promoter of the amylase gene, cloned from a mutant strain used for production of the enzyme, has been used. Linking this promoter to an aspartic protease from *Mucor miehei* gave yields of more that 3 g/l under appropriate fermentation conditions. In this experiment a DNA fragment from the 3'end of the glucoamylase gene from *A. niger* was used as a transcription terminator and polyadenylation site (Christensen *et al*, 1988).

This expression system has been used for the expression of other proteins including lipases, proteases, cellulases, xylanases and peroxidases. The enzymes originate from a number of different fungi including *Aspergillus niger*, *Humicola lanuginosa, Humicola insolens*, *Fusarium oxysporum*, *Mucor miehei*, *Candida antarctica* and *Coprinus cinnereus*.

Analysis of the heterologous enzymes showed that some of them are glycosylated by the *A. oryzae* host. This is the case for example for a lipase cloned from *H. lanuginosa*. The lipase is also glycosylated when synthesised in its original organism, but the amount of glycosylation is a little higher in *A. oryzae*. The two glycosylation forms have been examined, the specific activity of the lipase was no different, but some minor changes in thermostability were observed. At least some of the sugar residues are N-linked since they can be removed with endoglycosidase H. The size of endo-H treated lipase, estimated on a SDS-polyacrylamide gel, is the same whether the lipase is made in *H. lanuginosa* or *A. oryzae* (data not shown).

## PURE PRODUCTS

*A. oryzae* produces secreted proteins of its own which are undesirable in the final heterologous product; there are several ways to decrease the synthesis of these. Traditional mutagenesis could be used to inactivate the genes, or the gene(s) in question could be cloned and the cloned DNA could subsequently be used to delete or interrupt its chromosomal counterpart. Finally, fermentation conditions can be optimised to ensure maximal expression of the product and repression of synthesis of other secreted proteins.

The dominant secreted protein made by *A. oryzae* strain A1560 is the amylase. A1560 has three genes encoding this protein, which complicates both mutagenesis and gene disruption. In addition, as described above, the amylase promoter has often been used for synthesis of the heterologous product. This strategy is not compatible with fermentation conditions that repress amylase synthesis.

Attempts to inhibit synthesis of the endogenous amylase by the antisense mRNA technique were tried. *A. oryzae* was transformed with an expression plasmid that has a piece of the amylase gene inserted in the opposite of its normal orientation between the amylase promoter and the *A. niger* glucoamylase 3´end. The amylase promoter was chosen to direct synthesis of the amylase antisense transcript to ensure that the sense and the antisense RNA are coordinately regulated. The amylase gene fragment inserted in the expression plasmid is 636 bp long and covers the entire 5´end of the natural mRNA. This antisense expression plasmid was co-transformed into an A1560 sC mutant by co-transformation with the *A. nidulans* sC gene. 61 transformants were grown in test tubes in YP supplemented with 2% maltose. The fermentation broth was analysed by SDS-polyacrylamide gel electrophoresis, Figure 3 shows the result for some of the transformants. ToC710 and ToC711 had the lowest content of amylase in the medium. The two transformants and A1560 were subsequently fermented in 2 litre Kieler fermenters in a maltodextrin medium which induces amylase synthesis. A1560 produced 14, ToC710 produced 0.6 and ToC711 produced 0.4 amylase unit per ml. We concluded that it is possible to lower synthesis of endogenous amylase considerably by use of the antisense mRNA technique.

## STRAIN STABILITY

In *Aspergillus* transformants, the transforming DNA is normally integrated in a stable manner in the genome. This is a very important feature for production strains in order to obtain reliable fermentation results. *A. oryzae* transformants seem to be stable just as other *Aspergillus* transformants (Gomi *et al.*, 1987 and Cheevadhanarak *et al.*, 1991).

To examine the stability of *A. oryzae* strains under production conditions, DNA was isolated from a sample of mycelium taken at the end of a lipase production fermentation. This DNA was analysed by Southern blotting, in parallel with DNA prepared from a 10 ml test tube inoculated with the original transformant. The Southern blot was probed with a DNA fragment containing the structural gene for lipase. No rearrangement was observed in the lipase DNA after a production scale fermentation (data not shown). Lipase production started in 1988 from an *A. oryzae* transformant is still running. Reliable production results have proven that the strain is stable in large scale fermentations.

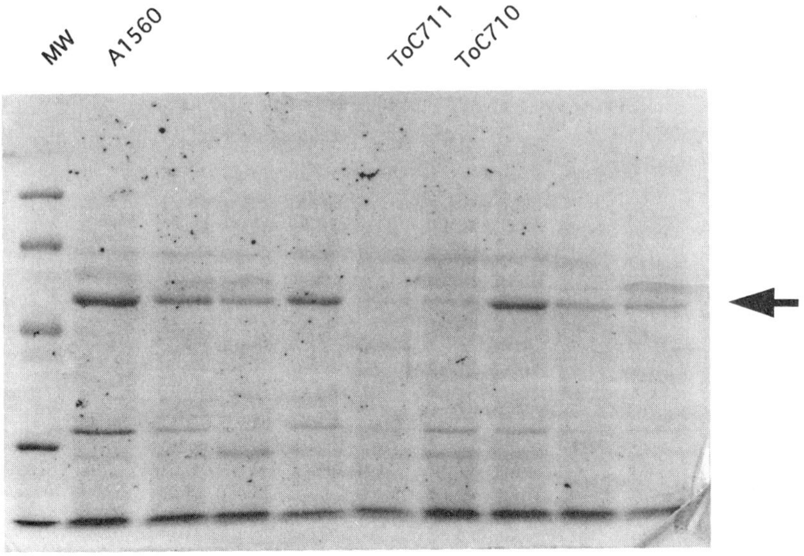


**Figure 3.** Amylase antisense transformants
8 independent transformants of the amylase antisense expression plasmid and A1560 were fermented in 10 ml YP with 2% maltose. A sample of the fermentation broth was applied to SDS-polyacrylamide gel electrophoresis on a 10% gel. The arrow indicate the position of the amylase. The MW lane contains a sample of the LMW molecular weight size marker from Pharmacia.

## CONCLUSION

The use of *A. oryzae* as a host for production of industrial enzymes has been explored. As part of this work several different markers for selection of transformants have been examined. A selection method was developed for the *benA*33 marker and it was shown that the temperature sensitive phenotype can be used for counter selection. In addition, the transformation frequency for the *amdS* marker has been enhanced by use of promoter mutants. The effect of different selection markers on expression from a co-transformed expression plasmid has also been studied.

A method using antisense RNA was effective in suppressing the expression of endogenous proteins made by *A. oryzae.*

To illustrate that the system is applicable for production of industrial enzymes, Table 3 shows trade names and the intended usage of five enzymes made with this expression system. None of these enzymes were commercially available earlier and have only become so via expression in the *A. oryzae* host.

**Table 3.** Commercial enzyme products made in *A. oryzae*.

| | |
|---|---|
| Lipolase™ | A lipase for detergents |
| Carezyme™ | A cellulase for detergents |
| Lipozyme™ | A lipase for transesterification |
| Resinase™ | A lipase for pitch removal in paper production |
| Novozym 435™ | A lipase for synthesis of esters |

## REFERENCES


Barbesgaard, P., Heldt-Hansen, H.P. and Diderichsen B. (1992) On the safety of *Aspergillus oryzae*: A review. Appl. Microbiol. Biotechnol. 36, 569-572.

Cheevadhanarak, S., Saunders, G., Renno, D.V. and Flegel, T.W. (1991) Transformation of *Aspergillus oryzae* with a dominant selectable marker. J. Biotech. 19, 117-122.

Christensen, T., Woeldike, H., Boel, E., Mortensen, S.B., Hjortshoej, K., Thim L. and Hansen M.T. (1988) High level expression of recombinant genes in Aspergillus oryzae. Biotechnology 6, 1419-1422.

Corrick, C. M., Twomey A P. and Hynes M.J. (1987) The nucleotide sequence of the *amdS* gene of *Aspergillus nidulans* and the molecular characterization of 5' mutations. Gene 53, 63-71.

Gomi, K., Iimura Y. and Hara S. (1987) Integrative transformation of *Aspergillus oryzae* with a Plasmid containing the *Aspergillus nidulans argB* gene. Agricol. Biol. Chem. 51, 2549-2555.

Gwynne, D.I., 1992, Foreign proteins, in "Applied Molecular Genetics of Filamentous Fungi" (Kinghorn, J.R. and Turner, G. Ed.), pp. 132-151. Blackie Academic & Professional, London.

Hynes, M.J., Corrick, C.M. and Littlejohn, K.J.M. and T.G. (1988) Identification of the sites of action for regulatory genes controlling the *amdS* gene of Aspergillus nidulans. Mol. Cell. Biol. 8, 2589-2596. Hynes, M.J., Corrick C.M. and King J.A. (1983) Isolation of genomic clones containing the *amdS* gene of *Aspergillus nidulans* and their use in the analysis of structural and regulatory mutations. Mol. Cell. Biol. 3, 1430-1439.

Katz, M., Saleeba, J.A., Sapats, S.I. and Hynes M.J. (1990) A mutation affecting *amdS* expression in *Aspergillus nidulans* contains a triplication of a cis-acting regulatory sequence. Mol. Gen. Genet. 220, 373-376.

Mattern, I.E., Unkles, S., Kinghorn, J.R., Pouwels P.H. and van den Hondel, C.A.M.J.J. (1987) Transformation of *Aspergillus oryzae* using the *A. niger pyrG* gene. Mol. Gen. Genet. 210, 460-461.

May, G.S., J. Gambino, W. J.A. and N.R. Morris (1985) Identification and functional analysis of beta-tubulin genes by site specific integrative transformation in *Aspergillus nidulans*. J. Cell. Biol. 101, 712-719.

Unkles, S.E., Campbell, E.I., de Ruiter-Jacobs, Y.M.J.T., Brockhuijsen, M., Macro, J.A., Carrez, D., Contreras, R., van den Hondel, C.A.M.J.J. and Kinghorn J.R. (1989) The development of a homologous transformation system for *Aspergillus oryzae* based on the nitrate assimilation pathway: A convenient and general selection system for filamentous fungal transformation. Mol. Gen. Genet. 218, 99-104.

# CURRENT SYSTEMATICS OF THE GENUS *ASPERGILLUS*

Robert A. Samson

Centraalbureau voor Schimmelcultures
P.O.Box 273, 3740 AG BAARN
The Netherlands

## INTRODUCTION

In their monograph of the genus *Aspergillus*, Raper and Fennell (1965) accepted 132 species subdivided into 18 groups. The generic and species concepts were well circumscribed and this monograph is still valuable and used. Several sections of the genus which remained problematic for identification have been reinvestigated, including the subgenus *Aspergillus* section *Aspergillus* (= the *Aspergillus glaucus* group = *Eurotium*) (Blaser, 1975; Pitt, 1985; Kozakiewicz, 1989), section *Nigri* (Al-Mussallam, 1980; Kozakiewicz, 1989 Kusters-van Someren *et al.*, 1990, 1991), section *Circumdati* (Christensen, 1982), section *Nidulantes* = *Emericella* (Horie, 1980; Christensen and States, 1982), section *Versicolores* (Klich, 1993), section *Flavi* (Murakami, 1971; Murakami *et al.*, 1982; Christensen 1981; Klich and Pitt, 1985, 1988*b*), section *Fumigati* (Frisvad and Samson, 1990; Samson *et al.*, 1990, Nielsen and Samson, 1992) and section *Restricti* (Pitt and Samson, 1990*b*). Samson (1979, 1992, 1993) provided a compilation of the species and varieties described since 1965 with a critical review of the validity of the published taxa. The genus now contains more than 180 species, with about 70 named teleomorphs.

The use of the non-morphological methods which could be important for *Aspergillus* taxonomy were already discussed 16 years ago by Fennell (1977). This review included cell wall composition, proteins (enzymes), pyrolysis products, nucleic acids, amino acid biosynthesis, hydrocarbon metabolism and inorganic elements. Of these techniques only the use of enzymes and nucleic acids are still considered to be relevant for taxonomic purposes.

The current systematics of *Aspergillus* were greatly influenced by the outcome of two NATO Advanced Research Workshops on *Penicillium* and *Aspergillus* Systematics (Samson and Pitt, 1986, 1990). Important contributions to the taxonomy of the genus came from multidisciplinary approaches including, biochemistry, molecular biology and serology. Furthermore the nomenclature has been stabilised by strict adherence to the rules of the International Code of Botanical Nomenclature (ICBN).

Today, species of *Aspergillus* are important organisms in industry and many fields of applied research and the need for clear taxonomic schemes are increasingly relevant. This paper reviews the current taxonomic concepts and discusses the impact of new approaches for modern *Aspergillus* systematics.

*The Genus Aspergillus*, Edited by Keith A. Powell *et al.*,
Plenum Press, New York, 1994

## MORPHOLOGICAL AND CULTURAL CRITERIA

The characteristic *Aspergillus* conidiophore, having an aseptate stipe terminating in a vesicle, on which the conidiogenous cells (phialides and metulae) are borne, are still primary morphological criteria for delimitation of the taxa. Except for the araneogenous genus *Gibellula*, which has *Cordyceps* and *Torrubiella* teleomorphs (Samson and Evans 1992) and septate stipes, there are no other genera in the Hyphomycetes with dry conidia which show a similar conidiophore structure. The colour of the conidial heads is also an important character, together with characteristics of Hülle cells and sclerotia. Many species produce sclerotia on agar media when freshly isolated but unfortunately the production of these vegetative bodies can be easily lost and therefore this, albeit a significant morphological character, is not often used.

*Aspergillus* has teleomorphs placed in at least nine genera (Table 1). Together with the morphological characters of the anamorph, the teleomorph species are easy to identify. Particularly, the ornamentation of the ascospores, which can be readily seen in both light microscopy and scanning electron microscopy (SEM). It is therefore not surprising that, for example, in the genera *Neosartorya* and *Emericella* (Table 2) several new species have been described, primarily based on unique ascospore ornamentation. In some species the teleomorphs are not known to produce an *Aspergillus* anamorph, but these fungi possess sufficient characters to justify their placement in these genera (Samson, 1992). The use of SEM can be useful for the examination of the ornamentation of thick-walled ascospores in the unfixed, air-dried state, but this must be applied with caution when examining conidia and other structures which are less resistant to shrinkage, artefacts may develop during fixation and examination. Pictures obtained from such SEM observations are sometimes difficult to intepret and the validity of a species concept based on this method as employed by

**Table 1**. Ascomycetous genera with an *Aspergillus* anamorph.

| |
|---|
| Section *Aspergillus*: *Eurotium* Link:Fr., *Dichlaena* Mont. & Durieu. |
| Section *Fumigati*: *Neosartorya* Malloch & Cain. |
| Subgenus *Ornati*: *Warcupiella* Subram., *Sclerocleista* Subram., *Hemicarpenteles* Sarbhoy & Elphick |
| Section *Nidulantes*: *Emericella* Berk. & Br. |
| Section *Flavipedes*: *Fennellia* Wiley & Simmons |
| Section *Circumdati*: *Petromyces* Malloch & Cain |
| Section *Cremei*: *Chaetosartorya* Subram. |

Kozakiewicz (1989) is debatable as discussed elsewhere (Samson, 1992). In addition, taxonomic conclusions drawn by Kozakiewicz (1989) on species in sections *Fumigati* and *Versicolores* based on DNA complementary measurements and cluster analysis of morphological data respectively, could not be confirmed by Peterson (1992) and Klich (1993).

Practical laboratory guides for identification of common aspergilli which are primarily based on morphological and colony characteristics have been published by Pitt and Hocking (1985), Samson and van Reenen-Hoekstra (1988), Klich and Pitt (1988*a*) and Tzean *et al.* (1990).

Most aspergilli can be readily identified when cultivated on Czapek and malt extract agars. Samson and Pitt (1986) recommended Czapek yeast agar (CYA) and malt extract agar (MEA) for most conidial species. For teleomorphic species, cornmeal agar, oatmeal agar and CBS malt agar were recommended. For xerophilic species, it is important to use low water activity media such as Czapek agar or CYA with 20% sucrose or malt yeast extract agar with 40% sucrose. Isolates of *Eurotium* or from the section *Restricti* have often been misidentified

**Table 2.** Species in the teleomorph genera *Emericella* and *Neosartorya*
Names indicated with an * are described since Raper and Fennell (1965)

| |
|---|
| *Emericella acristata* (Fennell & Raper) Horie |
| *Emericella astellata* (Fennell & Raper) Horie |
| *Emericella aurantio-brunnea* (Atkins *et al.*) Malloch & Cain |
| **Emericella bicolor* Christensen & States |
| **Emericella corrugata* Udagawa & Horie |
| *Emericella dentata* (Sandhu & Sandhu) Horie |
| **Emericella desertorum* Samson & Mouchacca |
| *Emericella echinulata* (Fennell & Raper) Horie |
| **Emericella falconensis* Horie *et al.* |
| **Emericella foeniculicola* Udagawa |
| **Emericella foveolata* Horie |
| *Emericella fruticulosa* (Raper & Fennell) Malloch & Cain |
| *Emericella heterothallica* (Kwon et al.) Raper & Fennell |
| *Emericella nidulans* (Eidam) Vuill. var. *lata* (Thom & Raper) |
| *Emericella parvathecia* (Raper & Fennell) Malloch & Cain |
| **Emericella purpurea* Samson & Mouchacca |
| *Emericella rugulosa* (C. Thom & Raper) C.R. Benjamin |
| **Emericella similis* Horie |
| **Emericella spectabilis* Christensen |
| *Emericella striata* (Rai et al.) Malloch & Cain) |
| **Emericella sublata* Horie |
| *Emericella quadrilineata* (C. Thom & Raper) C.R. Benjamin |
| **Emericella undulata* Kong & Qi |
| *Emericella unguis* Malloch & Cain |
| *Emericella variecolor* Berkeley & Broome |
| *Emericella violacea* (Fennell & Raper) Malloch & Cain |
| |
| *Neosartorya aurata* (Warcup) Malloch & Cain |
| *Neosartorya aureola* (Fennell & Raper) Malloch & Cain |
| **Neosartorya fennelliae* Kwon-Chung & Kim |
| *Neosartory fischeri* Wehmer |
| *Neosartorya glabra* (Fennell & Raper) Kozakiewicz |
| **Neosartorya hiratsukae* Udagawa *et al.* |
| **Neosartorya pseudofischeri* Petersen |
| *Neosartorya quadricincta* (Vuill) Malloch & Cain |
| **Neosartorya spathulata* Takada & Udagawa |
| *Neosartorya spinosa* (Raper & Fennell) Kozakiewicz |
| *Neosartorya stramenia* (E.H. Novak & Raper) Malloch & Cain |
| **Neosartorya tatenoi* Horie *et al.* |

due to the use of incorrect media. *Aspergillus* cultures should be incubated for 5-7 days at 25°C for anamorphic species and 10-14 days for most teleomorphic species.

Some specialised media are used for the rapid recognition of potentially aflatoxigenic aspergilli. The detection of *Aspergillus flavus* and *Aspergillus parasiticus* can be done within three days using *Aspergillus* Differential medium (Bothast and Fennell, 1974) or *Aspergillus flavus* and *parasiticus* agar (Pitt *et al.*, 1983) can also be used. A medium containing the antibiotic bleomycin helps to distinguish *A. parasiticus* from *Aspergillus sojae* (Klich and Mullaney, 1989). On this medium the growth of both species is somewhat reduced, but *A. sojae* produces very restricted colonies while *A. parasiticus* colonies are at least 3 mm in diameter within six days. These selective media are especially appropriate for the rapid detection of mycotoxinogenic isolates in food commodities.

## NEW APPROACHES

### Isozymes

Enzyme profiles were studied for a limited number of *Aspergillus* isolates by several investigators (Nealson and Garber, 1967; Nasuno, 1971; 1972*a,b*; 1974; Kurzeja and Garber, 1973; Cruickshank and Pitt, 1990; Sugiyama and Yamatoya, 1990; Yamatoya *et al.*, 1990). Saito *et al.* (1991) described a method for the identification of *A. flavus* and *A. parasiticus* by slab polyacrylamide gel electrophoresis of the alkaline proteinases, but it is not clear from their results if type isolates were used nor whether a good separation from *Aspergillus oryzae* and *A. sojae* was obtained. In the above-mentioned studies taxonomic relationships could be elucidated, but until now isozyme profiles have not provided a practical system for identification.

### Molecular studies

Several molecular studies on *Aspergillus* species have been carried out. Kozlowski and Stepien (1982) used restriction fragment length polymorphism (RFLP) to determine the relationship between several species of the genus *Aspergillus*. Using plasmids containing cloned fragments of *Emericella nidulans* mitochondrial (mt)DNA as probes, they concluded that this method was useful for the study of relationships between species, although surprisingly, they found that *A. oryzae* and *Aspergillus tamarii* were identical and that *Aspergillus awamori* and *Aspergillus niger* were not closely related to one another.

Several molecular biological studies have been conducted on members of the section. *Flavi* which are difficult to distinguish morphologically. Kurtzman *et al.* (1986) studied relative DNA reassociation values and could not distinguish between the closely related members of this group and therefore proposed to place the taxa in infraspecific categories. In contrast, RFLPs created with the restriction enzyme Sma I using total DNA, consistently differentiated the morphologically similar species *A. oryzae* and *A. flavus* (Klich and Mullaney, 1987). Gomi *et al.* (1989) could also distinguish four species of section *Flavi* using Sma I digests of total DNA. Moody and Tyler (1990*a,b*), using mitochondrial DNA restriction fragments found a clear separation of the three taxa, *A. flavus, A. parasiticus* and *Aspergillus nomius*. Kurtzman *et al.* (1987) however, when they used RFLPs, a limited correlation with geographic location of the isolates was observed. Kusters-van Someren *et al.* (1990, 1991) used pectin lyase production and pectin lyase probes to screen restriction enzyme DNA digests and ribosomal banding patterns of chromosomal digests to study the black aspergilli. In their studies they included the available type isolates of species of the *Aspergillus niger* complex and concluded that the isolates could be separated into two groups, centered around *A. niger* and *Aspergillus tubigensis* (see also under *A. niger* aggregate).

In a recent study S. W. Peterson (unpublished) found that *A. tamarii* of section *Flavi* was identical to *Aspergillus flavofurcatis* when nuclear DNA complementarities were compared with the ex-type cultures. *Aspergillus terricola*, a species placed in section *Wentii*, was closely related to *A. tamarii* with both species belonging to the section *Flavi* and not in a separate "*A. tamarii*-group" as proposed by Kozakiewicz (1989).

### Ubiquinones

The distribution of ubiquinone systems (especially Q-9 and Q-10) in *Aspergillus* have been used, often in combination with electrophoretic comparison of enzymes, by several researchers for taxonomic determinations (Kuraishi *et al.*, 1990; Matsuda *et al.*, 1992). These chemotaxonomic approaches can help clarify the relationship between important species. Matsuda *et al.* (1992) found, for example, that glucose-6-phosphate dehydrogenase and glutamate dehydrogenase were key enzymes for differentiating isolates of the pathogenic species *Aspergillus fumigatus*, however there was homogeneity between clinical and non-clinical strains.

## Secondary metabolites and mycotoxins

Profiles of secondary metabolites, including mycotoxins, are of great value for use in *Aspergillus* systematics. Samson *et al.* (1990) could clearly distinguish nine taxa within the genus *Neosartorya* by ascospore morphology and the profile of secondary metabolites. Frisvad and Samson (1990) carried out a biochemical and morphological study of 27 isolates of *A. fumigatus* and related taxa. The secondary metabolite production by *A. fumigatus* and described varieties was homogeneous: tryptoquivalins, fumigaclavine A, verruculogen, fumitremorgins and fumagillin were produced by all isolates tested; fumigatin and sulochrin-like metabolites were produced less frequently, while gliotoxin and helvolic acid were not detected (Table 3). The examined isolates of *Aspergillus brevipes, Aspergillus viridinutans, Aspergillus duricaulis* and *Aspergillus unilateralis* also produced unique profiles of secondary metabolites different from those found in *A. fumigatus*. This information supports the taxa of section *Fumigati* as proposed by Raper and Fennell (1965).

J.C. Frisvad and R.A. Samson (unpublished) investigated the species of section *Clavati* and found that they were closely related on the basis of their morphology, physiology and production of secondary metabolites. Clustering, principal component analysis and correspondence analysis have shown that in section *Clavati* four distinct species could be distinguished. A number of the secondary metabolites found have also been detected in species of the genus *Neosartorya* and the section *Fumigati*, which may indicate that sections *Fumigati* and *Clavati* are related.

Klich and Pitt (1988*b*) examined the relationship between current species concepts and secondary metabolite production in section *Flavi* and found a good correlation between the production of aflatoxins $B_1$ + $B_2$, $G_1$ + $G_2$, cyclopiazonic acid and the morphological characteristics of the species (see under *A. flavus* and *A. oryzae* below).

**Table 3.** Production of secondary metabolites in *Aspergillus* Section *Fumigati*.

| Species | Secondary metabolites |
|---|---|
| *A. brevipes* | Viriditoxin |
| *A. duricaulis* | Cyclopaldic acid, chromanol 1, 2, and 3 |
| *A. fumigatus* | Fumagillin, fumitoxins, fumigaclavine A, fumigatin, fumitremorgin A, B, and C, gliotoxin, monotrypacidin, tryptoquivalins, FUA, FUB, helvolic acid |
| *A. unilateralis* | Mycophenolic acid |
| *A. viridinutans* | Viriditoxin |

## Immunology

To enable the rapid detection of aspergilli sensitive enzyme immunoassay systems and monoclonal antibodies which identify unique epitopes are being developed. An ELISA (Enzyme linked immunosorbent assay) method and a latex agglutination method are available for the detection of heat-stable extracellular polysaccharides of certain aspergilli (Kamphuis *et al.*, 1989; Kamphuis and Notermans, 1992). A latex agglutination test used for the detection of aspergilli in the sera of patients with invasive aspergillosis has now been modified for the detection of food-borne aspergilli and penicillia (Stynen *et al.*, 1992) however until now there have been no immuno-assays reported which are specific for species or strains.

## NOMENCLATURE

Raper and Fennell (1965), in addition to other mycologists working with the genus *Aspergillus*, did not adhere to the rules of the ICBN (Greuter *et al.*, 1988). Problems with nomenclature were caused by neglecting the priority of older names and the failure to specify dried types for newly described species. The *Aspergillus* names used by Raper and Fennell (1965) were typified by Samson and Gams (1986) and Kozakiewicz (1989). Raper and Fennell (1965) subdivided *Aspergillus* into "Groups" however this infrageneric classification has no nomenclatural status under the ICBN, yet to stabilise the taxonomy, Gams *et al.* (1986) formally introduced names by subgenera and sections.

Many *Aspergillus* species have an ascomycete teleomorph (Table 1), but Raper and Fennell (1965) consistently retained the name *Aspergillus* for both teleomorph and anamorph, regarding *Aspergillus* as a genus of Ascomycetes, contrary to Article 59 of the ICBN. Samson and Gams (1986) and Kozakiewicz (1989) carried out the necessary nomenclatural changes (see also Pitt, 1989; Pitt and Samson, 1990a).

### Conservation and protection of names

Hundreds of *Aspergillus* names have been described since P.A. Micheli first gave the name to the genus in 1729. Many of these names were based on a single specimen being collected and described, but were not deposited in herbaria. In addition, herbaria throughout the world contain aspergilli cultures with very obscure names, not in common use. One of the rules of nomenclature for the ICBN is the priority of names, the earlier name should be adopted irrespective of the common usage of the name. For example in the revision of the black aspergilli Al-Musallam (1980) found that two older species names, *Aspergillus phoenicis* (Corda) Thom 1840 and *Aspergillus ficuum* (Reichardt) Hennings 1867, were synonymous with *A. niger*. Therefore following the rule about priority of names, the epithet *A. niger* should be replaced by one of the older names but the name *A. niger* has been widely used for a fungus used in the commercial production of citric and other organic acids around the world, and is clearly of major economic importance. In the case of *A. niger*, conservation of this name was suggested by Frisvad *et al.* (1990) and formally proposed by Kozakiewicz *et al.* (1992).

Another example of a case for conservation is the well-known name *Aspergillus nidulans*, a species used in a wide range of genetic studies. *A. nidulans* is threatened by the legitimate name *A. nidulellus* (Samson & W. Gams, 1986), introduced for the anamorph of *Emericella nidulans*, in strict adherence of Article 59 of the ICBN.

A procedure for stabilising *Aspergillus* nomenclature was recently developed to protect names in current use. On behalf of the International Commission on *Penicillium* and *Aspergillus* (ICPA, a commission of the International Union of Microbiological Societies) Pitt and Samson (1993) proposed a list of species names in current use in the family *Trichocomaceae* (Fungi, Eurotiales). This compilation was generated from discussions between the authors and other ICPA members. In order to protect *Aspergillus* names in current use from being threatened or displaced by names that are no longer in use, and in order to eliminate uncertainties regarding their application spelling, gender, typification date and place or valid publication, it has been proposed that published lists of names can be approved by an International Botanical Congress from 1993. Extensive discussions have taken place to ensure that the list finally generated represents the combined opinions of the experts involved. Important principles to be observed include: (1) where doubt exists about species circumscriptions, the "splitter's" view prevails, so to avoid omission of "narrower" names which might be in use by some; (2) where a species is well-defined, only a single name for each morph is accepted; and (3) varietal names are not considered, on the basis that, in these genera, conflicts over competing varieties within a single species have never occurred. The list of names of 186 aspergillli and 72 teleomorphs with an *Aspergillus* anamorph in current use were proposed to the XV International Botanical Congress in Tokyo for acceptance (Table 4).

**Table 4.** List of species names of *Aspergillus* and teleomorphs in current use proposed for protection (Pitt and Samson, 1993).

*Aspergillus* Fr. : Fr. (anamorphs)

*A. acanthosporus* Udagawa & Takada
*A. aeneus* Sappa
*A. albertensis* J. P. Tewari
*A. allahabadii* B. S. Mehrotra & Agnihotri
*A. alliaceus* Thom & Church
*A.amazonensis* (Henn.) Samson & Seifert (basionym: *Stilbothamnium amazonense* Henn.)
*A. ambiguus* Sappa
*A. amylovorus* Panas. ex Samson
*A. anthodesmis* Bartoli & Maggi
*A. appendiculatus* Blaser
*A. arenarius* Raper & Fennell
*A. asperescens* Stolk
*A. atheciellus* Samson & W. Gams
*A. aureolatus* Munt.-Cv&k. & Bata
*A. aureoluteus* Samson & W. Gams
*A. auricomus* (Guég.) Saito (basionym: *Sterigmatocystis auricoma* Guég.)
*A. avenaceus* G. Sm.
*A. awamorii* Nakaz.
*A. bicolor* M. Chr. & States
*A. biplanus* Raper & Fennell
*A. bisporus* Kwon-Chung & Fennell
*A. brevipes* G. Sm.
*A. bridgeri* M. Chr.
*A. brunneouniseriatus* Suj. Singh & B. K. Bakshi
*A. brunneus* Delacr.
*A. caesiellus* Saito
*A. caespitosus* Raper & Thom
*A. campestris* M. Chr.
*A. candidus* Link
*A.carbonarius* (Bainier) Thom (basionym: *Sterigmatocystis carbonaria* Bainier)
*A. carneus* Blochwitz
*A. cervinus* Massee
*A. chevalieri* L. Mangin
*A. chryseides* Samson & W. Gams *A. citrisporus* Höhn.
*A. clavatoflavus* Raper & Fennell
*A. clavatonanicus* Bat. *et al.*
*A. clavatus* Desm.
*A. compatibilis* Samson & W. Gams
*A. conicus* Blochwitz
*A. conjunctus* Kwon-Chung & Fennell
*A. coremiiformis* Bartoli & Maggi
*A. corrugatus* Udagawa & Y. Horie
*A. cremeoflavus* Samson & W. Gams
*A. cristatellus* Kozak.
*A. crustosus* Raper & Fennell
*A. curviformis* H. J. Chowdhery & J. N. Rai
*A. crystallinus* Kwon-Chung & Fennell
*A. deflectus* Fennell & Raper
*A. dimorphicus* B. S. Mehrotra & Prasad
*A. diversus* Raper & Fennell
*A. duricaulis* Raper & Fennell
*A. dybowskii* (Pat.) Samson & Seifert (basionym: *Penicilliopsis dybowskii* Pat.)
*A. eburneocremeus* Sappa
*A. egyptiacus* Moub. & Mustafa
*A. elegans* Gasperini
*A. ellipticus* Raper & Fennell
*A. elongatus* J. N. Rai & S. C. Agarwal
*A. erythrocephalus* Berk. & M. A. Curtis
*A. falconensis* Y. Horie
*A. fennelliae* Kwon-Chung & S. J. Kim
*A. flaschentraegeri* Stolk
*A. fischerianus* Samson & W. Gams
*A. flavipes* (Bainier & Sartory) Thom & Church (basionym: *Sterigmatocystis flavipes* Bainier & Sartory)
*A. flavofurcatus* Bat. & H. Maia
*A. flavus* Link
*A. floriformis* Samson & Mouch.
*A. foeniculicola* Udagawa
*A. foetidus* Thom & Raper
*A. foveolatus* Y. Horie
*A. fruticans* Samson & W. Gams
*A. fumigatus* Fresen.
*A. funiculosus* G. Sm.
*A. giganteus* Wehmer
*A. glaber* Blaser
*A. glaucoaffinis* Samson & W. Gams
*A. glauconiveus* Samson & W. Gams
*A. glaucus* Link
*A. globosus* H. J. Chowdhery & J. N. Rai
*A. gorakhpurensis* Kamal & Bhargava
*A. gracilis* Bainier
*A. granulosus* Raper & Thom
*A. halophilicus* C. M. Chr. *et al.*
*A. helicothrix* Al-Musallam
*A. heterocaryoticus* C. M. Chr. *et al.*
*A. h&eromorphus* Bat. & H. Maia

Table 4 (continued)

*A. hiratsukae* Udagawa *et al.*
*A. igneus* Kozak.
*A. insulicola* Montem. & A. R. Santiago
*A. intermedius* Blaser
*A. itaconicus* Kinosh.
*A. ivoriensis* Bartoli & Maggi
*A. janus* Raper & Thom
*A. japonicus* Saito
*A. kanagawaensis* Nehira
*A. lanosus* Kamal & Bhargava
*A. leporis* States & M. Chr.
*A. leucocarpus* Hadlok & Stolk
*A. longivesica* L. H. Huang & Raper
*A. lucknowensis* J. N. Rai *et al.*
*A. malodoratus* Kwon-Chung & Fennell
*A. maritimus* Samson & W. Gams
*A. medius* R. Meissn.
*A. melleus* Yukawa
*A. microcysticus* Sappa
*A. microthecius* Samson & W. Gams
*A. multicolor* Sappa
*A. navahoensis* M. Chr. & States
*A. neocarnoyi* Kozak.
*A. neoglaber* Kozak.
*A.nidulans* (Eidam) G. Winter (basionym: *Sterigmatocystis nidulans* Eidam)
*A. niger* Tiegh.
*A. niveus* Blochwitz
*A. nomius* Kurtzman *et al.*
*A. nutans* McLennan & Ducker
*A. ochraceoroseus* Bartoli & Maggi
*A. ochraceus* K. Wilh.
*A. ornatulus* Samson & W. Gams
*A. oryzae* (Ahlb.) Cohn
*A. ostianus* Wehmer
*A. paleaceus* Samson & W. Gams
*A. pallidus* Kamyschko
*A. panamensis* Raper & Thom
*A. paradoxus* Fennell & Raper
*A. parasiticus* Speare
*A. parvulus* G. Sm.
*A. penicillioides* Speg.
*A. petrakii* Vörös
*A. peyronelii* Sappa
*A. phoenicis* (Corda) Thom
*A. protuberus* Munt.-Cv&k.
*A. pseudodeflectus* Samson & Mouch.
*A. pulverulentus* (McAlpine) Wehmer (basionym:Sterigmatocystis pulverulenta McAlpine)
*A. pulvinus* Kwon-Chung & Fennell
*A. puniceus* Kwon-Chung & Fennell
*A. purpureus* Samson & Mouch.
*A. quadricingens* Kozak.
*A. raperi* Stolk
*A. recurvatus* Raper & Fennell
*A. reptans* Samson & W. Gams
*A. restrictus* G. Sm.
*A. robustus* M. Chr. & Raper
*A. rubrobrunneus* Samson & W. Gams
*A. rugulovalvus* Samson & W. Gams
*A. sclerotiorum* G. A. Huber
*A. sepultus* Tuthill & M. Chr.
*A. silvaticus* Fennell & Raper
*A. sojae* Sakag. & K. Yamada ex Murak.
*A. sparsus* Raper & Thom
*A. spathulatus* Takada & Udagawa
*A. spectabilis* M. Chr. & Raper
*A. spelunceus* Raper & Fennell
*A. spinosus* Kozak.
*A. stellifer* Samson & W. Gams
*A. striatulus* Samson & W. Gams
*A. stromatoides* Raper & Fennell
*A. sublatus* Y. Horie
*A. subolivaceus* Raper & Fennell
*A. subsessilis* Raper & Fennell
*A. sulphureus* (Fresen.) Wehmer (basionym: *Sterigmatocystis sulphurea* Fresen.)
*A. sydowii* (Bainier & Sartory) Thom & Church (basionym: *Sterigmatocystis sydowii* Bainier & Sartory)
*A. tamarii* Kita
*A. tardus* Bisset & Widden
*A. tatenoi* Y. Horie *et al.*
*A. terreus* Thom
*A. terricola* E. J. Marchal
*A. tetrazonus* Samson & W. Gams
*A. thermomutatus* (Paden) S. W. Peterson (basionym: *A. fischeri* var. *thermomutatus* Paden)
*A. togoensis* (Henn.) Samson & Seifert (basionym: *Stilbothamnium togoense* Henn.)
*A. tonophilus* Ohtsuki
*A. undulatus* H. Z. Kong & Z. T. Qi
*A. unguis* (Emile-Weil & L. Gaudin) Thom & Raper (basionym: *Sterigmatocystis unguis* Emile-Weil & L. Gaudin)
*A. unilateralis* Thrower
*A. ustus* (Bainier) Thom & Church (basionym: *Sterigmatocystis usta* Bainier)
*A. varians* Wehmer
*A. versicolor* (Vuill.) Tirab. (basionym: *Sterigmatocystis versicolor* Vuill.)
*A. violaceobrunneus* Samson & W. Gams
*A. violaceofuscus* Gasperini
*A. viridinutans* Ducker & Thrower
*A. vitellinus* (Massee) Samson & Seifert (basionym: *Sterigmatocystis vitellina* Ridl. ex Massee)
*A. vitis* Novobr.
*A. warcupii* Samson & W. Gams
*A. wentii* Wehmer
*A. xerophilus* Samson & Mouch.
*A. zonatus* Kwon-Chung & Fennell

***Chaetosartorya*** Subram. (holomorphs)

*Chaetosartorya chrysella* (Kwon-Chung & Fennell) Subram. (basionym: *Aspergillus chrysellus* Kwon-Chung & Fennell). [Anamorph: *Aspergillus chryseides* Samson & W. Gams].

*Chaetosartorya cremea* (Kwon-Chung & Fennell) . (basionym: *Aspergillus cremeus* Kwon-Chung & Fennell) [Anamorph: *Aspergillus cremeoflavus* Samson & W. Gams].
*Chaetosartorya stromatoides* B. J. Wiley & E. G. Simmons [Anamorph: *Aspergillus stromatoides* Raper & Fennell].

***Emericella*** Berk. (holomorphs)

*Emericella acristata* (Fennell & Raper) Y. Horie (basionym: *Aspergillus nidulans* var. *acristatus* Fennell & Raper)
*Emericella astellata* (Fennell & Raper) Y. Horie (basionym: *Aspergillus variecolor* var. *astellatus* Fennell & Raper)
*Emericella aurantiobrunnea* (G. A. Atkins *et al.*) Malloch & Cain (basionym: *Emericella nidulans* var. *aurantiobrunnea* G. A. Atkins *et al.*)
*Emericella bicolor* M. Chr. & States [Anamorph: *Aspergillus bicolor* M. Chr. & States].
*Emericella corrugata* Udagawa & Y. Horie [Anamorph: *Aspergillus corrugatus* Udagawa & Y. Horie].
*Emericella dentata* (D. K. Sandhu & R. S. Sandhu) Y. Horie (basionym: *Aspergillus nidulans* var. *dentatus* D. K. Sandhu & R. S. Sandhu)
*Emericella desertorum* Samson & Mouch.
*Emericella echinulata* (Fennell & Raper) Y. Horie (basionym: *Aspergillus nidulans* var. *echinulatus* Fennell & Raper)
*Emericella falconensis* Y. Horie et al. [Anamorph: *Aspergillus falconensis* Y. Horie].
*Emericella foeniculicola* Udagawa [Anamorph: *Aspergillus foeniculicola* Udagawa].
*Emericella foveolata* Y. Horie [Anamorph: *Aspergillus foveolatus* Y. Horie].
*Emericella fruticulosa* (Raper & Fennell) Malloch & Cain (basionym: *Aspergillus fructiculosus* Raper & Fennell). [Anamorph: *Aspergillus fruticans* Samson & W. Gams].
*Emericella heterothallica* (Kwon-Chung *et al.*) Malloch & Cain (basionym: *Aspergillus heterothallicus* Kwon-Chung *et al.*). [Anamorph: *Aspergillus compatibilis* Samson & W. Gams].
*Emericella navahoensis* M. Chr. & States [Anamorph: *Aspergillus navahoensis* M. Chr. & States].
*Emericella nidulans* (Eidam) Vuill. (basionym: *Sterigmatocystis nidulans* Eidam). [Anamorph: *Aspergillus nidulans* (Eidam) G. Winter].
*Emericella parvathecia* (Raper & Fennell) Malloch & Cain (basionym: *Aspergillus parvathecius* Raper & Fennell). [Anamorph: *Aspergillus microthecius* Samson & W. Gams].
*Emericella purpurea* Samson & Mouch. [Anamorph: *Aspergillus purpureus* Samson & Mouch.].
*Emericella quadrilineata* (Thom & Raper) C. R. Benj. (basionym: *Aspergillus quadrilineatus* Thom & Raper). [Anamorph: *Aspergillus tetrazonus* Samson & W. Gams].
*Emericella rugulosa* (Thom & Raper) C. R. Benj. (basionym: *Aspergillus rugulosus* Thom & Raper) [Anamorph: *Aspergillus rugulovalvus* Samson & W. Gams].
*Emericella similis* Y. Horie *et al.*
*Emericella spectabilis* M. Chr. & Raper [Anamorph: *Aspergillus spectabilis* M. Chr. & Raper].
*Emericella striata* (J. N. Rai *et al.*) Malloch & Cain (basionym: *Aspergillus striatus* J. N. Rai *et al.*). [Anamorph: *Aspergillus striatulus* Samson & W. Gams].
*Emericella sublata* Y. Horie [Anamorph: *Aspergillus sublatus* Y. Horie].
*Emericella undulata* H. Z. Kong & Z. T. Qi [Anamorph: *Aspergillus undulatus* H. Z. Kong & Z. T. Qi].
*Emericella unguis* Malloch & Cain [Anamorph: *Aspergillus unguis* (Emile-Weil & L. Gaudin) Thom & Raper].
*Emericella variecolor* Berk. & Broome [Anamorph: *Aspergillus stellifer* Samson & W. Gams].
*Emericella violacea* (Fennell & Raper) Malloch & Cain (basionym: *Aspergillus violaceus* Fennell & Raper). [Anamorph: *Aspergillus violaceobrunneus* Samson & W. Gams].

***Eurotium*** Link : Fr. (holomorphs)

*Eurotium amstelodami* L. Mangin [Anamorph: *Aspergillus vitis* Novobr.].
*Eurotium appendiculatum* Blaser [Anamorph: *Aspergillus appendiculatus* Blaser]. *Eurotium athecium* (Raper & Fennell) Arx (basionym: *Aspergillus athecius* Raper & Fennell) [Anamorph: *Aspergillus atheciellus* Samson & W. Gams].
*Eurotium carnoyi* Malloch & Cain [Anamorph: *Aspergillus neocarnoyi* Kozak.].
*Eurotium chevalieri* L. Mangin [Anamorph: *Aspergillus chevalieri* L. Mangin].
*Eurotium cristatum* (Raper & Fennell) Malloch & Cain (basionym: *Aspergillus cristatus* Raper & Fennell). [Anamorph: *Aspergillus cristatellus* Kozak.].
*Eurotium echinulatum* Delacr. [Anamorph: *Aspergillus brunneus* Delacr.].
*Eurotium glabrum* Blaser [Anamorph: *Aspergillus glaber* Blaser].
*Eurotium halophilicum* C. M. Chr. *et al.* [Anamorph: *Aspergillus halophilicus* C. M. Chr. *et al.*].
*Eurotium herbariorum* Link [Anamorph: *Aspergillus glaucus* Link].
*Eurotium intermedium* Blaser (basionym: *Aspergillus chevalieri* var. *intermedius* Thom & Raper) [Anamorph: *Aspergillus intermedius* Blaser].
*Eurotium leucocarpum* Hadlok & Stolk [Anamorph: *Aspergillus leucocarpus* Hadlok & Stolk].
*Eurotium medium* R. Meissn. [Anamorph: *Aspergillus medius* R. Meissn.].
*Eurotium niveoglaucum* (Thom & Raper) Malloch & Cain (basionym: *Aspergillus niveoglaucus* Thom &

Raper). [Anamorph: *Aspergillus glauconiveus* Samson & W. Gams].
*Eurotium pseudoglaucum* (Blochwitz) Malloch & Cain (basionym: *Aspergillus pseudoglaucus* Blochwitz). [Anamorph: *Aspergillus glaucoaffinis* Samson & W. Gams].
*Eurotium repens* de Bary [Anamorph: *Aspergillus reptans* Samson & W. Gams].
*Eurotium rubrum* Jos. König *et al.* [Anamorph: *Aspergillus rubrobrunneus* Samson & W. Gams].
*Eurotium tonophilum* Ohtsuki [Anamorph: *Aspergillus tonophilus* Ohtsuki].
*Eurotium xerophilum* Samson & Mouch. [Anamorph: *Aspergillus xerophilus* Samson & Mouch.].

***Fennellia*** B. J. Wiley & E. G. Simmons (holomorphs)

*Fennellia flavipes* B. J. Wiley & E. G. Simmons [Anamorph: *Aspergillus flavipes* (Bainier & Sartory) Thom & Church].
*Fennellia monodii* Locq.-Lin.
*Fennellia nivea* (B. J. Wiley & E. G. Simmons) Samson (basionym: *Emericella nivea* B. J. Wiley & E. G. Simmons). [Anamorph: *Aspergillus niveus* Blochwitz].

***Hemicarpenteles*** A. K. Sarbhoy & Elphick (holomorphs)

*Hemicarpenteles acanthosporus* Udagawa & Takada [Anamorph: *Aspergillus acanthosporus* Udagawa & Takada].
*Hemicarpenteles paradoxus* A. K. Sarbhoy & Elphick [Anamorph: *Aspergillus paradoxus* Fennell & Raper].

***Neosartorya*** Malloch & Cain (holomorphs)

*Neosartorya aurata* (Warcup) Malloch & Cain (basionym: *Aspergillus auratus* Warcup). [Anamorph: *Aspergillus igneus* Kozak.].
*Neosartorya aureola* (Fennell & Raper) Malloch & Cain (basionym: *Aspergillus aureolus* Fennell & Raper). [Anamorph: *Aspergillus aureoluteus* Samson & W. Gams].
*Neosartorya fennelliae* Kwon-Chung & S. J. Kim [Anamorph: *Aspergillus fennelliae* Kwon-Chung & S. J. Kim].
*Neosartorya fischeri* (Wehmer) Malloch & Cain (basionym: *Aspergillus fischeri* Wehmer). [Anamorph: *Aspergillus fischerianus* Samson & W. Gams].
*Neosartorya glabra* (Fennell & Raper) Kozak. (basionym: *Aspergillus fischeri* var. *glaber* Fennell & Raper). [Anamorph: *Aspergillus neoglaber* (Fennell & Raper) Kozak.].
*Neosartorya hiratsukae* Udagawa *et al.* [Anamorph: *Aspergillus hiratsukae* Udagawa *et al.*].
*Neosartorya pseudofischeri* S. W. Peterson [Anamorph: *Aspergillus thermomutatus* (Paden) S. W. Peterson].
*Neosartorya quadricincta* (E. Yuill) Malloch & Cain (basionym: *Aspergillus quadricinctus* E. Yuill). [Anamorph: *Aspergillus quadricingens* Kozak.].
*Neosartorya spathulata* Takada & Udagawa [Anamorph: *Aspergillus spathulatus* Takada & Udagawa].
*Neosartorya spinosa* (Raper & Fennell) Kozak. (basionym: *Aspergillus fischeri* var. *spinosus* Raper & Fennell). [Anamorph: *Aspergillus spinosus* Kozak.].
*Neosartorya stramenia* (R. Novak & Raper) Malloch & Cain (basionym: *Aspergillus stramenius* R. Novak & Raper). [Anamorph: *Aspergillus paleaceus* Samson & W. Gams].
*Neosartorya tatenoi* Y. Horie *et al.* [Anamorph: *Aspergillus tatenoi* Horie *et al.*].

***Petromyces*** Malloch & Cain (holomorphs)

*Petromyces albertensis* J. P. Tewari [Anamorph: *Aspergillus albertensis* J. P. Tewari].
*Petromyces alliaceus* Malloch & Cain [Anamorph: *Aspergillus alliaceus* Thom & Church].

***Sclerocleista*** Subram. (holomorphs)

*Sclerocleista ornata* (Raper *et al.*) Subram. (basionym: *Aspergillus ornatus* Raper *et al.*). [Anamorph: *Aspergillus ornatulus* Samson & W. Gams].
*Sclerocleista thaxteri* Subram. [Anamorph: *Aspergillus citrisporus* Höhn.].

***Warcupiella*** Subram. (holomorphs)

*Warcupiella spinulosa* (Warcup) Subram. (basionym: *Aspergillus spinulosus* Warcup). [Anamorph: *Aspergillus warcupii* Samson & W. Gams].

## TAXONOMY OF SOME IMPORTANT AND PROBLEMATIC *ASPERGILLUS* SPECIES

### *A. fumigatus*

*A. fumigatus* is morphologically more variable than described by Raper and Fennell (1965) with deviants described as *A. fumigatus* var. *acolumnaris* or *Aspergillus phialiseptus*. Clinical isolates, in particular, can be markedly abnormal (de Vries and Cormane, 1969; Leslie *et al.*, 1988) and are often more floccose with fewer conidia than type cultures. These isolates can not be separated from the typical saprobic isolates taken from substrates such as soil or feedstuffs (Debeaupuis *et al.*, 1990; Hearn *et al.*, 1990; Polonelli *et al.*, 1990). This was also confirmed in a recent study by Matsuda *et al.* (1992), who analysed the ubiquinones of many *A. fumigatus* strains and found that clinical and non-clinical isolates did not vary in this character.

The ex-type isolate of *A. fumigatus* var. *ellipticus* differs from the typcial isolates of *A. fumigatus* by its smooth-walled and ellipsoidal conidia, yet the isolate is biochemically identical to *A. fumigatus* var. *fumigatus* (Frisvad and Samson, 1990; Debeaupuis *et al.*, 1990). In addition Peterson (1992) made nuclear DNA complementary measurements and found a relationship of 93% NDNA between the type isolates of both varieties and concluded that the name *A. neoellipticus* proposed by Kozakiewicz (1989) should be treated as a synonym of *A. fumigatus*.

### *A. flavus and A. oryzae*

Members of section *Flavi* are important because of their biotechnological applications and their mycotoxin production. Taxonomy based purely on morphological criteria showed that isolates are often difficult to recognise and therefore several taxonomic studies have utilised non-conventional approaches. A number of these studies produced varied results suggesting unclear taxonomic conclusions. For example, on the basis of DNA complementarity, Kurtzman *et al.* (1986) reduced the aflatoxin-producing *A. parasiticus* to a subspecies of *A. flavus*. *A. oryzae* and *A. sojae* used in food fermentation were also reduced to varieties of the same species. Using electrophoretic comparison of enzymes and ubiquinone systems, Yamatoya *et al.* (1990) found that isolates of *A. flavus*, *A. oryzae*, *A. parasiticus* and *A. sojae* could be accommodated in two species: *A. flavus* and *A. parasiticus*.

*A. flavus*, *A. parasiticus*, *A. tamarii* and *A. nomius* were, however, clearly differentiated using enzyme electrophoretic patterns by Liljegren *et al* (1988) and Cruickshank and Pitt (1990). Klich and Pitt (1988*b*) also confirmed the study of Kozakiewicz (1982) where isolates of *A. flavus* were distinguished from those of *A. parasiticus* by conidial ornamentation. In addition they found that isolates of *A. flavus* produce aflatoxin $B_1$ and $B_2$ and/or cyclopiazonic acid, while *A. parasiticus* produced the aflatoxins $G_1$ ,$G_2$ , $B_1$ and $B_2$, but never cyclopiazonic acid. Other studies have shown that isolates of *A. flavus*, *A. parasiticus*, *A. nomius* and *A. sojae* consistently produce aspergillic acid, while isolates of *A. oryzae* never produce this toxin (Samson and Frisvad, 1991).

### The *A. niger* aggregate

One of the groups in Raper and Fennell's monograph not clarified was that of the black *Aspergillus* species, inspite of their importance in biotechnology. The taxonomy was primarily based on morphological criteria, and several taxa represent only modified industrial strains. Of the many species described, Raper and Fennell (1965) reduced the number of species to 12. The taxonomic distances between representative isolates of these taxa were assessed by Al-Musallam (1980) using cluster analysis involving all available morphological parameters. After both equal and iterative weighting of characters five readily distinguishable species and an *A. niger* aggregate which subdivided into seven varieties were recognised. Kusters-van Someren *et al.* (1990) separated the black aspergilli into *Aspergillus japonicus, Aspergillus carbonarius, Aspergillus heteromorphus, Aspergillus ellipticus* and an *A. niger* aggregate by using pectin lyase production and a pectin lyase probe to screen RFLPs. In a further study of the *A. niger* aggregate, Kusters-van Someren *et al.* (1991) studied rDNA banding patterns supported by Southern blots using several pectin lyase genes isolated from an *A. niger* isolate. The isolates, including many ex-(neo-)types, could be divided into two groups based around *A. niger* and *A. tubigensis* (Table 5). Since heterokaryon formation between the isolates of both groups was not possible, it was suggested that the two groups represented different species. The isolates cannot be distinguished on their morphology alone.

**Table 5.** Classification of species of the *A. niger* aggregate according to Kusters-van Someren *et al.* (1991)

| ***Aspergillus niger*** | ***Aspergillus tubigensis*** |
|---|---|
| *Aspergillus pseudocitricus* | *Aspergillus pseudoniger* |
| *Aspergillus awamori* | *Aspergillus satoi-kagoshimamaenis* |
| *Aspergillus foetidus* | *Aspergillus satoi* |
| *Aspergillus usamii* | *Aspergillus inuii* |
| *Aspergillus hennbergii* | *Aspergillus cinnamomeus* |
| *Aspergillus kawachi* | *Aspergillus schiemannii* |
| *Aspergillus aureus* | |

E. J.Mullaney and M. A. Klich (unpublished) verified the proposed *A. niger/A. tubigensis* division using *A. niger* phytase and acid phosphatase genes as probes. They also found that the RFLP patterns were not uniform for *A. ficuum* isolates, another species of the section *Nigri* which was not included in the investigations of Kusters-van Someren *et al.* (1991). Some of these *A. ficuum* isolates had a pattern distinct from either the reported *A. niger* or *A. tubigensis* pattern. In addition they found that isolates of *Aspergillus cinnamoneus* (NRRL 348) and *Aspergillus schiemanni* (NRRL 361) yielded the *A. niger* RFLP pattern when digested with Sma1 instead of the *A. tubigensis* pattern. The primary interest of these authors was to determine how widespread and similar the phytase and acid phosphatase genes were in section *Nigri* . If they were common to this group then the transfer of a modified phytase gene into one of the industrial strains in this group may further enhance the enzyme yield.

S. Peterson (unpublished) performed nuclear DNA reassociations and could segregate two major groups which were 40-50% related to each other. One group was composed mostly of isolates identified as *A. niger, A. awamori*, and *A. ficuum*; the second group contained mostly strains labelled as *A. phoenicis, A. pulverulentus*, and *A. tubingensis*. These molecular studies showed that morphological classification does not always reflect the genetic relationship of the strains. Although at the molecular level separation is clear, no correlated morphological differences was observed microscopically.

## NEEDS FOR FURTHER SYSTEMATIC RESEARCH

The genus *Aspergillus* is a relatively large taxon among the Hyphomycetes, and still many substrates and geographic areas have not been screened for their presence. Since aspergilli have a preference for (sub)tropical zones their presence in tropical soil, for example, could be abundant. One example is tropical subgenus *Stilbothamnium* (Hennings) Samson & Seifert, which was used by Samson and Seifert (1986) for studies on *Aspergillus* species with prominent synnematous taxa such as *A. togoensis* (= *Stilbothaminium nudipes*) (Roquebert and Nicot, 1986). These are subtropical or tropical species of *Aspergillus* which are rarely encountered and occur on seeds and are similar to members of section *Flavi* and may be significant producers of mycotoxin and other secondary metabolites. The subgenus now contains five species: *Aspergillus amazonenses*, *Aspergillus dybowskii*, *Aspergillus erythrocephalus*, *Aspergillus togoensis* and *Aspergillus vitellinus*, which are until now only known from herbarium specimens, with the exception of *A. togoensis* which has been isolated in culture.

Although the taxonomic concept of most *Aspergillus* species is clear and reliable identification is feasible, revisions of particular species or groups of species are still needed. This is especially the case for the sections *Nigri* and *Flavi*, which contain species of importance for biotechnological applications. Furthermore, other sections, such as *Circumdati* (= *A. ochraceus* group) with significant toxinogenic species, also urgently require revision.

Several molecular genetic studies on problematic but important aspergilli have been carried out and their potential for taxonomic use is apparent (Mullaney and Klich, 1990) and it is anticipated that the accurate and rapid detection of species and strains will be possible in the future. This is particularly required for the detection of strains which have undergone various modifications. These strains may no longer produce typical morphological structures and recognition might only be possible at the molecular level. However, at this time, the results from molecular genetic studies need careful evaluation, and should be compared with morphological observations. It is also essential that, wherever possible, type or authentic isolates of the species should be included in the genetic studies and that these are checked for their correct morphological features.

## REFERENCES


Al-Musallam, A. (1980) Revision of the black *Aspergillus* Species. Dissertation. Utrecht, Netherlands, University of Utrecht.

Blaser, P. (1975) Taxonomische und physiologische Untersuchungen über die Gattung *Eurotium* Link. ex Fr.Sydowia 28, 1-49.

Bothast, R.J. and Fennell, D.I. (1974) A medium for rapid identification and enumeration of *Aspergillus flavus* and related organisms. Mycologia 66, 365-369.

Christensen, M. (1981) A synoptic key and evaluation of species in the *Aspergillus flavus* group. Mycologia 73, 1056-1084.

Christensen, M. (1982) The *Aspergillus ochraceus* group: two new species from western soils and synoptic key. Mycologia 74, 210-225.

Christensen, M. and States, J.S. (1982) *Aspergillus nidulans* group: *Aspergillus navahoensis*, and a revised synoptic key. Mycologia 74, 226-235.

Cruickshanck R.H. and Pitt, J.I. (1990) Enzyme patterns in *Aspergillus flavus* and closely related taxa, in "Modern Concepts in *Penicillium* and *Aspergillus* Classification", (Samson, R.A. and Pitt, J.I., Eds), pp. 259-264. Plenum Press, New York.

Debeaupuis, J.P., Sarfati, P., Goris, A., Stynen, D. Diaquin, M. and Latgé, J.P. (1990) Exocellular polysaccharides of *Aspergillus fumigatus* and related taxa, in "Modern Concepts in *Penicillium* and *Aspergillus* Classification. (Samson, R.A. and J.I. Pitt, Eds), pp. 309-320. Plenum Press, New York.

Fennell, D.I. (1977) *Aspergillus* taxonomy, in "Genetics and Physiology of *Aspergillus*" (Smith, J.E. & Pateman, J.E., Eds), pp.1-21, Academic press, London.

Frisvad, J.C. and Samson, R.A. (1990) Chemotaxonomy and morphology of *Aspergillus fumigatus* and related taxa, in "Modern Concepts in *Penicillium* and *Aspergillus* Classification", (Samson, R.A. and Pitt, J.I., Eds), pp. 201-208. Plenum Press, New York.

Frisvad, J.C., Hawksworth, D.L, Kozakiewicz, Z., Pitt, J. I., Samson, R.A. and Stolk A.C. (1990) Proposal to conserve important species names in *Aspergillus* and *Penicillium*, in "Modern Concepts in *Penicillium* and *Aspergillus* systematics (Samson, R.A and Pitt, J.I., Eds), pp. 83-90. Plenum Press, New York.

Gams, W., Christensen, M., Onions, A.H.S., Pitt, J.I. and Samson, R.A. (1986)["1985"] Infrageneric taxa of *Aspergillus*, in "Advances in *Penicillium* and *Aspergillus* Systematics", (Samson, R.A. and Pitt, J.I., Eds) pp. 55-62. Plenum Press, New York.

Gomi, K., Tanaka, A., Iimura, Y. and Takahashi, K., (1989) Rapid differentiation of four related species of koji molds by agarose gel electrophoresis of genomic DNA digested units SMAI restriction enzyme. J. Gen. Appl. Microbiol. 35, 225-232.

Greuter, W. et al. (1988), International Code of Botanical Nomenclature adopted by the Fourteenth International Botanical Congress, Berlin, July-August, (1987). Koeltz Scientific Books, Konigstein, W. Germany.

Hearn, V. M., Moutaouakil, M. and Latgé, J.P., (1990) Analysis of components of *Aspergillus* and *Neosartorya* mycelial preparations by gel electrophoresis and Western blotting techniques, in "Modern Concepts in *Penicillium* and *Aspergillus* Classification". (Samson, R.A. and J.I. Pitt, Eds), pp. 309-320. Plenum Press, New York.

Horie, Y. (1980) Ascospore ornamentation and its application to the taxonomic re-evaluation in *Emericella*. Trans. mycol. Soc. Japan 21, 483-493.

Kamphuis, H.J., Notermans, S., Veeneman, G.H., Boom, J.H.van and Rombouts, F.M. (1989) A rapid and reliable method for the detection of moulds in foods: Using the latex agglutination assay. J. Food Protect. 52, 244-247.

Kamphuis, H. and Notermans, S. (1992) Development of a technique for immunological detection of fungi, in "Modern Methods in Food Mycology" (Samson, R.A., A.C. Hocking, J.I. Pitt and A.D. King, Eds.), pp. 197-203. Elsevier, Amsterdam.

Klich, M. A. (1993) Morphological studies of *Aspergillus* section *Versicolores* and related species. Mycologia 85, 100-107.

Klich, M.A. and Mullaney, E.J. (1987) DNA restriction enzyme fragment polymorphism as a tool for rapid differentiation of *Aspergillus flavus* from *Aspergillus oryzae*. Exp. Mycol. 11, 170-175.

Klich, M.A. and Mullaney, E.J. (1989) Use of bleomycin-containing medium to distinguish *Aspergillus parasiticus* from *A. sojae*. Mycologia 81, 159-160.

Klich, M.A. and Pitt, J.I. (1985) The theory and practice of distinguishing species of the *Aspergillus flavus* group. In "Advances in *Penicillium* and *Aspergillus* Systematics". (Samson, R.A. and J.I. Pitt, Eds.) pp. 211-220. Plenum Press, New York.

Klich, M.A. and Pitt, J.I. (1988*a*) A laboratory guide to common *Aspergillus* species and their teleomorphs. North Ryde, NSW, CSIRO Division of Food Processing.

Klich, M.A. and Pitt, J.I. (1988*b*) Differentiation of *Aspergillus flavus* from *A. parasiticus* and other closely related species. Trans. Br. Mycol. Soc. 91, 99-108.

Kozakiewicz, Z. (1982) The identity and typification of *Aspergillus parasiticus*. Mycotaxon 15, 293-305.

Kozakiewicz, Z. (1989) *Aspergillus* species on stored products. Mycol. Papers 161, 1-188.

Kozakiewicz, Z., Frisvad, J.C.. Hawksworth, D.L., Pitt, J.I., Samson, R.A. and Stolk, A.C. (1992) Proposal for nomina specifica conservanda and rejicienda in *Aspergillus* and *Penicillium*. Taxon 41, 109-113.

Kozlowski, M. and Stepien, P.R. (1982) Restriction enzyme analysis of mitochondrial DNA of members of the genus *Aspergillus* as an aid in taxonomy. J. gen. Microbiol. 128, 471-476, 1982.

Kuraishi, H., Itoh, M., Tsuzaki, N., Katayama, Y., Yokoyama, T. and Sugiyama, J. (1990) Ubiquinone system as a taxonomic tool in *Aspergillus* and its teleomorphs, in "Modern Concepts in *Penicillium* and *Aspergillus* Classification", (Samson, R.A. and Pitt, J.I., Eds.), pp. 407-420. Plenum Press, New York.

Kurtzman, C.P., Smiley, M.J., Robnett, C.J. and Wicklow, D.T. (1986) DNA relatedness among wild and domesticated species in the *Aspergillus* flavus group. Mycologia 78, 955-959.

Kurtzman, C.P., Horn, B.W. and Hesseltine, C.W. (1987) *Aspergillus nomius*, a new aflatoxin-producing species related to *Aspergillus flavus* and *Aspergillus tamarii*. Antonie van Leeuwenhoek 53, 147-158.

Kurzeja, K.C. and Garber, E. D. (1973) A genetic study of electrophoretically variant extracellular amylolytic enzymes of wild-type strains of *Aspergillus nidulans*. Can. J. Gen. Cytol. 15, 275-287.

Kusters-van Someren, M., H.C.M. Kester, Samson, R.A. and Visser, J. (1990) Variation in pectinolytic enzymes of the black Aspergilli: a biochemical and genetic approach, in "Modern Concepts in *Penicillium* and *Aspergillus* Classification", (Samson, R.A. and Pitt, J.I., Eds), pp. 321-334. Plenum Press, New York.

Kusters-van Someren, M.A., Samson, R.A. and Visser, J. (1991) The use of RFLP analysis in classification of the black Aspergilli: reinterpretation of the *Aspergillus niger* aggregate. Current Genetics 19, 21-26l.

Leslie, L.E., Flannigan, B., and Milner, L.J.R. (1988) Morphological studies on clinical isolates of *Aspergillus fumigatus*. J. Med. Vet. Mycology 26, 335-341.

Liljegren, K., Svendsen, A. and Frisvad, J.C. (1988) Mycotoxin and exoenzyme production by members of *Aspergillus* section *Flavi*: An integrated taxonomic approach to their classification. Proceedings Jap. Assoc. Mycotoxicol. Suppl. 1, 35-36.

Matsuda, H., Kohno, S., Maesaki, S., Yamada, H., Koga, H., Tamura, M., Kuraishi, H. and Sugiyama, J. (1992) Application of ubiquinone systems and electrophoretic comparison of enzymes to identification of clinical isolates of *Aspergillus fumigatus* and several other species of *Aspergillus*. J. Clinic. Microbiol. 3, 1999-2005.

Moody, S.F. and Tyler, B.M. (1990*a*) Restriction enzyme and mitochondrial DNA of the *Aspergillus flavus* group, *Aspergillus flavus, Aspergillus parasiticus* and *Aspergillus nomius*. Appl. Environ. Microbiol. 56, 2441-2452.

Moody, S.F. and Tyler, B.M. (1990*b*) Use of nuclear DNA restriction length polymorphisms to analyze the diversity of the *Aspergillus flavus* group, *Aspergillus flavus, Aspergillus parasiticus* and *Aspergillus nomius*. Appl. Environ. Microbiol. 56, 2453-2461.

Mullaney, E.J. and Klich, M.A. (1990) A review of molecular biological techniques for systematic studies of *Aspergillus* and *Penicillium*, in "Modern Concepts in *Penicillium* and *Aspergillus* Classification". (Samson, R.A. and J.I. Pitt, Eds.), pp. 301-307. Plenum Press, New York.

Murakami, H. (1971) Classification of the koji mold. J. Gen. Appl. Microbiology 17, 281-309.

Murakami, H. Hayashi, K. and Ushijima, S. (1982) Useful key characters separating three *Aspergillus* taxa: *A. sojae, A. parasiticus*, and *A. toxicarius*. J. Gen. Appl. Microbiology 28, 55-60.

Nasuno, S. (1971) Polyacrylamide gel disc electrophoresis of alkaline proteinases from *Aspergillus* species. Agric. Biol. Chem. 35, 1147-1150.

Nasuno, S. (1972*a*) Differentiation of *Aspergillus sojae* from *Aspergillus oryzae* by polyacrylamide gel disc electrophoresis. J. Gen. Microbiol. 71, 29-33.

Nasuno, S. (1972*b*) Electrophoretic studies of alkaline proteinases from strains of *Aspergillus flavus* group. Agric.Biol. Chem. 36, 684-689.

Nasuno, S. (1974) Further evidence on differentiations of *Aspergillus sojae* from *Aspergillus oryzae* by electrophoretic patterns of cellulase, pectin-lyase, and acid proteinase. Canad. J. Microbiol. 20, 413-416.

Nealson, K. H. and Garber, E. D. (1967) An electrophoretic survey of esterases, phosphatases, and leucine amino-peptidases in mycelial extracts of species of *Aspergillus*. Mycologia 59, 330-336.

Nielsen, P.V. and Samson, R.A., (1992) Differentiation of food-borne taxa of *Neosartorya*, in "Modern Methods in Food Mycology (Samson, R.A., Hocking, A.C., Pitt, J.I. and A.D. King, Eds), pp. 159-168. Elsevier, Amsterdam.

Peterson, S. W. (1992) *Neosartorya pseudofischeri* sp. nov. and its relationship to other species in *Aspergillus* section *Fumigati*. Mycol. Res. 96, 547-554.

Pitt, J.I. (1986)["1985"]. Nomenclatorial and taxonomic problems in the genus Eurotium, in "Advances in *Penicillium* and *Aspergillus* Systematics". (Samson, R.A. and Pitt, J.I., Eds) pp. 383-395. Plenum Press, New York.

Pitt, J.I. (1989) Recent developments in the study of *Penicillium* and *Aspergillus* systematics. J. Appl. Bact. , Symp. Suppl. 37S-45S.

Pitt, J.I. and Hocking, A.D. (1985) *Fungi and Food Spoilage*. Sydney, Academic Press.

Pitt, J.I., Hocking, A.D. and Glenn, D.R. (1983) An improved medium for the detection of *Aspergillus* flavus and *A. parasiticus*. J. Appl. Bact. 54, 109-114.

Pitt, J.I. and Samson, R.A. (1990*a*). Approaches to *Penicillium* and *Aspergillus* systematics. Stud. Mycol. Baarn 32, 77-91.

Pitt, J.I. and Samson, R.A. (1990*b*) Taxonomy of *Aspergillus* Section Restricta, in "Modern Concepts in *Penicillium* and *Aspergillus* Classification, (Samson, R.A. and Pitt, J.I., Eds), pp. 249-257. Plenum Press, New York.

Pitt, J.I. and Samson, R.A. (1993) Species names in current use in the Trichocomaceae (Fungi, Eurotiales). Regnum Vegetabile 128, 13-57.

Polonelli, L., Conti, A., Campani, L. and Fanti, F. (1990) Biotyping of *Aspergillus fumigatus* and related taxa by the yeast killer system, in "Modern Concepts in *Penicillium* and *Aspergillus* Classification". (Samson, R.A. and J.I. Pitt, Eds), pp. 309-320. Plenum Press, New York.

Raper, K.B. and Fennell, D.I. (1965) *The Genus Aspergillus*. Baltimore, Maryland, Williams and Wilkins.

Roquebert, M.F. and Nicot, J. (1986)["1985"]. Similarities between the genera Stilbothaminium and *Aspergillus*, in "Advances in *Penicillium* and *Aspergillus* Systematics" (Samson, R.A. and Pitt, J.I., Eds), pp. 221-228. Plenum Press, New York.

Saito, M., Kusumoto, K. and Kawasumi, T. (1991) A simple method for identification of *Aspergillus flavus* and *Aspergillus parasiticus* by slab polyacrylamide gel electrophoresis of alkaline proteinases. Rep. Natl. Food Res, inst. 55, 49-51.

Samson, R.A. (1979) A compilation of the Aspergilli described since (1965). Stud. Mycol., Baarn 18, 1-40.

Samson, R.A. (1992) Current taxonomic schemes of the genus *Aspergillus* and its teleomorphs in "*Aspergillus* in "The biology and Industrial applications" (Bennett, J. W. and M. A. Klich, Eds), pp. 353-388. Butterworth Publishers.

Samson, R.A. (1993) Current concepts of *Aspergillus* systematics In "Biotechnology Handbooks vol.20. *Aspergillus*" (J.E. Smith, Ed.). Plenum, New York, in press.

Samson, R.A. and Evans, H.C. (1992) New species of *Gibellula* on spiders (Araneida) from South America. Mycologia 84, 300-314.

Samson, R.A. and Frisvad, J.C. (1991*a*). Taxonomic species concepts of Hyphomycetes related to mycotoxin production. Proc. Japan Assoc. Mycotoxicol. 32, 3-10.

Samson, R.A. and Gams, W. (1986)["1985"] Typification of the species of *Aspergillus* and associated teleomorphs, in Advances in *Penicillium* and *Aspergillus* Systematics, (Samson, R.A. and Pitt, J.I., Eds) pp. 31-54. Plenum Press, New York.

Samson, R.A. and Pitt, J.I. (1986)["1985"]. Advances in *Penicillium* and *Aspergillus* Systematics". Plenum Press, New York.

Samson, R.A. and Pitt, J.I. (1990) "Modern Concepts in *Penicillium* and *Aspergillus* Classification". Plenum Press, New York.

Samson, R.A. and van Rreene-Hoekstra, E.S. (1988) "Introduction to Food-borne Fungi". Third edition. Centraalbureau voor Schimmelcultures, Baarn.

Samson, R.A. and Seifert, K. A. (1986)["1985"]. The ascomycete genus Penicilliopsis and its anamorphs, in *Advances in* Penicillium *and* Aspergillus *Systematics*, (Samson, R.A. and Pitt, J.I., Eds) pp. 397-426. Plenum Press, New York.

Samson, R.A., Nielsen, P.V. and Frisvad, J.C. (1990) The genus *Neosartorya*: differentiation by scanning electron microscopy and mycotoxin profiles, in "Modern Concepts in *Penicillium* and *Aspergillus* systematics" (Samson, R.A. and Pitt, J.I., Eds.), pp. 455-467. Plenum Press, New York.

Stynen, D., Meulemans, L., Goris, A., Braendlin, N. and Symons, N. (1992) Characteristics of a Latex Agglutination Test, Based on Monoclonal Antibodies, for the Detection of Mould Antigen in Foods, in "Modern Methods in Food Mycology", (Samson, R.A., A.C. Hocking, J.I. Pitt and A.D..D. King, Eds.), pp. 213-219. Elsevier, Amsterdam.

Sugiyama, J. and Yamatoya, T. (1990) Electrophoretic comparison of enzymes as a chemotaxonomic aid among *Aspergillus* taxa (1). *Aspergillus* sects. *Ornati* and *Cremei*, in "Modern Concepts of *Penicillium* and *Aspergillus* Classification" (Samson, R.A. and Pitt, J.I., Eds), pp. 385-394. Plenum Press, New York.

Tzean, S.S. Chen J.L., Liou, G.Y., Chen, C.C. and Hsu, W.H. (1990) *Aspergillus* and related teleomorphs of Taiwan. Mycological Monograph no. 1. Culture Collection and Research Center, Taiwan.

Vries, G.A. de and Cormane, R.H. (1969) A study on the possible relationships between certain morphological and physiological properties of *Aspergillus fumigatus* Fres. and its presence in, or on, human and animal (pulmonary) tissue. Mycol. Mycopath. Appl. 39, 241-253.

Yamatoya, T., Sugiyama, J. and Kuraishi, H. (1990) Electrophoretic comparison of enzymes as a chemotaxonomic aid among *Aspergillus* taxa (2). *Aspergillus* sect. *Flavi*, in "Modern Concepts of *Penicillium* and *Aspergillus* Classification" (Samson, R.A. and Pitt, J.I., Eds), pp. 395-406. Plenum Press, New York.

# APPLICATION OF RFLPS IN SYSTEMATICS AND POPULATION GENETICS OF ASPERGILLI

James H. Croft[1] and János Varga[2]

[1]School of Biological Sciences, University of Birmingham, P.O. Box 363 Birmingham B15 2TT, UK
[2]Department of Microbiology, Attila József University, P.O. Box 533, Szeged H-6701, Hungary

## INTRODUCTION

The classical methods for systematic studies of the genus *Aspergillus* have been very successful and have provided one of the better classifications among fungal genera. However, the genetic basis of a majority of characters used for these studies is not known and as a result the phylogenetic and evolutionary relationships between taxa and even between most sections of the genus are uncertain since it is probable that some characters which carry a high taxonomic significance have only a low significance in phylogenetic and evolutionary terms. This is especially so for anamorphic taxa. Consequently, it is possible to argue that direct comparison of the nucleotide sequence and organisation of nucleic acids is likely to be the most sensitive discriminator between individuals and is most likely to indicate clearly their phylogenetic and evolutionary relationships.

Polymorphism in DNA can be detected by a number of methods which range from measurement of physical parameters of the molecule to nucleotide sequence analysis. The maximum information could be obtained from an extensive nucleotide sequence analysis. This is not feasible at the present time though comparisons of the sequences of small regions or of individual genes, such as those for 5S rRNA or for part of the 18S rRNA, are proving very informative (Gniadkowski *et al.*, 1991; Sugiyama *et al.*, 1991). However, restriction fragment length polymorphism (RFLP) analysis provides a method by which the nucleotide sequence of the whole genome is sampled. Type II restriction endonucleases have specific recognition sequences in DNA and cleave the molecule at those sites. Sequence changes in genomic DNA can lead to the appearance or disappearance of particular cleavage sites, thus altering the sizes of fragments generated from a given region. These changes in sequence may result from single base substitutions, additions or deletions of small fragments, or large chromosomal changes such as inversions and translocations. Differences in the sizes of fragments resulting from digestion of the corresponding region of DNA from homologous chromosomes have been termed restriction fragment length polymorphisms (Gusella, 1986).

The steps involved in the detection of RFLPs are as follows:

(i) purification of the DNA;
(ii) digestion of the DNA with a type II restriction endonuclease;
(iii) size fractionation of the fragments by agarose gel electrophoresis;
(iv) transfer of the fragments to nylon or nitrocellulose membranes (ie. Southern blot);
(v) molecular hybridisation of these fragments to a suitably labelled DNA 'probe';
(vi) detection of the hybridised fragments by autoradiography or other suitable method.

Alternative procedures, such as hybridisation directly to the gel or the comparison of the restriction fragments produced by cloned homologous sequences, have been used less frequently. In some laboratories the restriction enzyme digests of the total DNA samples of different fungal species or strains have been compared successfully directly on the gel. Repetitive sequences can be observed in this way. Normally such sequences may originate from the mitochondrial DNA (mtDNA) or the ribosomal repeat unit (rDNA) of the strains. The restriction enzyme *Sma*I has the recognition sequence 5'-CCCGGG-3'. This sequence is very rare in the AT-rich mtDNA, and only moderate numbers of recognition sites are usually found in the nuclear DNA of fungi. Electrophoretic separation of the fragments generated results in repetitive DNA bands which have been shown to correspond to parts of the ribosomal repeat unit. Such analyses were carried out with *Fusarium* (Coddington *et al.*, 1987), *Candida* (Scherer and Stevens, 1987, 1988; Smith *et al.*, 1989), *Phytophthora* (Panabiéres *et al.*, 1989), and *Aspergillus* (discussed below).

Mitochondrial DNA (mtDNA) is an attractive source of RFLPs because of its small size, relatively large copy number and the fairly simple procedures available for its purification. MtDNA RFLPs can be detected by purifying mtDNA from the strains and comparing their patterns observed on agarose gels after digestion with restriction endonucleases. Detailed analysis of these patterns lead to the preparation of restriction fragment maps. DNA hybridisation with known mtDNA genes can also be carried out to identify some of the fragments, and correlate the genetic and physical maps. Another possibility has been described recently for observing mtDNA RFLPs more quickly, enabling the researcher to examine large numbers of isolates. This involves the use of restriction enzymes which have GC-rich 4-base recognition sites such as *Hae*III or *Hha*I (Spitzer *et al.*, 1989). The nuclear DNA digests to small fragments which migrate to the anodic end of the gel leaving the larger AT-rich mtDNA fragments clearly isolated towards the cathodic end (Spitzer *et al.*, 1989). These fragments has been shown to be of mitochondrial origin by comparison with digests of purified mtDNA (Varga *et al.*, 1993) or by hybridisation to total mtDNA.

## RFLP STUDIES IN ASPERGILLI

As yet comparatively few studies of RFLP analysis in *Aspergillus* have been described though other studies are current in several laboratories. The published work will now be reviewed according to the section of the genus (Gams *et al.*, 1985) in which these studies have been performed.

### Section *Fumigati*

Published RFLP studies within this section focus on *Aspergillus fumigatus*. The main aims of these studies concern the detection of this species in mixed clinical samples (Gabal, 1989), and its classification into discrete types (Denning *et al.*, 1990, 1991; Spreadbury *et al.*, 1990).

Gabal (1989) used the total DNA of an *A. fumigatus* strain as a probe to detect this species in the sputum of patients. He could prove the presence of *A. fumigatus* even if other aspergilli and penicillia, yeasts or bacteria were present in the same specimen. The probe cross-hybridised with *Aspergillus* and *Penicillium* DNAs, but the pattern obtained was different. The pattern found to be characteristic to *A. fumigatus* was the same as the repetitive DNA pattern observed on gels when *A. fumigatus* DNA was digested with *Eco*RI and the fragments were separated by electrophoresis. These bands have been shown to correspond to mitochondrial and ribosomal DNA fragments (J. Varga unpublished).

The *A. fumigatus* isolates could be distinguished readily from the other *Aspergillus* species by using a *Saccharomyces cerevisiae* rDNA probe to hybridise to the *Eco*RI digested total DNA preparations of *Aspergillus* species (Bainbridge *et al.*, 1990; Spreadbury *et al.*, 1990). Spreadbury *et al.* (1990) also observed intraspecific RFLPs in the ribosomal DNA repeat units of *A. fumigatus* isolates when *Eco*RI-digested total DNA patterns of the strains were probed with part of the intergenic spacer region of the rRNA gene complex of *Aspergillus nidulans*.

Denning *et al.* (1990, 1991) worked out a DNA fingerprinting method for typing *A. fumigatus* isolates. The authors prepared total DNA from mycelia of the strains by a protoplast lysis method. The high molecular weight DNA (60 kb) obtained was digested with *Sal*I and *Xho*I restriction endonucleases, electrophoresed, stained and differences in the mobilities of 10 to 50 kb bands were used to distinguish between isolates. The 45 isolates revealed 30 different patterns (DNA types). The results indicated widespread dispersal of some clones and very restricted localisation for others. The precise nature of these high molecular weight bands is not clear though they are probably of nuclear origin.

Burnie *et al.* (1992) also used RFLP analysis to distinguish between clinical isolates of *A. fumigatus*. They could classify 21 isolates into six types by using *Xba*I digestion of total cellular DNA and scoring the brightest bands in the patterns obtained. They could not find variation among these isolates by using *Eco*RI digestion.

Extensive studies have been carried out recently in this laboratory on *A. fumigatus* strains of different origin (Varga and Rinyu, 1991; J. Varga, E. Rinyu and L. Ferenczy unpublished). The strains were compared by their morphological features, isoenzyme analysis was carried out, and mitochondrial and ribosomal DNA patterns of the strains were investigated after digesting the total DNA preparations of the strains with restriction enzymes. The strains showed the same pattern when *Eco*RI was used. No mitochondrial RFLPs were observed among the 60 *A. fumigatus* strains after digestion with *Hae*III. Mitochondrial DNA RFLPs were not observed even when DNA samples of some of the strains representing type cultures of *A. fumigatus* and its 'subspecies' were digested with different combinations of restriction enzymes (*Hae*III-*Bgl*II, *Hae*III-*Eco*RI, *Hae*III-*Hin*dIII). These results support the earlier proposal, based on the examination of the micromorphology and secondary metabolite profiles (Frisvad and Samson, 1990) and DNA reassociation kinetics (Peterson, 1992) of strains of *A. fumigatus* and its subspecies (*A. fumigatus* var. *ellipticus* and var. *acolumnaris*), that these subspecies are morphological variants of *A. fumigatus* and do not deserve the subspecies status. The results obtained by isoenzyme analysis also strengthen this proposal. Some more restriction enzymes should be tested on purified mtDNA preparations to find out if there is any variability in the mtDNAs of these strains. Restriction enzymes with AT-rich recognition sequences were found to be useful in characterising mtDNAs of strains in section *Flavi* (Moody and Tyler, 1990*a*); these enzymes might also be useful in revealing mtDNA RFLPs in *A. fumigatus* strains. The *Hae*III-generated mtDNA pattern characteristic to *A. fumigatus* was compared to those of other strains representing related species from section *Fumigati* (see Table 1 and Figure 1). The mtDNA patterns were specific for most of the species studied but *A. fumigatus* showed the same pattern as *Neosartorya fischeri* strains. Intraspecific variability was observed only in *Aspergillus viridinutans*. The examination of the ribosomal repeat unit (rDNA) was carried out by digesting the total DNA with *Sma*I, separating the fragments by electrophoresis and directly observing the repetitive DNA patterns on the gels. All except one of the examined *A. fumigatus* strains showed the same repetitive DNA pattern.

**Table 1.** MtDNA types of species of section *Fumigati*

| Strain | Origin | mtDNA type[a] |
|---|---|---|
| *Aspergillus brevipes* | NRRL 2439 | 8 |
| *Aspergillus citrisporus* | NRRL 4735 | 2 |
| *Aspergillus fumigatus* var. *acolumnaris* | NRRL 5587 | 4 |
| *Aspergillus fumigatus* | ATCC 32722 | 4 |
| *Aspergillus fumigatus* | NRRL 163 | 4 |
| *Aspergillus fumigatus* var. *ellipticus* | NRRL 5109 | 4 |
| *Aspergillus fumigatus?* | FRR 1266 | 16 |
| *Aspergillus unilateralis* | NRRL 577 | 3 |
| *Aspergillus viridi-nutans* | NRRL 576 | 9 |
| *Aspergillus viridi-nutans* | NRRL 6106 | 10 |
| *Neosartorya aurata* | NRRL 4378 | 2 |
| *Neosartorya aurata* | NRRL 4379 | 2 |
| *Neosartorya aureola* | NRRL 2244 | 7 |
| *Neosartorya aureola* | NRRL 2391 | 7 |
| *Neosartorya fennelliae* | NRRL 5534 | 3 |
| *Neosartorya fennelliae* | NRRL 5535 | 3 |
| *Nosartorya fischeri* | NRRL 181 | 4 |
| *Neosartorya fischeri* | NRRL A-7223 | 4 |
| *Neosartorya glabra* | NRRL 183 | 1 |
| *Neosartorya glabra* | NRRL 2163 | 1 |
| *Neosartorya hiratsukae* | NHL 3008 | 2 |
| *Neosartorya hiratsukae* | NHL 3009 | 2 |
| *Neosartorya pseudofischeri* | NCAIM F 00757 | 3 |
| *Nosartorya pseudofischeri* | NRRL 20748 | 3 |
| *Neosartorya pseudofischeri* | NRRL 3496 | 3 |
| *Neosartorya quadricincta* | NRRL 2154 | 6 |
| *Neosartorya quadricincta* | NRRL 2221 | 6 |
| *Neosartorya spathulata* | NHL 2947 | 5 |
| *Neosartorya spathulata* | NHL 2948 | 5 |
| *Neosartorya spathulata* | NHL 2949 | 5 |
| *Neosartorya spinosa* | NRRL 3435 | 3 |
| *Neosartorya spinosa* | NRRL 5034 | 3 |
| *Neosartorya straminea* | NRRL 4652 | 2 |

[a] for mtDNA types, see Figure 1

When the *A. nidulans* rDNA was hybridised to the filters obtained after blotting the polymorphic band did not show homology to this probe and the same hybridisation pattern was observed in all *A. fumigatus* strains. Similar results were obtained for other species of section *Fumigati*. A high proportion of the polymorphic bands observed on the gels as repetitive bands did not hybridise to the *A. nidulans* ribosomal repeat unit, consequently the hybridisation patterns were similar for most of the species. This result could mean that the fragments not showing homology to the probe used are part of the intergenic spacer region. From the RFLP analysis of these species, possible close relationship of *A. citrisporus*, *Neosartorya aurata* and *Neosartorya straminea*, and *A. fumigatus* and *N. fischeri*, respectively, is proposed. This idea is further supported by the results obtained by isoenzyme and RAPD analyses of these strains.

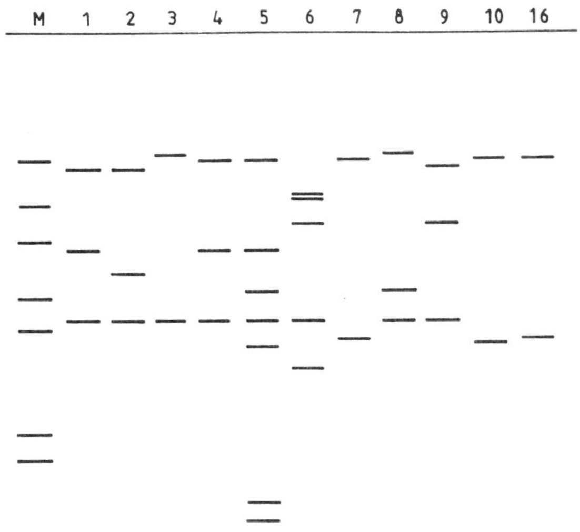


**Figure 1.** Mitochondrial DNA patterns observed in species of section *Fumigati* (for species showing these patterns, see Table 1). M, *Hin*dIII digested lambda DNA

## Section *Flavi*

Klich and Mullaney (1987) and Gomi *et al.* (1989) used the *Sma*I digestion patterns of nuclear DNAs to distinguish between species of this section. Strains belonging to the species *Aspergillus sojae, Aspergillus oryzae, Aspergillus flavus* and *Aspergillus parasiticus* were differentiated successfully even if the strain examined showed degenerated morphology. The differences in the patterns were found to be due to a missing *Sma*I site in the rDNAs of *A. flavus* and *A. parasiticus* compared to those of *A. oryzae* and *A. sojae*.

Moody and Tyler (1990*a*, 1990*b*) carried out extensive RFLP studies on three species of this section, namely *A. flavus, A. parasiticus* and *Aspergillus nomius*. Their aim was to correlate mitochondrial and nuclear RFLPs with morphological and biochemical features characteristic of the species examined and to investigate the sequence divergence between these fungi. They found very small intra- or interspecific variability in the mitochondrial genomes of these strains. Only restriction enzymes with AT rich recognition sequences detected RFLPs. *A. flavus* and *A. parasiticus* were shown to have mtDNAs of the same size (32 kb), while *A. nomius* has a slightly larger mitochondrial genome (33 kb) due to insertions at two sites. The gene order was the same in all three species as that found in *A. nidulans*. RFLPs generated by *Hin*fI or *Rsa*I made it possible to distinguish all the three species. During the study of the nuclear DNA of the species (Moody and Tyler, 1990*b*), highly conserved cloned genes were used as heterologous probes. Very little inter- or intraspecific polymorphism was detected in the spacer region of the ribosomal repeat unit of these species. Hybridisation of *Neurospora crassa* ribosomal DNA to the *Taq*I or *Nco*I digested DNA samples of the strains did not result in species-specific patterns. This finding is in contrast with the results of Klich and Mullaney (1987) and Gomi *et al.* (1989) who found the *Sma*I generated repetitive rDNA patterns to be species-specific. The other probes used (β-tubulin gene and a ribosomal protein gene of *N. crassa*) made it possible to classify isolates into one of the three species examined. The statistical analysis of the data obtained showed a closer relationship between *A. flavus* and *A. nomius*, than between *A. flavus* and *A. parasiticus*. This is in contrast to the results obtained during the analysis of the mtDNA of these strains; more

*A. nomius* isolates should be examined to estimate accurately the relatedness of this species to *A. flavus* and *A. parasiticus*. From the very small variation observed in the mtDNA and rDNA of these species, the authors concluded that the three species are closely related and must have emerged very recently. Limited correlation was found between the mitochondrial and nuclear RFLPs and the geographical origin, while no RFLPs were found to correlate with mycotoxin production of the strains.

## Section *Circumdati*

Examination of the mtDNA and rDNA patterns of some collection strains of the section has been carried out in our laboratory (Table 2, Figures 2 and 3). The mtDNA patterns observed were species-specific, except for *Aspergillus ochraceus* and *Aspergillus petrakii*. These strains showed the same pattern when *Hae*III-digested mtDNA and *Sma*I-generated rDNA patterns were compared (see Figures 2 and 3). MtDNA patterns 10 and 13 are identical to each other as are rDNA patterns 12 and 15. Examination of the rDNA patterns of the other strains revealed possible relationships between some species, but further studies including more strains from each species are necessary to draw taxonomic conclusions.

## Section *Nigri*

Because of their importance in fermentation industries, the taxonomy of black aspergilli has been well established. Mosseray (1934; cf. Raper and Fennell, 1965) described 35 species of the black aspergilli. Raper and Fennell (1965) reduced the number of species accepted within the *Aspergillus niger* group to twelve. Al-Musallam (1980) revised the taxonomy of the *A. niger* group by taking mainly morphological features into account. She recognised seven species within this group and described *A. niger* itself as an aggregate consisting of seven varieties and two formae. Several attempts have been made to clarify the problematic inter-relationships between species belonging to the *A. niger* aggregate and between these and other black *Aspergillus* species by using nuclear or mitochondrial DNA RFLPs.

The first report came from Kozlowski and Stepien (1982), who found only a distant relationship between an *A. niger* and an *Aspergillus awamori* strain by comparing their *Hin*dIII-*Eco*RI double-digested mtDNA patterns. The similarity was less than that observed between *A. niger* and *A. oryzae*.

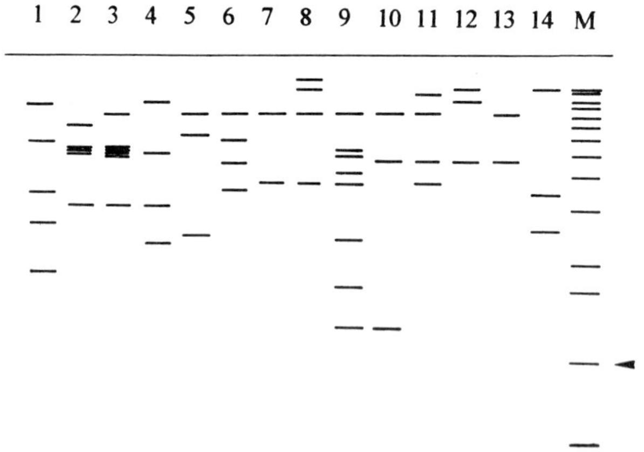


**Figure 2.** Mitochondrial DNA patterns observed in species of section *Circumdati* (for species showing these patterns, see Table 2). M, 1 kb DNA ladder (the arrow indicates the 1 kb band).

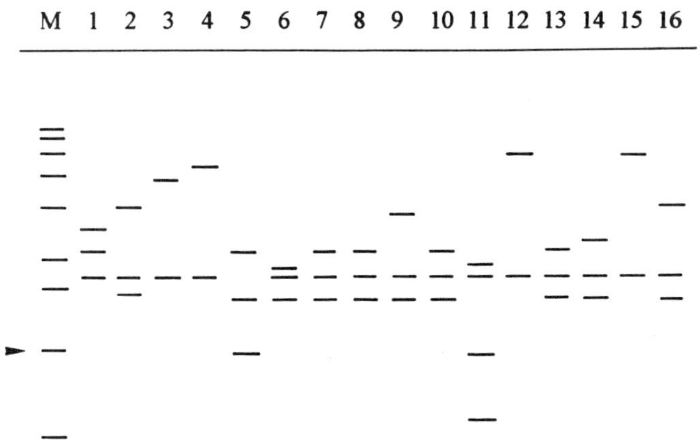


**Figure 3.** Ribosomal DNA hybridisation patterns of species belonging to section *Circumdati*. M, 1 kb DNA ladder (the arrow indicates the 1 kb band).

**Table 2.** MtDNA and rDNA types of strains belonging to section *Circumdati* (one of each of *A. candidus* and *A. wentii* strains were also included)

| Species | Origin | mtDNA[a] pattern | rDNA[b] pattern |
|---|---|---|---|
| *A. auricomus* | IMI 172277 | 12 | 14 |
| *A. bridgeri* | IMI 259060 | 7 | 8 |
| *A. campestris* | IMI 259099 | 1 | 2 |
| *A. candidus* | IMI 091889 | 14 | 16 |
| *A. dimorphicus* | IMI 131553 | 4 | 5 |
| *A. elegans* | IMI 133962 | 5 | 6 |
| *A. elegans* | NRRL 407 | 5 | 6 |
| *A. lanosus* | IMI 226007 | 2 | 3 |
| *A. ochraceoroseus* | IMI 223071 | ND | 1 |
| *A. ochraceus* | IMI 016247 | 10 | 12 |
| *A. ochraceus* | NRRL 405 | 10 | 12 |
| *A. ostianus* | IMI 015960 | 3 | 4 |
| *A. petrakii* | IMI 172291 | 13 | 15 |
| *A. petrakii* | NRRL 416 | 13 | 15 |
| *A. quercinus* | IMI 235600 | 6 | 7 |
| *A. robustus* | IMI 216610 | 9 | 11 |
| *A. sclerotiorum* | IMI 056673 | 8 | 10 |
| *A. sulphureus* | IMI 211397 | 11 | 13 |
| *A. wentii* | IMI 017295 | ND | 9 |

[a] for mtDNA patterns, see Figure 2.
[b] for rDNA patterns, see Figure 3.

Kirimura *et al.* (1992*a*) prepared the physical map of the mtDNA of an *A. niger* strain. The size of the mtDNA was 32.6 kb. The gene order observed in this strain was the same as that of *A. nidulans*. Later Kirimura *et al.* (1992*b*) found mtDNA of a different size (31.9 kb) in another *A. niger* strain.

Kusters-van Someren *et al.* (1990) used Western blotting and DNA hybridisation with a pectin lyase (*pelD*) gene to find out if these methods can be used for rapid strain identification. Only limited correlation was found between the results obtained by Western and Southern blotting. The DNA hybridisation experiments showed that the *pelD* gene is conserved in all isolates belonging to the *A. niger* aggregate. Hybridisation was also observed to DNAs of an *A. carbonarius* and all *A. foetidus* strains (the *A. carbonarius* strain was later shown to be misclassified, and belongs to the *A. niger* aggregate). The authors established three groups within the *A. niger* aggregate on the basis of presence or absence of three other bands which hybridised strongly to the *pelD* gene, and probably resulted from heterologous hybridisation to other *pel* genes. Other species of black aspergilli did not show homology to *pelD*. As a continuation of this work, Kusters-van Someren *et al.* (1991) carried out a more extensive study on nuclear DNA RFLPs of several black *Aspergillus* collection strains. Two groups of strains were distinguished according to their *Sma*I-generated repetitive DNA patterns. The fragments were shown to correspond to ribosomal DNA fragments. The two groups were also clearly distinguishable by their hybridisation patterns when pectin lyase genes (*pelA*, *pelB*) and the pyruvate kinase (*pki*) gene were used as probes in DNA hybridisation experiments. The two groups found were proposed to represent different species, namely *A. niger* and *Aspergillus tubigensis*. This idea was further strengthened by the observation that heterokaryon formation was not possible between isolates belonging to the different species mentioned above. Examination of other species not belonging to the *A. niger* aggregate was also carried out. *Aspergillus foetidus* strains, classified into a different species by Al-Musallam (1980), showed the same nuclear DNA RFLPs as *A. niger*. *Aspergillus helicothrix* was found to represent only a morphological variant of *Aspergillus ellipticus*, and *Aspergillus aculeatus* should only ranked to subspecies status as it showed the same *Sma*I-digested rDNA pattern as the *Aspergillus japonicus* strains examined.

Mégnégneau *et al.* (1993) examined collection strains of black aspergilli by applying several approaches (examination of nuDNA RFLPs, isoenzyme analysis, separation of chromosomal DNAs by CHEF, RAPD analysis). They established four rDNA types within the *A. niger* aggregate based on the rDNA patterns of the strains obtained by using *Sma*I, *Eco*RI and *Pst*I restriction enzymes and hybridising the *A. nidulans* ribosomal repeat unit to the filters obtained after blotting. The strains showing the four types belonged to two main groups corresponding to the two species proposed by Kusters-van Someren *et al.* (1991). The two species were also distinguishable by the other methods applied.

Varga *et al.* (1993) studied the mtDNA RFLPs of collection strains of the *A. niger* aggregate. The two main groups established according to the mtDNA patterns obtained by using *Eco*RI, *Bgl*II, *Pvu*II and *Hae*III restriction enzymes coincided with the two groups based on the *Sma*I-digested rDNA patterns of the strains, representing *A. niger* and *A. tubigensis* respectively. MtDNA variation was observed within these species. *A. foetidus* strains could not be distinguished by their mtDNA or rDNA RFLPs from *A. niger* strains. The size of the mtDNA was about 30-32 kb depending on the mtDNA type the given strain belongs to.

Recently, J. Varga, F. Kevei, Z. Kozakiewicz and J. H. Croft (unpublished) have carried out a more extensive study on field isolates belonging to the *A. niger* aggregate. The method using *Hae*III-*Bgl*II double digestion of total DNA samples described earlier (Varga *et al.*, 1993) was used to characterise mtDNAs of the strains. The bands observed on the gels were shown to correspond to mtDNA fragments (Varga *et al.*, 1993). Most of the isolates were classifiable as *A. niger* or *A. tubigensis* according to their mtDNA and *Sma*I generated rDNA patterns. A high level of variation was observed in the mtDNAs of the strains; the 150 strains examined showed 12 different mtDNA patterns (see Figure 4). The *Sma*I generated rDNA patterns remained specific to *A. niger* or *A. tubigensis*. The geographical distribution of the mtDNA types was uneven. Almost all the strains collected in Mallorca or Indonesia showed mtDNA patterns characteristic of *A. niger*, while Australian isolates showed an unexpectedly high degree of variability (the 15 isolates showed five different mtDNA types, three of which were shown only by these strains). Hybridisation experiments were carried out with cloned *A. nidulans* and *A. niger* mtDNA fragments. The hybridisation patterns revealed that the two species are more distantly related than concluded earlier (Varga *et al.*, 1993) since bands which appeared common to both major classes of mtDNA in the *Hae*III-*Bgl*II digests were shown to be non-homologous in the hybridisation experiments.

Six of the examined 14 Brasilian isolates revealed a mtDNA pattern which was shown only by these strains. The strains also showed different rDNA patterns from those characteristic to *A. niger* or *A. tubigensis*. These were the only isolates which could not be classified into one of the species proposed by Kusters-van Someren *et al.* (1991). Classical taxonomic examination of these strains is in progress.

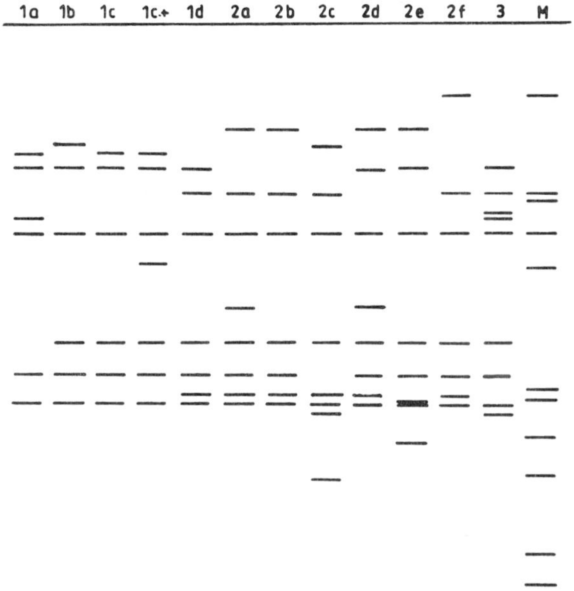


**Figure 4.** Mitochondrial DNA patterns observed in isolates belonging to the *A. niger* aggregate. M, lambda DNA double digested with *Eco*RI and *Hin*dIII.

F. Kevei, Zs. Hamari, Z. Kozakiewicz, J. H. Croft and J. Varga (unpublished) compared several *A. japonicus* and *A. carbonarius* strains by their mtDNA patterns obtained by using *Eco*RI, *Pvu*II, *Bgl*II and *Hae*III restriction enzymes. The size of the mtDNAs was 42-50 kb and 40-43 kb in the cases of *A. japonicus* and *A. carbonarius*, respectively. The 30 *A. japonicus* strains examined belong to seven different RFLP groups, while all except one of the 11 examined *A. carbonarius* strains showed the same pattern with all the enzymes used. Hybridisation experiments with cloned *A. nidulans* mtDNA fragments made it possible to establish subgroups within the *A. japonicus* species. The *Sma*I-generated rDNA patterns of the strains were specific to each species. Several molecular hybridisation experiments between probes derived from *A. niger* and *A. japonicus* DNA have all given weak hybridisation suggesting only low homology. This has also been observed by Mullaney (1993) who used a region of the phytase gene derived from a member of the *A. niger* aggregate (*Aspergillus ficuum* SRRC 265) as a probe and obtained only very weak hybridisation with a strain of *A. japonicus*. These results suggest a distant relationship between the *A. niger* aggregate and *A. japonicus*.

## Section *Nidulantes*

The section *Nidulantes* has a particularly good taxonomy. It contains about 30 species, many of which have well defined teleomorphs (*Emericella* spp.) but others are anamorphic taxa and are perhaps less well defined. *Aspergillus nidulellus* (teleomorph *Emericella nidulans*), more generally known as *A. nidulans*, has been extensively studied genetically. Using the techniques developed during these genetical studies a large collection of natural isolates has been investigated in order to determine the extent and nature of natural variation within the species and to establish the genetic structure of the natural population (Croft and Jinks, 1977; Croft, 1987). The results of this work show that the natural population of *A. nidulans* consists of a series of vegetative or heterokaryon compatibility (h-c) groups of strains. Members of each h-c group are genetically closely related and clearly represent asexually dispersed strains. However, it is apparent that there is a considerable amount of genetic variation in the population for characters such as linear growth rate, penicillin production or cleistothecial production and this is seen as differences between h-c groups.

In spite of the obvious genetic variation in general phenotypic characters found within *A. nidulans* very little variability can be demonstrated within the species by RFLP analysis. The species appears to be tightly defined at the DNA level. No RFLPs have yet been detected in the mtDNA of *A. nidulans*. Similarly, random genomic clones revealed only minor RFLPs between strains when used as probes in Southern blot experiments. Those that were present were found as simple differences between h-c groups. Preliminary results from RAPD analysis using several primers has also shown only very minor polymorphism within the species. Other preliminary experiments in which phage M13 has been used as a probe in Southern blot experiments have revealed the presence of some between h-c group polymorphism and possibly also of a small amount of intra-group polymorphism, but much more information is required to extend and confirm these observations. The presence of some further intraspecific polymorphism is indicated by the observation that the penicillin non-producing strains which comprise one h-c group (h-c F) all contain a deletion of the cluster of three genes which are responsible for penicillin synthesis (MacCabe *et al.*, 1990). Inter h-c group molecular polymorphism in *A. nidulans* has been studied recently by D. M. Geiser, M. L. Arnold and W. E. Timberlake (unpublished). They demonstrated that 14 out of 42 randomly chosen clones from two chromosome specific cosmid libraries showed polymorphisms when used as probes in Southern blot experiments against restriction digests of DNA from isolates representing seven h-c groups. The average size of each clone was 35kb and the 42 clones thus represent about 5% of the total *A. nidulans* genome. Eleven of these polymorphic clones were then used as probes against strains from 20 h-c groups and the resultant 11 locus haplotypes obtained were found to be specific to each group. It is of interest to note that the polymorphisms detected by these clones were complex and involved rearrangements in or flanking the region covered by the probe. Multiple 'allelic' forms of these probes were present in most cases.

When other sexual taxa from the section, such as *Aspergillus quadrilineatus* or *A. nidulans* var. *echinulatus*, were examined clear interspecific polymorphisms were revealed. It has been possible to locate several RFLPs to linkage groups by the analysis of haploid chromosome substitution lines derived from an allodiploid strain isolated following the protoplast fusion of genetically marked strains of *A. nidulans* and *A. quadrilineatus*. Again, little or no variation was detected within each species. This was so both for RFLPs detected by random probes and in mtDNA.

The comparison of the mitochondrial genome of these species is of interest since it can be made in some detail. Kozlowski and Stepien (1982) reported that there was a high degree of similarity in the restriction patterns produced by the digestion of the mtDNA of *A. nidulans* and *A. nidulans* var. *echinulatus*. However, since the sequence of the mtDNA of one strain of *A. nidulans* (NRRL 194) is almost fully known a very complete and accurate restriction site map is available. Consequently, the restriction site maps of the mtDNA of *A.*

*quadrilineatus* and *A. nidulans* var. *echinulatus* were simple to produce. This revealed that the major differences between the mitochondrial genome of these three taxa were due to the presence or absence of introns within some genes. The common parts of the molecules were nearly identical in all three taxa, there being only a few restriction site differences. Comparison of the nucleotide sequence of a short region from *A. nidulans* var. *echinulatus* with the homologous region in *A. nidulans* showed only 0.17% base substitution (Jadayel, 1986).

By contrast three other members of section *Nidulantes* are clearly distantly related both to each other and to *A. nidulans* (Jadayel, 1986). The restriction maps of the mtDNAs of *A. heterothallicus, A. stellatus* and *A. unguis* all show considerable divergence from each other and from *A. nidulans* and, based on the assumption that the bands of similar size represented common bands, estimates of base substitution by comparison with *A. nidulans* of from 11% to 16% were made. When a short nucleotide sequence from *Aspergillus heterothallicus* was compared to the homologous sequence from *A. nidulans* an actual base substitution of 12% was seen. In *A. heterothallicus*, but not in the other two species, the order of the genes in the mtDNA was rearranged.

Random probes derived from nuclear DNA of *A. nidulans* failed completely to hybridise to DNA of these three species under the conditions of stringency used. This was of interest because these probes hybridised strongly to the DNA of *Aspergillus terreus*. This observation was supported by the work of Mullaney and Klich (1987) who surveyed some *Aspergillus* species for homology with *A. nidulans* developmental genes (*brlA*, *spoC* and *tubC*). They found that all of the species tested have homologous regions to all three genes, but the intensity of the autoradiogram bands generated by Southern hybridisation analysis of restriction enzyme digests of *A. fumigatus* DNA indicated a lower level of homology between this species and *A. nidulans* as compared to *A. terreus* and *A. nidulans*.

## DISCUSSION

Although the use of RFLP data in the study of the taxonomy of *Aspergillus* is limited at present to only a small part of the genus, it is clear from the results achieved so far that the technique is very useful for this purpose. Relatively impure DNA preparations are suitable for many of the experimental techniques involved and simple methods for the extraction of DNA from quite small amounts of mycelium are available. Much information can be obtained directly from the digestion of crude preparations of total DNA and the separation of the resultant fragments by electrophoresis. Mitochondrial genomes can be studied in this way by the digestion of total DNA preparations with restriction enzymes which recognise GC-rich four base pair sequences such as *Hae*III, either alone or in combination with other enzymes. Similarly, repetitive nuclear DNA (particularly the ribosomal RNA genes) can be studied by digestion of total DNA preparations with enzymes such as *Sma*I or *Eco*RI. This method makes it possible to survey large numbers of strains. The accuracy of this method is greatly improved if it is used in conjunction with Southern blot DNA hybridisation experiments thus making the identification of DNA fragments more certain, but this of course requires that suitable probes are available. In general it is probable that RFLP analysis will be most informative when used in conjunction with other methods such as nucleotide sequence analysis or DNA amplification methods but alone it can provide much information of use to the taxonomist.

The results reviewed here have shown that in *Aspergillus* mtDNA RFLP analysis is a very informative method for the grouping of field isolates and that mtDNA is very useful for the identification of particular species. In Sections *Fumigati* and *Nidulantes* there appears to be no intraspecific mitochondrial DNA polymorphism (with the possible exception of *Aspergillus viridinutans*). This is so in both sections even though the individual species (for example *A. fumigatus* and *A. nidulans*) are known to be genetically very variable for many morphological and physiological characters. This intraspecific molecular constancy is confirmed by the analysis of the nuclear rDNA. It is interesting that such a population structure is present in both an asexual and a sexual species. The approach is also useful for detecting some unexpected possible close relationships. For example, in Section *Fumigati,* the asexual species *A. citrisporus* and the two sexual species, *N. aurata* and *Neosartorya. straminea,* have identical RFLP patterns for both mtDNA and rDNA, as have *A. fumigatus* and *N. fischeri.*

The population structure revealed in Section *Nigri* is in complete contrast. Several distinct species are present in this group. These are *A. carbonarius, A. ellipticus, A. heteromorphus* and *A. japonicus* in addition to the *A. niger* aggregate. Polymorphism in mtDNA has been demonstrated in *A. japonicus* dividing the species into seven groups, but there was no variation in the rDNA patterns. The *A. niger* aggregate has been shown by RFLP analysis to consist of two genetically related groups, each containing considerable polymorphism. It was proposed by Kusters-van Someren *et al.* (1991) on evidence derived from nuclear DNA that these two groups should be given species status and called *A. niger* and *A. tubigensis.* This division is supported by the analysis of mtDNA but much more polymorphism is present in mtDNA than was revealed in nuclear DNA. Also, the presence of a third distinct type among this aggregate was demonstrated among isolates from Brasil. The whole of the *A. niger* aggregate is very difficult to distinguish using the methods of classical taxonomy and thus the separation of this aggregate into two (possibly three) distinct species is being proposed purely on the basis of RFLP data.

## ACKNOWLEDGEMENTS

We would like to thank Zofia Kozakiewicz (IMI) both for supplying many strains and for her help with strain identification. Part of the study referred to in this review was carried out while J. Varga was in receipt of a joint one-year postdoctoral fellowship of the Royal Society and the Hungarian Academy of Sciences, sponsored by the Soros Foundation.

## REFERENCES

Al-Musallam, A. (1980) Revision of the black *Aspergillus* species. Ph.D. Thesis, Utrecht.

Bainbridge, B.W., Spreadbury, C.L., Scalise, F.G. and Cohen, J. (1990) Improved methods for the preparation of high molecular weight DNA from large and small scale cultures of filamentous fungi. FEMS Microbiol. Lett. 66, 113-118.

Burnie, J.P., Coke, A. and Matthews, R.C. (1992) Restriction endonuclease analysis of *Aspergillus fumigatus* DNA. J. Clin. Pathol. 45, 324-327.

Coddington, A., Matthews, P.M., Cullis, C. and Smith, K.H. (1987) Restriction digest patterns of total DNA from different races of *Fusarium oxysporum* f. sp. *pisi* an improved method for race classification. J. Phytopathol. 118, 9-20.

Croft, J.H. (1987) Genetic variation and evolution in *Aspergillus*, in "Evolutionary biology of the fungi" (Rayner, A.D.M., Braiser, C.M. and Moore, D., Eds.), pp. 311-323. Cambridge University Press, Cambridge.

Croft, J.H. and Jinks, J.L. (1977) Aspects of the population genetics of *Aspergillus nidulans*, in "Genetics and physiology of *Aspergillus*" (Smith, J.E. and Pateman, J.A., Eds), pp. 339-360. Academic Press, London.

Denning, D.W., Clemons, K.V., Hanson, L.H. and Stevens, D.A. (1990) Restriction endonuclease analysis of total cellular DNA of *Aspergillus fumigatus* isolates of geographically and epidemiologically diverse origin. J. Infect. Dis. 162, 1151-1158.

Denning, D.W., Shankland, G.S. and Stevens, D.A. (1991) DNA fingerprinting of *Aspergillus fumigatus* isolates from patients with aspergilloma. J. Med. Veter. Mycol. 29, 339-342.

Frisvad, J.C. and Samson, R.A. (1990) Chemotaxonomy and morphology of *Aspergillus fumigatus* and related taxa, in "Modern concepts in *Penicillium* and *Aspergillus* classification"(Samson, R.A. and Pitt, J.I., Eds.), pp. 201-208. Plenum Press, New York.

Gabal, M.A. (1989) Development of a chromosomal DNA probe for the laboratory diagnosis of aspergillosis. Mycopathologia 106, 121-129.

Gams, W., Christensen, M., Onions, A.H.S., Pitt, J.I. and Samson, R.A. (1985) Infrageneric taxa of

*Aspergillus*, in "Advances in *Penicillium* and *Aspergillus* systematics" (Samson, R.A. and Pitt, J.I., Eds.), pp. 55-61. Plenum Press, New York.

Gniadkowski, M., Fiett, J., Borsuk, P., HoffmanZacharska, D., Stepien, P.P. and Bartnik, E.(1991) Structure and evolution of 5S rRNA genes and pseudogenes in the genus*Aspergillus*. J. Mol. Evol. 33, 175-178.

Gomi, K., Tanaka, A., Iimura, Y. andTakahashi,K. (1989) Rapid differentiation of four related species of koji moulds byagarosegel electrophoresis of genomic DNA digested with SmaI restriction enzyme. J. Gen. Appl. Microbiol. 35, 225-232.

Gusella, J.F. (1986) DNA polymorphism and human disease. Ann. Rev. Biochem. 55, 831-854.

Jadayel, D.M. (1986) Variation in the organization and structure of the mitochondrial DNA of species of *Aspergillus*. Ph. D. Thesis, University of Birmingham, UK.

Kirimura, K., Fukuda, S., Abe, H., Kanayama, S. and Usami, S. (1992*a*) Physical mapping of the mitochondrial DNA from *Aspergillus niger*. FEMS Microbiol. Lett. 90, 235-238.

Kirimura, K., Fukuda, S., Sarngbin, S., Kanayama, S. and Usami, S. (1992*b*) The mitochondrial genome of *Aspergillus niger*. Abstracts of 1st European Conference on Fungal Genetics, Nottingham, P1/02.

Klich, M.A. and Mullaney, E.J. (1987) DNA restriction enzyme fragment polymorphism as a tool for rapid differentiation of *Aspergillus flavus* from *Aspergillus oryzae*. Exp. Mycol. 11, 170-175.

Kozlowski, M. and Stepien, P.P. (1982) Restriction enzyme analysis of mitochondrial DNA of members of the genus *Aspergillus* as an aid in taxonomy. J. Gen. Microbiol. 128, 471-476.

Kusters-van Someren, M.A., Kester, H.C.M., Samson, R.A. and Visser, J. (1990) Variation in pectinolytic enzymes in black *Aspergilli*: a biochemical and genetic approach, in "Modern concepts in *Penicillium* and *Aspergillus* classification" (Samson, R.A. and Pitt, J.I., Eds.), pp.321-334. Plenum Press, New York

Kusters-van Someren, M.A., Samson, R.A. and Visser, J. (1991) The use of RFLP analysis in classification of the black *Aspergilli*: reinterpretation of *Aspergillus niger* aggregate. Curr. Genet. 19, 21-26.

MacCabe, A.P., Riach, M.B.R., Unkles, S.E. and Kinghorn, J.R. (1990) The *Aspergillus nidulans npeA* locus consists of three contiguous genes required for penicillin biosynthesis. EMBO J. 9, 279-287.

Mégnégneau, B., Debets, F. and Hoekstra, R.F. (1993) Genetic variability and relatedness in the complex group of black *Aspergilli* based on random amplification of polymorphic DNA. Curr. Genet. 23, 323-329.

Moody, S.F. and Tyler, B.M. (1990*a*) Restriction enzyme analysis of mitochondrial DNA of the *Aspergillus flavus* group: *A. flavus*, *A. parasiticus*, and *A. nomius*. Appl. Environ. Microbiol. 56, 2441-2452.

Moody, S.F. and Tyler, B.M. (1990*b*) Use of nuclear DNA restriction fragment length polymorphisms to analyze the diversity of the *Aspergillus flavus* group: *A. flavus*, *A. parasiticus*, and *A. nomius*. Appl. Environ. Microbiol. 56, 2453-2461.

Mullaney, E.J. (1993) Survey of the *Aspergillus niger* group for DNA sequences cross hybridizing to the five prime region of the fungal phytase gene. Mycologia 85, 71-73.

Mullaney, E.J. and Klich, M.A. (1987) Survey of representative species of *Aspergillus* for regions of DNA homology to *Aspergillus nidulans* developmental genes. Appl. Microbiol. Biotechnol. 25, 476-479.

Panabiéres, F., Marais, A., Trentin, F., Bonnet, P. and Ricci, P. (1989) Repetitive DNA polymorphism analysis as a tool for identifying *Phytophtora* species. Phytopathol 79, 1105-1109.

Peterson, S.W. (1992) *Neosartorya pseudofischeri* sp. nov. and its relationship to other species in *Aspergillus* section *Fumigati*. Mycol. Res. 96, 547-554.

Raper, K.B. and Fennell, D.I. (1965) "The genus *Aspergillus*." Williams & Wilkins: Baltimore.

Scherer, S. and Stevens, D.A. (1987) Application of DNA typing methods to epidemiology and taxonomy of *Candida* species. J. Clin. Microbiol. 25, 675-679.

Scherer, S. and Stevens, D.A. (1988) A *Candida albicans* dispersed, repeated gene family and its epidemiologic applications. Proc. Natl. Acad. Sci., USA 85, 1452-1456.

Smith, R.A., Hitchcock, C.A., Evans, E.G.V., Lacey, C.J.N. and Adams, D.J. (1989) The identification of *Candida albicans* strains by restriction fragment length polymorphism analysis of DNA. J. Med. Veter. Mycol. 27, 431-434.

Spitzer, E.D., Lasker, B.A., Travis, S.J., Kobayashi, G.S. and Medoff, G. 1989) Use of mitochondrial and ribosomal DNA polymorphisms to classify clinical and soil isolates of *Histoplasma capsulatum*. Infect. Immun. 57, 1409-1412.

Spreadbury, C.L., Bainbridge, B.W. and Cohen, J. (1990) Restriction fragment length polymorphisms in isolates of *Aspergillus fumigatus* probed with part of the intergenic spacer region from the ribosomal RNA gene complex of *Aspergillus nidulans*. J. Gen. Microbiol. 136, 1991-1994.

Sugiyama, J., Rahayu, E.S., Chang, J.-M. and Oyaizu, H. (1991) Chemotaxonomy of *Aspergillus* and associated teleomorphs. Jpn. J. Med. Mycol. 32, Suppl. 39-60.

Varga, J. and Rinyu, E. (1991) Characterization of *Aspergillus fumigatus* strains by isoenzyme analysis. Acta Microbiol. Hung. 38, 225.

Varga, J., Kevei, F., Fekete, Cs., Coenen, A., Kozakiewicz, Z and Croft, J.H. (1993) Restriction fragment length polymorphisms in the mitochondrial DNAs of the *Aspergillus niger* aggregate. Mycol. Res. 97, 1207-1212.

# MODERN APPROACHES TO THE TAXONOMY OF *ASPERGILLUS*

Brian W. Bainbridge

Microbiology Group, Life Sciences Division
Kings College London, Campden Hill Road
London W8 7AH, UK

## INTRODUCTION

Techniques from molecular biology have provided a series of new tools for the analysis of diversity in the fungi. These techniques have been applied to a variety of fungal groups but only rather limited work has been done on the genus *Aspergillus*, which is surprising considering the economic importance of species within the genus. The availability of a detailed molecular genetic systems in *Aspergillus nidulans* has been of considerable help in providing a scientific base, but it appears on the whole that molecular geneticists have not been very interested in the taxonomy of the genus. However, a need to study the epidemiology, detection, diagnosis, identification, classification, characterisation and quantification of *Aspergillus* has resulted in an increasing interest in the taxonomic basis for differences within the genus. This has made it essential that closer links are forged between molecular biologists and taxonomists.

Recent techniques are based on variation in the nucleotide sequence of nucleic acids which is frequently analysed indirectly by hybridisation with nucleic acid probes. These may be based on repetitive DNA such as the ribosomal RNA gene complex (rDNA) which has both conserved and variable sequences (see below). Hybridisation or annealing of short oligonucleotide primers to a target sequences is also the basis of amplification using the polymerase chain reaction (PCR) which can be used to allow the specific detection of DNA or to produce a genetic fingerprint. Hybridisation is often combined with restriction enzyme analysis to produce restriction fragment length polymorphisms (RFLPs) but the reliability of this method for phylogenetic analysis has been questioned (Swofford and Olsen, 1990). This is because the fragments obtained may not fulfil the crucial test that characters are independent of each other. A new site may evolve between two pre-existing sites and one longer fragment may disappear and two new fragments will appear. Thus two organisms may have two restriction sites in common with only a single base pair different and yet have no fragments in common. Combined with other analyses, however conserved RFLPs have some value but further discussion of the value of RFLP data will not be given as it covered in the chapter by Croft .

Alternatively, complete or partial nucleotide sequences of target molecules can provide data for a more accurate phylogenetic comparison. Such sequences also provide useful data for the development of general or specific nucleic acid probes which can be used for RFLP

*The Genus Aspergillus*, Edited by Keith A. Powell *et al.*,
Plenum Press, New York, 1994

analysis, for slot blot hybridisation or for specific amplification. Analysis of chromosomal structure by use of pulse field gel electrophoresis (PFGE) can also reveal information about taxonomic relationships and some details of this technique will be found in the chapter by van der Bos.

Other taxonomic methods have been based on ubiquinones, antigenic properties, monoclonal antibodies, cell wall structure, isozymes polymorphisms and analysis of metabolic pathways. A detailed analysis of these methods is not possible in a chapter of this length although reference will be made to some examples.

Ideally a good classification system should be based on evolutionary descent and therefore a decision may need to be made on which characters are primitive and which are derived. This would allow a cladistic analysis based on natural relationship and assumptions about ancestors but such an approach is difficult in a filamentous fungus like *Aspergillus* where questions about whether imperfect species are primitive or advanced may be unhelpful. A more useful approach may be to apply phenetic or numerical analysis on a number of characters and to look for similarity making no assumptions about evolution. Total or partial base sequences of nucleic acids can however be analysed by cladistic methods. It is important, however,that molecular data are used as an extra character combined with conventional taxonomic methods, such as morphological and metabolic properties. Analysis of, for example nucleic acids alone, should not be expected to produce a reliable classification. A molecular phylogenetic analysis of the genus *Aspergillus* should provide us with data for use in a variety of new and exciting approaches to the analysis of genome diversity in industrial, agricultural, medical and environmental isolates of *Aspergillus*.

The scope of this chapter will include current and possible future applications of these approaches to the classification of *Aspergillus* into subgenera and species, as well as to methods for producing fine distinctions between strains, races and isolates using fingerprinting techniques.

## CHOICE OF A MOLECULAR METHOD

A variety of factors will affect the choice of a method for the analysis of the taxonomic status of a range of *Aspergillus* strains. Ideally as many characters as possible should be analysed and molecular techniques should be combined with an analysis using a standard key as used by Klich and Pitt (1988). The approach then may be to make an epidemiological study of a large number of closely related isolates, or the requirement may be to analyse the phylogenetic relationships between a range of subgenera or species. Other factors such cost, speed, sensitivity and specificity will also influence the choice of methods. Fine distinctions between closely related strains will obviously need a finger printing techniques which shows subtle differences between strains. These techniques may be based on repetitive DNA or an amplification techniques using mismatch priming of polymerase chain reactions. Genetic fingerprinting is often an extension of RFLP analysis and this may be based on repetitive DNA or random probes. Attempts have been made to repeat the success of human genetic fingerprinting (Jefferys *et al*., 1991) using bacteriophage M13 probes and synthetic repeat oligonucleotides but only limited data is available for *Aspergillus* spp. (Meyer *et al*., 1991). These will need to be correlated with functional characters such as pathogenicity, vegetative compatibility and toxin production. Alternatively a fundamental phylogenetic study should analyse sequence data from an appropriate molecule which is expected to correlate with natural relationships between taxa. An example is the 16S-like ribosomal RNA subunit which has been analysed extensively in bacteria (Pace *et al*., 1986). Disagreement between the two methods would need to be resolved by the accumulation of further data as well as checking of the data already obtained.

Another approach depends on targeting a single gene or a pathway which has been shown to correlate with taxonomic groupings. A good example of this is the potential use of a probe based on a cloned gene for an enzyme in the pathway for the synthesis of aflatoxin in *Aspergillus flavus* (Payne, *et al.*, 1993). Alternatively, messenger RNA for the gene could be targeted by in situ hybridisation. This general approach can also be done on an empirical basis simply by preparing a genomic gene library and then taking sequences at random for use as a nucleic acid probe against DNA extracted from the strains under study. It may be possible to sequence such probes later and to identify the function of the gene by comparisons with nucleotide data bases. Finally to show fine distinctions between related strains it may be possible to produce genomic and cDNA libraries from different strains and to identify specific probes by subtraction hybridisation.

## THE RIBOSOMAL RNA GENE COMPLEX (rDNA)

Fungal rDNA is particularly useful as a target for hybridisation, as a source of nucleic probes and a source of sequence variation for taxonomic analysis. This is because the complex consists of about 100 copies which are tandemly repeated head to tail (Garber *et al.*, 1988) in which there are highly conserved regions in the coding regions and highly

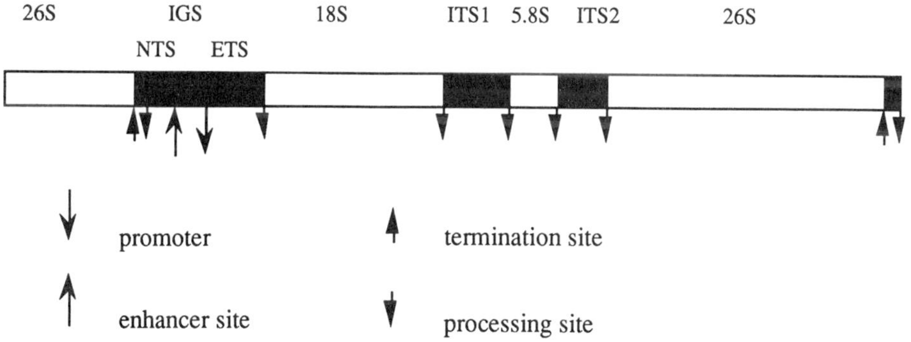


**Figure 1.** Diagramatic representation of the ribosomal RNA complex in filamentous fungi to illustrate the gene order and distribution of processing sites in the spacer regions. Note particularly the diversity of sites in the intergenic spacer region (IGS). NTS is the nontranscribed spacer, ETS is the external transcribed spacer and ITS1 and 2 are the two internal transcribed spacers.

variable sequences in the control and spacer regions (Sogin and Gunderson, 1987; Metzenberg, 1991). The multiple copies mean that hybridisation techniques are automatically one hundred times more sensitive than techniques where the target is a single copy gene. The conserved regions mean that rDNA probes can be used for the analysis of RFLPs in a variety of fungi, for example, probes from *Saccharomyces cerevisiae* rDNA will hybridise to *Aspergillus* spp. (Bainbridge *et al.*,1990). In general however homologous probes are to be preferred as fragments may not be detected due to poor homology in the variable regions. These variable or nonconserved regions can provide sequences for specific probes as well as giving sequence data for phylogenetic comparisons.

In *Aspergillus* spp., as in other filamentous fungi, the complex is organised as shown in Figure 1 with the order intergenic spacer (IGS), 18S, internal transcribed spacer (ITS1), 5.8S, ITS2 and 26S while the 5S subunit gene is located elsewhere in the genome (Lockington *et al.*, 1982). The IGS region can be subdivided further into the nontranscribed spacer (NTS) and the external transcribed spacer (ETS). The NTS contains transcription termination signals and enhancer sequence sites while the promoter region is located at the upstream end of the ETS region. A single polygenic RNA molecule is produced from the promoter, and processing of this molecule produces the corresponding ribosomal RNA subunits.

It is also possible to use probes and sequences based on the large (23S) and small (16S) ribosomal subunits from mitochondrial DNA but very few of these are available for analysis (Neefs *et al.*, 1991; Rijk *et al.*, 1992)

## Intergenic Spacer Region (IGS)

It is generally accepted that the IGS region is the most variable region of the complex and a number of RFLPs have their origin here (Spreadbury *et al.*, 1990; Metzenberg 1991; B. W. Bainbridge unpublished). The region is expected by analogy with higher organisms to contain species-specific sequences which interact with RNA polymerase I, transcription factors, enhancers and terminators (Dover and Flavell, 1984; Bell *et al.*, 1989). The origin of this variation is likely to be that there has been relatively recent co-evolution between the specific sequences and compensating alterations in the proteins which interact with these regions. Similar variation is unlikely to occur in the coding regions as there will be much stronger constraints on the interaction between the large number of ribosomal proteins and their subunit RNAs.

Heterogeneity of the IGS region has been detected in clinical isolates of *Aspergillus fumigatus* as minor RFLPs which were revealed after hybridisation to an 1.2kb IGS probe from *Aspergillus nidulans* (Spreadbury *et al.*, 1990). This probe probably contained 100 base pairs from the 3' end of the 26S and the rest was in the IGS region. We suggested that this showed that there was heterogeneity in the length of the IGS and that different copies of the complex in different isolates had a variable numbers of a 200 base pair repeat unit. Three different patterns could be detected in eleven different isolates allowing epidemiological surveys to be made. Length heterogeneity of the IGS region has also been demonstrated in *Yarrowia lypolitica* (Fournier *et al.*, 1986) and in *Pythium ultimum* (Buckho and Klassen, 1990). As the region is variable, it has proved possible to use nucleic acid probes from the IGS region for relatively specific hybridisation to DNA from *Aspergillus* species in slot blot experiments. The 1.2kb probe mentioned above hybridised to *Aspergillus* spp., and to *Penicillium* spp. but not to a range of other fungi including *Candida*, *Fusarium*, *Metarhizium*, *Rhizopus* or *Ustilago* spp. (C. L. Spreadbury unpublished). Evidentally some of the sequences, probably the 26S region, are shared between *Aspergillus* and *Penicillium*, and are sufficiently closely related to show cross hybridisation. It should prove possible to remove these sequences to obtain genus and species specific probes. Verma and Dutta (1987) have also found that nucleic acid probes taken from the IGS region are useful for studying phylogenetic relationships between species of *Neurospora*.

The IGS region has been found to have multiple promoters in higher organisms such as *Xenopus*, and *Triticum* and *Drosophila* (Tautz *et al.*, 1987). In the fungi, only limited sequence data is available for the IGS region in *Neurospora* ( Dutta and Verma 1990), *A. fumigatus* (C. L. Spreadbury, J. Cohen and B. W. Bainbridge unpublished), *Saccharomyces cerevisiae* (Skryabin *et al.*, 1984) and *Penicillium hordei* (T. P. Roberts and B. W. Bainbridge unpublished), but from this there is no evidence for multiple promoters. As more sequences become available it should prove possible to study evolutionary divergence between promoters and spacers in species of *Aspergillus* as has been done by Dover's group (Tautz *et al.*, 1987). However there is evidence for some conservation of processing signals such as termination sequences (CTCCC) which are similar in human, yeast, *Neurospora*, *A. fumigatus* and *P. hordei* IGS DNA (Dutta and Verma, 1990, T. P. Roberts, C. L. Spreadbury and B. W. Bainbridge unpublished). A comparison of the nucleotide sequences of the IGS regions of *A. fumigatus* and *P. hordei* has been used to design oligonucleotide primers for specific amplification and genetic fingerprinting of *Aspergillus* DNA (see below).

**Small Subunit Ribosomal RNA Sequences (18S)**

Bacterial taxonomists have argued successfully for the use of the 16S-like ribosomal RNA sequences as a basis for demonstrating phylogenetic and evolutionary relationships between a wide variety of organisms and organelles (Pace *et al.*, 1986). Basically they have argued that the ideal molecule for molecular taxonomy should not be transferred horizontally between organisms, it should have a constant function, it should be universally distributed, it should be large enough to show conserved and variable regions, the genes involved should provide a slow and reliable molecular clock which relates to genealogical time and, finally, it should be technically feasible to analyse. The 16S-like ribosomal RNA fulfilled these criteria and an analysis of this molecule has had a major impact on bacterial taxonomy. Predictions from the analysis of 16S RNA have been confirmed when metabolic pathways and wall structure of bacterial groups have been analysed. The sequences can be obtained directly from rRNA using reverse transcriptase, or the DNA can be cloned or amplified and then sequenced (see below).

In recent compilations of the complete sequences of fungal 16S-like ribosomal subunit (Neefs *et al.*, 1991; Rijk *et al.*, 1992) only 52 out of 927 sequences were from the fungi and only two species of *Aspergillus* were represented, the mitochondrial small subunit from *A. nidulans* and the 18S from *A. fumigatus*. A comparison has been made between the complete 18S sequences of *Candida* spp. and *A fumigatus* (Barns *et al.*, 1991), and between partial 18S sequences in *Aspergillus* species and their teleomorphs (Chang *et al.*, 1991) The latter group analysed 558 nucleotides in positions 384-563, 942-119 and 1419-1623. Their conclusions were that there was a major taxonomic break between *Aspergillus* and *Penicillium* and that the structural difference between an "aspergillum and a penicillus " was taxonomically significant (See Figure 2). Two analyses will illustrate their approach: firstly they used cluster analysis by the unweighted pair group method with arithmetic averages (UPGMA) which is not affected by adding outlying distantly related groups. Secondly they used a neighbor-joining (NJ) method which is influenced by using different outlying groups but which is considered to be a closer reflection of true evolutionary distances between organisms (Swofford and Olsen, 1990). *A. flavus*, *Aspergillus oryzae* and *Eurotium repens* formed a single cluster by both methods, and Chang *et al.*, (1991) suggested that *Aspergillus raperi* should be considered to be closer to *A. nidulans* rather than to the *Aspergillus versicolor* group as originally proposed. This is an interesting study but needs to be extended to a greater range of species before more general conclusions can be drawn.

The method used to compare sequences is crucial as variation in results can occur using different methods (Swofford & Olsen, 1990). Sequences are usually first lined up by eye using conserved sequences as markers. Obvious deletion and insertions are eliminated and comparisons made between the remaining bases. Percentage similarity or mutation fixed per base are then calculated by computer programmes. Hypervariable regions may be omitted although this is controversial. These pairwise comparisons are transformed into a matrix and various programmes are used to construct dendrograms and trees. Parsimony methods, using the smallest numbers of assumptions about changes in sequence, are often preferred and the aim is to produce a minimum tree length (Sober, 1989).

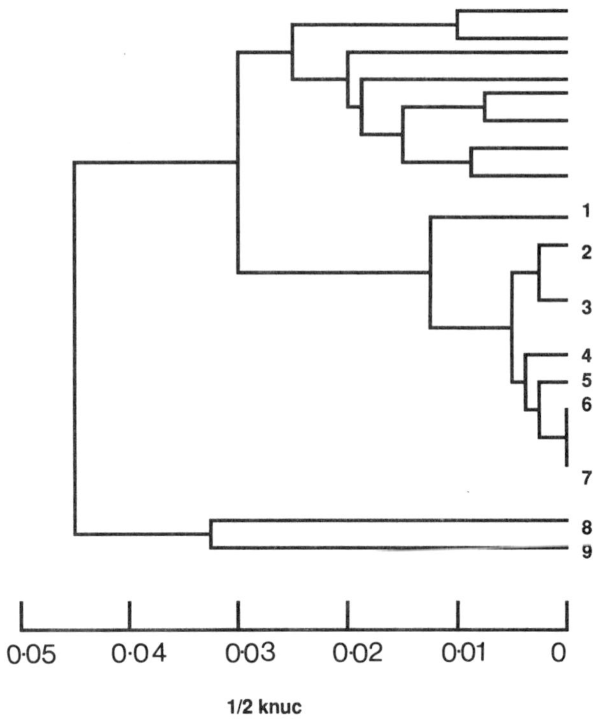


**Figure 2.** Phylogenetic tree for species and teleomorphs of *Aspergillus* and *Penicillium*. Based on partial nucleotide sequences from 18S ribosomal RNA subunit analysed by UPGMA. 1/2 knuc is the relative evolutionary distance. Modified and redrawn from Chang *et al.*, (1991). Key 1. *Warcupiella* (Ornati) ; 2. *Aspergillus raperi* ; 3. *Emericella nidulans* (Nidulantes); 4. *Sclerocleista ornata* (Ornati); 5. *Aspergillus fumigatus* (Fumigati) ; 6. *Aspergillus oryzae*; 7. *Aspergillus flavus* (Circumdati); 8. *Neurospora crassa*; 9. *Saccharomyces cerevisiae*

## Internally Transcribed Spacers (ITS) and 5.8S RNA

The length of the 18S region, normally about 1.8kb, means that accumulating total sequence data is time consuming. For this reason, some groups have chosen to study the ITS1-5.8S-ITS2 region which is shorter at about 900bp in length. The region also has the advantage that it has two variable spacer regions flanking a relatively conserved 5.8S region. The latter helps to align the sequences and aids comparisons as well as providing some limited taxonomic data. Unfortunately few sequences appear to be generally available although groups are known to be working on this region (Moens, 1992). One disadvantage of the region for the analysis of fine variation is that very few control and processing regions will be present in the ITS regions (see Figure 1). This is because processing sites are only required for removal of the ITS region during the production of the ribosomal RNA sub units. There will be no termination sites, enhancer sites or promoters, as are found in the IGS. There is evidence however from other fungi such as *Verticillium* spp. that there is sufficient variation to design species-specific primers for DNA amplification.

## SPECIFIC AMPLIFICATION OF DNA USING THE POLYMERASE CHAIN REACTION (PCR)

Many of the approaches already discussed have been revolutionised by PCR. The identification of conserved regions means that pairs of compatible primers can be used to amplify the ITS1-5.8S-ITS2 regions (White *et al.*, 1990) or the IGS region (B. W. Bainbridge unpublished see Figure 3). RFLPs can then be detected directly on gels following digestion of the amplified fragments with restriction enzymes. Alternatively they can be cloned and sequenced, or sequenced directly by asymmetric PCR. The error rate of Taq polymerase and of sequencing reactions inevitably means that there will be some errors and ideally amplifications and sequencing should be repeated and compared to give a consensus sequence (Eckert and Kunkel, 1991).

A more direct approach is to use a conserved primer and a nonconserved primer to directly amplify DNA from the target fungus. This has been done for *A. fumigatus* both from pure cultures and from clinical samples (Spreadbury *et al.*, 1993). There was no amplification from most other fungi tested. The reliability of this method will depend on the specificity of the primers used and in this case the primers were designed on the basis of a comparison of sequences from the 5' end of the IGS from *A. fumigatus* and *P. hordei* (C. L Spreadbury, T. P. Roberts, J. Cohen and B. W. Bainbridge unpublished data). A further check of the specificity of this method was made by blotting the amplified bands obtained, followed by hybridisation with a specific, homologous IGS probe. The method showed considerable specificity and potential as a clinical diagnostic method. In principle the method could be applied to detecting and identifying a variety of different species but this will depend on the availability of more sequence data.

## GENETIC FINGERPRINTING BY PCR AMPLIFICATION

The methods discussed so far have depended to varying extents on sequence data, restriction enzyme digestion, Southern transfer and hybridisation. The last two years has seen the discovery of an empirical method which can be rapidly applied to any fungus without previous knowledge of sequence data, gene probes or cloned DNA. The method generates useful fragment length polymorphisms (genetic fingerprints) which can analysed directly on agarose gels stained with ethidium bromide. Typically, this method relies on low temperature (30°- 45°) annealing of a single short primer (10-20 bases, with greater than 60% G+C) which produces mismatching on the target DNA on opposite strands. Multiple bands patterns are produced which have found a variety of uses from production of genetic maps to identification of fine distinctions between closely related strains.

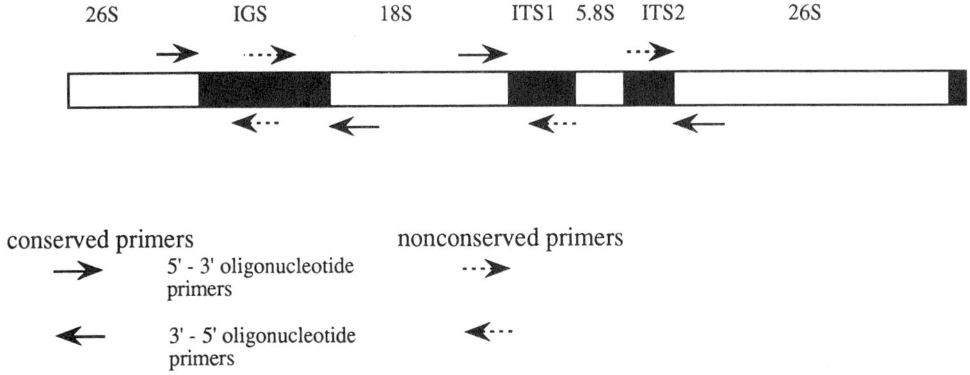


**Figure 3.** Diagram of the ribosomal RNA gene complex to illustrate the regions used for amplification of spacer regions. Non specific amplification can be obtained using conserved pairs of primers whereas specific amplification is obtained if at least one primer is nonconserved and specific to the target DNA.

The method has variously been called randomly amplified polymorphic DNA (RAPDs) (Williams *et al*., 1991) or arbitrarily primed polymerase chain reaction (AP-PCR) (Welsh *et al*., 1991). Neither of these terms are completely accurate particularly where the sequence of the primer is based on a known sequence and therefore the term amplified fragment length polymorphisms (AFLPs) will be used for any polymorphisms produce by these approaches.

Initially scientists using this method have worked through genuinely random sets of primers which are available commercially from companies such as Operon Inc. (USA). Typically only one out of three of these primers will give a useful number of fragments usually between four and ten bands of 300 bp to 4 kb in length. The origin of these bands is unknown but must relate to specific sequences in the genome. These could be conserved, repetitive or unique sequences. Some of these sequences may have taxonomic or epidemiological significance but this can only be ascertained by detailed study. The AFLPs on their own have no significance and it is only when a correlation can be shown, for example between the fingerprints and pathogenicity of a particular species, that they have value.

A variety of primers from the IGS region of *P. hordei* have been studied in this laboratory and several of these have been found to produce useful AFLPs. Primers initially developed to specifically amplify *P. hordei* DNA (T. P. Roberts and B. W. Bainbridge unpublished) have been shown to produce useful AFLPs at 37°- 45°C in a variety of species of *Aspergillus* (see Figure 4, J. Madeira and B. W. Bainbridge unpublished). It can be seen that all of the *A. flavus* isolates are very similar and show a similar pattern to *A. oryzae* . In addition three isolates of *A. parasiticus* have similar patterns whereas the other species show different fingerprints. The same primers have also been shown to be useful in *Penicillium* and *Fusarium* spp. (A. Kelly, J. B. Heale, E. R. Neto and B. W. Bainbridge unpublished). Aufavre-Brown *et al*. (1992) have shown the application of this technique to distinguish clinical isolates of *A. fumigatus*. Twenty one out of forty four decamers gave AFLPs which were difficult to interpret or were unreliable whereas one primer of sequence GTATTGCCCT gave different AFLPs with nine independent isolates.

Little work seems to have been done on assessing the potential of AFLPs for taxonomic or phylogenetic analysis. The fragments produced may or may not represent independent characters as required for taxonomic analysis as discussed earlier. For example two fragments may have one site in common with a variable second site giving perhaps 100% homology over a length of DNA but also a length polymorphism. The two fragments amplified would not strictly speaking be totally independent. Some workers have weighted brighter bands on the assumption that they must have a higher homology with the primer. Assuming that AFLPs can be treated as independent, similarity coefficients can be calculated from presence/absence matrices. Relationships between, for example, pathotypes has been analysed in *Leptosphaeria maculans* by Goodwin and Annis (1991). It is obviously essential that these analyses are compared with the results from traditional taxonomic methods so that an assessment can be made of their general reliability. Ideally the AFLPs should be based on known sequences with known phylogenetic significance. Although this method is simple and quick it is inevitably sampling only a small proportion of the total genome. Therefore a variety of primers need to be chosen with great care if reliable taxonomic data are to be obtained.

## FUTURE POTENTIAL OF MOLECULAR TAXONOMIC METHODS

The future potential of molecular taxonomic methods for *Aspergillus* can be assessed by looking at successes in other groups. For example it has proved possible to generate AFLPs in *Fusarium oxysporum* f.sp. *curcurbitae* which could distinguish between mating populations and races (Crowhurst *et al*., 1991). DNA fragments which differentiated between the mating types were used as gene probes and were found to hybridise specifically to one mating type or the other. We have also isolated symptom-associated probes from AFLPs generated in *F. oxysporum* f.sp.*ciceris* (A. Kelly, J. B. Heale and B. W. Bainbridge, unpublished). It follows that at least some of the bands in AFLPs may be related to species-specific or subgenera-specific groupings. A wider screening of AFLPs and derived gene probes will be required to test this idea.

An alternative approach with a less empirical base is to build up a data base of sequences for target molecules such as the 18S, IGS and ITS1-5.8S-ITS2 regions. Computer comparisons of these regions from a variety of *Aspergillus* spp. should allow the design of primers or gene probes for whatever level of distinction is required. The approach is basically that of Giovannoni *et al.* (1988) and Amann *et al.* (1990) who have designed oligonucleotides at a variety of levels of discrimination particularly for *in situ* hybridisation to cells. This method could also be extended either to taxonomic amplification of DNA using specific primers or to hybridisation of end-labelled oligonucleotides to slot blots using fungal genomic DNA.

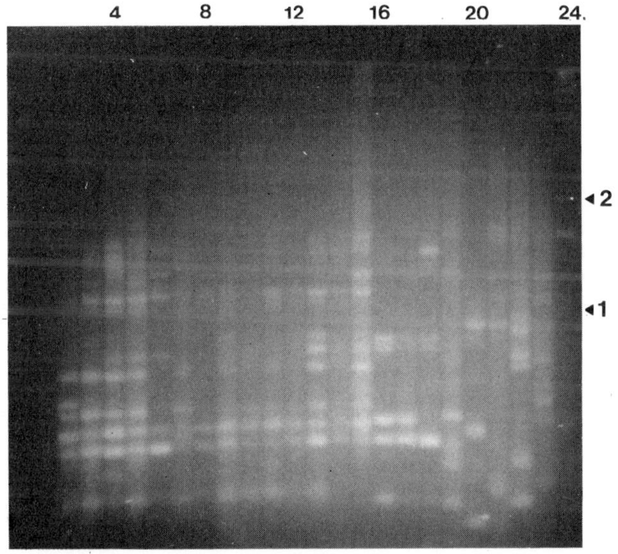


**Figure 4.** Genetic fingerprints produced by amplified fragment length poylmorphisms (AFLPs) or RAPDs. Lane 1 is a no DNA control; lanes 2 - 14 different isolates of *A.flavus*, Lane 15 *A.oryzae*; lanes 16 - 18 *A.parasiticus*; lane 19 *A.amstelodami*.; lane 20 *A.candidus*; lane 21 *A.fumigatus*; lane 22 *A.nidulans*; lane 23 *A.terreus* and lane 24 1kb ladder. The primer used was P2 based on the IGS region of *Penicillium hordei* (T. P. Roberts and B. W. Bainbridge, unpublished data). (Photograph by courtesy of Jovita Madeira).

A final illustration of the success of molecular methods is the analysis of *Pneumocystis carnii*, an unculturable organism associated with pneumonia in AIDS patients. Initially this organism was thought to be a protozoan but sequencing of the 18S region showed closer relationships with the fungi (Edman *et al.*,1988). More recently this study has been extended to a taxonomic grouping by amplifying and sequencing the large ribosomal RNA subunit from the mitochondrial genome of *P.carnii* and a wide variety of other fungi (Wakefield *et al.*, 1992). The organisms showed a close relationship with the ustomycetous red yeasts. An extension of this approach seems very promising and will become easier as the sequence data base is increased.

# REFERENCES


Amann, R.I., Krumholz, L. and Stahl, D.A. (1990) Fluorescent-oligonucleotide probing of whole cells for determinative, phylogenetic and environmental studies in microbiology. J. Bact. 172, 762-770.

Aufauvre-Brown, A., Cohen, J. and Holden, D.W. (1992) Use of randomly amplified polymorphic DNA markers to distinguish isolates of *Aspergillus fumigatus*. J. Clinical Micro. 30, 2991-2993.

Bainbridge, B.W., Spreadbury, C.L., Scalise, F.G. and Cohen, J. (1990) Improved methods for the preparation of high molecular weight DNA from large and small scale cultures of filamentous fungi. FEMS Micro. Lett. 66, 113-118.

Barns, S.M., Lane, D.J., Sogin, M.L., Bibneau, C. and Weisberg, W.G . (1991) Evolutionary relationships among *Candida* species and relatives. J. Bact. 173, 2250-2255.

Bell, S.P., Pikaard, C.S., Reeder, R.H. and Tjian, R. (1989) Molecular mechanisms governing species-specific transcription of ribosomal RNA. Cell 59, 489-497.

Buchko, J. and Klassen, G.R. (1990) Detection of length heterogeneity in the ribosomal DNA of *Pythium ultimum* by PCR amplification of the intergenic region. Curr. Genet. 18, 203-205.

Chang, J., Oyaizu, H. and Sugiyama, J (1991) Phylogenetic relationships among eleven selected species of *Aspergillus* and associated teleomorphic genera estimated from 18S ribosomal RNA partial sequences. J. Gen. Appl. Microbiol. 37, 289-308.

Crowhurst, R.N., Hawthorne, B.T., Rikkerink, E.H.A. and Templeton, M.D. (1991) Differentiation of *Fusarium solani* f.sp. *cucurbitae* races 1 and 2 by random amplification of polymorphic DNA. Curr. Genet. 20, 391-396.

Dover, G.A. and Flavell, R.B. (1984) Molecular co-evolution : DNA divergence and the maintenance of function. Cell 38, 622-623.

Dutta, S.K. and Verma, M. (1990) Primary structure of the nontranscribed spacer region and flanking sequences of the ribosomal DNA of *Neurospora crassa* and comparison with other organisms. Biochem. Biophys. Res. Comm. 170, 187-193.

Eckert, K.A. and Kunkel, T.A. (1991) The fidelity of DNA polymerases used in the polymerase chain reactions in "PCR a practical approach" (McPherson, M.J., Quirke, P. and Taylor, G.R. Eds.) pp 225-244. IRL Press, Oxford. Edman, J.C.

Edman, J. C., Kovacs, J. A., Masur, H., Santi, D. V., Elwood, H. J. and Sogin, M. L. (1988) Ribosomal RNA sequence shows *Pneumocystis carnii* to be a member of the fungi. Nature, 334, 519-522.

Fournier, P., Gaillardin, C., Persuy, M., Klootwijk, J. and van Heerikhuizen, H. (1986) Heterogeneity in the ribosomal family of the yeast *Yarrowia lipolytica* : genomic organisation and segregation analysis. Gene 42, 273-282.

Garber, R.C., Turgeon, B.G., Selker, E.U. and Yoder, O.C. (1988) Organisation of the ribosomal RNA genes in the fungus *Cochliobolus heterotrophus*. Curr. Genetics 14, 573-582.

Giovannoni, S.J., DeLong, E.F., Olsen, G.J. and Pace, N.R. (1988) Phylogenetic group-specific oligodeoxynucleotide probes for identification of single microbial cells. J. Bact. 170 720-726.

Goodwin, P.H. and Annis, S.L. (1991) Rapid identification of Genetic Variation and pathotype of *Leptosphaeria maculans* by Random Amplified Polymorphic DNA assay. Appl. Environ. Micro. 57, 2482-2486.

Jefferys, A.J., MacLeod, A., Tanaki, K., Neil, D.L. and Monckton, D.G. (1991) Minisatellite repeat coding as a digital approach to DNA typing. Nature, 354, 201-209.

Klich, M.A. and Pitt, J.I. (1988) A laboratory guide to common *Aspergillus* species and their teleomorphs. Commonwealth Scientific and Industrial Research Organisation, North Ryde, New South Wales, Australia.

Kovacs, J.A., Masur, H., Santi, D.V., Elwood, H.J. and Sogin, M.L. (1988) Ribosomal RNA sequence shows *Pneumocystis carnii* to be a member of the fungi. Nature, 334, 519-522.

Lockington, R.A., Taylor, G.G., Winther, M., Scazzocchio, C. and Davies, R.W. (1982) A physical map of the ribosomal DNA repeat unit of *Aspergillus nidulans*. Gene 20, 135-137.

Metzenberg, R.L. (1991) Benefactors' lecture: the impact of molecular biology on mycology. Mycol. Res. 95, 9-13.

Meyer, W., Koch, A., Niemann, C., Beyermann, B., Epplen, J.T. and Borner,T. (1991) Differentiation of species and strains among filamentous fungi by DNA fingerprinting. Curr. Genet. 19, 239-242.

Moens, W. (1992) Fast design of Fungal PCR Markers. Proceedings of the Ist European Conference on Fungal Genetics at Nottingham August 1992 P2/58.

Neefs, J., de Peer, Y., de Rijk, P., Goris, A. and de Wachter, R. (1991) Compilation of small ribosomal subunit RNA sequences. Nuc. Acids Res. 19 suppl., 1987-2015.

Pace, N.R., Olsen, G.J. and Woese, C.R. (1986) Ribosomal RNA phylogeny and the primary lines of evolutionary descent. Cell 45, 325- 326.

Payne, G.A., Nystrom, G.J., Bhatnagar, D. Cleveland, T.E. and Woloshuk, C.P. (1993) Cloning of the afl-2 gene involved in aflatoxin biosynthesis from *Aspergillus flavus*. Appl. Environ. Micro. 59, 156-162.

Rijk, de ,P., Neefs, J., Van de Peer, Y. and de Wachter, R. (1992) Compilation of small ribosomal RNA sequences. Nuc. Acids Res. 20 suppl. 2075-2089.

Skryabin, K.G., Eldarov, M.A., Larionov, V.I., Bayev, A.A., Klootwijk, J., Regt, V.C.H.F., Veldman, G.M., Planta, R.J., Georgiev, O.I. and Hadjiiolov, A.A. (1984) Structure and function of the nontranscribed spacer region of yeast rDNA. Nuc. Acid. Res.12, 2955-2968.

Sober, E. (1989) Reconstructing the past: Parsimony, evolution, and inference. MIT Press, Cambridge, MA, USA. Sogin, M.L. and Gunderson, J.H. (1987) Structural diversity of eukaryotic small subunit ribosomal RNAs. Evolutionary implications. Annals N.Y. Acad.Sciences 503, 125-139.

Spreadbury, C.L., Bainbridge, B.W. and Cohen, J. (1990) Restriction fragment length polymorphisms in isolates of *Aspergillus fumigatus* probed with part of the intergenic spacer region from the ribosomal RNA gene complex of *Aspergillus nidulans* . J. Gen. Microbiol. 136, 1991-1994.

Spreadbury, C.L., Holden, D., Aufauvre-Brown, A., Bainbridge, B.W. and Cohen, J. (1993) Detection of *Aspergillus fumigatus* by the polymerase chain reaction. J. Clin. Microbiol. 31, 615-621.

Swofford, D.L. and Olsen, G.J. (1990) Phylogeny reconstruction in "Molecular Systematics" (Hillis, D.M. and Moritz, C., Eds.) pp. 411-501. Sinhauer, Sunderland, Mass, U.S.A.

Tautz, D., Tautz, C., Webb, D. and Dover, G.A. (1987) Evolutionary divergence of promoters and spacers in the rDNA family of four *Drosophila* species. Implications for molecular co-evolution in multigene families. J. Mol. Biol. 195, 525-542.

Verma, M. and Dutta, S.K. (1987) Phylogenetic implications of the heterogeneity of the nontranscribed spacer of rDNA repeating unit in various *Neurospora* and related fungal species. Curr. Genet. 11, 309-314.

Wakefield, A.E., Peters, S.E., Banjeri, S., Bridge, P.D., Hall, G.S., Hawksworth, D.L., Guiver, L.A., Allen, A.G. and Hopkin, J.M. (1992) *Pneumocystis carnii* shows DNA homology with the usomycetous red fungi. Molec. Microbiol. 6, 1903-1911.

Welsh, J., Petersen, C. and McClelland, M. (1991) Polymorphisms generated by arbitrarily primed PCR in the mouse: application to strain identification and genetic mapping. Nucl. Acid Res. 19, 303- 306.

White, T.J., Bruns, T., Lee, S. and Taylor, J. (1990) Amplification and direct sequencing of fungal ribosomal RNA genes for phylogenetics, in "PCR Protocols: a Guide to Methods and Applications" (Innis, M.A., Gelfand, D.H., Sninsky, J.J. and White, T.J. Eds.) pp. 315-322. Acad. Press, Inc. San Diego.

Williams, J.G.K., Kubelik, A.R., Livak, K.J., Rafalski. J.A. and Tingey, S.V. (1991) DNA polymorphisms amplified by arbitrary primers are useful as genetic markers. Nuc. Acid Res. 18, 6531-6535.

# *ASPERGILLUS* TOXINS AND TAXONOMY

Zofia Kozakiewicz

International Mycological Institute
Bakeham Lane
Egham, Surrey TW20 9TY

## INTRODUCTION

Members of the genus *Aspergillus* and its close relative *Penicillium*, are the dominant fungal contaminants of stored products, foods and feedstuffs. Both genera produce harmful mycotoxins and therefore, correct identification at the species level is of paramount importance.

Literature on mycotoxins is vast, with emphasis being placed on toxicology and chemistry rather than the identification of toxin-producing fungal species. A common mistake which has resulted, is that when fungal isolations and mycotoxin analyses are made on a particular product, the fungi isolated therein are assumed to be responsible for the mycotoxin recovered (Table 1). Where names have been used, very few such identifications have been sent for verification by an expert.

Much of the literature on mycotoxins stems from studies undertaken in the laboratory, and although *Aspergillus* species produce a range of toxins *in vitro*, but how many of these mycotoxins actually occur *naturally*? In order to answer such a question, a comprehensive multidisciplinary approach is required to firmly establish the correlation between *Aspergillus* species and the toxins which they produce. Fortunately, the last decade has seen a revolution in new approaches to *Aspergillus* taxonomy (Samson and Pitt, 1985, 1990; Kozakiewicz, 1989). The combination of using traditional morphological characters with results from the new techniques has given some groups a more stable taxonomy.

## NEW APPROACHES TO *ASPERGILLUS* TAXONOMY

Since these new approaches to *Aspergillus* taxonomy will be discussed in greater detail elsewhere in this book, only a brief summary will be given here.

Several biochemical and physiological techniques have been introduced to improve *Aspergillus* taxonomy. These include isoenzyme patterns (Cruickshank and Pitt, 1990; Yamatoya *et al.*, 1990), ubiquinone systems (Kuraishi *et al.*, 1990), molecular techniques

*The Genus Aspergillus*, Edited by Keith A. Powell *et al.*,
Plenum Press, New York, 1994

**Table 1.** Species of *Aspergillus* and teleomorphs reported to produce sterigmatocystin.

| | |
|---|---|
| *A. ustus** | *A. versicolor* |
| *Eurotium chevalieri** | *Emericella nidulans* |
| *E. repens** | *E. quadrilineata* |
| *E. rubrum** | *E. rugulosa* |
| | *E. unguis* |

* Species not known to produce the toxin.

(Kozlowski and Stepien, 1982; Croft *et al.*, 1990; Moody and Tyler, 1990*a,b*; Mullaney and Klich, 1990; Kusters van Someren *et al.*, 1991; Varga *et al.*, 1993), secondary metabolites (Frisvad, 1985, 1989), latex agglutination tests (Stynen *et al.*, 1992) and scanning electron microscopy (SEM) (Udagawa and Takada 1985; Kozakiewicz, 1989; Samson *et al.*, 1990). Selective media have been devised for isolation and direct identification and notable amongst these for aspergilli is AFPA (*Aspergillus flavus* and *Aspergillus parasiticus* agar, Pitt *et al.*, 1983).

The value of such multidisciplinary approaches is gaining acceptance and is improving our understanding of mycotoxigenic species delimitations. Indeed, an international working group "The International Commission on *Penicillium* and *Aspergillus*" (ICPA), was established in 1986 and is undertaking studies of these two important genera. Participants include taxonomists using morphological techniques, as well as specialists from the fields of physiology and biochemistry.

However, despite these numerous advances, misidentifications and mistakes still occur, usually because the researchers concerned have been inadequately trained in the fields of taxonomy and mycotoxicology; misidentifications have also been made by experts.

## MISIDENTIFICATIONS

The ability to produce cyclopiazonic acid (CPA) has been reported in some *Penicillium* species, notably *Penicillium commune* and *Penicillium griseofulvum, Aspergillus tamarii* and *A. flavus* (Kurata, 1978; Dorner, 1983). In addition, this mycotoxin was attributed to *Aspergillus versicolor* (Ohmomo *et al.*,1973). However, Domsch *et al.* (1980) showed that Ohmomo's isolate was a misidentification. According to Domsch *et al.* (1980) the correct identification was *Aspergillus oryzae*, a known producer of CPA. Unfortunately, the mistake continues to be cited in the literature, even in substantial texts (Davis and Diener, 1987; Golinski, 1991; Smith and Ross, 1991).

An important group within the aspergilli are the aflatoxin producers, *A. flavus* and *A. parasiticus.* Isolates of these are maintained in all the major world collections and are used extensively for reference and as verified isolates for mycotoxin research. The integrity of isolate labels associated with such collections is rarely if ever questioned.

*A. parasiticus* was originally described from material isolated from sugar cane grown in Hawaii (Speare, 1912). Subsequently his original isolate was subcultured and distributed around the world. Scanning electron microscopy of all extant isolates revealed conidia of two distinct ornamentations or morphs (Figures 1*a* and 1*b*). Isolates always consisted of one or other of these two morphs but never mixtures of both (Kozakiewicz, 1982). When these morphs were attributed to their respective cultures, a sharp dichotomy was revealed. One form occurs in the isolate derived from the type of *A. parasiticus* (see Figure 1*a)* whilst the other has been established as that of *A. flavus* (see Figure 1*b*) (Kozakiewicz, 1982). In other words, cultures from the type of *A. parasiticus* held in three major world collections are in fact *A. flavus*. There has been no confirmation that this mistake has been corrected other than at the IMI, Egham, Surrey, England. Misidentifications of these two species are very prevalent in cultures held in many collections. In routine examinations of IMI cultures ten isolates have been re-identified to date.

Such mistakes must lead to invalid conclusions and confusion concerning research into mycotoxin production. A pertinent example from the literature (Mirocha and Christensen, 1982) supports this argument:

> "In the United States, *A. flavus* is commonly assumed to be found on field corn and to be responsible predominantly for the production of $B_1$ and $B_2$. *A. parasiticus* is assumed to be usually found on peanuts and to produce the full complement of aflatoxin $B_1$, $B_2$, $G_1$ and $G_2$. Some investigators do not fully agree with this, claiming that *A. parasiticus* is also found on maize and produces its corresponding derivatives."

Misidentifications of *A. flavus* on the scale mentioned above may well account for such conflicts of opinion. A more serious situation must operate outside the areas of taxonomic expertise where not only will the level of competent identification be lower but sadly such inaccuracies may lead to a total disassociation with the importance of the species as a biological unit. Again Mirocha and Christensen (1982) provide a suitable corroboration:

> "Most scientists other than mycologists simply identify this species [*A. parasiticus*] as *A. flavus* without bothering to determine any fine structure ..."

Then speaking, presumably as chemists, they go on to say:

> ".....and for most determinations this is sufficient."

The recent description of a new species, *Aspergillus nomius*, has further complicated the situation (Kurtzman *et al.*, 1987). Morphologically it resembles *A. flavus*, but differs by producing smaller more elongate sclerotia, those in *A. flavus* being more globose, and by the production of aflatoxin $B_1$, $B_2$, $G_1$, $G_2$ and a unique metabolite, nominine (Klich and Pitt, 1988; Liljegren *et al.*, 1988; Samson and Frisvad, 1991). *A. flavus* produces only aflatoxin $B_1$ and $B_2$. An examination of *A. flavus* cultures in IMI showed one to be attributable to *A. nomius* (Klich and Pitt, 1988). Others must exist.

The secondary metabolites which distinguish *A. flavus* and each of its related species are listed in Table 2. Included are *A. oryzae, A. sojae,* and *A. tamarii*, species used in the food fermentation industry and considered to be "safe" in that they do not produce aflatoxins. *A. oryzae* is thought to be a "domesticated" form of *A. flavus*, and *A. sojae* a "domesticated" form of *A. parasiticus* (see Figures 1*a*,1*b*,1*c*,1*d*), since they are morphologically identical but do not produce the corresponding aflatoxins (Kozakiewicz, 1985). *A. tamarii* (Figure 1e) is a distinct species which produces kojic and cyclopiazonic acids, but additionally produces aspirochlorin, canadensolide and fumigaclavine A (Frisvad and Samson, 1991).

A classic misidentification within the section *Flavi* concerns the attribution of the production of aflatoxin to an isolate of *A. oryzae* (El-Hag and Morse, 1976). A protracted published argument ensued between the authors of this statement and the curator of the donor collection (Fennell, 1976). On subsequent examination of a sample returned to the donor institute, the curator declared it to be a pure culture of *A. parasiticus* heavily infested with mites, thus accounting for the aflatoxin production.

Nevertheless, the argument remained unresolved and continued to appear in the literature (Morse, 1976), the mycotoxin researchers insisting they were working with a mycotoxin producing variant of *A. oryzae*. An SEM examination of an isolate grown from the batch culture in dispute revealed the presence of two conidial forms (Kozakiewicz, 1986). One was identifiable as *A. parasiticus* and the other as *A. oryzae* (Figure 1). Thus, the isolate was found to be mixed, the question is entirely resolved.

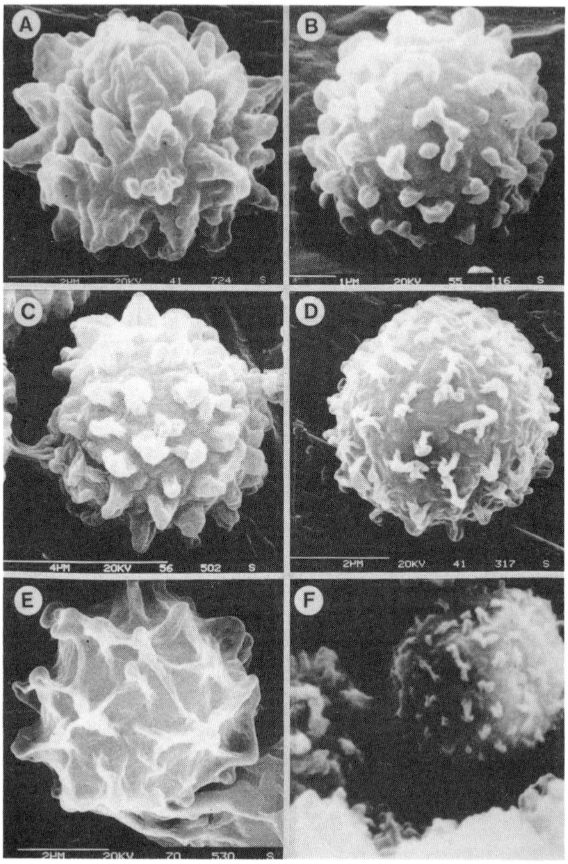


**Figure 1.** Scanning electron micrographs illustrating conidial ornamentation in *A. flavus* and related species.
**A**. *A. parasiticus* **IMI** 15957*vi*; **B**. *A. flavus* **IMI** 124930; **C**. *Aspergillus sojae* **IMI** 191300; **D**. *Aspergillus oryzae* **IMI** 126842; **E**. *Aspergillus tamarii* **IMI** 91888; **F**. Contaminated culture of *A. oryzae* **NRRL** 1988

## NATURAL OCCURRENCE OF MYCOTOXINS

Despite the fact that aspergilli readily produce a variety of secondary metabolites on artificial media in the laboratory, few of these occur naturally on food or feed products. A list of important *Aspergillus* species and their mycotoxins is provided (see Table 3).

**Table 2.** Metabolites produced by *A. flavus* and related species (Samson and Frisvad 1991)

| Species | Afl.B | Afl.G | CPA | KA | Nom. |
|---|---|---|---|---|---|
| *A. flavus* | + | - | + | + | - |
| *A. parasiticus* | + | + | - | + | - |
| *A. nomius* | + | + | - | + | + |
| *A. oryzae* | - | - | (+) | + | - |
| *A. sojae* | - | - | - | + | - |
| *A. tamarii* | - | - | + | + | - |

Afl. B = Aflatoxin $B_1$, $B_2$
Afl. G = Aflatoxin $G_1$ , $G_2$
Nom. = nominine
CPA = cyclopiazonic acid
KA = kojic acid

Aflatoxins are common on products high in oils, such as groundnuts, Brazil nuts, pistachio, almonds, walnuts and other edible nuts, cotton seed, copra and cereals from warmer climates such as maize, sorghum, and millet (Moss, 1991). Aflatoxin is not found in significant quantities on soybeans (Moss, 1991).

Sterigmatocystin, produced mainly by *A. versicolor* and *Emericella nidulans*, is found in small grains such as wheat, barley, and rice, in animal feedstuffs and wheat and oat based breakfast cereals (Yoshizawa, 1991), spices and acid treated bread (Frisvad, 1988).

Ochratoxins, most commonly ochratoxin A, have been isolated world-wide from many products including cereals (maize, barley, oats, rye and sorghum), also animal feedstuffs (Smith and Ross, 1991), breakfast cereals, Japanese rice (Yoshizawa, 1991) and European lagers (Payne *et al.*, 1983). Ochratoxin A is produced by *Aspergillus ochraceus* mainly in hot climates, and by *Penicillium verrucosum* in temperate climates. It has been implicated in Balkan endemic nephropathy (Pavlovic *et al.*, 1979).

CPA is produced by *A. flavus* and most strains of *A. tamarii* (Frisvad and Samson, 1991). It has been detected in nuts, spices, meat and eggs (Frisvad, 1988), and naturally contaminated agricultural products, where it has been noted to co-occur with aflatoxin (Smith and Ross, 1991).

Patulin is produced by *Aspergillus clavatus* and *Aspergillus terreus*. *A. clavatus* is a common contaminant of malting barley and is the causal agent of the allergic respiratory disease known as malt worker's lung (Flannigan, 1986).

## CONCLUSIONS

Contamination of food and feedstuffs by *Aspergillus* species and their toxic metabolites is a serious problem world-wide. They have adverse effects on animal and human health and cause economic problems for international trade, in particular that of developing countries. Despite the wealth of literature on this topic, much more basic survey and research work needs to be done in order to determine the true extent of fungal and mycotoxin contamination. Therein lies the problem. Such research programmes require adequately trained personnel. Unfortunately, fungal taxonomy is not seen as a "high tech" research area and therefore does not attract adequate funding.

**Table 3.** Important mycotoxins produced by *Aspergillus* species and their teleomorphs.

| Species | Mycotoxins |
|---|---|
| *A. clavatus* | cytochalasin E, patulin, tryptoquivalins |
| *A. fumigatus* | fumigaclavines, fumigatin, fumitoxins, fumitremorgins, gliotoxin, tryptoquivalins, verruculogen |
| *A. restrictus* | mitogillin |
| *Eurotium amstelodami* | physcions |
| *E. chevalieri* | physcions |
| *E. repens* | physcions |
| *E. rubrum* | physcions |
| *A. terreus* | citreoviridin, citrinin, patulin, territrems |
| *A. versicolor* | nidulotoxin, sterigmatocystin |
| *Emericella nidulans* | nidulotoxin, sterigmatocystin |
| *A. candidus* | terphenyllin, xanthoascin |
| *A. flavus* | aflatoxin $B_1$, $B_2$, aflatrem, kojic acid maltoryzin, 3-nitropropionic acid |
| *A. nomius* | aflatoxin $B_1$, $B_2$, $G_1$, $G_2$, kojic acid |
| *A. parasiticus* | aflatoxin $B_1$, $B_2$, $G_1$, $G_2$, kojic acid 3-nitropropionic acid |
| *A. tamarii* | cyclopiazonic acid, fumigaclavine A, kojic acid |
| *A. niger* | malformins, naphthopyrones |
| *A. ochraceus* | ochratoxin, penicillic acid, secalonic acid A, xanthomegnin, viomellein |

The few specialists are to be found in Europe and North America, and not in developing countries where many of the mycotoxin problems occur. In order that misidentifications on the scale discussed in this paper do not continue to occur in the future, the following is proposed:

1. Surveys conducted in order to assess species-toxin relationships for a wide variety of commodities.

2. The methodology for identification and enumeration of mycotoxigenic fungi in foods and feedstuffs to be standardised.

3. Fungal isolates used in taxonomic, biochemical and toxicological studies to be deposited in recognised culture collections, where they can be kept under optimal conditions.

4. The identification of such isolates to be verified by a specialist.

5. ICPA to regularly publish a list of misidentifications taken from the literature.

6. Information on isolates and their properties to be stored in databases.

7. Isolates of commercial value to be safety-deposited.

8. The creation of small research collections in developing countries, suitable to the local needs and resources.

9. Adequate funding to be available to maintain such collections where they already exist, so that important isolates are not lost.

10. Personnel running such collections to be properly trained.

11. Communication between laboratories in developing and developed countries to be improved.

## INTERNATIONAL COMMISSION ON *PENICILLIUM* AND *ASPERGILLUS* (ICPA)

ICPA is already carrying out collaborative taxonomic studies and revising culture collection names. In addition, it is collating a set of reference cultures for *Aspergillus* and *Penicillium* toxigenic species with morphological and secondary metabolite profile descriptions. Such publications will be of use to all mycologists, chemists and toxicologists who need to accurately identify species within these two important genera.

## REFERENCES


Croft, J.H., Bhattacherjee, V. and Chapman, K.E. (1990) RFLP analysis of nuclear and mitochondrial DNA and its use in *Aspergillus* systematics, in "Modern Concepts in *Penicillium* and *Aspergillus* Systematics" (Samson, R.A. and Pitt, J.I., Eds.), pp. 309-320. Plenum Press, New York.

Cruickshank, R.H. and Pitt, J.I. (1990) Isoenzyme patterns in *Aspergillus flavus* and closely related species, in "Modern Concepts in *Penicillium* and *Aspergillus* Systematics" (Samson, R.A. and Pitt, J.I., Eds.), pp. 259-265. Plenum Press, New York.

Davis, N.D. and Diener, U.L. (1987) Mycotoxins, in "Food and Beverage Mycology" (Beuchat, L.R., Ed.), pp. 517-570. Avi, New York.

Domsch, K.H., Gams, W. and Anderson, T.H. (1980) "Compendium of Soil Fungi," pp. 76-124. Academic Press, London.

Dorner, J.W. (1983) Production of cyclopiazonic acid by *Aspergillus tamarii* Kita. Appl. environ. Microbiol. 46, 1435-1437.

El-Hag, N. and Morse, R.E. (1976) Aflatoxin production by a variant of *Aspergillus oryzae* (NRRL 1988) on cowpeas (*Vigna sinensis*). Science 192, 1345-1346.

Fennell, D.I. (1976) *Aspergillus oryzae* (NRRL 1988): A clarification. Science 194, 1188.

Flannigan, B. (1986) Mycotoxins and the fermentation industries, in "Spoilage and Mycotoxins of Cereals and other Stored Products" (Flannigan, B., Ed.), pp. 109-114. CAB International, Wallingford.

Frisvad, J.C. (1985) Secondary metabolites as an aid to *Emericella* classification, in "Advances in *Penicillium* and *Aspergillus* Systematics" (Samson, R.A. and Pitt, J.I., Eds.), pp. 437-444. Plenum Press, New York.

Frisvad, J.C. (1988) Fungal species and their specific production of mycotoxins, in "Introduction to Food-borne Fungi" (Samson, R.A. and Reenen-Hoekstra, E.S., Eds.), pp. 239-249. CBS, Baarn.

Frisvad, J.C. (1989) The connection between the penicillia and aspergilli and mycotoxins with special emphasis on misidentified isolates. Arch. Environ. Contam. Toxicol. 18, 452-467.

Frisvad, J.C. and Samson, R.A. (1991) Mycotoxins produced by species of *Penicilllium* and *Aspergillus* occurring in cereals, in "Cereal Grain Mycotoxins, Fungi and Quality in Drying and Storage" (Chelkowski, J., Ed.), pp. 441-476. Elsevier, Amsterdam.

Golinski, P. (1991) Secondary metabolites (mycotoxins) produced by fungi colonizing cereal grain in store - structure and properties, in "Cereal Grain Mycotoxins, Fungi and Quality in Drying and Storage" (Chelkowski, J., Ed.), pp. 355-403. Elsevier, Amsterdam.

Klich, M.A. and Pitt, J.I. (1988) Differentiation of *Aspergillus flavus* from *Aspergillus parasiticus* and other closely related species. Trans. Br. mycol. Soc. 91, 99-108.

Kozakiewicz, Z. (1982) The identity and typification of *Aspergillus parasiticus*. Mycotaxon 15, 293-305.

Kozakiewicz, Z. (1985) Solutions to some problems in *Aspergillus* taxonomy using the scanning electron microscope, in "Advances in *Penicillium* and *Aspergillus* Systematics" (Samson, R.A. and Pitt, J.I., Eds.), pp. 351-361. Academic Press, New York.

Kozakiewicz, Z. (1986) New developments in the accurate identification of aspergilli in stored products, in "Spoilage and Mycotoxins of Cereals and other Stored Products" (Flannigan, B., Ed.), pp. 115-122. CAB International, Wallingford.

Kozakiewicz, Z. (1989) "*Aspergillus* Species on Stored Products." Mycological Pap. 161, pp. 1-161. CAB International, Wallingford.

Kozlowski, M. and Stepien, P.R. (1982) Restriction enzyme analysis of mitochondrial DNA of members of the genus *Aspergillus* as an aid in taxonomy. J. gen. Microbiol. 128, 471-476.

Kuraishi, H., Itoh, M., Tsuzaki, N., Katayama, Y., Yokoyama, T. and Sugiyama, J. (1990) Ubiquinone systems as a taxonomic tool in *Aspergillus* and its teleomorphs, in "Modern Concepts in *Penicillium* and *Aspergillus* Systematics" (Samson, R.A. and Pitt, J.I., Eds.), pp. 407-421. Academic Press, New York.

Kurata, H. (1978) Current scope of mycotoxin research from the viewpoint of food mycology, in "Toxicology, Biochemistry and Pathology of Mycotoxins" (Uraguchi, K. and Yamazaki, M., Eds.), pp. 13-17. John Wiley and Sons, New York.

Kurtzman, C.P., Horn, B.W. and Hesseltine, C.W. (1987) *Aspergillus nomius*, a new aflatoxin-producing species related to *Aspergillus flavus* and *Aspergillus tamarii*. Antonie van Leeuwenhoek 53, 147-158.

Kusters van Someren, M.A., Samson, R.A. and Visser, J. (1991) The use of RFLP analysis in classification of the black aspergilli - reinterpretation of the *Aspergillus niger* aggregate. Curr. Gen. 19, 21-26.

Liljegren, K., Svendsen, A. and Frisvad, J.C. (1988) Mycotoxin and exoenzyme production by members of *Aspergillus* section *Flavi* : An integrated taxonomic approach to their classification. Proc. Jap. Assoc. Mycotoxicol. Suppl.1, 35-36.

Mirocha, J.C. and Christensen, C.M. (1982) Mycotoxins, in "Storage of Cereal Grains and their Products" (Christensen, C.M., Ed.), pp. 241-290. Amer. Assoc. Cer. Chem. Inc., St. Paul, Minnesota.

Moody, S.F. and Tyler, B.M. (1990*a)* Restriction enzyme and mitochondrial DNA of the *Aspergillus flavus* group, *Aspergillus flavus, Aspergillus parasiticus* and *Aspergillus nomius*. Appl. environ. Microbiol. 56, 2441-2452.

Moody, S.F. and Tyler, B.M. (1990*b)* Use of nuclear DNA restriction length polymorphisms to analyze the diversity of the *Aspergillus flavus* group, *Aspergillus flavus, Aspergillus parasiticus* and *Aspergillus nomius*. Appl. environ. Microbiol. 56, 2453-2461.

Morse, R.E. (1976) *Aspergillus oryzae* (NRRL 1988): A clarification. Science 194, 1188.

Moss, M.O. (1991) The environmental factors controlling mycotoxin formation, in "Mycotoxins and Animal Foods" (Smith, J.E. and Henderson, R.S., Eds.), pp. 37-56. CRC Press, Boca Raton.

Mullaney, E.J. and Klich, M.A. (1990) A review of molecular biological techniques for systematic studies of *Aspergillus* and *Penicillium*, in "Modern Concepts in *Penicillium* and *Aspergillus* Classification" (Samson, R.A. and Pitt, J.I., Eds.), pp. 301-307. Plenum Press, New York.

Ohmomo,S., Sugita, M. and Abe, M. (1973) Production of alkaloids and related substances by fungi. II. Isolation of cyclopiazonic acid, cyclopiazonic acid imine and bissecodehydrocyclopiazonic acid from the culture of *Aspergillus versicolor* (Vuill.) Tiraboschi. J. Agric. Chem. Soc. Jap. 47, 57-63.

Pavlovic, M., Plestina, R. and Krogh, P. (1979) Ochratoxin A contamination of foodstuffs in an area with Balkan (endemic) nephropathy. Acta Pathol. Microbiol. Scand. Sect. B, 87, 243-246.

Payen, J., Girard, T., Gaillardin, M. and Lafont, P. (1983) Sur la presence de mycotoxines dans des bières. Micro. Alim. Nutr. 1, 143-146.

Pitt, J.I., Hocking, A.D. and Glenn, D.R. (1983) An improved medium for the detection of *Aspergillus flavus* and *Aspergillus parasiticus*. J. Appl. Bacteriol. 54, 109-114.

Samson, R.A. and Frisvad, J.C. (1991) Current taxonomic concepts in *Penicillium* and *Aspergillus*, in "Cereal Grain Mycotoxins, Fungi and Quality in Drying and Storage" (Chelkowski, J., Ed.), pp. 405-439. Elsevier, Amsterdam.

Samson, R.A. and Pitt, J.I. (1985) "Advances in *Penicillium* and *Aspergillus* Systematics," pp. 1-483. Plenum Press, New York.

Samson, R.A. and Pitt, J.I. (1990) "Modern Concepts in *Penicillium* and *Aspergillus* Classification," pp. 1-478. Plenum Press, New York.

Samson, R.A., Nielsen, P.V. and Frisvad, J.C. (1990) The genus *Neosartorya* differentiation by scanning electron microscopy and mycotoxin profiles, in "Modern Concepts in *Penicillium* and *Aspergillus* Classification" (Samson, R.A. and Pitt, J.I., Eds.), pp. 455-467. Plenum Press, New York.

Smith, J.E. and Ross, K. (1991) The toxigenic aspergilli, in "Mycotoxins and Animal Foods" (Smith, J.E. and Henderson, R.S. Eds.), pp. 101-118. CRC Press, Boca Raton.

Speare, A.T. (1912) Fungi parasitic upon insects injurious to sugar cane. Hawaii. Sugar Plrs' Assoc. Exp. Stat. Path. Phys. Series Bull. 12, 1-62.

Stynen, D., Meulemans, L., Goris, A., Braedlin, N. and Symons, N. (1992) Characteristics of a latex agglutination test based on monoclonal antibodies for the detection of fungal antigens in foods, in "Modern Methods in Food Mycology" (Samson, R.A., Hocking, A.D., Pitt, J.I. and King, A.D. Eds.), pp. 213-219. Elsevier, Amsterdam.

Udagawa, S. and Takada, M. (1985) Contributions to our knowledge of *Aspergillus* teleomorphs: some taxonomic problems, in "Advances in *Penicillium* and *Aspergillus* Systematics" (Samson, R.A. and Pitt, J.I. Eds.), pp. 429-435. Plenum Press, New York.

Varga, J., Kevei, F., Fekete, C., Coenen, A., Kozakiewicz, Z. and Croft, J.H. (1993) Restriction fragment length polymorphisms in the DNA's of the *Aspergillus niger* aggregate. Mycol. Res. (in press).

Yamatoya, K., Sugiyama, J. and Kuraishi, H. (1990) Electrophoretic comparison of enzymes as a chemotaxonomic aid among *Aspergillus* taxa: (2) *Aspergillus* sect. *Flavi*, in "Modern Concepts in *Penicillium* and *Aspergillus* Classification" (Samson, R.A. and Pitt, J.I. Eds.), pp. 395-405. Plenum Press, New York.

Yoshizawa, T. (1991) Natural occurrence of mycotoxins in small grain cereals (wheat, barley, rye, oats, sorghum, millet, rice), in "Mycotoxins and Animal Foods" (Smith, J.E. and Henderson, R.S. Eds.), pp. 301-324. CRC Press, Boca Raton.

# FORMS OF ASPERGILLOSIS

C.K. Campbell

PHLS Mycological Reference Laboratory
Public Health Laboratory
Myrtle Road
Bristol, BS2 8EL

## INTRODUCTION

Aspergilli are among the most numerous and most abundant of the saprophytic moulds, occurring on a wide variety of substrates, wherever vegetation decomposes. From these habitats their conidia disseminate on air currents. Opportunities are thus abundant for animal infection and it must be assumed that all terrestrial animals are continually exposed to all the commonest species of *Aspergillus* in their environment. Despite this only a handful of species commonly cause aspergillosis and it is tempting to look for "virulence factors" which might explain this feature.

Among warm-blooded animals, the vast majority of cases of deep infection are due to one species, *A.fumigatus*, a thermotolerant fungus known to occur naturally on decomposing, self-heating plant material. A minority of infections are due to other thermotolerant saprophytes such as *A.flavus*, *A.niger*, *A.terreus* and *A.nidulans*, all of which are able to grow well at blood temperatures. Outside this caucus of recognised "pathogens" are those aspergilli only rarely seen in invasive disease. Rippon (1988) lists 17 additional causes of deep aspergillosis, by fungi representing most of the major species groups. An overlapping but not identical range of species is involved in infections of the more superficial sites such as the outer ear, nails, etc.,

Even the apparently strongly pathogenic species such as *A.fumigatus*, *A.flavus*, etc., have been unable to adapt to the parasitic life, since aspergillosis is an ecological "dead-end", with no escape back to the saprophytic environment. It is therefore most unlikely that any apparent virulence factors that we can identify are evolutionary adaptations to life in the environment of the host tissues. Probably, as is the case with thermotolerance, they are really attributes associated with saprophytic life on the substrates they normally inhabit. Fungal products like toxins and anti-inflammatory compounds have possibly evolved to deter saprophytic competitors.

Assuming that no adaptive parasitism has arisen in the aspergilli, the pathogenicity of the most virulent species may be due to chance conformity of their physiological requirements with the conditions in host tissue. Certainly the morphology of the fungal

*The Genus Aspergillus*, Edited by Keith A. Powell *et al.*,
Plenum Press, New York, 1994

hyphae in the various forms of aspergillosis appears to reflect differences in the degree of resistance exerted by the host.

Some authors have suggested the small size of conidia of *A.fumigatus*, allowing ease of penetration into the lung alveoli, may explain its prevalence as the main pathogenic species (Cohen, 1982). Larger spores are more likely to be impacted onto the sides of the air-passages before reaching the deeper recesses of the lungs (Hatch, 1961). This may go some way to explain why *A.niger* and *A.flavus* are more frequently found infecting the upper airways than the lung parenchyma. However, there are other *Aspergillus* species with spores of the same general size (3µm.) as those of *A.fumigatus* (*A.candidus*, *A.flavipes*, *A.versicolor*) but they do not feature in deep lung invasion. Another species with small spores *A.clavatus*, causes alveolar allergy (Riddle *et al.*, 1968) showing that its spores can reach the alveoli, but that it is physiologically unable to invade the tissue.

**Early Infection**

The mammalian and avian bodies have a range of non-specific defences against the ingress of micro-organisms, and these undoubtedly reduce the risk that conidia of fungi may lodge in the lungs long enough to establish a mycelium. Healthy lungs are supplied with an upwardly directed flow of mucus at the surface of the air passages. This is maintained by action of ciliated epithelial cells and has the tendency to physically remove any airborne particle that is inhaled. In addition, phagocytic macrophages are active in the mucus and can engulf and destroy many types of microbes. Tests have shown that the conidia of aspergilli can be destroyed by these cells (Waldorf *et al.*, 1984; Levitz & Diamond, 1985).

In situations where there is constant inhalation of more conidia such as with cattle housed indoors in an atmosphere heavily laden with spores from hay and straw, at least some of the conidia are retained in the lung long enough to enable them to germinate and produce a short hyphal cell. This develops no further in healthy individuals, as it is held in check by the secretions of the surrounding host response to its presence. It remains viable, encased in a thick envelope of antibody-rich material and is visible in histological preparations as an"asteroid body" (Austwick, 1962; Matsui *et al.*, 1985).

To invade the non-pulmonary tissues the fungus must either get past these formidable barriers in the lung, or take advantage of breaks in the skin at wound sites. Wounds may be traumatically produced or they may be the result of medical procedures such as surgery or catheterisation. Even with exposed living tissues, the conidium has little chance of progressing beyond the germination stage unless the cellular immune system is seriously impaired (Carlile *et al.*, 1978).

It has been shown that the removal of cellular defences following the administration of corticosteroid drugs to animals experimentally infected with *Aspergillus* conidia results in a progressive and usually fatal infection, in contrast to control animals in which the normal immunity is able to contain and eventually eliminate the conidia (Sidransky *et al.*, 1972).

## ASPERGILLOSIS IN MAN

For over a hundred years following the first account of human aspergillosis by Bennett (1844), the infection was a rarity and usually took the form of aspergilloma or fungus ball in the lungs. The infection was often an unexpected finding at post-mortem examinations. The more superficial forms of *Aspergillus* infection such as *otitis externa* and nail invasion were undoubtedly common but the pathogenic role of the fungus was (and to some extent still is) disputed by clinicians.

The realisation that aspergilli were capable of fulminant infection in any deep organ of the body with often a rapidly fatal outcome is a feature of the second half of the twentieth century. It came as a result of the striking successes in prolongation of life in people with various forms of cancer and in patients undergoing new surgical procedures involving immunosuppressants.

**Aspergilloma**

This type of aspergillosis, often called "fungus ball", typically occurs in human patients with physical abnormalities of the airways. An air-filled cavity left by destruction of lung parenchyma in old tuberculous lesions or by surgery, may form the focus for this infection. In such places the normal mucus stream is ineffectual, and the inhaled conidium is able to germinate and grow for many months or years. The result is a compact ball of fungal mycelium which grows to partially fill the available air space. In these conditions the fungus is often able to sporulate and the resulting conidia may germinate and contribute to the mass of the fungal body.

Aspergilloma patients usually have intact immune responses and a strong cellular infiltrate effectively seals off the ball and prevents hyphae penetrating into the surrounding respiratory tissue. The cells also produce a strong anti-*Aspergillus* antibody response and these allow differentiation from similar focal lesions of non-fungal etiology.

Diagnosis is primarily a result of interpretation of chest radiography, supplemented with serological determination of these antibodies, and occasionally the culture of the fungus from fragments of the mycetoma coughed up in sputum. The patient is usually not in acute danger except that in a proportion of cases the lesion involves a major blood vessel and rupture of this can lead to massive blood loss (Rippon, 1988: Kwon-Chung & Bennett, 1992).

**Invasive aspergillosis**

When the host resistance to infection is particularly impaired, the aspergilli can cause a rapidly developing acute infection, which may have a dramatic and devastating effect on the organs and frequently results in death (Young *et at.*, 1970). Typically invasive aspergillosis is the result of absence of or abnormal functioning of the leukocytic series of blood cells. Leukocytes may be poorly produced in some naturally occurring conditions, such as the various forms of leukaemia, and following infection with human immunodeficiency virus (HIV). Artificially induced leukocyte depletion occurs in patients on anti-cancer chemotherapy with cytotoxic drugs and radiotherapy, and on immunosuppressants following organ transplants.

Most cases of invasive aspergillosis begin as pulmonary infections, involving areas of lung infarction and lobar pneumonia. Extensive damage to the respiratory parenchyma occurs with eventual necrosis at a distance from the invading hyphae, suggesting the action of secreted toxic metabolites of the fungus. In addition, focal lesions such as abscesses or aspergilloma-like infections may develop in some patients.

Pulmonary aspergillosis in these immunosuppressed patients may spread from the lungs to any part of the body with corresponding symptomatology and pathology. Thus fatal invasion of the viscera, including the heart, liver, kidneys and other abdominal organs are all too common, as are dissemination to the facial tissues, eyes and brain.

In all these tissues the histological appearance of the invading hyphae suggests the fungus is growing as if uninhibited and often resemble hyphae grown in pure culture, very different from hyphae in chronic infections in patients with an intact immune response (Rippon, 1988; Kwon-chung & Bennett, 1992). Histological visualisation of the hyphae in biopsy specimens is the "gold standard" among diagnostic methods in that absolute proof of the presence of the fungus is achieved and other infectious agents can be discounted. However, in these patients biopsy is seldom possible owing to their extreme susceptibility to any invasive procedure, and much effort has gone into developing serological diagnostic methods. Despite more than 20 years of attempts to use antibody detection to give early warning of invasive aspergillosis, no satisfactory test has emerged. The very immunosuppression which predisposes to aspergillosis also precludes effective antibody response, and so tests often fail to detect antibody even in cases of fulminant disease (Young & Bennett, 1970; Holmberg *et al.*, 1980).

More recently attempts have been made to identify antigenic fungal derivatives in patients' sera and other body fluids such as urine, with some success. It seems, however, that even in proven cases of aspergillosis, such derivatives are detectable only in brief pulses. Successful detection thus requires the processing of frequent sequential samples. (Andrews & Wiener, 1982; Wiener *et al.*, 1983)

### Chronic Necrotising Aspergillosis

In attempts to fully describe the spectrum of disease types in human aspergillosis, chronic necrotising aspergillosis was suggested for the less acutely invasive infection of patients with low degrees of immunosuppression often in combination with physical abnormality of the lung (Binder *et al.*, 1982).

This is a slowly progressive disease with less immediate implications than with invasive aspergillosis. Typically there is extensive growth of aspergilli in the airways, with some destruction of functional respiratory tissue. With time the necrotic lesions may cavitate and become the site of a fungus ball. Unlike classical aspergilloma the cavity itself is the direct result of the fungal infection. These patients usually raise antibodies to the infecting agent.

### Paranasal Aspergilloma

As its name suggests this is essentially a mass of fungal mycelium and reaction tissue growing in the air-filled sinuses of the head. Infected patients may have no obvious predisposing condition. The ethmoid and maxillary sinuses are the most frequent sites. It is not a common disease anywhere, but significant numbers, caused by *A.flavus*, have been reported from Sudan (Milosev *et al.*, 1969; Rudwan & Sheikh, 1976).

The infection is not rapidly invasive, but in prolonged cases the orbit of the eye may be involved, with consequent proptosis. As in aspergillosis involving other air spaces, the organism may produce conidia at the air surface. Some reports even describe the ascosporic state of *A.nidulans* in this material (Mitchell *et al.*, 1987).

## ASPERGILLOSIS IN OTHER MAMMALS

Lesions reminiscent of all the human forms of aspergillosis have been described in other mammals, though the presentation is often different. Again it is possible to recognise the importance of the effectivity of the host resistance.

Aspergilloma-like lesions are evidently rare (Austwick, 1968; Gupta, 1978). More noticeable are the common sinus infections described in both dogs (Lane & Warnock, 1977; Sharp *et al.*, 1986; Sullivan *et al.*, 1986) and horses (Lane, 1993). In both these animals the sinusitis presents as an excessive drainage of mucus from the nose. Radiographs of the nasal area, at least in dogs, shows destruction of bone and obstruction of the airways. Endoscopy of the sinuses can give precise location of the lesions and facilitate direct application of antifungal drugs to the affected area.

Similar infections of the equine guttural pouch (an air-filled diverticulum of the eustachian canal) can be more serious as a fatal loss of blood can follow the rupture of the carotid artery which lies close under the epithelium lining the pouch (Cook, 1968; Lane, 1989).

Not surprisingly, as animals are not yet subjected to the extreme degrees of medical management predisposing to human invasive aspergillosis, examples of acute disease in other mammals are associated with more natural types of lowered immunity. One of these is seen in outbreaks (presumably from a common source) of acute pulmonary aspergillosis in very young pigs (Austwick, 1965) and sheep (Austwick *et al.*, 1960; Young, 1970). In these the only known reason for susceptibility to infection was their very young age (less than 3 weeks). In all cases the lung was filled with very many small nodules of necrosis, each centred around a germinating spore of *Aspergillus*.

Another form of acute aspergillosis occurs in mycotic placentitis, usually presenting as abortion of the foetus. This is classically a disease of cattle (Austwick & Venn, 1961; Sheridan *et al.*, 1985), though it has been described in sheep, horses and pigs (Leash *et al.*, 1968; Mason, 1971). Besides aspergilli, a wide range of moulds, (notably the thermotolerent Mucoraceous species) and yeasts cause essentially similar mycotic abortion, strongly suggesting that general susceptibility of the placenta and foetus are more important than specific virulence of the infecting aspergilli.

In this disease the mother animal typically shows no sign of pulmonary aspergillosis or other focal infection from which it may have spread to the placenta. The mode of entry of the fungus to the uterus is thus unknown, though haematogenous spread from asteroid lesions in the lung is a possibility. The foetal infection takes the form of skin, gut and lung lesions, presumably following carriage from the placenta tissue via the amniotic fluid, from which the fungus can be isolated.

Finally a disseminated form of invasive aspergillosis seems to be peculiar to adult dogs. Isolated reports have occurred since 1978 (Wood *et al.*, 1978), and an interesting cluster was recorded by Day *et al.* (1985; 1986). In these investigations the infections were caused by *A.terreus* and almost all the subjects were in the German Shepherd breed. The infection occurred as focal lysis of bones in both spine and limbs, and as infections of the soft organs, particularly spleen and kidneys. In another series (Jang *et al.*, 1986) German Shepherd dogs were similarly infected, this time with *A.deflectus*. Evidently there is some unknown factor at work in this breed which is exposing the tissues to the lower levels of protection comparable with human immunosuppression.

## ASPERGILLOSIS IN BIRDS

Aspergilli have long been known to infect the lungs and air-sac systems of birds of all types (Ainsworth & Austwick, 1973). Two major clinical diseases can be distinguished; acute pneumonia in young chicks and chronic pulmono-visceral disease in older birds.

The acute form, also called brooder pneumonia, has caused devastating loss of birds in hatcheries where thousands of chicks are reared by intensive farming methods (Ainsworth & Austwick, 1973). In these outbreaks the source of the fungus has sometimes been traced to contaminated litter. In addition broken eggs may become infected before hatching and act as ideal substrates for sporulation. In consequence nearby newly-hatched chicks are readily supplied with a massive inoculum of conidia.

Each of the inhaled spores initiates a cellular response in the lung, with the result that the air passages are soon obliterated by a mixture of cells and hyphae. Death by asphyxiation is the inevitable rapid outcome.

Young age has been shown to be a factor in the extreme susceptibility shown in acute avian pneumonia. Chicks are considerably more resistant to infection only 2 weeks after hatching (Campbell, 1970; Klimes & Rosa, 1966).

In older birds the usual form of aspergillosis is a chronic infection of the lungs and air-sacs, leading slowly to reduction in respiratory function, and often to dissemination among the other viscera. This is particularly a disease of captive wild birds, especially the larger species, but enough reports exist to show that it also occurs naturally among free-living birds. It is reasonable to suppose that the stresses of captivity explain its prevalence in zoological gardens and other wild bird collections.

In a typical case, several to many slowly enlarging nodules of cellular reaction occur scattered throughout the lung tissue, reducing gaseous exchange (Ainsworth & Austwick, 1973). The extensive system of air-sacs, essential for proper inhalation and exhalation become infected and fail to draw fresh air into the lungs. Their normally thin, elastic walls become thickened with reaction cells and irreversibly rigid. In advanced cases the fungus sporulates in abundance on the air-surface of the sac wall. In the disseminated form, which is always preceded by pulmonary infection, similar necrotic nodules to those in the lungs develop in the liver, gut wall and other viscera.

## REFERENCES


Ainsworth, G.C. and Austwick, P.K.C. (1973) "Fungal Diseases of Animals" p. 39. 2nd.Ed.Commonwealth Agricultural Bureaux, Farnham Royal, U.K.

Andrews, C.P. and Weiner,M.H. (1982) *Aspergillus* antigen detection in bronchoalveolar lavage fluid from patients with invasive aspergillosis and aspergillomas. Amer. J. Medicine 73, 372-380.

Austwick, P.K.C. (1962) The presence of *Aspergillus fumigatus* in the lungs of dairy cows. Laboratory Invest. 11, 1065-1072.

Austwick, P.K.C. (1965) "The Genus *Aspergillus*" (Raper, K.B. and Fennell, D.I.) p. 92. Williams & Wilkins Co., Baltimore.

Austwick, P.K.C. (1968) Mycotic Infections "Diseases of free-living wild animals."Symposia of the Zoological Society of London 24, pp. 249-271 Academic Press.

Austwick, P.K.C., Gitter, M. and Watkins, C.V. (1960) Pulmonary aspergillosis in lambs. Veterinary Record 72, 19-21.

Austwick, P.K.C. and Venn, J.A.J. (1961) Mycotic Abortion in England and Wales 1954-1960. Proceedings IVth Internat. Congress of Animal Reproduction, The Hague, pp. 562-568.

Bennett, J.H. (1844) On the parasitic vegetable structures found growing in living animals. Trans. Royal Soc. Edinburgh 15, 277-294.

Binder, R.E., Faling, L.J., Pugatch, R.D.,Mahasaen, C. and Snider, G.L. (1982) Chronic necrotizing pulmonary aspergillosis, a discrete clinical entity. Medicine 61, 109-124.

Campbell, C.K. (1970) Electron microscopy of aspergillosis in fowl chicks. Sabouraudia 8, 133-140.

Carlile, J.R., Millet, R.E., Cho, C.T. and Vats, T.S. (1978) Primary cutaneous aspergillosis in a leukaemic child. Arch. of Dermatol. 114, 78-80.

Cohen, J. (1982) Aspergillosis in the compromised patient. "Fungal Infection in the Compromised Patient" (Warnock, D.W.W. and Richardson, M.V., Eds.), p. 121. J. Wiley & Sons Ltd., Chichester.

Cook, W.R. (1968) The clinical features of guttural pouch mycosis in the horse. Vet. Record 83, 336-345.

Day, M.J., Eger, C.E., Shaw, S.E. and Penhale, W.J. (1985) Immunologic study of systemic aspergillosis in German Shepherd dogs. Vet. Immunology and Immunopathology 9, 335-347.

Day, M.J., Penhale, W.J., Eger, C.E., Shaw, S.E., Kabay, M.J., Robinson, W.F., Huxtable, C.R.R., Mills, J.N. and Wyburn, R.S. (1986) Disseminated aspergillosis in dogs. Austral. Vet. J. 63, 55-59.

Gupta, B.N. (1978) Pulmonary aspergillosis in a rat. J. Amer. Vet. Medical. Assoc. 173, 1196-1197.

Hatch, T.F. (1961) Distribution and deposition of inhaled particles in the respiratory tract. Bacteriol. Rev. 25, 237-240.

Holmberg, K., Berdischewsky, M. and Young, L.S. (1980) Serological immunodiagnosis of invasive aspergillosis. J. Infect. Dis. 141, 656-664.

Jang, S.S., Dorr, T.E., Bilberstein, E.L. and Wong, A. (1986) *Aspergillus deflectus* infection in four dogs. J. Med. Vet. Mycol. 24, 95-104.

Klimes, B. and Rosa, L. (1964) Die altersresistenz von kuken gegenuber *Aspergillus fumigatus*. Berliner Munchener Tierartzliche Wochenschrift 77, 125-126.

Kwon-Chung, K.J. and Bennett, J.E. (1992) "Medical Mycology" pp. 201-247. Lea & Febiger,Philadelphia.

Lane, J,G. (1989) The management of guttural pouch mycosis. Equine Vet. J. 21, 321-324.

Lane, J.G. (1993) The management of sinus disorders of horses-Part 1. Equine Vet. Educ. 5, 5-9.
Lane, J.G. and Warnock, D.W.W. (1977) The diagnosis of *Aspergillus fumigatus* infection of the nasal chambers of the dog with particular reference to the value of the double diffusion test. J.Small Anim. Practice 18, 169-177.
Leash, A.M., Sachs, S.D.,Abrams and Limbert, R. (1968) Control of *Aspergillus fumigatus* infection in fetal sheep. Lab. Animal Care 18, 407-408.
Levitz, S.M. and Diamond, R.D. (1985) Mechanisms of resistance of *Aspergillus fumigatus* to killing by neutrophil *in vitro*. J. infect. Dis. 152, 33-42.
Mason, R.W. (1971) Porcine mycotic abortion caused by *Aspergillus fumigatus*. Austral. Vet. J. 47, 18-19.
Matsui, T., Taguchi-Ochi, S., Takano, M., Kuroda, S., Tapiyama, H. and Ono,T. (1985) Pulmonary aspergillosis in apparently healthy young rabbits. Vet. Pathol. 22, 200-205.
Milosev, B., Mahgoub, El S., Aal, S. and Hassan, A.M.El (1969) Primary aspergilloma of the paranasal sinuses in the Sudan. Brit. J. Surg. 56, 132-137.
Mitchell, R.G., Chaplin, A.J. and Mackenzie, D.W.R. (1987) *Emericella nidulans* in a maxillary sinus fungal mass. J. Med. Vet. Mycol. 25, 339-341.
Rippon, J.W. (1988) "Medical Mycology" 3rd Ed. pp. 618-650 W.B. Saunders Company, Philadelphia.
Riddle, H.V.F., Channell, S., Blyth, W., Weir, D.M., Lloyd, M., Amos, W.M.G. and Grant, I.W.B. (1968) Allergic alveolitis in a maltworker. Thorax 23, 271-280.
Rudwan, M.A. and Sheikh, H.A. (1976) Aspergilloma of paranasal sinuses, a common cause of unilateral proptosis in Sudan. Clin. Radiol. 27, 497-502.
Sharp, N., Burrell, M.H., Sullivan, M. and Cervantes-Olivares, R.A. (1984) Canine nasal aspergillosis: serology and treatment with ketoconazole. J. Small Anim. Pract. 25, 149-158.
Sheridan, J.J., White, D.S.C. and McGarvie, Q.D. (1985) The occurrence of and organisms concerned with bovine mycotic abortion in some counties of Ireland. Vet. Research Communications 9, 221-226.
Sidransky, H., Epstein, S.M., Verney, E. and Horowitz, C. (1972) Experimental visceral aspergillosis. Amer. J. Pathol. 69, 55-70.
Sullivan, M., Lee,R., Jakovlgevic, S., Sharp,N.J.M. (1986) The radiological features of aspergillosis in the nasal cavity and frontal sinus in the dog. J. Small Anim. Prac. 27, 167-180.
Waldorf, A.R., Levitz, S.M. and Diamond, R.D. (1984) *In vivo* bronchoalveolar macrophage defense against *Rhizopus oryzae* and *Aspergillus fumigatus* J. Inf. Dis. 150, 752-760.
Weiner, M.H., Talbot, G.H., Gerson, S.L., Felice, G. and Cassileth, P.A. (1983) Antigen detection in the diagnosis of invasive aspergillosis. Ann. Internal Med. 99, 777-782.
Wood, G.L., Hirsch,D.C., Selcer, R.R., Rinaldi, M.G. and Boorman, G.A. (1978) Disseminated aspergillosis in a dog. J. Amer. Vet. Med. Ass. 172, 704-707.
Young, N.E. (1970) Pulmonary aspergillosis in the lamb. Vet. Record 86, 790.
Young, R,C. and Bennett, J.E. (1971) Invasive aspergillosis. Absence of detectable antibody response Am. Rev. Resp. Dis. 104, 710-716.
Young, R.C., Bennett, J.E., Vogel, C.L., Carbone, P.P. and De Vita, V.T. (1971) Aspergillosis. The spectrum of the disease in 98 patients. Medicine 49, 147-173.

# EXOANTIGENS OF *ASPERGILLUS FUMIGATUS*:

# SERODIAGNOSIS AND VIRULENCE

J.P. Latgé[1], S. Paris[1], J. Sarfati[1], J.P. Debeaupuis[1] and M. Monod[2]

[1] Mycologie, Institut Pasteur, 75015 Paris, France
[2] Dermatologie, CHUV 1011 Lausanne, Switzerland

## INTRODUCTION

*Aspergillus fumigatus* commonly grows in damp environments such as soil, decaying vegetation and organic debris from which it releases a high number of spores into the atmosphere (Raper and Fennell, 1965). The number of *A. fumigatus* conidia present in the air usually varies from 0.1 to 10 spores/m$^3$, but may reach 10$^6$/m$^3$ under certain environmental conditions (Mullins *et al.*, 1976; Bodey and Vartivarian, 1989; Gumowski *et al.*, 1992). Due to their small size (2-3 $\mu$m) and presence in the air, *A. fumigatus* conidia are inhaled by all individuals and transit through the upper respiratory tract, bronchia and bronchiola to end in the alveoli (Cole and Samson, 1984). The pulmonary epithelium is the largest epithelial surface exposed to the external environment (40-80 m$^2$/individual). It possesses several lines of defence capable of expelling *A. fumigatus* from the lung (Seaton *et al.*,1989). Impairment of these defence mechanisms can lead to the establishment of the fungus in the lung. This impairment can be of genetical (cystic fibrosis, chronic granulomatous disease) or clinical origin (tuberculosis, chemo- and corticotherapy) (Cohen *et al.*, 1981; Wagner, 1984; Schönheyder *et al.*, 1988; Bodey and Vartivarian, 1989; Cohen, 1991). There are three clinical manifestations of *A. fumigatus* infections: 1) allergic bronchopulmonary aspergillosis (ABPA) resulting from repeated inhalation of conidia with limited fungal growth; 2) aspergilloma where colonisation of a pre-existing pulmonary cavity forms a fungal ball; 3) invasive pulmonary aspergillosis where *A. fumigatus* invades lung parenchyma and disseminates to other organs. The latter form of the disease occurs in immunocompromised hosts and is usually life threatening in the absence of early diagnosis.

The small size of the conidia, their prevalence in the atmosphere and an ability to grow at 37°C are not sufficient to explain the pathogenic development of *A. fumigatus*. Other fungal species with the same biological characteristics are not pathogenic. This

*The Genus Aspergillus*, Edited by Keith A. Powell *et al.*,
Plenum Press, New York, 1994

suggests that *A. fumigatus* possesses some specific virulence factors which are responsible for the germination of the conidia and the growth of mycelial filaments in lung tissues. Until now, these factors have not been precisely identified.

Molecules located intracellularly in the cytoplasm and in the cell wall or secreted in the surrounding environment are known to play a role in the pathogenesis of *A. fumigatus*. The role of constitutive wall components during *Aspergillus* infections is presented in this volume and will not be discussed here. The present Chapter only deals with *A. fumigatus* molecules passively and actively released from conidia and mycelium. After reviewing the lung defence mechanisms, the role of exocellular fungal molecules in the inhibition of these defences will be discussed. Special emphasis will be then directed towards exoantigens which can be used in the serodiagnosis of aspergillosis and/or facilitate the establishment of *A. fumigatus* in the lung.

## THE LUNG MILIEU AND PULMONARY HOST DEFENCES AGAINST MICROORGANISMS

To elucidate the sequential events resulting in the development of an aspergillosis, it is essential to apprehend the lung environment and the role of host defence mechanisms acting against the fungus (Cole and Wilson, 1989; Seaton *et al.*, 1989; Reynolds, 1991).

### Conducting Airways

The airway surface of the respiratory tract is covered by vibratory cilia which bath in a mucus containing 95% water, 3% proteins (mainly mucins and IgA), carbohydrates and lipids (Figure 1). Following their deposition on mucosal surfaces, particles are mechanically removed from the respiratory tract, by their entrapment and transport via the mucociliary escalator. This mechanism is responsible for the clearance of most microorganisms which will then not reach the alveoli. Alterations of the ciliary machinery due to the presence of very tenacious secretions found for example in cystic fibrosis, or due to exoproducts such as bacterial toxins that paralyse or slow ciliary beating, may allow local microbial colonisation to proceed. The primary function of IgA is to counteract these toxins and to reduce the adherence of microorganisms to airways epithelial cells.

### Alveoli

When a microorganism reaches the alveolar surface it encounters a different cellular and nutritional environment (Figure 1). The alveolar wall consists of type I and II pneumocytes, basement membrane, interstitial tissue with bundles of elastin and collagen, capillary endothelial cells. Macrophages are found in the interstitial space as well as in the alveolar lumen and represent 80-90 % of the cells found in broncho-alveolar lavages (BAL). T lymphocytes ($<$ 10 %), mainly T helpers are also present in the lumen. The alveolar milieu is of lipoproteic nature. The main lipophilic acellular compound is the surfactant, which is manufactured by type II alveolar pneumocytes and is mainly composed of dipalmitoyl lecithin and dipalmitoyl phosphatidyl ethanolamine.

The first event in the microbial host defence reaction is the opsonisation of the microorganism for optimal phagocytic uptake. Non-immune and immune proteins present in the alveolus can opsonise microorganisms. Amongst these proteins are proteins of the surfactant, fibronectin fragments, C-reactive protein, complement components and IgG immunoglobulins (mainly IgG2). The second axis of defence involves the macrophages which are the principal phagocytes on the air exchange surface of the lung. Cell mediated immunity is required to activate the macrophages. This activation step is under the control of T helper cells and interferons. In turn, macrophages can influence lymphocytes through an array of different cytokines that these cells can produce. Secretion of these chemotactic factors is essential to initiate the influx of neutrophils (PMN) which is the third participant to the lung defence. PMNs recruitment occurs in conditions favouring the proliferation of a microorganism such as a large inoculum or a virulent strain. The mobilisation of inflammatory cells from the intravascular compartment into the lung is critical for an effective pulmonary host defence. Although a small number of PMN are usually found in

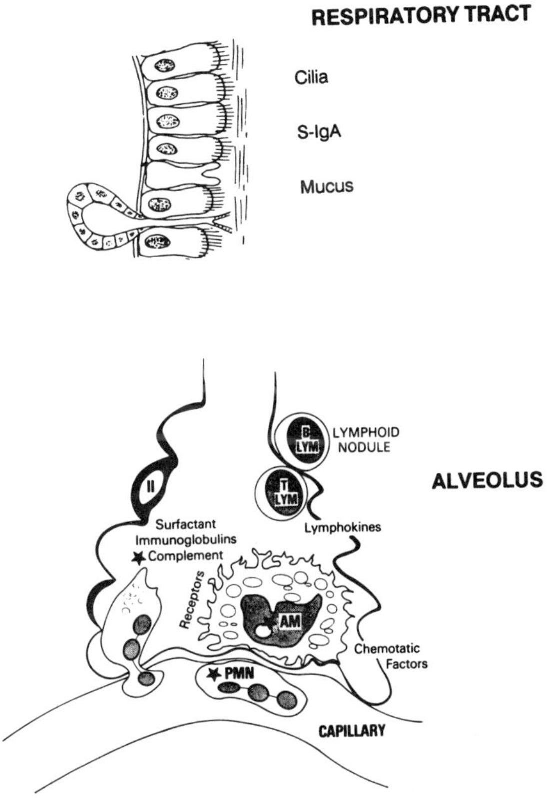


**Figure 1.** Schematic representation of the respiratory tract conducting airways and alveoli presenting the main cellular and humoral pulmonary defence mechanisms (adapted from Reynolds, 1991). The factors and cells which play a role in the phagocytosis of *A. fumigatus* are marked with an asterisk.

BAL, the actual number of PMN is probably greater since most of them are located in the pulmonary vasculature as a transient and dynamic population.

Pharmacological therapy has adverse effects on pulmonary host defence reactions (Mahler and Kinsella, 1991; Nelson and Summer, 1991). Cytotoxic drugs or whole body irradiations decrease the number of alveolar macrophages and PMNs. Immunosuppressive drugs inhibit macrophage chemotactic responsiveness and their killing potential. These drugs also decrease the recruitment of PMNs to the airways. In contrast, immunosuppression does not seem to interfere with microorganism opsonisation. All these pharmacological factors that impair PMN and macrophage responsiveness will be predictably associated with infections from airborne pathogens such as *Aspergillus*.

## INHIBITION OF HOST DEFENCE REACTIONS BY EXOCELLULAR *A. FUMIGATUS* MOLECULES

### Respiratory Tract

The role of mucus, secretory IgA and cilia during entrapment of *A. fumigatus* conidia in the pseudostratified ciliated epithelium has not been studied. However, some biological traits of *A. fumigatus* suggest that this pathogenic organism may interfere with these defences.

Upon contact on a surface, rodlet material from the outer wall layer of the conidia, can be detached from the wall (Figure 2). Moreover, after a short incubation in different physiological media, conidia of *A. fumigatus* release a significant amount of soluble proteins and sugars (unpublished). The role of the fungal material released from the conidia in the adhesion to mucins or to the ciliary apparatus has not been investigated. In a way similar to what has been shown during bacterial infections, toxins secreted by *A. fumigatus* and conidial enzymes may play an essential role in the impairment of ciliary beating (Steinfort *et al.*, 1989; Wilson *et al.*, 1991). More than 10 different mycotoxins secreted by *A. fumigatus* have been demonstrated (Turner and Aldridge, 1983; Seiglemurandi *et al.*, 1990). Inhibition of ciliary beating can be also promoted by antigens released from the surface of the conidia. This phenomenon can occur specially in atopic patients sensitised to fungal allergens (Bouziane *et al.*, 1988). IgG and IgA response against conidial diffusates have been demonstrated (Figure 3), but the role of secretory IgA against *A. fumigatus* in the respiratory tract is not known. A more comprehensive study of the conidial factors interacting with the surface of the respiratory tract would make a new insight in the understanding of aspergillosis in cystic fibrosis patients.

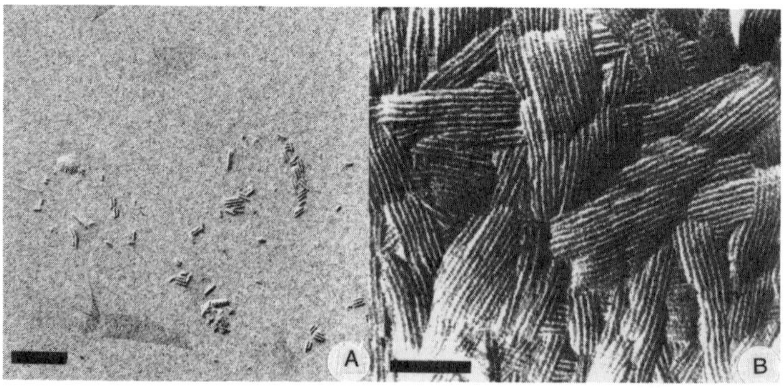


**Figure 2.** Carbon-platinum replicas showing rodlets detached from conidia of *A. fumigatus* after impact on a mica sheet (A). Organisation of the rodlets on the surface of the conidium (B) Bars represent 0.2 $\mu$m (A) and 0.1 $\mu$m (B).

## Alveoli

In contrast to the lack of information on the role of the conducting airways against *Aspergillus*, the primary involvement of complement, alveolar macrophages and PMNs during the phagocytosis of *A. fumigatus* conidia and germ tubes has been repetitively established (Schaffner, 1989; Waldorf, 1989; Sturtevant and Latgé, 1992) (Figure 1).

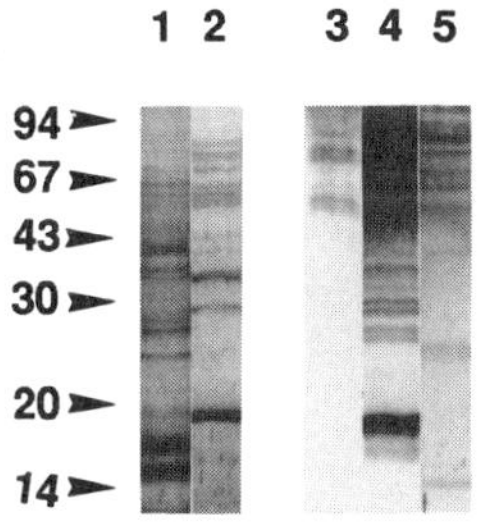


**Figure 3**. SDS-PAGE protein patterns and immunoblot analysis of material released from conidia of *A. fumigatus*. Silver nitrate staining of water soluble extract of disrupted conidia (lane 1) and proteins released from intact conidia (lane 2). Immunoblot analysis of conidial diffusates probed with a pool of sera from aspergilloma patients and anti-human IgG (lanes 3, 4) and anti-human IgA (lane 5) peroxydase conjugates. Conidial diffusates are obtained after incubation of spores in a NaCl (0.9 %) + $NaHCO_3$ (0.3 %) solution (lane 3) or mild ultrasonication of a conidial suspension in water (lanes 2, 4, 5). Aspergilloma patient sera were diluted 1/1000 (lanes 3, 4) or 1/100 (lane 5).

**Molecules Interfering with Opsonisation**. Unlike IgG, complement is one of the major factors involved in *A. fumigatus* phagocytosis. However, *A. fumigatus* releases compounds that can counteract complement efficacy during phagocytosis. It has recently been found that C3 can be cleaved by protease(s) which are released from conidia (Sturtevant and Latgé, 1992*b*). As suggested for bacteria (Hostetter, 1986), cleavage of C3 could contribute to the pathogenicity of *A. fumigatus*. The ability or inability of non-pathogenic *Aspergillus* species to degrade C3 should be investigated as well as the nature of protease(s) released by *A. fumigatus* conidia.

This conidial proteolytic component is not responsible for the complement inhibitory activity isolated from mycelial cultures by Washburn *et al.* (1986, 1990). This extracellular inhibitory fraction of the alternative complement pathway was not purified but proved to be extremely hydrophobic. These authors suggest that the inhibitory activity is due to phospholipids (most probably phosphatidyl serine/phosphatidyl inositol and phosphatidyl ethanolamine). The mechanism of action of this lipidic complement inhibitor of *A. fumigatus* is different from other complement inhibitory fractions isolated from non-pathogenic fungi or from bacteria or parasites . How important is the interference of these complement inhibitors in the production of chemotaxins C5a and in opsonophagocytosis is unknown.

Other lipophilic molecules may interact with plasma factors involved in host defence. C-reactive substances binding in a $Ca^{++}$ dependent fashion to C-reactive protein have been described in 1971 by Pepys and Longbottom. The molecules involved in this binding were never isolated and it was suggested by Borglum Jensen *et al.* (1986) that C-reactive substances of *A. fumigatus* are heterogenous. Phospholipids seems to be involved in the binding of *A. fumigatus* to CRP since a significant amount of C-reactive substances are retained in an hydrophobic column and react positively with an anti-phosphoryl choline monoclonal antibody. Another part of CRP-binding activity could be due to a proteic doublet recently identified by western blotting (J.P. Latgé, unpublished data).

**Phagocytosis Inhibitors**. Alveolar macrophages kill the phagocytosed conidia by a non-oxidative killing system. Germ tube killing is due to neutrophils which secrete actively reactive oxygen immediately after contact with mycelium (Levis and Diamond, 1985; Robertson *et al.*, 1989).

Inhibition of host immune reactions by *A. fumigatus* seems to be due to low molecular weight molecules. In a series of papers, Robertson and collaborators (1987*a*, *b*, *c*; 1988) have described a conidial diffusate which affects both phagocytosis and the release of superoxide anion and hydrogen peroxide. The inhibitor diffuses from the conidia as soon as they are put into suspension and is not produced as the result of germination. The effector mechanism of the conidial diffusate seems a reduction of the migration ability of phagocytes due to its interaction with the contractile elements of the cell cytoplasm. Its effect on cells looks similar to cytochalasin B but it has not been chemically identified. The inhibitory activity of the diffusate was removed by dialysis indicating that the molecular weight of the active substance in the diffusate is less than 10 kDa. It is different from gliotoxin which has also an antiphagocytic activity but which is released once germination has taken place.

Gliotoxin is an epipolythiodioxopiperazine produced *in vitro* and *in vivo* by *A. fumigatus* (Müllbacher and Eichner, 1984; Eichner *et al.*, 1986). It inhibits phagocytosis by macrophages, proliferation of T cells after stimulation by mitogens, release of interleukin by activated T cells and generation of cytotoxic cells. It induces DNA fragmentation of macrophages and activated T cells in a pattern characteristic of apoptosis, a programmed form of cell death (Eichner *et al.*, 1988; Waring *et al.*, 1988). However, induction of apoptosis seems unrelated to the antiphagocytic properties of gliotoxin.

## EXOANTIGENS AND LUNG TISSUE COLONISATION

### Exoantigens and Serodiagnosis

During *in vivo* and *in vitro* growth, *A. fumigatus* releases a complex mixture of polysaccharides and proteins whose antigenicity has been recognised for a long time. However, it is only recently that the isolation and purification of these "metabolic" antigens has begun with a view to developing standardised antigens for diagnosis.

**Polysaccharides**. Two different polysaccharides are secreted by *A. fumigatus*: a galactomannan (GM) and a polymer composed of galactosamine and *N*-acetyl galactosamine (Latgé *et al.*, 1991*a*). Glucan can also be found associated with GM depending on the composition of the culture medium. The composition of the polygalactosamine is unknown and its antigenicity has not been tested. In contrast, the galactomannan has been studied in detail because of its high immunogenicity in man and animals.

While the presence of acid-labile galactofuranose residues in GM preparation is generally accepted, opinions differ on other structural elements (Azuma *et al.*, 1971; Reiss and Lehmann, 1979; Barreto-Bergter *et al.*, 1981; Bennett *et al.*, 1985; van Bruggen-Van der Lugt *et al.*, 1992). These variations may be partly due to the different culture, extraction and purification protocols used in each laboratory. In our group, galactomannan is chemically purified by a double sequential hydrazine-nitrous acid treatment. Methylation-acetylation analysis of sugar derivatives by GLC/MS and $^{1}H$ and $^{13}C$ NMR studies of intact and acid hydrolysed galactomannan (16h at 100°C at pH 2) have been performed. The data obtained and comparisons with NMR spectra published by Unkefer and Gander (1979, 1990) showed that the mannan core has a linear configuration containing α(1-2) and (1-6) linked residues in the ratio of 3 to 1. Side chains are composed solely of ß(1-5) galactofuranosyl residues with an average degree of polymerisation of 5 (Figure 4). Unanswered questions concern the degree of polymerisation of all galactosyl residues in all side chains and the repartition of the linkages anchoring the galactofuranose on the mannosyl residues.

Monoclonal antibodies (MAb) have been selected against *Aspergillus* galactomannan (Stynen *et al.*, 1992). Immunochemical studies with synthetic ß-galactofuranose oligosaccharides have shown that the shortest chain that could efficiently inhibit all MAbs selected is a tetra ß(1-5) galactofuranoside. Longer chains were as efficient as the tetramer. This result suggests that the immunodominant epitope on the galactomannan is a tetra galactofuranoside with the chemical configuration presented in Figure 4.

With regard to GM, *A. fumigatus* behaves *in vivo* as *in vitro*, i.e. it secretes GM which can be recovered in the biological fluids of patients with invasive aspergillosis.

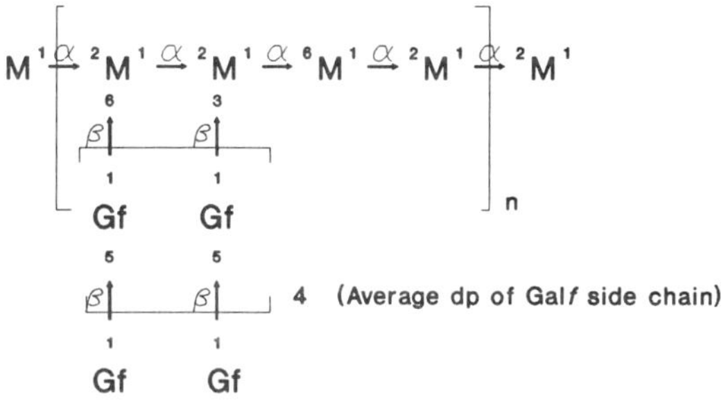


**Figure 4**. Structure of the galactomannan (GM) secreted by *A. fumigatus*. GM is purified by a double hydrazine-nitrous acid treatment and analysed by $^1$H and $^{13}$C NMR and GLC/MS after methylation-acetylation of the GM.

Since its detection could be used in the diagnosis of invasive aspergillosis this molecule has been of particular interest to medical mycologists for the last 20 years (de Repentigny, 1989; Rogers *et al.*, 1990; Buckley *et al.*, 1992). Anti-galactomannan MAbs have been used in antigen detection in the sera and urine of patients with invasive aspergillosis. The recent development of a sandwich ELISA has lowered the detection threshold to 0.5 ng/ml (J.P. Latgé, J. Sarfati and D. Stynen, unpublished data). In a retrospective study, circulating GM has been detected in patients more than 30 days before their death. This early detection is very encouraging for the diagnosis of invasive aspergillosis at a stage where the disease can be stopped by an appropriate antifungal therapy. Monitoring the GM concentration in the biological fluids of patients will be also used to assess the efficacy of the antifungal treatment on the development of *A. fumigatus in situ*.

The antigenic capacity of GM is well established. However, in contrast to *Candida* mannan for which immunomodulating activity has been studied (Domer *et al.*, 1989), the role of *A. fumigatus* GM in cell mediated immunity during the course of an infection is totally unknown.

**Proteins.** Around forty antigenic proteins produced in culture media have been identified in the literature on the basis of their molecular mass (Mr) after SDS-PAGE; but less than 10 have been purified (Table 1). The majority of the secreted proteins with high Mr are glycosylated as indicated by their reaction with anti-galactofurane MAb and concanavalin A (Figure 5). The antigenicity of these proteins is due in most cases to their galactofuranose containing sugar moiety. Treatment of immunoblots with periodic acid removed the reactivity of these glycoproteins with patient IgG without altering the antigenicity of the low Mr non-glycosylated proteins (unpublished). Depending on the composition of the culture medium proteins secreted have distinct Mr (Hearn, 1992). In spite of the variation in the Mr and glycosylation of exoantigens, IgG from aspergilloma patients react with the majority of the proteins secreted in the different culture media (Figure 6). Reactivity of each individual serum towards a single crude antigenic extract is also different (Figure 7) (Latgé *et al.*, 1991*b*). Only a few antigens present in all media are indeed detected by all patient sera. These results may be the reason for the absence of any standardised antigen used for the detection of aspergillosis.

**Table 1.** Purified proteic exoantigens of *Aspergillus fumigatus*[1].

| Antigens[2] | Function | Identification[3] | Purification | Ig binding |
|---|---|---|---|---|
| 18 kDa [4]<br>pI 9.5 | RNAse | WB, LISA | Cation exchange chromatography + gel filtration | IgE, IgG |
| 20 kDa (GP) [5]<br>pI 6.5 | | X E | Preparative IEF + ConA-affinity chromatography | IgG, IgE |
| 32-33 kDa [6]<br>pI 8.0 | Protease | WB, LISA | Anion exchange chromatography + gel filtration | IgG |
| 36 kDa (GP) [7]<br>pI 4.8 | | XIE WB | Gel filtration + immuno affinity chromatography | IgG, IgE |
| 40 kDa (GP) [8]<br>pI 5.5 | Protease | WB | Hydroxyapatite + cation exchange chromatography | IgG |
| 58 kDa (GP) [9] | | WB | Con A affinity + immunoaffinity chromatography | IgG |
| 70 kDa (GP) [10] | Protease ? | XIE, ELISA | Gel filtration + immunoaffinity chromatography | IgG |
| 88 kDa (GP) [11]<br>pI 5.8 | | WB, ELISA | Anion exchange chromatography + gel filtration | IgG |

[1] Antigens from commercial extracts and non-purified exo-antigens identified by Bernstein *et al.* (1990), Kurup (1989), Kurup *et al.* (1988, 1989) Latgé *et al.* (1991*b*) and Taylor and Longbottom (1988) are not included. [2] Molecular masses calculated after SDS-PAGE in reducing conditions; GP: glycoprotein. [3] XIE: Crossed immunoelectrophoresis; WB: Western blot. [4] Longbottom (1986, 1989); Arruda *et al.* (1990); Latgé *et al.* (1991). [5] Kurup *et al.* (1986). [6] Reichard *et al.* (1990); Monod *et al.* (1991); Frosco *et al.* (1992). [7] Harvey and Longbottom (1986). [8] Monod *et al.* (1993*b*). [9] Fratamico and Buckley (1991). [10] Harvey and Longbottom (1987). [11] Kobayashi *et al.* (1993).

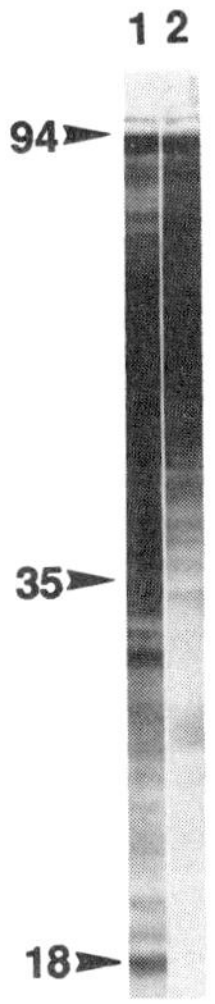


**Figure 5**. Glycosylation of *A. fumigatus* exoantigens produced in a 2 % glucose + 1 % peptone medium. Western blots of an ethanol precipitated culture filtrate are incubated with a pool of aspergilloma patient sera (lane 1) and an anti-galactomannan monoclonal antibody (lane 2).

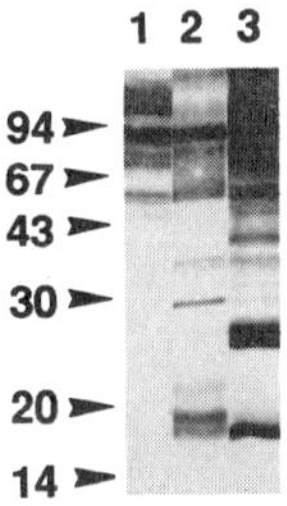


**Figure 6**. Immunoblot analysis of exoantigens produced in three different media : 0.2 % collagen (lane 1), 1 % yeast extract (lane 2), 2 % glucose + 1 % peptone (lane 3). All blots are incubated with the same pool of aspergilloma patient sera diluted 1/1000.

Taking into account these considerations, we recently purified an 88-kDa antigen which could be a good candidate for the serodiagnosis of aspergilloma (Kobayashi *et al.*, 1993). This antigen is a glycoprotein with a 9-kDa *N*-linked sugar moiety composed of mannose : glucose : galactose (16:10:1). It reacts negatively with anti-galactomannan monoclonal antibody. The antigenicity is associated with the polypeptide part of the molecule. It is recognised by more than 95% of sera from aspergilloma patients (Figure 8). Besides its antigenic property, this antigen could play a role in the development of the mycosis since sera from mice surviving to intravenous or intranasal infection of *A. fumigatus* react almost exclusively with this antigen (J.P. Latgé, D. Boucias and M. Moutaouakil, unpublished data). The protective role of this antigen against an *Aspergillus* infection which is suggested by these results has not yet been evaluated. Peptide sequencing has revealed that this 88-kDa antigen does not show any homology to known heat shock proteins (HSP) and should be different from an HSP-related antigen with similar Mr recently described by Burnie and Matthews (1991).

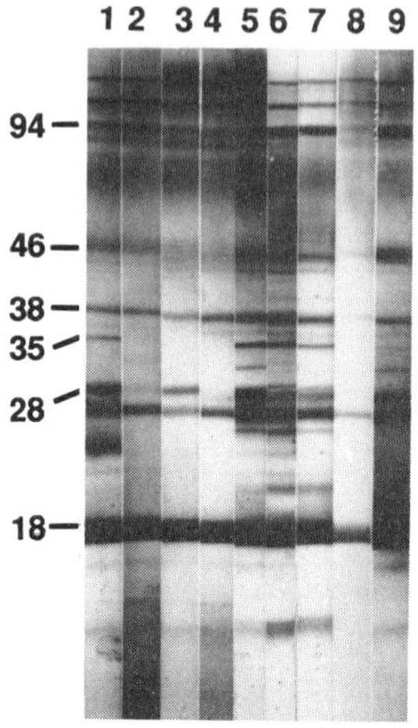


**Figure 7.** Differential reactivity of individual sera from aspergilloma patients with an ethanol precipitate of a culture filtrate of *A. fumigatus* grown in a 2 % glucose + 1 % peptone medium. Western blots are incubated with patient sera diluted 1/1000.

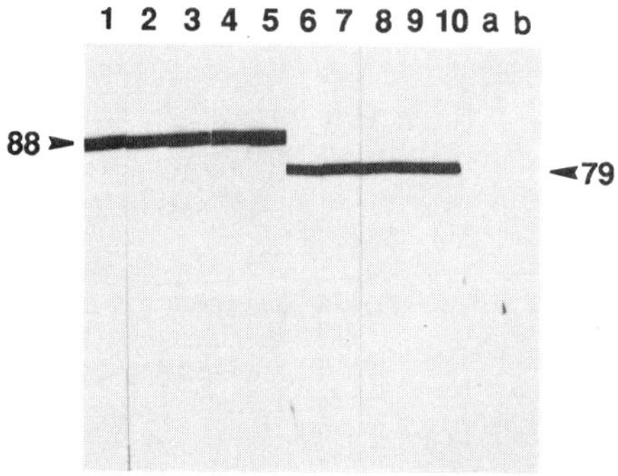


**Figure 8.** Immunoblot reactivity of the native 88-kDa antigen (lanes 1-5) and its deglycosylated form (lanes 6-10) with 10 individual sera from aspergilloma patients (lanes 1-10) and 2 sera from patients with candidiasis (lanes a, b). All sera were diluted 1/1000 and probed with anti-human IgG (H+L) peroxidase conjugate.

Antigenic proteins have also been detected in the biological fluids of patients and animals with invasive aspergillosis (Phillips and Radigan, 1989; Haynes *et al.*, 1990; Yu *et al.*, 1990). Most antigens detected *in vivo* have low Mr (11, 13, 14, 18, 20, 21, 27, 29, 33, 38, 40 kDa). Antigens with Mr >40 kDa (44, 45, 50, 80, 84-88 kDa) are seldom found and are most often glycosylated. Several of these antigens produced *in vivo* had similar Mr to antigens secreted *in vitro*. A basic protein of 18 kDa (also called AspFI and Ag3) was recognised as a major antigen binding to IgE and IgG of allergic and aspergilloma patients (Arruda *et al.*, 1990; Longbottom, 1986). The recent purification of this protein produced *in vitro* and the subsequent selection of monospecific antibodies directed towards this antigen has allowed us to demonstrate that the 18-kDa antigen produced *in vitro* and the protein secreted in the urine of patients with invasive aspergillosis were indeed the same molecule (Latgé *et al.* 1991*b*). The potential of this antigen in the diagnosis of invasive aspergillosis is currently being assessed using a sandwich ELISA. With the exception of this 18-kDa antigen, the identity between other antigens with identical Mr present both in the biological fluid of patients and in culture media has not been demonstrated in absence of purification of these antigens.

**Exoantigens and Virulence**

With the exception of their Ig binding properties, very little is known on the function of the different antigenic proteins. Only three proteins with an enzyme activity have been purified. It is an 18-kDa ribonuclease, a 33-kDa serine protease and a 40-kDa metalloprotease. Other enzymatic activities which may play a role during the pathogenic development of *A. fumigatus* have been detected in culture filtrates (e.g. catalase, phospholipase) (Biguet *et al.*, 1964; V. Choumet, J.P. Debeaupuis, J.P. Latgé and C. Bon, unpublished data), but the protein possessing these activities has not been identified. Two of the antigenic enzymes, i.e. the 18-kDa and the 33-kDa proteins, have been characterised genetically and their putative role during infection is discussed in the following section.

**The 33- and 40-kDa Proteases.** Proteolytic degradation of the lung tissue has been suggested to be one of the key events involved in the pathogenesis of *A. fumigatus*. Virulence of this species has been correlated with proteolytic activity of the fungus (Miyaji and Nishimura, 1977; Kothary *et al.*, 1984). An alkaline protease (ALP) of the subtilisin family (elastase) has been isolated recently by several research groups (Reichard *et al.*, 1990, Monod *et al.*, 1991, Frosco *et al.*, 1992, Larcher *et al.*, 1992). This 32-33-kDa serine protease is totally inhibited by PMSF, antipain, chymostatin and ß-2-macroglobulin, a pattern characteristic of a subtilisin. The ability of the 33-kDa enzyme to degrade elastin and collagen, two main proteins of the lung tissue and its detection *in vivo* in an experimental murine aspergillosis (Moutaouakil *et al.*, 1993) were two sufficient reasons to consider this protein as a putative virulent factor of *A. fumigatus*. The gene encoding for this protease has been cloned from both genomic and cDNA libraries of *A. fumigatus*. It contains three exons, codes for a 403 amino acid proenzyme with a leader sequence of 121 amino acids (Jaton-Ogay *et al.*, 1992). This gene was inactivated by gene disruption after transformation of a pathogenic strain of the fungus with a linear DNA fragment carrying the gene from which the central part was replaced by the selectable *E. coli* hygromycin B dominant resistance marker (Tang *et al.*, 1992; Monod *et al.*, 1993*a*). Both isogenic ALP producing and ALP non producing hygromycin resistant strains were tested for virulence in immunosuppressed mice (Figure 9). Surprisingly, no difference of pathogenicity was observed between isogenic ALP positive and ALP negative strains. The mortality rate and the incidence of lung tissue invasion by mycelium of the two strains were similarly high (Monod *et al.*, 1993*a*). These results show that ALP is not required by *A. fumigatus* to invade the lung tissues.

An analysis of the proteolytic activity of the ALP- mutant has shown that these mutants were not able to degrade elastin but were still able to lyse collagen. The residual proteolytic activity is due to a 40-kDa metalloprotease (MEP) which is inhibited by EDTA, phosphoramidon and 1,10 phenantroline (Monod *et al.*, 1993b). The presence of this other proteolytic activity, not due to ALP, is sufficient to allow mycelial development *in vitro* in a collagen medium and *in vivo* in the lung tissues. The absence of an essential role of elastase during lung colonisation is confirmed by a recent anatomopathological study of invasive aspergillosis where no overall loss of elastic tissues was seen in blood vessel walls infiltrated by hyphae (Denning *et al.*, 1992). The authors of this study concluded that if

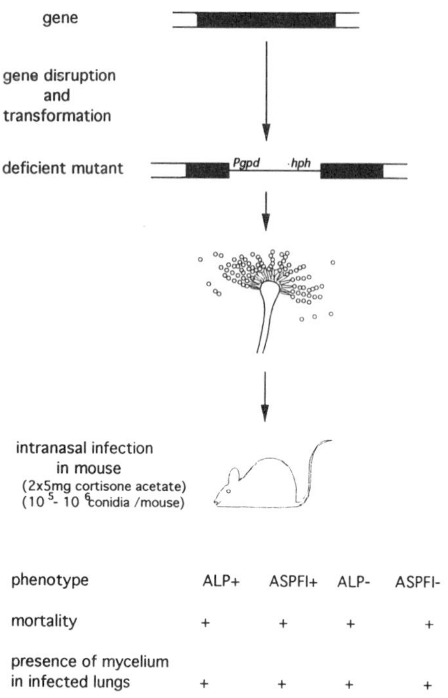


**Figure 9.** Strategy used in *A. fumigatus* to disrupt the genes of the 33-kDa serine protease and 18-kDa ribonuclease and to assess the virulence of deficient mutants in a mouse infection model.

elastolysis of vessel walls occurs it must be very localised and/or transient. In contrast to anti-ALP-antibodies, anti-MEP antibodies are produced in high amount in infected patients and are discriminatory for aspergilloma sera. These data suggested that MEP is secreted in higher amount than ALP during lung invasion by *A. fumigatus* and that in fact MEP may be the essential protease in the colonisation of the lung tissue by *A. fumigatus*. This 40-kDa metalloprotease is currently being cloned and gene disruption will be undertaken to verify its role as a virulence factor in *A. fumigatus*.

**The 18-kDa Antigen.** The 18-kDa antigen (ASPFI) is a very specific intra-ribosomal RNA nuclease with cleaves a single phosphodiester bond in a highly conserved region and releases a 400-base fragment from the 3' end of the large ribosomal RNA (Lamy and Davies, 1991). By doing so it interferes with the elongation factors and inactivates ribosome and protein synthesis in the cell. On the basis of its enzymatic function and amino acid sequence, this protein has been shown to be identical to restrictocin, a ribotoxin described in early 1960s (Lamy *et al.*, 1991). The gene of this protein, designated under the different names of restrictocin (Lamy and Davies, 1991), 18-kDa antigen (Lamy *et al.*, 1991), ASPFI (Arruda *et al.*, 1992) and allergen I/a (Moser *et al.*, 1992), has been cloned and sequenced by 4 independent research groups. The protein is composed of 149 amino acids with a 27 amino acid leader sequence. The coding sequence is interrupted by a short intron of 52 nucleotides. The earlier interest of restrictocin was due to its acute toxicity towards different tumour cells. Penetration of this protein synthesis inhibitor seems only possible if the integrity of the cell membrane is disturbed (Munoz *et al.*, 1985). Internalisation of this protein can be also achieved by the construction of immunoconjugates, prepared by the attachment of the ribotoxin to monoclonal antibodies specific of receptors present on the surface of the target cells

(Conde *et al.*, 1989; Wawrzynczak *et al.*, 1991). The occurrence of this ribosome inactivating protein in the course of an *Aspergillus* infection and its presence on the conidial surface has prompted several groups to consider this molecule as a potential virulence factor in aspergillosis (Lamy *et al.*, 1991; Arruda *et al.*, 1992; Brandhorst and Kenealy, 1992).

To demonstrate the contribution of ASPFI in the pathogenesis of *A. fumigatus* related diseases, the ASPFI gene was disrupted using a strategy identical to that used for the disruption of the alkaline protease (Figure 9). The ASPFI mutant strain and the parental wild type strain produced a similar mortality after nasal injection of cortisone-treated mice consecutive to mycelial growth in the lung tissues (Paris *et al.*, 1993). These negative results demonstrated that, at least in our mouse model, ASPFI does not play a prominent role in helping *A. fumigatus* to colonise the lung tissues.

Other related animal experiments have shown that, although very toxic *in vitro* when cells are permeabilised, ASPFI does not have an acute toxicity *in vivo*. First of all, injection to rabbits of 0.5 mg of pure ASPFI protein during immunisation experiments did not result in any toxicity to animal. Guinea pig tracheal epithelial cells are insensitive to 6 hr exposure of high concentration of ASPFI (up to 100 $\mu$g/ml). Cell death was not observed (measured by Cr51 release) even if the tracheal cells are primed with a polycation such as polyarginin, which at 10 $\mu$g/ml disturb partially the integrity of these cells (M.-A. Nahori, J.P. Debeaupuis, J.P. Latgé, B. Vargaftig, unpublished data). Similarly, airway responsiveness evaluated in a rat model, was not modified by the injection of 80 $\mu$g of ASPFI to the rat (A.J. Coyle, J.P. Debeaupuis, J.P. Latgé, unpublished data). Western and Southern blot experiments have shown that the production of this protein is independent of the pathogenicity to man of the species producing ASPFI (Table II). It is worth mentioning that two species of *Aspergillus* viz. *A. giganteus* which secretes $\alpha$-sarcin, a ribotoxin with 90% amino acid identity with ASPFI or *A. oryzae* which produces the ribonuclease TI closely related to ASPFI, are not human pathogens (Lamy *et al.*, 1991).

**Table 2.** Ribotoxin producers in *Aspergillus* and their pathogenicity to man[1]

| Species | Pathogenicity | Production of ASPFI | |
|---|---|---|---|
| | | Western blot | Southern blot |
| *A. fumigatus* | + | + | + |
| *A. flavus* | + | - | - |
| *A. niger* | + | - | - |
| *A. terreus* | + | - | - |
| *A. nidulans* | + | - | - |
| *Neosartorya fischeri* | - | + | + |
| *N. aureola* | - | + | ND |
| *N. straemenia* | - | + | ND |
| *A. clavatus* | - | + | ND |
| *A. ochraceus* | - | + | ND |
| *A. restrictus* | - | + | + |

[1] Compiled data from Arruda *et al.* (1990, 1992); Hearn *et al.* (1990); Lamy and Davies (1991); Latgé *et al.* (1991*b*); Monod, Paris and Latgé (unpublished).

All these data suggest that ASPFI is not a major virulence factor in the establishment of aspergillosis. However, ASPFI could play a role in association with other *A. fumigatus* secreted molecules. For example, associated with gliotoxin, ASPFI impairs lymphocyte stimulation in mixed lymphocyte cultures (P.I. Gumowski, F. Goergen, J.P. Debeaupuis, J.P. Latgé, unpublished data).

One of the molecular biological problems related to ASPFI is to understand how *A. fumigatus* protects itself from the highly toxic activity of this molecule. Because of its extremely potent inhibitory activity against protein synthesis, due to its catalytic mode of action, one molecule of this protein would be lethal to the cell. Moreover, it is known that the ribosomes of the producing organism are sensitive to the toxin (Lamy *et al.*, 1991). The genomic organisation of the ASPFI gene have prompted Lamy and Davies (1991) to suggest that ASPFI would be secreted as an inactive protoxin which becomes active after the cleavage of a signal peptide of 27 amino acids residues in the Golgi apparatus. Recent studies by Yang and Kenealy (1992), however, have shown that this hypothesis was wrong. The toxicity of genetically engineered ASPFI was not reduced by the presence of the 27 amino acid leader sequence or a 7-8 amino acid putative leader sequence. On the basis of sequence similarities between ASPFI and RNAse U2, the same authors have identified the amino acid histidine 136 as the putative active site of ribotoxin. Replacement of histidine 136 by leucine resulted in an inactive protein. In contrast to native ASPFI, the mutated inactive protein associated to low concentration of gliotoxin looses immunosuppressive activity in lymphocyte transformation tests (P.I. Gumowski, F. Goergen, J.P. Debeaupuis, R. Kenealy and J.P. Latgé, unpublished data) (Figure 10). The transcription signals and regulation factors involved in the secretion of this fascinating antigen can now be investigated as well as its potential role as an immunomodulator of the response towards *A. fumigatus*.

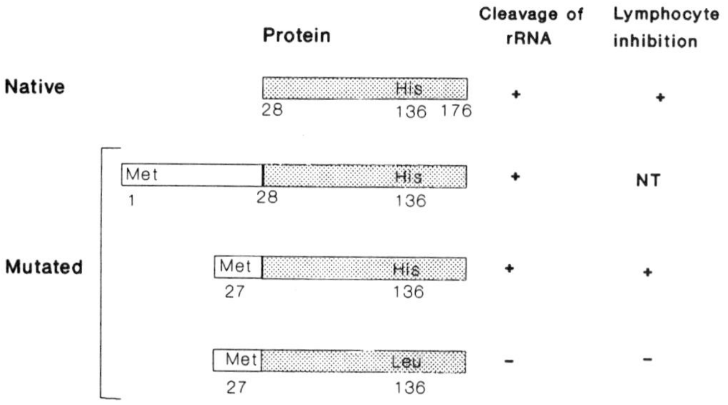


**Figure 10.** Effect of amino-terminal extension and specific mutations on the ribonuclease activity and lymphocyte inhibition in mixed lymphocyte cultures preincubated with stimulatory concentration of gliotoxin adapted from Lamy and Davies (1991); Yang and Kenealy (1992); P.I. Gumowski *et al.* (unpublished data).

## CONCLUSION

Exocellular molecules which help anchoring *A. fumigatus* in the lower respiratory tract or which interact with phagocytosis have been poorly studied. Disturbance of the alternative complement pathway and inhibition of the phagocytic uptake have been demonstrated. However, most of these data have been obtained during *in vitro* studies and are too limited to certify that the molecules responsible for the inhibition of these defence reactions play a significant role in the establishment of the fungus in the lung.

Due to the interest in serodiagnosis of most researchers studying *A. fumigatus*, the bulk of information on this fungus concerns the antigens produced during mycelial growth. Numerous antigens or semi-purified antigenic fractions have been described in the literature but only a few have been purified and characterised biochemically and immunologically. Amongst these purified antigens which can be used today for the serodiagnosis of aspergillosis are a galactomannan in invasive aspergillosis, a glycoprotein of 88-kDa for aspergilloma and a 18-kDa protein for aspergilloma and ABPA.

The role of exocellular enzymes in the establishment of the mycosis was tested by reverse genetics for an 18-kDa ribonuclease and a 33-kDa serine protease. In both cases, it was proven that these proteins with enzyme activities which could logically be involved in the development of the disease *in vivo*, were not essential in preventing the fungus to colonise the lung tissues. These results demonstrate that either these proteins do not have any role in the pathogenicity of *A. fumigatus* or more probably that lung colonisation is under the control of several enzymatic proteins acting either sequentially or simultaneously. In the later case, the deletion of one factor is not sufficient to prevent the installation of the disease when the other virulence determinants are functional.

Two pitfalls may buffer the results obtained until now on the role of antigenic enzymes in the virulence of *A. fumigatus*. First of all, the pathogenicity studies were performed in a mouse model which does not mimic the human situation. In immunocompromised patients, invasive aspergillosis is seen in patients with long-term granulocytopenia who inhale quantities of spores likely to be many orders of magnitude lower than those used in the mouse experiments. Levels of $10^1$ to $10^2$ conidia, daily, reach the lung in a human individual exposed to normal aerial concentration of spores (1-10 spores/$m^3$ and 10-20, 000 l of air processed daily by the lung). In contrast, in our experimental model, the mouse lungs received only one spore injection but with a high dose of $10^5$-$10^6$ conidia. Secondly, all antigens studied have been obtained from *in vitro* grown cultures and are not necessarily expressed in lung tissues. Recent experiments with infected animals have shown that most antigens found in infected kidneys are also produced *in vitro*. However, the number of antigens detected *in vitro* is lower than *in vivo*. Moreover, major antigens are different *in vivo* and *in vitro* (J. Sarfati, D. Boucias, J.P. Latgé, manuscript in preparation). These data and the results obtained with ALP suggest that the amount and nature of the proteases and other virulence factors secreted by *A. fumigatus* in infected tissues and in culture media may be different.

In the absence of true non-pathogenic strains or at least strains with lower virulence due to restricted growth in the lung tissue, it is difficult to assess the role of exoantigens of *A. fumigatus* in the establishment of the mycosis. Non-pathogenic strains have sometimes been reported in the literature, but in our hands with in our mouse model, all strains tested are able to kill mice following lung invasion by mycelium.

If some progress has been made recently in the isolation of pure polysaccharidic and proteic antigens useful in the serodiagnosis of aspergillosis, the role of the exoantigens of *A. fumigatus* in the immune regulation of the disease is poorly recognised. Very little is known indeed on their role in the different T and B cells and macrophage populations and in the production of the different cytokins regulating the immune response.

In conclusion, a lot remains to be done to understand the pathogeny of aspergillosis which is without doubt a mycosis of the future due to the continuous increase of immunosuppressive therapies in modern medicine.

## ACKNOWLEDGEMENTS

This research was partly funded by INSERM grant #900313.

## REFERENCES

Arruda L.K., Mann B.J., and Chapman M.D. (1992) Selective expression of a major allergen and cytotoxin, Asp fI, in *Aspergillus fumigatus*. Implications for the immunopathogenesis of *Aspergillus*-related diseases. J. Immunol. 149, 3354-3359.

Arruda L.K., Platts-Mills T.A., Fox J.W., and Chapman M.D. (1990) *Aspergillus fumigatus* allergen I, a major IgE-binding protein, is a member of the mitogillin family of cytotoxins. J. Exp. Med. 172,1529-1532.

Azuma I., Kimura H., Hirao F., Tsubura E., Yamamura Y., and Misaki A. (1971) Biochemical and immunological studies on *Aspergillus*. III: chemical and immunological properties of glycopeptide obtained from *Aspergillus fumigatus*. Jpn J. Microbiol. 15, 237-246.

Barreto-Bergter E., Gorin P.A.J., and Travassos L.R. (1981) Cell constituents of mycelia and conidia of *Aspergillus fumigatus*. Carbohydr. Res. 95, 205-218.

Bennett J.E., Bhattacharjee A.K., and Glaudemans C.P.J. (1985) Galactofuranosyl groups are immunodominant in *Aspergillus fumigatus* galactomannan. Mol. Immunol. 22,251-254.

Bernstein J.A., Zeiss C.R., Greenberger P.A., Patterson R., Marhoul J.F., Smith L.L., (1990), Immunoblot analysis of sera from patients with allergic bronchopulmonary aspergillosis: correlation with disease activity. J Allergy Clin. Immunol. 86, 532-539.

Biguet J., Tran van Ky P., Andrieu S., and Fruit J. (1964) Analyse immunoélectrophorétique d'extraits cellulaires et de milieux de culture d' *Aspergillus fumigatus* par des immunsérums expérimentaux et des sérums de malades atteints d'aspergillome bronchopulmonaire, Ann. Inst. Past. 107, 73-97.

Bodey G.P., and Vartivarian S. (1989) Aspergillosis. Eur. J. Clin. Microbiol. Infect. Dis. 8, 413-437.

Borglum-Jensen T.D., Schoenheyder H., Andersen P., and Stenderup A. (1986) Binding of C-reactive protein to *Aspergillus fumigatus* fractions. J. Med. Microbiol. 21, 173-177.

Bouziane H., Latgé J.P., Prévost M.C., Chevance L.G., and Paris S. (1988) Nasal allergy in guinea pigs. Mycopathologia 101, 181-186.

Brandhorst T.T., and Kenealy W.R. (1992) Production and localization of restrictocin in *Aspergillus restrictus*. J. Gen. Microbiol. 138, 1429-1435.

van Bruggen-Van der Lugt A.W., Kamphuis H.J., De Ruiter G.A., Mischnick P., Van Boom J., and Rombouts F.M. (1992) New structural features of the antigenic extracellular polysaccharides of *Penicillium* and *Aspergillus* species revealed with exo-beta-*D*-galactofuranosidase. J. Bacteriol. 174, 6096-6102.

Buckley H.R., Richardson M.D., Evans E.G.V., and Wheat L.J. (1992) Immunodiagnosis of invasive fungal infection. J. Med. Vet. Mycol. 30 (Suppl 1), 249-260.

Burnie J.P., and Matthews R.C. (1991) Heat-shock protein 88 and *Aspergillus* infection, J. Clin. Microbiol. 29, 2099-2106.

Cohen M.S., Isturiz R.E., Malech H.L., Root R.K., Wilfert C.M., Gutman L., and Buckley R.H. (1981) Fungal infection in chronic granulomatous disease. The importance of the phagocytes in defense against fungi. Amer. J. Med. 71, 59-66.

Cohen J. (1991) Clinical manifestations and management of aspergillosis in the compromised patient, in: "Fungal infection in the compromised patient" (Warnock D.W., Richardson M.D. Eds.), 2nd edn, pp. 117-152. John Wiley & Sons, New York.

Cole G.T., and Samson R.A. (1984) The conidia, in: "Mould allergy", (Al-Doory Y., Domson J.F. Eds.) pp. 66-103. Lea & Febiger, Philadelphia.

Cole P., and Wilson R. (1989) Host microbial interrelationships in respiratory infections. Chest 95, 217-221S.

Conde F.P., Orlandi R., Canevari S., Mezzanzanica D., Ripamonti M., Munoz S.M., Jorge P., and Colnaghi M.I. (1989) The *Aspergillus* toxin restriction is a suitable cytotoxic agent for generation of immunoconjugates with monoclonal antibodies directed against human carcinoma cells. Eur. J. Biochem. 178, 795-802.

Denning D.W., Ward P.N., Fenelon L.E., and Benbow E.W. (1992) Lack of vessel wall elastolysis in human invasive pulmonary aspergillosis. Infect. Immun. 60, 5153-5156.

Domer J.E., Garner R.E., and Befidi-Mengue R.N. (1989) Mannan as an antigen in cell-mediated immunity (CMI) assays and as a modulator of mannan-specific CMI. Infect. Immun. 57, 693-700.

Eichner R.D., Al Salmi M., Wood P.R., and Müllbacher A. (1986) The effects of gliotoxin upon macrophage function. Int. J. Immunopharmac. 8, 789-797.

Eichner R.D., Waring P., Geue A.M., Braitwaite A.W., Müllbacher A. (1988) Gliotoxin causes oxidative damage to plasmid and cellular DNA. J. Biol. Chem. 263, 3772-3777.

Fratamico P.M., Long W.K. and Buckley H.R. (1991) Production and characterisation of monoclonal antibodies to a 58-kilodalton antigen of *Aspergillus fumigatus*. Infect. Immun. 59, 316-322.

Frosco M., Chase T., McMillan J.D. (1992) Purification and properties of the elastase from *Aspergillus fumigatus*. Infect. Immun. 60, 728-734.

Gumowski P.I., Dunoyer-Geindre S., Goergen F., Latgé J.P., Beffa T., Aragno M., Selldorf P., and Gandolla M. (1992) Evaluation of occupational risk factors for workers in municipal composting facilities. Eur. Resp. J. 5(S15), 406S-407S.

Harvey C., and Longbottom J.L. (1986) Characterization of a major antigenic component of *Aspergillus fumigatus*. Clin. Exp. Immunol. 65, 206-214.

Harvey C., and Longbottom J.L. (1987) Characterization of a second major antigen Ag 13 (antigen C) of *Aspergillus fumigatus* and investigation of its immunological reactivity. Clin. Exp. Immunol. 70, 247-254.

Haynes K.A., Latgé J.P., and Rogers T.R. (1990) Detection of *Aspergillus* antigens associated with invasive infection. J. Clin. Microbiol. 28, 2040-2044.

Hearn V.M. (1992) Antigenicity of *Aspergillus* species. J. Med. Vet. Mycol. 30, 11-25.

Hearn V.M., Moutaouakil M., and Latgé J.P. (1990) Analysis of components of *Aspergillus* and *Neosartorya* mycelial preparations by gel electrophoresis and western blotting procedures, in: "Modern concepts in *Penicillium* and *Aspergillus* classification" (Samson, R.A. and Pitt, J.I. Eds.), pp. 235. Plenum Press, London, New York.

Hostetter M.K. (1986) Serotypic variations among virulent *Pneumococci* in deposition and degradation of covalently bound C3b: implications of phagocytosis and antibody production. J. Infect. Dis. 153, 682.

Jaton-Ogay K., Suter M., Crameri R., Falchetto R., Fatih A., and Monod M. (1992) Nucleotide sequence of a genomic and a cDNA clone encoding an extracellular alkaline protease of *Aspergillus fumigatus*. FEMS Microbiol. Lett. 92, 163-168.

Kothary M.H., Chase T. jr, and McMillan J.D. (1984) Correlation of elastase production by some strains of *Aspergillus fumigatus* with ability to cause pulmonary invasive aspergillosis in mice. Infect. Immun. 43, 320-329.

Kurup V.P. (1989) Murine monoclonal antibodies binding to the specific antigens of *Aspergillus fumigatus* associated with allergic bronchopulmonary aspergillosis. J. Clin. Lab. Anal. 3, 116-121.

Kurup V.P., Greenberger P.A., and Fink J.N. (1989) Antibody response to low-molecular-weight antigens of *Aspergillus fumigatus* in allergic bronchopulmonary aspergillosis. J. Clin. Microbiol. 27, 1312-1316.

Kurup V.P., John K.V., Resnick A., and Fink J.N. (1986) A partially purified glycoprotein antigen from *Aspergillus fumigatus*. Int. Archs. Allergy. Appl. Immun. 79, 263-269.

Kurup V.P., Ramasamy M., Greenberger P.A., and Fink J.N. (1988) Isolation and characterization of a relevant *Aspergillus fumigatus* antigen with IgG- and IgE-binding activity. Int. Archs Allergy. Appl. Immun. 86, 176-182.

Lamy B., and Davies J. (1991) Isolation and nucleotide sequence of the *Aspergillus restrictus* gene coding for the ribonucleic toxin restrictocin and its expression in *A. nidulans*: the leader sequence protects producing strains from suicide. Nucl. Acids Res. 19, 1001-1006.

Lamy B., Moutaouakil M., Latgé J.P., and Davies J. (1991) Secretion of a potential virulence factor, a fungal ribonucleotoxin, during human aspergillosis infections. Mol. Microbiol. 5, 1811-1815.

Lamy B., Schindler D., and Davis J. (1991) The *Aspergillus* ribonucleolytic toxins (ribotoxins), in "Genetically engineered toxins" (Frankel A.E. Ed.) Marcel Dekker Inc., New York.

Larcher G., Bouchara J.P., Annaix V., Symoens F., Chabasse D., and Tronchin G. (1992) Purification and characterization of a fibrinogenolitic serine protease from *Aspergillus fumigatus* culture filtrate. FEBS Lett. 308, 65-69.

Latgé J.P., Debeaupuis J.P., Moutaouakil M., Diaquin M., Sarfati J., Prévost M.C., Wieruszeski J.M., Leroy Y., and Fournet B. (1991*a*) Galactomannan and the circulating antigens of *Aspergillus fumigatus*, in "Fungal cell wall and immune response" (Latgé, J.P. and Boucias, D.J. Eds.), pp. 143-151. Springer Verlag, Heidelberg.

Latgé J.P., Moutaouakil M., Debeaupuis J.P., Bouchara J.P., Haynes K., and Prévost M.C. (1991*b*) The 18-kilodalton antigen secreted by *Aspergillus fumigatus*. Infect. Immun. 59, 2586-2594.

Levitz S.M. and Diamond R.D (1985) Mechanisms of resistance of *Aspergillus fumigatus* conidia to killing by neutrophils *in vitro*. J. Infect. Dis. 152, 33-42.

Longbottom J.L. (1986) Antigens and allergens of *Aspergillus fumigatus*. II: their further identification and partial characterization of a major allergen (Ag 3). J. Allergy Clin. Immunol. 78, 18-24.

Longbottom J.L. (1989) Fungal allergens: their characterization with special reference to *Aspergillus fumigatus*. Adv. Biosci. 74, 43-56.

Mahler P.A., and Kinsella T.J. (1991) Radiation and cytotoxic drug effects on lung natural defence mechanisms, in "Respiratory disease in the immunosuppressed host" (Shelhama J., Pizzo P.A., Parrillo J.E., Masur H. Eds.), pp.30-36. J.B. Lippincott Company, Philadelphia.

Miyaji M., and Nishimura K. (1977) Relationship between proteolytic activity of *Aspergillus fumigatus* and the fungus' invasiveness of mouse brain. Mycopathologia 62, 161-166.

Monod M., Paris S., Sarfati J., Jaton-Ogay K., Ave P., and Latgé J.P. (1993*a*) Virulence of alkaline protease-deficient mutants of *Aspergillus fumigatus*. FEMS Microbiol. Lett. 106, 39-46.

Monod M., Togni G., Rahalison L., and Frenk E. (1991) Isolation and characterization of extracellular alkaline protease of *Aspergillus fumigatus*. J. Med. Microbiol. 35, 23-28.

Moser M., Crameri R., Menz G., Schneider T., Dudler T., Virchow C., Gmachl M., Blaser K., and Suter M. (1992) Cloning and expression of recombinant *Aspergillus fumigatus* allergen I/a (rAsp f 1/a) with IgE binding and type I skin test activity. J. Immunol. 149, 454-460.

Moutaouakil M., Monod M., Prévost M.C., Bouchara J.P., and Latgé J.P. (1993) Identification of the 33-kDa alkaline protease of *Aspergillus fumigatus in vitro* and *in vivo*. J. Med. Microbiol. (in press).

Müllbacher A., and Eichner R.D. (1984) Immunosuppression *in vitro* by a metabolite of a human pathogenic fungus. Proc. Natl Acad. Sci. U.S.A. 81, 3835-3837.

Mullins J., Harvey R., and Seaton A. (1976) Sources and incidences of airborne *Aspergillus fumigatus*. Clin. Allergy 6, 209-217.

Munoz A., Castrillo J.L., and Carrasco L. (1985) Modification of membrane permeability during Semliki Forest Virus infection. Virology 146, 203-212.

Nelson S., and Summer W.R. (1991) Adverse effects of pharmacologic therapy on pulmonary host defenses, in "Respiratory disease in the immunosuppressed host" (Shelhama J., Pizzo P.A., Parrillo J.E., Masur H. Eds), pp. 15-29. J.B. Lippincott Company, Philadelphia.

Pepys J., and Longbottom J.L. (1971) Antigenic and C-substance activities of related glycopeptides from fungal, parasitic and vegetable sources. Int. Archs Allergy Appl. Immun. 41, 219-221.

Phillips P., and Radigan G. (1989) Antigenemia in a rabbit model of invasive Aspergillosis. J. Infect. Dis. 159, 1147-1150.

Raper K.B., and Fennell D.I. (1965) *Aspergillus fumigatus* group, in "The genus *Aspergillus*", (Raper K.B. and Fennell, D.I. Eds), pp. 238-268. William & Wilkins, Baltimore.

Reichard U., Büttner S., Eiffert H., Staib F., and Rüchel R. (1990) Purification and characterisation of an extracellular serine proteinase from *Aspergillus fumigatus* and its detection in tissue. J. Med. Microbiol. 33, 243-251.

Reiss E., and Lehmann P.F. (1979) Galactomannan antigenemia in invasive aspergillosis. Infect. Immun. 25, 357-365.

Repentigny L. de, (1989) Serological techniques for diagnosis of fungal infection. Eur. J. Clin. Microbiol. Infect. Dis. 4, 362-375.

Reynolds H.Y. (1991) Pulmonary host defenses, in "Respiratory disease in the immunosuppressed host" (Shelhama J., Pizzo P.A., Parrillo J.E., and Masur H. Eds.), pp. 3-14. J.B. Lippincott Company, Philadelphia.

Robertson M.D., Brown D.M., MacLaren W.M., and Seaton A. (1988) Fungal handling by phagocytic cells from asthmatic patients sensitised and non-sensitised to *Aspergillus fumigatus*. Thorax 43, 224.

Robertson M.D., Kerr K.M., Seaton A. (1989) Killing *Aspergillus fumigatus* spores by human lung macrophages: a paradoxical effect of heat-labile serum component. J. Vet. Med. Mycol. 27, 295-302.

Robertson M.D., Seaton A., Milne L.J.R., and Raeburn J.A. (1987) Resistance of spores of *Aspergillus fumigatus* to ingestion by phagocytic cells. Thorax 42, 466-472.

Robertson M.D., Seaton A., Milne L.J.R., and Raeburn J.A. (1987) Suppression of host defense by *Aspergillus fumigatus*. Thorax 42, 19-25.

Robertson M.D., Seaton A., Raeburn J.A., and Milne L.J.R. (1987) Inhibition of phagocyte migration and spreading by spore diffusates of *Aspergillus fumigatus*. J. Med. Vet. Mycol. 25, 389-396.

Rogers T.R., Haynes K.A., and Barnes R.A. (1990) Value of antigen detection in predicting invasive pulmonary aspergillosis. Lancet 336, 1210-1213.

Schaffner A. (1989) Experimental basis for the clinical epidemiology of fungal infections. A review. Mycoses 32, 499-515.

Schönheyder H., Jensen T., Hoiby N., and Koch C. (1988) Clinical and serological survey of pulmonary aspergillosis in patients with cystic fibrosis. Int. Archs Allergy Appl. Immun. 85, 472-477.

Seaton A., Seaton D., and Leitch A.G. (1989) "Crofton and Douglas's respiratory diseases", pp. 1215. Blackwell Scientific Publ., Oxford.

Seiglemurandi F., Krivobok S., Steiman R., and Marzin D. (1990) Production, mutagenicity, and immunotoxicity of gliotoxin. J. Agric. Fd Chem. 38, 1854-1856.

Steinfort C., Wilson R., Mitchell T., Feldman C., Rutman A., Todd H., Sykes D., Walker J., Saunders K., Andrew P.W., Bournois G.J., and Cole P. (1989) Effect of *Streptococcus pneumoniae* on human respiratory epithelium *in vitro*. Infect. Immun. 57, 2006-2013.

Sturtevant J., and Latgé J.P. (1992*a*) Interactions between conidia of *Aspergillus fumigatus* and human complement component C3. Infect. Immun. 60, 1913-1918.

Sturtevant J., and Latgé J.P. (1992*b*) Participation of complement in the phagocytosis of the conidia of *Aspergillus fumigatus* by human polymorphonuclear cells. J. Infect. Dis. 166, 580-586.

Stynen D., Sarfati J., Goris A., Prévost M.C., Lesourd M., Kamphuis H., Darras V., and Latgé J.P. (1992) Rat monoclonal antibodies against *Aspergillus* galactomannan. Infect. Immun. 60, 2237-2245.

Tang C.M., Cohen J., and Holden D.W. (1992) An *Aspergillus fumigatus* alkaline protease mutant constructed by gene disruption is deficient in extracellular elastase activity. Mol. Microbiol. 6, 1663-1671.

Taylor M.L., and Longbottom J.L. (1988) Partial characterization of rapidly released antigenic/allergenic component (Ag5) of *Aspergillus fumigatus*. J. Allergy Clin. Immunol. 81, 548-557.

Turner W.B., and Aldridge D.C. (1983) "Fungal metabolites", II, pp. 631. Academic Press, London.

Unkefer C.J., and Gander J.E. (1979) The 5-0-beta-*D*-galactofuranosyl-containing glycopeptide from *Penicillium charlesii*. Carbon-13 nuclear magnetic resonance studies. J. Biol. Chem. 254, 12131-12135.

Unkefer C.J., and Gander J.E. (1990) The 5-0-beta-*D*-galactofuranosyl-containing peptido-phospho-galactomannan of *Penicillium charlesii*. Characterization of the mannan by 13-C NMR spectroscopy. J. Biol. Chem. 265, 685-689.

Wagner G.E. (1984) Bronchopulmonary aspergillosis and aspergilloma, in "Mould allergy" (Al-Doory Y., Domson J.F. Eds.) pp. 202-216. Lea & Febiger, Philadelphia.

Waldorf A.R. (1989) Pulmonary defense mechanisms against opportunistic fungal pathogens, *in:* "Immunology of fungal diseases" Immunol series 47 (Kurstak E., Marques G., Auger P., de Repentigny L., Montplaisir S. Eds.) pp. 243-270. Marcel Dekker, New York.

Waring P., Eichner R.D., Müllbacher A., and Sjaarda A. (1988) Gliotoxin induces apoptosis in macrophages unrelated to its antiphagocytic properties. J. Biol. Chem. 263, 18493-9.

Washburn R.G., DeHart D.J., Agwu D.E., Bryant-Varela B.J., and Julian N.C. (1990) *Aspergillus fumigatus* complement inhibitor: production, characterization, and purification by hydrophobic interaction and thin layer chromatography. Infect. Immun. 58, 3508-3515.

Washburn R.G., Hammer C.H., and Bennett J.E. (1986) Inhibition of complement by culture supernatants of *Aspergillus fumigatus*. J. Infect. Dis. 154, 944-951.

Wawrzynczak E.J., Henry R.V., Cumber A.J., Parnell G.D., Derbyshire E.J., and Ulbrich N. (1991) Biochemical, cytotoxic and pharmacokinetic properties of an immunotoxin composed of a mouse monoclonal antibody Fib75 and the ribosome-inactivating protein alpha-sarcin from *Aspergillus giganteus*. Eur. J. Biochem. 196, 203-209.

Wilson R., Read R., Thomas M., Rutman A., Harison K., Lund V., Cookson B., Goldman W., Lambert H., and Cole P. (1991) Effects of *Bordetella pertussis* infection on human respiratory epithelium *in vivo* and *in vitro*. Infect. Immun. 59, 337-345.

Yang R., and Kenealy W.R. (1992) Effects of amino-terminal extensions and specific mutations on the activity of restrictocin. J. Biol. Chem. 267, 16801-16805.

Yu B., Niki Y., and Armstrong D. (1990) Use of immunoblotting to detect *Aspergillus fumigatus* antigen in sera and urines of rats with experimental invasive aspergillosis. J. Clin. Microbiol. 28, 1575-1579.

# CELL WALL IMMUNOCHEMISTRY AND INFECTION

Veronica M. Hearn

PHLS Mycology Reference Laboratory
Department of Microbiology
University of Leeds
Leeds LS2 9JT

## INTRODUCTION

The fungal wall performs many functions, giving the cell its shape, serving a protective role and regulating the flow of molecules, including digestive enzymes, out of the cell (Chang and Trevithick, 1974; Pugh and Cawson, 1977). Protection is of particular importance with fungal pathogens. It has been postulated that primary pathogens differ from opportunistic pathogens such as *Aspergillus* in the greater resistance of the former to microbicidal agents produced by the phagocytic cells of an invaded host (Diamond, 1991).

The cell wall is not a static structure, it is flexible to allow for conidial growth and expansion with differentiation to the hyphal form. It is also necessary for the organism to be able to adapt to its environment, where any changes may be reflected by corresponding changes in the wall (Venkateswerlu and Stotzky, 1986; Hazan, 1989).

*Aspergillus* is a saprophyte in nature whose natural habitat is soil however, under certain conditions, a limited number of species can act as opportunistic pathogens of man, causing a variety of lung-related diseases. Much of the detailed analysis of *Aspergillus* wall structure has focused on relatively few species, all of them pathogenic *viz.*, *A.fumigatus, A.nidulans* and *A.niger*. The need for diagnostic tests for aspergillosis led to immunochemical studies involving the use of defined antibodies as probes to detect *in situ* antigens on germ-tubes and hyphal structures (Drouhet *et al.*, 1972 and Schønheyder *et al.*, 1982). With the advent of monoclonal antibodies, additional studies have resulted as to the antigenic nature of the cell wall and attempts to localise the so-called immunodominant antigens. Studies of cellular immunity, where *Aspergillus* spores and hyphae were allowed to interact with cellular defence systems, has given additional information on surface characteristics (Levitz and Diamond, 1985).

With the major pathogenic species of *Aspergillus*, the changes that occur during growth are reflected by changes in host response. Thus, in addition to studying the chemical composition, the architecture and the biosynthetic processes involved in the

*The Genus Aspergillus*, Edited by Keith A. Powell *et al.*,
Plenum Press, New York, 1994

assembly of the cell wall, by analysing its immunochemistry and the immunological response of the host, insight may be gained into how this complex, dynamic structure works. This contribution will examine some of the features of differentiation of *Aspergillus* wall both as a function of age, as it progresses from spore to hypha, and as a function of speciation. It is of interest to establish how this affects the organism's survival and how it affects the host response in pathogenesis. Initially, the more recent chemical and morphological data available on *Aspergillus* will be presented followed by immunochemical and immunological analysis of the major pathogenic species.

## MORPHOLOGY

In a paper published in 1969, Hess and Stocks analysed surface structure of conidia from ten species of *Aspergillus* using freeze-etching procedures. As with *Penicillium* species, the surfaces of all species of *Aspergillus* conidia studied were covered with "rodlets" which consist of linear arrays of particles approximately 50°A in diameter. Individual "rodlets" are identical to those reported for *Penicillium* conidia but the patterns in which they are arranged appear to be species specific. The conidial wall of *A.niger* was investigated by Cole and Pope (1981). Non-ionic detergent extracted an amorphous, hydrophobic layer (thought to contain melanin and protein). "Rodlets" remained *in situ* but were extractable with dilute alkali and were shown to be lipoprotein structures. The authors reported surface ultrastructure of *A.fumigatus* conidia as comparable to that of *A.niger* (Cole *et al.*, 1983).

The changes that occur in *A.nidulans* conidial wall during germination were monitored in an elegant study (Border and Trinci, 1970). Wall layers present in dormant conidia rupture as spores swell and expand. As germination proceeds, new inner wall layers are formed and one of these emerges to form the germ tube wall. The most rapid changes to the fungal wall are seen during spore germination. Vesicles have been reported present in the region of germ tube emergence. They are abundant in hyphal tips, and appear to function as transporters of wall precursors and/or hydrolytic and synthetic enzymes to allow for plasticity during cell growth by cell wall extension (Girbardt, 1969). Enzymic lysis is indicated during active hyphal extension by high autolysis of wall galactans and mannans in newly-formed wall of *A.nidulans* (Rosenberger, 1979). Vesicles and hyphal walls of *A.nidulans* contain a mixture of glucanases, proteinases and chitinase (Polacheck and Rosenberger, 1978). However, it is difficult to determine which enzymes are truly mural and which are located in the periplasmic space. These aspects of fungal growth have recently been reviewed (Bartnicki-Garcia, 1990; Gooday and Gow, 1990).

Specific staining of conidial surfaces with wheat germ agglutinin (which binds preferentially to N-acetyl-glucosamine residues) strongly supports the contention that chitin is exposed on *A.fumigatus* wall (Kan and Bennett, 1988). In hyphae the major location of chitin deposition is the hyphal tip, with all the evidence pointing towards the cell membrane as the site of chitin polymerisation (Gooday, 1977). Only (1-3)-ß-D-glucan is found in the newly-formed wall at the apex. In sub-apical regions, incorporation into the wall of a (1-3)-ß-/(1-6)-ß-D-glucan occurs (Sietsma *et al.*, 1985). Fluorescein-conjugated lectins, specific for N-acetyl-galacosamine and ß-D-galactose, gave a strong staining reaction, when used as a probe, with hyphal sections of *A.fumigatus* (Stoddart and Herbertson, 1978).

## CHEMISTRY

*Aspergillus* belongs to the group of fungi characterised by chitin-glucan walls walls(Bartnicki-Garcia, 1973). The hyphal wall consists of amorphous matrix components *viz.*, $\alpha$-glucans and glycoproteins, which include peptido-galactomannans. Glycoproteins perform a variety of cellular functions, they are involved in recognition systems, as enzymes, as antigenic determinants and as immunomodulators. In most glycoproteins the protein moiety is thought to represent the functional part whereas the carbohydrate is thought to contribute to tertiary structure and so modify the molecule, making it more stable and possibly facilitating secretion. Skeletal compounds are polysaccharidic, highly crystalline, and include chitin and ß-linked glucans. Depolymerisation of chitin with $HNO_2$ renders ß-glucan partly water-, partly alkali- soluble, as has been demonstrated for *A.niger* (Stagg and Feather, 1973). Walls of several fungal species are sensitive to $HNO_2$ without prior deacetylation, indicating that chitin is present in a partly-deacetylated form. A fragment containing N-acetylglucosamine, glucose and some amino acids, predominantly lysine, is thought to be involved in the chitin-glucan link (Sietsma and Wessels, 1979).

The detailed analysis of the galactomannans (GMs) of *Aspergillus* is dealt with elsewhere (see Latgé *et al.*, this volume) and will be mentioned only briefly here. Alkai-extracted *A.fumigatus* conidial and mycelial cell wall showed polysaccharides of different structure. Mycelial GM contained a core of $\alpha$-D-mannopyranosyl (man*p*) units 1,6-linked with ß-D-galactofuranosyl units (gal*f*) 1,6-linked to the mannan joining side chains of approximately 6 units which are 1,5 interlinked. The glucan, co-solubilised by alkali, is thought to be glycogen. In the conidial wall polysaccharide chains consist of single units of ß-D-gal*f* linked 1,6 to man*p*. In contrast to mycelium the alkali-soluble glucan was not glycogen. The conidial wall contained much less protein than did mycelial wall (Barreto-Bergter *et al.*, 1981). That the oligogalactosyl side chains of *A.fumigatus* GM may not be entirely ß-linked has been suggested by a study where the lectin BS-1 from *Bandeiraea simplicifolia* has been used to probe extracts of *A.fumigatus*. This lectin with specificity for $\alpha$-D-gal*p* units bound to some components which were also capable of binding the mAb (monoclonal antibody) to GM (Ste-Marie *et al.*, 1990).

Of interest are the common features shared by *Aspergillus* species as well as their differences, which can be considerable (Reiss, 1986). Alkali extraction of hyphal wall material of *A.niger* yielded a hetero-polysaccharide *viz.*, galactosaminogalactan, a linear molecule of gal*p* units joined mainly in $\alpha$-1,4 linkage which also contains approximately 20% of galactosamine (Bardalaye and Nordin, 1976). These $\alpha$-galactosamine-galactans form a family of polysaccharides which seem characteristic of *Aspergillus* species (Gomez-Miranda and Leal, 1981). Differences among the Aspergilli include the presence in *A.nidulans* of significantly larger quantities of galactosamine than is found in the majority of fungal walls. *A.niger*, unusually among the Aspergilli, contains arabinose and nigeran, a glucan with alternating $\alpha$-1,3 and $\alpha$-1,4 linkages which is also found in *A.awamori* but not in *A.fumigatus* (Johnson, 1965; Bull, 1970; Bobbitt *et al.*, 1977).

The GM of *A.niger* also shows differences from that of *A.fumigatus* in that the $\alpha$-1,2 man*p* backbone is thought to be linked in a relatively short series of 5-9 hexose units through $\alpha$-1,6 bonds with side chains consisting of 2-3 internalised D- gal*p* units and nonreducing terminal gal*f* residues (Bardalaye and Nordin, 1977). Differences between conidial and hyphal wall structures have been shown by the action of a mixture of ß-1,3-glucanase and chitinase, present in a Streptomyces culture filtrate. Hyphal walls of *A.phoenicis* were extensively digested by these enzymes while conidial wall was resistant due to the presence of melanin (Bloomfield and Alexander, 1967). A similar enzyme system was shown to be active in lysing *A.oryzae* and *A.nidulans* (Skujins *et al.*, 1965; Bull, 1970). It was found that these enzymes were sensitive to melanin with chitinase the more susceptible (Bull, 1970). $\alpha$-1,3-glucan is a major constituent of the outer cell wall and appears to protect young, non-melanised hyphae from the action of hydrolytic enzymes (Polacheck and Rosenberger, 1977). Melanin fulfills a protective function, similar to that of $\alpha$-1,3-glucan in old or differentiated hyphae which, at least in *A.nidulans*, no longer appears to contain α-1,3-glucan (Zonneveld, 1974). Species differences have also been revealed when five species of *Aspergillus* were treated with a variety of proteinases and polysaccharidases. Differences in susceptibility of the intact organisms showed inter-species differences in structural organisation of the cell wall (Hearn, 1984).

## IMMUNOCHEMISTRY

A GM-peptide with immediate and delayed-type hypersensitivity responses was obtained from both *A.fumigatus* spores and mycelium. The immediate reactivity was associated with the GM while the delayed reaction was linked with the protein (Azuma *et al.*, 1968; Suzuki and Hayashi, 1975). The similar structure of oligo-mannoside side chains of *A.fumigatus* GM and *Candida* mannan offers a plausible explanation for the observed reactivity of anti- *Candida* antiserum with *A.fumigatus* GM and the binding of antiserum against *A.fumigatus* mycelial antigen with *C.albicans* mannan (Suzuki *et al.*,1967; Marier *et al.*, 1979). Of the two mAbs isolated by Ste-Marie *et al.* (1990), one was a non-precipitating antibody that cross-reacted with *Candida* mannan; the other did not recognise *Candida* mannan but recognised immunodominant oligogalactoside side chains of *A.fumigatus* GM. The gal*f* residues are immunodominant in rabbits and most likely account for cross-reactivity with GMII of dermatophytes (Azuma, *et al.*, 1968). Monoclonal antibodies of the IgM class (EB-A1 to EB-A7), raised in rats to *A.fumigatus* GM, bound to an epitope located on the ß(1-5)- galactofuranose- containing side chains of GM (Stynen *et al.*, 1992). The mAb EB-A1 has been reported as being specific for the detection of *Aspergillus* species and *Penicillium marneffei* in fixed tissue section. It revealed hyphae, remnants of filaments and what the authors interpreted as hyphal debris included by macrophages (Pierard *et al.*, 1991).

The presence of a heat-labile, non-ConA-binding antigen/allergen on the germ tube surface of *A.fumigatus* has been demonstrated. The same molecule is rapidly excreted into the medium from resting spores (Taylor and Longbottom, 1988). Incubation of intact germ tubes and branched hyphae with serum from patient's with antibodies specific for *Aspergillus*, showed marked binding, in each case, to surface antigens. Probing of germ tubes and hyphae, before and after treatment with hydrolytic enzymes, with antisera raised in rabbits to isolated fractions of *A.fumigatus*, revealed distinct differences in wall architecture between the two morphological forms (John *et al.*, 1984).

Recent EM work, using *A.fumigatus* in thin section, has shown that many of the highly-reactive cellular antigens appear surface located in both conidia and mycelium (Hearn *et al.*, 1991; Reijula *et al.*, 1991). While mycelial surface antigens are readily elutable with aqueous solvents (Leung *et al.*, 1988; Hearn *et al.*, 1991), aqueous extracts of spore surface components demonstrate low biological and low immunological activity (Kauffman *et al.*, 1984). However, following alkaline extraction, the spore wall of *A.clavatus* yielded both precipitinogenic and alveolitis-inducing molecules. Using a mouse model, it was found that delipidated, dead spores provoked more severe allergic alveolitis than did live spores (Blyth, 1978).

Membrane-bound acid protease has been described for *A.oryzae* by Tsujita and Endo (1976). Membrane vesicles, obtained from *A.fumigatus* mycelium by extraction with Triton X-100, gave two cross-reacting acid proteinases which bound to immunoglobulin (Ig) from both allergic bronchopulmonary aspergillosis and aspergilloma patient's sera. They were both shown to be glycoprotein molecules containing 93% and 85% carbohydrate, respectively (Piechura *et al.*, 1990). Extraction of a crude wall fraction of *A.fumigatus*, also with Triton X-100, yielded an extract having both proteinase and catalase components with antigenic activity (Hearn, 1991). Recently, an alkali-soluble, water-insoluble glycoprotein fraction was isolated from the cell wall of *A.fumigatus*. This

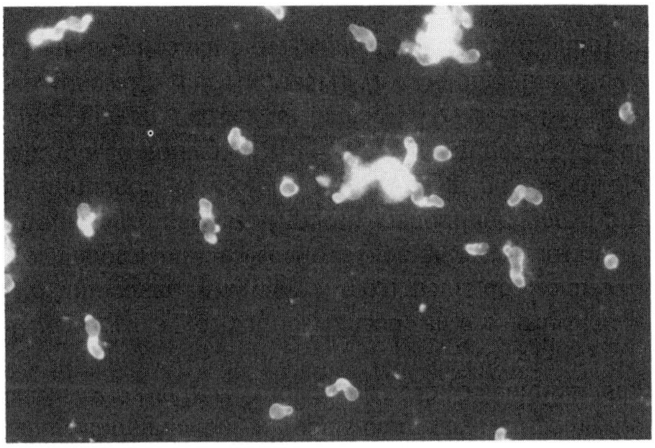

**Figure 1.** Indirect immunofluorescence of germinating conidia of *A.fumigatus* with EB-A2. The conidia were only weakly stained (arrowheads), whereas the mycelial germ tubes were more reactive. Bar, 25$\mu$m. (C) The arrowhead indicates a more brightly stained, young conidium. Bar, 5$\mu$m. (From Stynen *et al.*, 1992, Infection & Immunity, Am. Soc. Microbiol. With permission).

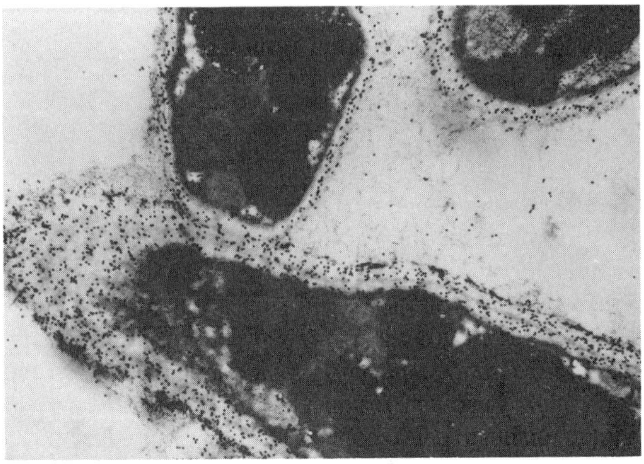

**Figure 2.** Immunoelectron microscopy of germinating spores of *A.fumigatus*. The section was incubated with a serum from a patient with ABPA followed by goat antihuman IgG and rabbit antigoat IgG conjugated with colloidal gold (original magnification x 13,990). (From Reijula *et al.*, 1991, J. Allergy Clin. Immunol., Mosby Year Book, Inc. With permission).

fraction binds to antibodies raised in rabbits to wall material and to antibodies present in the serum of patient's with aspergilloma. Antigenicity appears to reside in the sugar moiety (V.M. Hearn and J.H. Sietsma, ms in preparation). An additional peptido-GM component with immunological activity has been obtained from intact mycelium of *A.fumigatus* by aqueous extraction followed by fractionation with Cetavlon (E.M. Barreto-Bergter, personal communication). The relationship among these various glycoprotein antigens remains to be elucidated.

## HOST-FUNGUS INTERACTIONS

Conidial and mycelial extracts of *A.nidulans* have been shown to share common antigenic zones while other antigenic zones were found to be common to all pathogenic species of *Aspergillus* tested. The authors suggested that the degree of conservation of antigenic proteins is higher in the wall and membrane fractions of different growth phases and strains than in the cytosol (Puenta *et al.*, 1991). While *Aspergillus* species possess shared antigenic epitopes, structural differences and differences in host response to invasion are quite evident. The resistance of *A.fumigatus* conidia to ingestion and killing by phagocytes has been widely demonstrated (Lehrer and Jan, 1970; Kurup, 1984). It has been emphasised that while conidia of *Penicillium ochrochloron* are readily ingested by phagocytes, a substantial number of *A.fumigatus* conidia appear to remain on the surface of the cell (Robertson *et al.*, 1987). It has further been shown that *A.fumigatus* and *A.flavus* possess a far greater ability to germinate in the presence of alveolar macrophages than does *A.niger* (Schaffner *et al.*, 1983; Kurup, 1984). Resting and swollen spores are both susceptible to phagocytosis by neutrophils but the former stimulates production of significantly less superoxide anion and induces far less myeloperoxidase-dependent iodination by neutrophils than does pre-swollen, *A.fumigatus* conidia. The authors postulated that this reaction might result from greater expression or exposure of certain surface antigens (Levitz and Diamond, 1985). Alveolar macrophages have the ability to kill metabolically-active *A. fumigatus* conidia, perhaps because of increased permeability, when compared with resting spores (Levitz *et al.*,1986).

Specific effects of serum opsonins on phagocytosis may be due to their deposition on the fungal cell wall. *A.fumigatus* is a potent activator of the complement system (Marx and Flaherty, 1976). The deposition of complement component C3 onto all three life forms of *A.fumigatus* has been measured. Complement activation by resting conidia was mediated by the alternative pathway. In contrast, there was a progressive dependence on the classical pathway as the fungus matured into swollen conidia and then hyphae (Kozel *et al.*, 1989). A separate study has confirmed the importance of the alternative complement pathway in *Aspergillus* infections. Human complement component C3 and/or a C3 fragment/s is bound by glycoproteins of apparent molecular mass of 54 and 58 kDa, present on *A.fumigatus* conidial cell surface. Deposited C3 is rapidly degraded to lower molecular weight fragments. Evidence suggests that these fragments are produced as a result of proteolytic activity present in the outer conidial wall layer (Sturtevant and Latgé, 1992 *a*). Other studies have suggested a role for complement in phagocytosis of *A.fumigatus* by human polymorphonuclear cells (Waldorf and Diamond, 1985). Optimal cell-cell interaction appears dependent on an active alternative complement pathway (Sturtevant and Latgé, 1992 *b)*.

Host responses to *Aspergillus* can be affected by other serum components including fibrinogen which fosters adherence of conidia to phagocytic cells (Bouchara *et al.*, 1988). Uncertainty still exists as to the part played by humoral immunity in host defence against *Aspergillus* (Forman *et al.*, 1978; Bennett, 1988). A role for antibody-dependent cell-mediated cytotoxicity against *A. fumigatus* antigen-coated target cells has been suggested in patients with *Aspergillus* infections (Kurup, 1985).

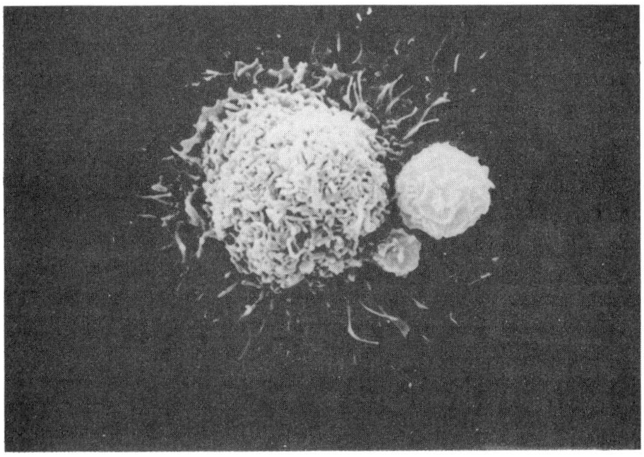

**Figure 3a.** Scanning electron micrograph of the cell association of fungal conidia following the interaction *in vivo* of mouse peritoneal exudate cells and conidia of *A.fumigatus*. O, attached conidia (From Robertson *et al.*, 1987, Thorax, BMJ Publishing Group. With permission).

The importance of carbohydrates and lectins in recognition and attachment, before phagocytosis, of fungal yeast cells to host cells has been well documented (Warr, 1980; Oda *et al.*, 1983). In competitive binding experiments sugars, including chitotriose, N-acetylglucosamine, D-mannose and L-fucose were effective inhibitors of binding of *A.fumigatus* conidia to pulmonary macrophages. The effectiveness of chitotriose as an inhibitor suggests the presence of another lectin (in addition to the mannosyl-fucosyl receptor) on macrophage membrane that recognises chitin components of *A.fumigatus* conidial cell wall (Kan and Bennett, 1988). Again, where pre-swollen spore of *A.fumigatus* were sensitive to the fungicidal activity of neutrophil cationic peptides, this activity could be inhibited by the chitin fragments, chitobiose and chitotriose (Levitz *et al.*, 1986).

## CONCLUSIONS

What most concerns the medical mycologist is the considerable ability of *A.fumigatus* (in comparison with other air-borne organisms) to function as a significant pathogen in man. While many similarities exist among the Aspergillie, differences are what determine the ability of *A.fumigatus* to survive, usually at the expense of the host. Among the features which help promote its pathogenicity are small spore size, thermotolerance and an innate resistance to various killing mechanisms of the host, accomplished both by its surface composition and secreted molecules. In all of these processes, the wall is of prime importance.

Certain lines of investigation may prove particularly fruitful. The use of mutant strains has already given valuable information concerning *Aspergillus* cellular structure (Katz and Rosenberger, 1970; Zonneveld, 1974; Bainbridge *et al.*, 1979; Latgé *et al.*, this volume). A mutant strain of *A.fumigatus* has recently been produced which lacks the gene for alkaline proteinase, a major secretory product, (Monod, *et al.*, 1993).

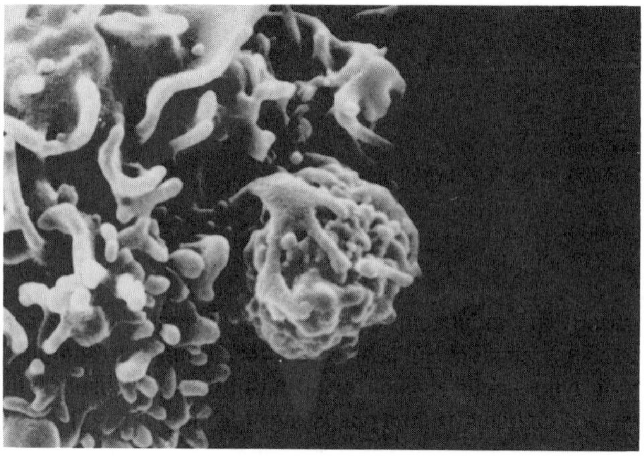

**Figure 3b.** Scanning electron micrograph of the cell association of fungal conidia following the interaction *in vivo* of mouse peritoneal exudate cells and conidia of *P. ochrochloron*. *O*, ingested conidia (From Robertson *et al.*, 1987, Thorax, BMJ Publishing Group. With permission).

Further studies on the regeneration of *Aspergillus* protoplasts could give us additional insight into wall synthesis and the morphogenetic changes that occur (Douglas *et al.*, 1984; Peberdy, 1990). Experiments involving animal models of disease plus *in vitro* systems studying cell-cell interactions represent another line of research. The multi-functional roles of the fungal cell wall, growth-associated changes and its ability to adapt to its environment make it a highly complex structure. Thus, while considerable knowledge has been gained in recent years on the fine structure of *Aspergillus* cell wall, much remains to be elucidated.

## REFERENCES


Azuma, I., Kimura, H. and Yamamura, Y. (1968) Purification and characterization of an immunologically active glycoprotein from *Aspergillus fumigatus*. J. Bacteriol. 96, 272-273.

Azuma, I., Kimura, H., Hirao, F., Tsubura, E., Yamamura, Y. and Misaki, A. (1971) Biochemical and immunological studies on *Aspergillus* 3. Chemical and immunological properties of glycopeptide obtained from *Aspergillus fumigatus*. Jap. J. Microbiol. 15, 237-246.

Bainbridge, B.W., Valentine, B.P. and Markham, P. (1979) The use of temperature-sensitive mutants to study wall growth, in "Fungal Walls and Hyphal Growth" (Burnett, J.H. and Trinci, A.P.J., Eds.), pp. 71-91. Cambridge Univ. Press, London.

Bardalaye, P.C. and Nordin, J.H. (1976) Galactosaminogalactan from cell walls of *Aspergillus niger*. J. Bacteriol. 125, 655-669.

Bardalaye, P.C. and Nordin, J.H. (1977) Chemical structure of the galactomannan from the cell wall of *Aspergillus niger*. J. Biol. Chem. 252, 2584-2591.

Barreto-Bergter, E.M., Gorin, P.A. and Travassos, L.R. (1981) Cell constituents of mycelia and conidia of *Aspergillus fumigatus*. Carbohydr. Res. 95, 205-218.

Bartnicki-Garcia, S. (1973) Fundamental aspects of hyphal morphogenesis, in "Microbial Differentiation" (Ashworth, J.M. and Smith, J.E., Eds.), pp. 245-268. Cambridge Univ. Press, London.

Bartnicki-Garcia, S. (1990) Role of vesicles in apical growth and a new mathematical model of hyphal morphogenesis, in "Tip Growth in Plant and Fungal Cells" (Heath, I.B., Ed.), pp. 211-232. Acad. Press, London.

Bennett, J.E. (1988) Role of the phagocyte in host defence against aspergillosis, in "*Aspergillus* and aspergillosis" (Van den Bossche, H., Mackenzie, D.W.R. and Cauwengergh, G., Eds.), pp 115-119. Plenum Press, New York.

Bloomfield, B.J. and Alexander, M. (1967) Melanins and resistance of fungi to lysis. J. Bacteriol. 93, 1276-1280.

Blyth, W. (1978) The occurence and nature of alveolitis-inducing substances in *Aspergillus clavatus*. Clin. exp. Immunol. 32, 272-282.

Bobbitt, T.F., Nordin, J.H., Roux, M, Revol, J.F. and Marchessault, R.H. (1977) Distribution and conformation of crystalline nigeran in hyphal walls of *Aspergillus niger* and *Aspergillus awamori*. J. Bacteriol. 132, 691-703.

Border, D.J. and Trinci, A.P.J. (1970) Fine structure of the germination of *Aspergillus nidulans* conidia. Trans. Br. Mycol. Soc. 54, 143-146.

Bouchara, J.P., Bouali, A., Tronchin, G., Robert, R., Chabasse, D. and Senet,J.M. (1988) Binding of fibrinogen to the pathogenic *Aspergillus* species. J. Med. Vet. Mycol. 26, 327-334.

Bull, A.T. (1970) Inhibition of polysaccharases by melanin: enzyme inhibition in relation to mycolysis. Arch. Biochem. Biophys. 137, 345-356.

Chang, P.L.Y. and Trevithick, J.R. (1974) How important is secretion of exoenzymes through apical cell walls of fungi? Arch. Microbiol. 101, 281-293.

Cole, G.T. and Pope, L.M. (1981) Surface wall components of *Aspergillus niger* conidia, in "The Fungal Spore: Morphogenetic controls" (Turian, G. and Hohl, H., Eds.), pp. 195-215. Academic Press, London.

Cole, G.T., Sun, S.H. and Huppert, M. (1983) Isolation and ultrastructural examination of conidial wall components of *Coccidioides* and *Aspergillus*. Scan. Elect. Microscopy 4, 1677-1685.

Diamond, R.D. (1991) The Fungal Cell Wall and Vertebrate Phagocytosis, in "Fungal Cell Wall and Immune Response" (Latge, J.P. and Boucias, D., Eds.), pp. 331-340. Springer Verlag, Heidelberg.

Douglas, C.M., Synan, T.R., Bobbitt, T.F. and Nordin, J.H. (1984) Nigeran synthesis by regenerating protoplasts of *Aspergillus awamori* correlates with formation of hyphae. Exper. Mycol. 8, 146-160.

Drouhet, E., Camay, L. and Segretain, G. (1972) Valeur de l'immunoprecipitation et de l'immunofluorescence indirecte dans les Aspergilloses bronchopulmonaires. Ann. Inst. Pasteur (Paris) 123, 379-395.

Forman, S.R., Fink, J.N., Moore, V.L., Wang, J. and Patterson, R. (1978) Humoral and cellular immune responses in *Aspergillus fumigatus* pulmonary disease. J. Allergy Clin. Immunol. 62, 132-140.

Girbardt, M. (1969) Die Ultrastruktur der Apikalregion von Pilzhyphen. Protoplasma 67, 413-441.

Gomez-Miranda, B. and Leal, J.A. (1981) Extracellular and cell wall polysaccharides of *Aspergillus alliaceus*. Trans. Br. Mycol. Soc. 76, 249-253.

Gooday, G.W. (1977) Biosynthesis of the fungal wall-mechanisms and implications. J. Gen. Microbiol. 99, 1-11.

Gooday, G.W. and Gow, N.A.R. (1990) Enzymology of tip growth in fungi, in "Tip Growth in Plant and Fungal Cells" (Heath, I.B., Ed.), pp. 31-58. Acad. Press, London.

Hazen, K. (1989) Yeast cell surface hydrophodbicity in adherence of *Candida albicans*. Infect. Immun. 57, 1894-1900.

Hearn, V.M. (1984) Surface antigens of intact *Aspergillus fumigatus* mycelium: their localisation using radiolabelled Protein A as marker. J. Gen. Microbiol. 130, 907-917.

Hearn, V.M. (1991) Glycoproteins of *Aspergillus fumigatus* cell wall, in "Fungal Cell Wall and Immune Response" (Latgé, J.P. and Boucias, D., Eds.), pp. 219-228.

Hearn, V.M., Latgé, J-P. and Prevost,M.-C. (1991) Immunolocalisation of *Aspergillus fumigatus* mycelial antigens. J. Med. Vet. Mycol. 29, 73-81. Springer Verlag, Heidelberg.

Hess, W.M. and Stocks, D.L. (1969) Surface characteristics of *Aspergillus* conidia. Mycologia 61, 560-571.

John, J., Wilson, E.V. and Hearn, V.M. (1984) Analysis of *Aspergillus fumigatus* germ tube surface structures by an immunofluourescent labelling technique. Mykosen 27, 485-497

Johnson, I.R. (1965) The composition of the cell wall of *Aspergillus niger*. Biochem. J. 96, 651-658.

Kan, V.L. and Bennett, J.E. (1988) Lectin-like attachment sites on murine pulmonary alveolar macrophages bind *Aspergillus fumigatus* conidia. J. Infect. Dis. 158, 407-414.

Katz, D. and Rosenberger, R.F. (1970) A mutation in *Aspergillus nidulans* producing hyphal walls which lack chitin. Biochem. Biophys. Acta 208, 452-460.

Kauffman, H.F., van der Heide, S., Beaumont, F., de Monchy, J.G.R. and de Vries, K. (1984) The allergenic and antigenic properties of spore extracts of *Aspergillus fumigatus*: a comparative study of spore extracts with mycelium and culture filtrate extracts. J. Allergy Clin. Immunol. 73, 567-573.

Kozel, T.R., Wilson, M.A., Farrell, T.P. and Levitz, S.M. (1989) Activation of C3 and binding to *Aspergillus fumigatus* conidia and hyphae. Infect. Immun. 57, 3412-3417.

Kurup, V.P. (1984) Interaction of *Aspergillus fumigatus* spores and pulmonary alveolar macrophages of rabbits. Immunobiol. 166, 53-61.

Kurup, V.P. (1985) Serum antibodies and their role in antibody-dependent cell-mediated cytotoxicity in aspergillosis. Immunobiol. 169, 362-371.

Lehrer, R.I. and Jan, R.G. (1970) Interaction of *Aspergillus fumigatus* spores with human leukocytes and serum. Infect. Immun. 1, 345-350.

Leung, P.S.C., Gershwin, M.E., Coppel, R., Halpern, G., Novey, H. and Castles, J.J. (1988) Localisation, molecular weight and immunoglobulin subclass response to *Aspergillus fumigatus* allergens in acute bronchopulmonary aspergillosis. Int. Archs. Allergy appl Immunol. 85, 416-421.

Levitz, S.M. and Diamond, R.D. (1985) Mechanisms of resistance of *Aspergillus fumigatus* conidia to killing by neutrophils *in vitro*. J. Infect. Dis. 152, 33-42.

Levitz, S.M., Selsted, M.E., Ganz, T., Lehrer, R.I. and Diamond, R.D. (1986) *In vitro* killing of spores and hyphae of *Aspergillus fumigatus* and *Rhizopus oryzae* by rabbit neutrophil cationic peptides and bronchoalveolar macrophages. J. Infect. Dis. 154, 483-489.

Marier, R., Smith, W., Jansen, M. and Andriole, V.T. (1979) A solid phase radioimmunoassay for the measurement of antibody to *Aspergillus* in invasive aspergillosis. J. Infect. Dis. 140, 771-779.

Marx, J.J and Flaherty, D.K. (1976) Activation of the complement sequence by extracts of bacteria and fungi associated with hyperasensitivity pneumonitis. J. Allergy clin. Immunol. 57, 328-334.

Monod, M., Paris, S., Sarfati, J., Jaton-Ogay, K., Ave, P. and Latgé, J.P. (1993) Virulence of alkaline protease-deficient mutants of *Aspergillus fumigatus*. FEMS Microb. letters 106, 39-46.

Oda, L.M., Kubelka, C.F., Alviano, C.S. and Travassos, L.R. (1983) Ingestion of yeast forms of *Sporothrix schenckii* by mouse peritoneal macrophages. Infect. Immun. 39, 497-504.

Peberdy, J.F. (1990) Fungal Cell Walls - A Review, in "Biochemistry of Cell Walls and Membranes in Fungi" (Kuhn, P.J., Trinci, A.P.J., Jung, M.J., Goosey, M.W. and Copping, L.G., Eds.), pp. 5-30. Springer Verlag, Heidelberg.

Piechura, J.E., Kurup, V.P. and Daft, L.J. (1990) Isolation and immunochemical characterization of fractions from membranes of *Aspergillus fumigatus* with protease activity. Can. J. Microbiol. 36, 33-41.

Pierard, G.E., Estrada, J.A., Pierard-Franchimont, C., Thiry, A. and Stynen, D. (1991) Immunohistochemical expression of galactomannan in the cytoplasm of phagocytic cells during invasive aspergillosis. Am. J. Clin. Path. 96, 373-376.

Polacheck, I. and Rosenberger, R.F. (1977) *Aspergillus nidulans* mutant lacking $\alpha$-(1,3)-glucan, melanin, and cleistothecia. J. Bacteriol. 132, 650-656.

Polacheck, I. and Rosenberger, R.F. (1978) The distribution of autolysins in hyphae of *Aspergillus nidulans*: existence of a lipid mediated attachment to hyphal walls. J. Bacteriol. 135, 741-754.

Puente, P., Ovejero, M.C., Fernandez, N. and Leal, F. (1991) Analysis of *Aspergillus nidulans* conidial antigens and their prevalence in other *Aspergillus* species. Infect. Immun. 59, 4478-4485.

Pugh, D. and Cawson, R.A. (1977) The cytochemical localisation of acid hydrolases in four common fungi. Cell. Mol. Biol. 22, 125-132.

Reijula, K.E., Kurup, V.P. and Fink, J.N. (1991) Ultrastructural demonstration of specific IgG and IgE antibodies binding to *Aspergillus fumigatus* from patients with aspergillosis. J. Allergy clin. Immunol. 87, 683-688.

Reiss, E. (1986) *Aspergillus*, in "Molecular Immunology of Mycotic and Actinomycotic Infections" pp.129-156. Elsevier, New York.

Robertson, M.D., Seaton, A., Milne, L.J.R. and Raeburn, J.A. (1987) Resistance of spores of *Aspergillus fumigatus* to ingestion by phagocytic cells. Thorax 42, 466-472.

Rosenberger, R.F. (1979) Endogenous lytic enzymes and wall metabolism, in "Fungal Walls and Hyphal Growth" (Burnett, J.H. and Trinci, A.P.J., Eds.), pp. 265-277. Cambridge Univ. Press, London.

Schaffner, A., Douglas, H., Braude, A.I. and Davis, C.E. (1983) Killing of *Aspergillus* spores depends on the anatomical source of the macrophage. Infect. Immun. 42, 1109-1115.

# *ASPERGILLUS* AND AEROBIOLOGY

John Mullins

Asthma and Allergy Unit
Sully Hospital
Penarth, S.Glam.

## INTRODUCTION

Aerobiology is the study of the organisms and biologically significant particles which are transported through the atmosphere. This includes pollens, spores of mosses and ferns, and fungal spores which may be actively or passively discharged into the atmosphere. In fact the air we breathe contains a great variety of organisms. The fungal spores alone have resulted in a daily average spore count of over 100,000 spores in a cubic metre of air in Cardiff during the summer. The airspora varies in content from season to season, day to day and even hour to hour.

In the nineteenth century Pasteur sampled the air by drawing it through plugs of gun cotton which were later dissolved to allow microscopic examination of the sediment, but soon abandoned the method noting that it could be improved and used to study the effects of seasons and locations. Thereafter the airspora was investigated for the causes of infectious diseases and the work was largely dominated by bacteriologists (Gregory, 1973).

Studies on pollens as a cause of hayfever were undertaken by Blackley (1873) who noticed in passing the severe effect that fungal spores had upon his health, but work on the fungal spore component of the airspora was not resumed until this century with the recognition of fungal spores as allergens by Cadham (1924) and Feinberg (1935). Since then many methods have been used to assess the spore content of the atmosphere.

## METHODS

### Qualitative Methods

The method originally most widely used was the 'gravity slide technique' in which a glass slide with an adhesive coating was simply exposed to the atmosphere on a surface and pollens and spores were deposited by gravity and then identified under a microscope. Alternatively a culture plate was exposed on a horizontal surface and incubated to allow the moulds to grow so that they could be identified. The obvious disadvantage of these

*The Genus Aspergillus*, Edited by Keith A. Powell *et al.*,
Plenum Press, New York, 1994

techniques is that the rate at which spores are deposited from the air will be dependent on the size and density of the spores so that large spores will be deposited more readily than small spores and therefore there is a tendency for small spored fungi to be under represented in any census of the airspora using gravity slides or plates.

## Quantitative Methods

The method most widely used in the continual census of the airspora is the Automatic Volumetric Spore Trap and its successor the Burkard Trap developed by Hirst (1952). It samples continuously by drawing air through a slit onto a glass slide with an adhesive coating of Vaseline or glycerine jelly. The trap constantly faces into the wind to avoid problems of eddies and the sample obtained is a sample of the whole 24 hours and it is possible to discern variations in spore concentrations through the course of the day. Ideally the trap should sample at the same speed as the wind but this is not practical.

The Cascade Impactor (May, 1945) samples onto four glass slides in series, and at each stage the width of the sampling orifice is reduced to increase the impaction speed of the air striking the slide, so that as the air speed increases smaller spores are deposited on the slides, giving a gradation of different spore types on the slides. The disadvantage is that the sampler can only be used for short periods of time or the spore deposit becomes too dense for microscopy.

The 'Rotorod sampler' (Perkins, 1957) operates by whirling glass slides so that the spores are impacted by the speed of the slide as it travels through the air. Sampling is onto the edge of the slide for examination under the microscope but again this can only be used for short periods of time or the sample becomes overloaded.

It is possible to draw air through a filter such as a membrane filter and either examine the filter directly by microscopy or wash the spores off and culture them.

For culturable moulds the Cassella slit sampler (Bourdillon *et al,* 1941) has been widely used followed by the Andersen sampler (Andersen, 1958). The advantage of the latter is that it is based on the same philosophy as the cascade impactor. Air is drawn successively over six culture plates and the velocity of the air is increased at each stage so that the smaller spores, which have a higher impaction velocity, are deposited selectively on the later stages of the sampler. This sampler is considered very efficient at removing all particles from the sampled air. However, a disadvantage lies in the fact that the sampler can only be run for a short period of time or the plates become overcrowded and impossible to interpret.

## Results

In Cardiff the Automatic Volumatic Spore Trap (AVST) has been used to monitor the airspora since 1954, and whereas it is relatively straightforward to recognise many conidia, ascospores and basidiospores the problem in sampling for *Aspergillus* is that the spores are comparatively small so that when seen under the microscope with the spores of other fungi sampled from the air it is not possible to distinguish them from any other small spored genus. Accordingly the classification of 'Aspergillaceae' used by aerobiologists is potentially a very heterogeneous group of fungal spores and cannot with any confidence be said to represent *Aspergillus*, and as a category the 'Aspergillaceae' comprises less than 1% of the total spore count. However the picture presented by the AVST has been used as an indication of the seasonal incidence of *Aspergillus* (Figure 1).

Numerous surveys of the air have been carried out using culture plates, (Mullins, 1974) and have tended to indicate that in this country *Aspergillus* spp. are more abundant during the spring and winter, and that they have a proportionately higher incidence in the air of cities than in rural areas. *Aspergillus* usually occupies the position of 5th or 6th in order of frequency and makes up around 1 to 4% of the total catch of culturable moulds. Surveys around the world show considerable variability in the incidence of *Aspergillus* from being the commonest mould sampled to being very rare. Indoor surveys have tended to indicate higher levels of *Aspergillus* inside houses than outside, (Lumpkins and Corbit, 1976), but where identification has been taken to species level the results for different species are more variable (Garcia-Pantaleon *et al*, 1992).

The importance of fungal infection in immuno-compromised patients has resulted in a particular interest in the incidence of *Aspergillus* in the air of hospitals and operating theatres (Noble and Clayton 1963, Mullins 1974, Harvey *et al* 1988, Nolard-Tintigner and Beguin 1990). In studies of operating theatres in Cardiff the incidence of *A.fumigatus* in the air of these units was generally found to be a pale reflection of the levels found in

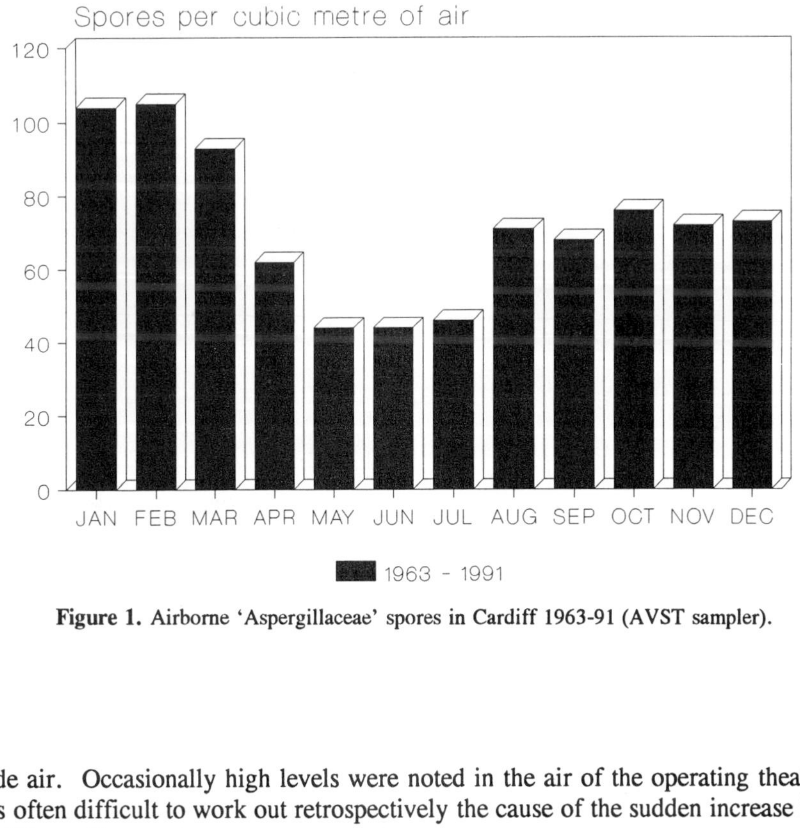


**Figure 1.** Airborne 'Aspergillaceae' spores in Cardiff 1963-91 (AVST sampler).

outside air. Occasionally high levels were noted in the air of the operating theatres and it was often difficult to work out retrospectively the cause of the sudden increase in spore level (Figure 2).

One possibility investigated in our studies was the influence of filter changing in the air conditioning system. It was found that during the procedure the *A.fumigatus* spore content of the air rose dramatically (Figure 3) and investigation of the filters which were in place for one month showed them to have an *A.fumigatus* spore content of more than $7.5 \times 10^6$. As the filters were changed from within the operating suite, the possibility of dislodging spores into the air was high.

Mycologists are usually called in to investigate a problem which may have occurred some months previously by which time the situation may have changed and it is often difficult at that stage to find the cause.

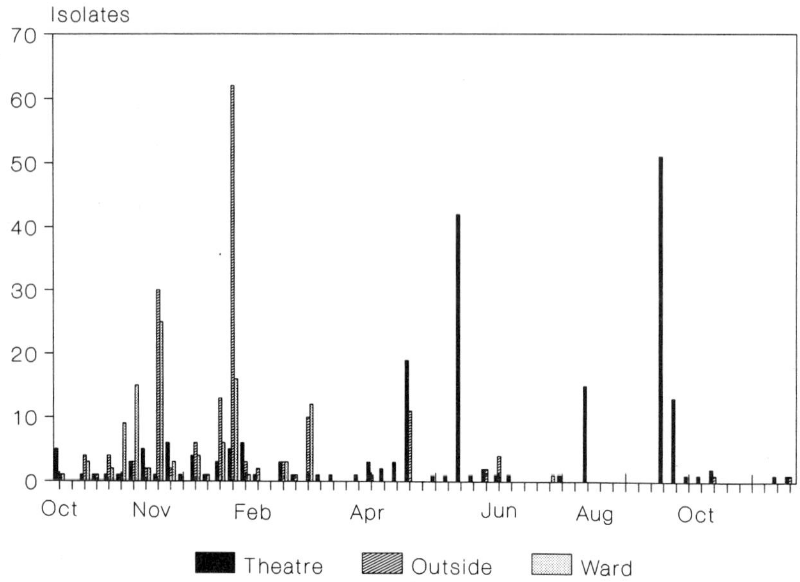


**Figure 2.** Isolations of *A.fumigatus* from an operating theatre, a ward and outside air.

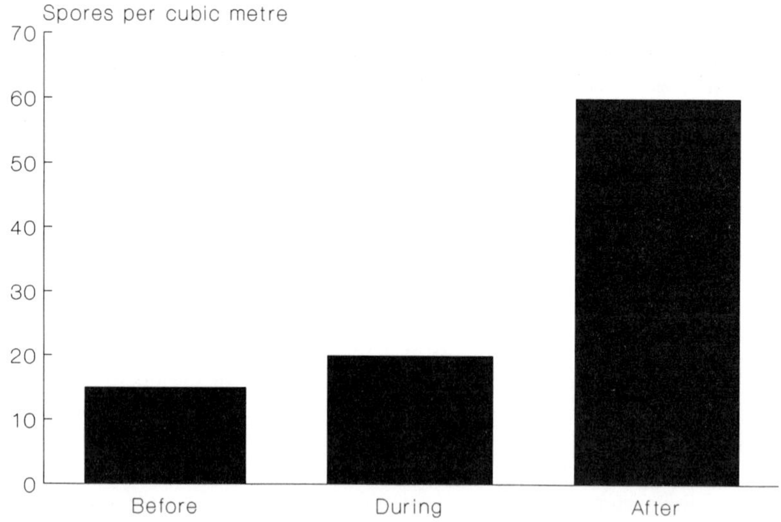


**Figure 3.** *A.fumigatus* levels during filter changing in an operating suite.

Quantitative sampling of the air has been carried out far less frequently. The equipment necessary for quantitative sampling is comparatively expensive and it is difficult to run a survey which requires samples to be taken from several different sites synchronously. The Andersen sampler also requires the use of six culture plates at a time. The length of sampling varies dependent upon the spore load of the sir, but sampling cannot usually be continued for periods of longer than a few minutes, or the plates

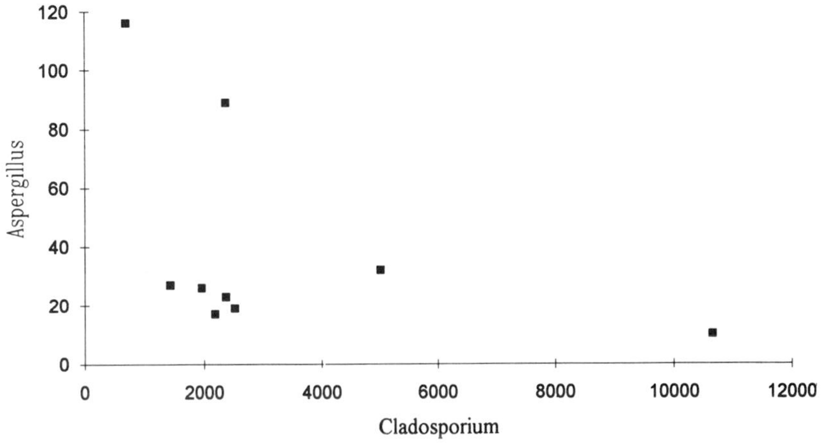


**Figure 4.** *Aspergillus* isolates plotted against *Cladosporium* isolates.

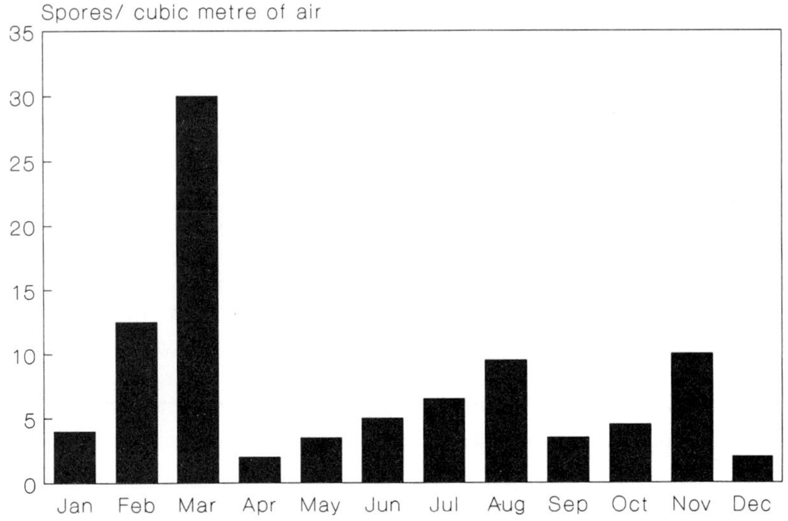


**Figure 5.** *A.fumigatus* levels in the centre of Cardiff (Andersen sampler).

become overcrowded. This highlights a particular problem with *Aspergillus* spp. in that, as a mould which accounts for such a small proportion of the airspora, it tends not to compete particularly well on culture plates with the moulds which are dominant in the air during the summer. It is particularly noticeable from the studies of Richards (1953) that *Aspergillus* was more frequent in cities than in rural areas, and the converse was found for *Cladosporium* (Figure 4) which must suggest the possibility that the results of general surveys of moulds can be greatly influenced by competition on the culture plate.

Hudson (1969) carried out a survey of the air in Cambridge specifically for *Aspergillus* using the Andersen sampler and coped with the problem of overcrowding by reducing the sampling time to 0.5 min in the summer, removing fast growing colonies on the second day of incubation and subculturing colonies of the *A.glaucus* group. By this method he recorded 14 different species of *Aspergillus*, namely *A.amstelodami, A. candidus, A.chevalieri. A.echinulatus, A.fumigatus, A.gracilis, A.nidulans, A.niger, A. ochraceus, A.oryzae, A.repens, A.restrictus, A.ruber* and *A.versicolor*. The incidence of *Aspergillus* in Cambridge again indicated the tendency for spore incidence to be higher during the spring and winter.

Studies of *Aspergillus* in Cardiff were restricted to *A.fumigatus*, where sampling is simplified because it is possible to take advantage of the thermotolerant nature of the organism. Evans (1972) studied thermophilous fungi in the air at Keele by incubating the culture plates at 45°C and found that *A.fumigatus* made up 45% of the total catch. In Cardiff the culture plates were incubated at 37°C and this species made up more than 90% of the catch. By using this method both mesophilic and thermophilic fungi were selectively avoided and overcrowding of the culture plates never became a problem. (Mullins *et al* 1976). The incidence of *A.fumigatus* in the air at the centre of Cardiff is shown in Figure 5.

As seen in studies of the Aspergilli in general, the fungus was more prevalent in the air during the spring and winter. Air sampling was carried out using both gravity plate surveys and an Andersen sampler using Malt extract agar.

Similar results were obtained from surveys of the outside air with the Andersen sampler a few miles from the hospital centre (Figure 6) and from a gravity plate survey of the air carried out inside and outside a hospital ward every week for two years. The results showed that spore concentrations inside the ward were just a lower reflection of the spore levels outside (Figure 7), and there were no high counts to compare with the very high levels reported by Noble and Clayton (1963) in a London hospital.

A comparative survey of the air in Cardiff and St Louis, Missouri, U.S.A. (Mullins *et al* 1984) was carried out 3 days/week over a 12 month period using Andersen samplers which were controlled by time clocks to sample for periods of one or two minutes every hour to try and get a more representative sample of the air and avoid any possibility of the influence of diurnal periodicity. Both sites showed seasonal variations with the highest concentrations during the winter months. Other surveys of *A.fumigatus* spore levels in the air in the U.S.A. (Solomon and Burge 1975, Solomon *et al* 1978) had indicated relatively low concentrations particularly when compared to the levels recorded by Noble and Clayton (1963), however in our surveys, the average spore concentration was $13.5/m^{-3}$ in St Louis and $11.3/m^{-3}$ in Cardiff. Both sites showed similar seasonal variations with highest concentrations recorded during the winter months (Figure 8).

## DISCUSSION

The census of *Aspergillus* spores in the air has been relatively neglected as the spores are not distinguishable under the light microscope and on culture plates it is easily overlooked as it represents such a small proportion of the airspora. To gain a true

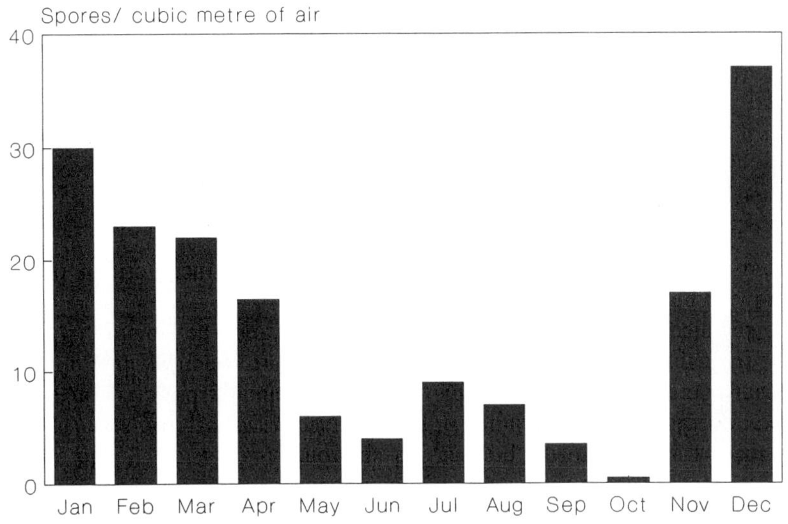


**Figure 6.** *A.fumigatus* levels in hospital grounds (Andersen sampler).

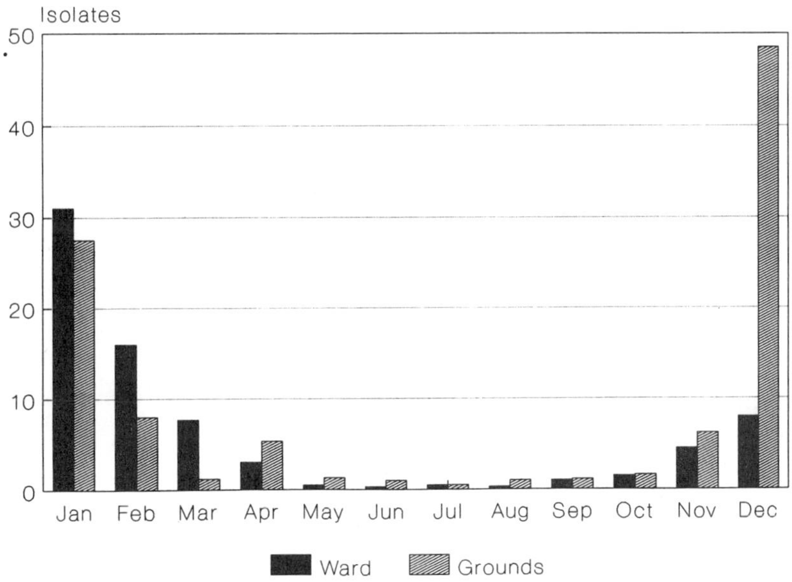


**Figure 7.** *A.fumigatus* isolates from hospital ward and grounds (gravity plates).

impression of the incidence of *Aspergillus* in the air requires that the survey should be carried out specifically for *Aspergillus* using selective culture techniques wherever possible to avoid the overcrowding of plates. The survey carried out by Hudson (1969) shows the benefits of selective sampling and enabled him to report *Aspergillus* concentrations averaging $27/m^{-3}$ with highest concentrations recorded during the winter months.

Surveys for *A.fumigatus* have the advantage that incubation temperatures can be used that exclude mesophilic and thermophilic fungi, and from these surveys average spore levels of 13.5/m$^{-3}$ have been recorded with concentrations highest during the winter months. Indoor levels were found to be a reflection of the counts in outdoor air and there was no indication that spore levels were higher in the city than at a rural site. Although samples taken alongside a compost heap did indicate higher spore levels they did not really reflect the massive spore levels within the compost heap which presumably are unlikely ever to become airborne (Mullins *et al*, 1976).

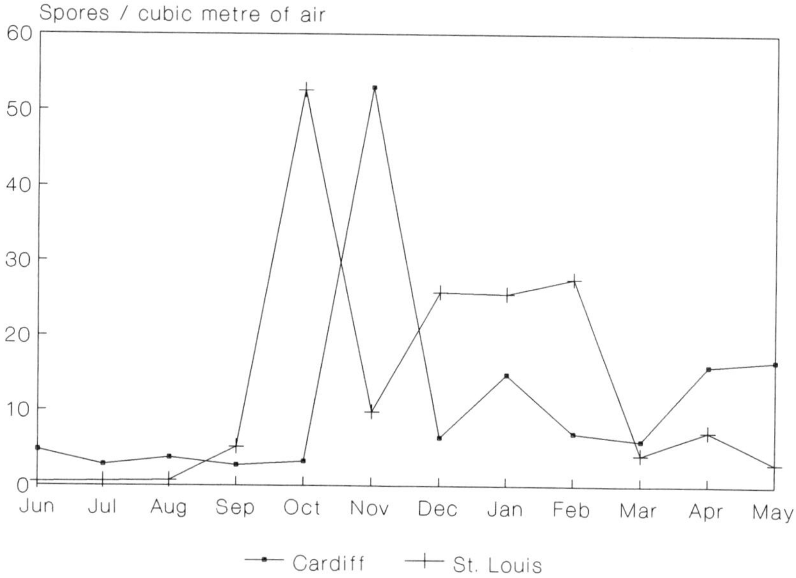


**Figure 8.** Comparative incidence of *A.fumigatus* in the air of Cardiff and St. Louis.

## REFERENCES


Andersen A.A. (1958) New sampler for the collection, sizing and enumeration of viable airborne particles. J.Bact. 76, 471-484.

Blackley C.H. (1873) Experimental researches on the cause and nature of Catarrhus Aestivus (Hay fever or hay asthma) Ballière, Tindall and Cox, London.

Bourdillon R.B., Lidwell, O.M. and Thomas, J.C. (1941) A slit sampler for collecting and counting airborne bacteria. J.Hyg.,Camb. 41, 197-224.

Cadham F.T. (1924) Asthma due to grain rusts. J. Am. Med. Ass. 83, 27.

Evans H.C. (1972) Thermophilous fungi isolated from the air. Trans. Brit. Mycol. Soc. 59, 516-519.

Feinberg S.M. (1935) Mold allergy: its importance in asthma and hay fever. Wisconsin Med. J. 34, 254.

Garcia-Pantaleon F.I., Soldevilla,C.G., Vilches, E.D., Romero, J.A. and Molina, A.M. (1992) Airspora microfungi in dwellings of south of Spain. Aerobiologia 8, 245-253.

Gregory P.H. (1973) The microbiology of the atmosphere. 2nd Ed. Leonard Hill, Aylesbury.

Harvey I.M., Leadbeatter S., Peters T.J., Mullins J., Philpot, C.M. and Salaman J.R. (1988) An outbreak of disseminated aspergillosis associated with an intensive care unit. Community Med. 10, 306-313.

Hirst J.M. (1952) An automatic volumetric spore trap. Ann. Appl. Biol. 39, 257-265.

Hudson H.J. (1969) Aspergilli in the air-spora at Cambridge. Trans. Br. Mycol. Soc. 52, 153-159.

Lumpkins E.D. and Corbit S.L. (1976) Airborne fungi survey II. Culture plate survey of the home environment. Ann Allerg. 36, 40-44.

May K.R. (1945) The cascade impactor: An instrument for sampling coarse aerosols. J. Sci Instrum. 22, 187-195.

Mullins J. (1974) The ecology of *Aspergillus fumigatus* (Fres.) PhD Thesis, University of Wales.

Mullins J., Harvey R. and Seaton A. (1976) Sources and incidence of airborne *Aspergillus fumigatus* (Fres.) Clin. Allergy 6, 209-217.

Mullins J., Hutcheson P. and Slavin R.G. (1984) *Aspergillus fumigatus* spore concentrations in outside air: Cardiff and St. Louis compared. Clin. Allergy 14, 351-254.

Noble W.C. and Clayton Y.M. (1963) Fungi in the air of hospital wards. J. Gen Microbiol. 32, 397-402.

Nolard-Tintigner, N. and Beguin, H. (1990) Epidemiology of *Aspergillus* infection in the hospital environment. Bull. de la Soc. Francaise de Mycol.Med. 19, 125-130.

Perkins W.A. (1957) The rotorod sampler. 2nd Semiannual Rep Aerosol Lab Dep. Chemistry and Chem Engng., Stanford Univ. CML. 186, 66pp.

Richards M. (1953) An investigation of the identity and incidence of airborne fungal spores as a basis for studies on fungi as allergens. PhD Thesis University of Wales.

Solomon W.R. and Burge H.P.(1975) *Aspergillus fumigatus* levels in and out of doors in urban air. J. All. Clin Immunol. 57, 46-51.

Solomon W.R., Burge H.P. and Boise, J.P. (1978) Airborne *Aspergillus fumigatus* levels outside and within a large clinical center. J. All. Clin. Immunol. 62, 56-60.

# INTERACTIONS OF FUNGI WITH TOXIC METALS

G.M. Gadd

Department of Biological Sciences
University of Dundee
Dundee, DD1 4HN

## INTRODUCTION

Interactions of toxic metals with fungi have long been of importance since metal toxicity still is the basis of many fungicidal preparations while in an environmental context, accelerating pollution of the natural environment has led to increased interest because of the sometimes dominant presence of fungi in metal-polluted habitats, the translocation of toxic metals and radionuclides to fruit bodies of edible higher fungi, and the significance of mycorrhizal fungi in the amelioration of metal phytotoxicity (Colpaert and Van Assche, 1987). Furthermore, the use of fungal (and other microbial) biomass for the detoxification of metal/radionuclide-containing industrial effluents is of biotechnological potential (Gadd, 1990, 1992*a*).

Metals are directly and/or indirectly involved in all aspects of fungal growth, metabolism, reproduction and differentiation. While many metals are essential, e.g. K, Na, Mg, Ca, Mn, Fe, Cu, Zn, Co, Ni, many others have no essential function, e.g. Rb, Cs, Al, Cd, Ag, Au, Hg, Pb. However, all can interact with fungal cells, and be accumulated by physico-chemical mechanisms and transport systems of varying specificity (Gadd, 1988). The majority of essential and inessential metals exhibit toxicity above a certain concentration, which will vary depending on the organism, the physico-chemical properties of the metal, and environmental factors, and this may necessitate expression of a detoxification mechanism if the organism is to survive (Gadd, 1992*b*).

Many attempts have been made to define metals in relation to biological effects using their physical, chemical and biological properties. Such classifications of metals may be useful in defining or predicting certain interactions with organisms and are also of use in assessing metal speciation in environmental samples or growth media (Hughes and Poole, 1991). Some elements not considered to be true metals according to chemical definitions may have metallic attributes and exhibit varying degrees of toxicity as well as accumulation by fungi, e.g. metalloids (germanium, arsenic, selenium and tellurium) and the actinides and lanthanides. Alkali metals are not usually included in discussions on metal toxicity but nevertheless exhibit a dramatic range of biological properties as exemplified by $K^+$, $Na^+$ and $Cs^+$ as respectively representing an essential non-toxic, an essential but potentially-toxic, and

*The Genus Aspergillus*, Edited by Keith A. Powell *et al.*,
Plenum Press, New York, 1994

an inessential potentially-toxic metal ion. Definitions are particularly variable in the context of "heavy metals". These are often defined as a group of approximately 65 metallic elements, of density greater than 5, with the general ability to exert toxic effects on microbial and other life forms (Gadd and Griffiths, 1978). However, this imprecise definition includes elements with diverse physical, chemical and biological properties (Nieboer and Richardson, 1980). Organometallic compounds are defined as a compound containing at least one metal-carbon bond. These are of significance as environmental pollutants, as fungicides, and because some may be synthesized by fungi as a result of exposure to and transformation of the "parental" metal species.

## METAL MYCOTOXICITY

### Mechanisms

Toxic metals can exert harmful effects in many ways although major mechanisms of toxicity are frequently a result of strong coordinating abilities. Toxic effects include the blocking of functional groups of biologically important molecules, e.g. enzymes, and transport systems for essential nutrients and ions, the displacement and/or substitution of essential metal ions from biomolecules and functional cellular units, conformational modification, denaturation and inactivation of enzymes and disruption of cell and organellar membrane integrity (Ochiai, 1987). Because of the wide spectrum of potentially toxic interactions between metals and fungi, almost every aspect of metabolism, growth and differentiation may be affected, depending on the organism, metal species and concentration and physico-chemical factors (Ross, 1975; Gadd, 1986; Gadd and White 1989*a*;). Organometals are generally more toxic towards fungi than corresponding inorganic metal species and the toxicity of organometal compounds varies with the number and identity of the organic groups (Cooney and Wuertz, 1989). Major effects of organotins and organoleads are disruption of plasma and mitochondrial membranes and action as $Cl^-/OH^-$ ionophores depolarizing electrochemical gradients (Cooney and Wuertz, 1989).

### Resistance and Tolerance

Fungal survival mainly depends on intrinsic biochemical and structural properties, physiological and/or genetical adaptation, including morphological changes, and environmental modification of metal speciation, availability and toxicity (Gadd and Griffiths, 1978; Gadd, 1992*b,c*, 1993). It is probably more appropriate to define "metal resistance" as the ability of an organism to survive metal toxicity by means of a mechanism produced in direct response to the metal species concerned, e.g. metallothionein or γ-glutamyl peptide synthesis (Mehra and Winge, 1991). "Metal tolerance" may be defined as the ability of an organism to survive metal toxicity by means of intrinsic properties and/or environmental modification of toxicity (Gadd, 1992*b,c*, 1993). Intrinsic properties that can determine survival include possession of impermeable pigmented cell walls, extracellular polysaccharide, and metabolite excretion, especially where this leads to detoxification of the metal species by, e.g. binding or precipitation. However, distinctions are difficult in many cases because of the involvement of several direct and indirect physico-chemical and biological mechanisms in survival. Biological mechanisms implicated in fungal survival (as distinct from environmental modification of toxicity) include extracellular precipitation, complexation and crystallization, transformation of metal species by, e.g. oxidation, reduction, methylation and dealkylation, biosorption to cell walls, pigments and extracellular polysaccharide, decreased transport or impermeability, efflux, intracellular compartmentation, precipitation and/or sequestration (Ross, 1975; Gadd and Griffiths, 1978; Gadd, 1988, 1990, 1992*a,b,c*, 1993; Mehra and Winge, 1991). A given organism may directly and/or indirectly rely on several survival strategies (see Figure 1).

**METAL(LOID) SPECIES**
- **charge, ionic radius, ligand preferences organometal(loid)s, actinides, lanthanides metal particulates, radionuclides**

↓

**ENVIRONMENTAL FACTORS**
- **physico-chemical and biological**

↓

**EXTRACELLULAR FUNGAL PRODUCTS**
- **polysaccharide, melanins, pigments, siderophores, organic acids, other metabolites and gases, e.g. $H_2S$; precipitation, binding, complexation, transformation**

↓

**CELL WALL COMPONENTS**
- **structural components, melanins, pigments, metabolites, wall-associated enzymes; precipitation, binding, complexation, transformation**

↓

**PLASMA MEMBRANE**
- **transport processes, permeability changes, reductase and other enzymes**

↓

**INTRACELLULAR**
- **cytosol, e.g. metallothioneins, metal γ-glutamyl peptides, precipitation, transformations**
- **vacuole, e.g. compartmentation, polyphosphate precipitation, $S^{2-}$ incorporation into metal γ-glutamyl peptides, amino acids**
- **other organelles, e.g. precipitation, reduction**

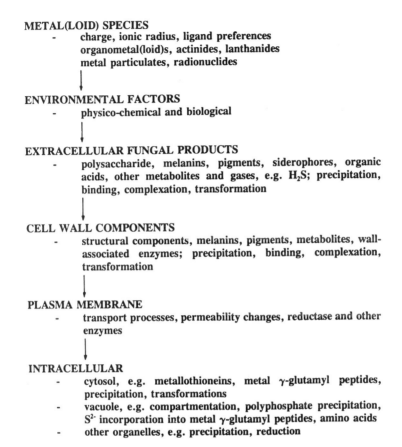

**Figure 1.** Diagrammatic representation of important fungal interactions with toxic metals. All the phenomena shown can be involved in metal detoxification and survival. A given organism may interact at several levels; external physico-chemical factors influence all responses.

## Environmental Influence on Metal Toxicity

The physico-chemical properties of a given environment, or growth medium, determine metal speciation, and therefore biological availability and toxicity, as well as other essential and inessential metal-organism interactions. Where such factors as pH, $E_h$, the presence of other anions and cations, particulate and soluble organic matter, and clay minerals decrease biological activity, toxicity may be reduced (Gadd and Griffiths, 1978). Physico-chemical attributes of the environment will also determine aspects of fungal growth and physiology which will influence responses to metals (Collins and Stotzky, 1989, 1992) (Figure 2).

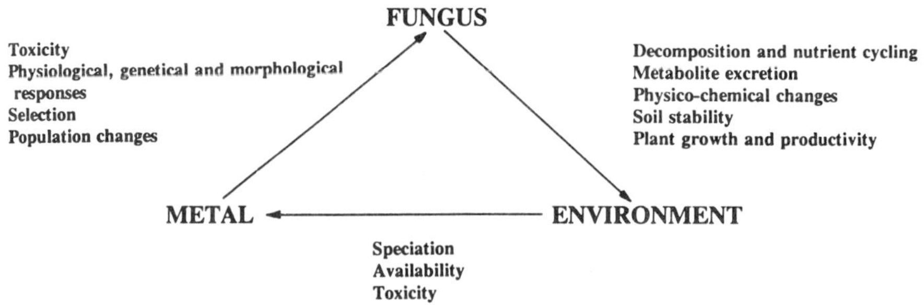


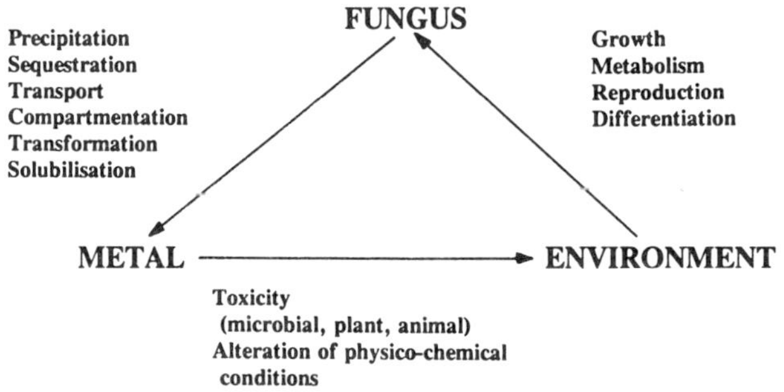


**Figure 2.** Diagrammatic representation of the dynamic interrelationships between fungi, toxic metals (and related substances), and the environment. The environment (natural, synthetic or laboratory) can affect fungal growth and metabolism and metal speciation, toxicity and availability. Metal species may positively or negatively affect fungal growth, metabolism and differentiation depending on the element, concentration and speciation, and alter the physico-chemical characteristics of an environment as well as affecting other life forms. Fungi may solubilise metals from insoluble compounds, remove them from solution by physico-chemical and biological mechanisms, and transform them into other forms, including organometals.

## Fungi in Polluted Habitats

A range of fungi from all the major groups may be found in metal-polluted habitats and the ability to survive and grow in the presence of potentially toxic concentrations is frequently encountered. In general terms, toxic metals are believed to affect fungal populations in terrestrial environments by reducing abundance and species diversity and selecting for a resistant/tolerant population (Babich and Stotzky, 1985; Duxbury, 1985). However, the effect of toxic metals on microbial abundance in natural habitats varies with the metal species and organisms present and environmental factors (Gadd and Griffiths, 1978; Duxbury, 1985). Metal pollution of plant surfaces is widespread and while there may be significant decreases in total microbial numbers (including bacteria) on phylloplanes, numbers of filamentous fungi and non-pigmented yeasts are generally little affected and if metal-supplemented isolation media are used only fungi may be isolated (Bewley, 1980; Bewley and Campbell, 1980; Mowll and Gadd, 1985).

It must be stressed that it may be difficult to separate metal effects from those of other environmental components. Phylloplane mycoflora may be subject to the influence of other potential toxicants, e.g. $SO_2$ (Babich and Stotzky, 1980) while polluted soils may be nutrient-poor, of variable pH and also contain additional toxicants (Gadd and Griffiths, 1978). Furthermore, while species diversity may be reduced in certain cases, resistance/tolerance can be exhibited by fungi from both polluted and non-polluted habitats (Freedman and Hutchinson, 1980; Arnebrant *et al.*, 1987). Such studies indicate that in certain cases, survival must be dependent on intrinsic properties of the organisms, rather than adaptive changes (which are generally studied under laboratory conditions), and physico-chemical properties of that environment which may modulate toxicity and affect species composition (Gadd, 1986, 1992*b,c*, 1993).

## CELLULAR INTERACTIONS BETWEEN TOXIC METALS AND FUNGI

### Extracellular Precipitation and Complexation

Many extracellular fungal products can complex or precipitate heavy metals. Citric acid can be an efficient metal-ion chelator, and oxalic acid can form insoluble metal oxalates (Sutter *et al.*, 1983). The production of $H_2S$ by yeasts can result in metal precipitation as insoluble sulphides (Minney and Quirk, 1985). Many fungi release high affinity Fe-binding siderophores (Winkelmann, 1992). Externally-formed $Fe^{3+}$ chelates may subsequently be taken up into the cell. In several fungi, the excretion of iron-binding molecules is markedly stimulated by Fe deficiency, and such compounds may bind other metals, e.g. Ga (Adjimani and Emery, 1987).

### Metal Binding to Cell Walls

The wall is the first cellular site of interaction with metal species and metal removal from solution may be rapid and include such phenomena as ion exchange, adsorption, complexation, precipitation and crystallization. A variety of potential sites may be involved in metal sequestration including carboxyl, amine, hydroxyl, phosphate and sulphydryl groups (Strandberg *et al.*, 1981) although primary interactions are likely to involve binding to carboxyl and phosphate groups which may be enhanced by electrostatic attraction to other negatively charged functional groups (Tobin *et al.*, 1990). Metabolism-independent biosorption is frequently rapid and unaffected over moderate ranges of temperature, e.g. 4-30°C (Norris and Kelly, 1977; De Rome and Gadd, 1987). In *Rhizopus arrhizus*, biosorption was related to ionic radius for, e.g. $La^{3+}$, $Mn^{2+}$, $Cu^{2+}$, $Zn^{2+}$, $Cd^{2+}$, $Ba^{2+}$, $Hg^{2+}$, $Pb^{2+}$, $UO_2^{2+}$ and $Ag^+$, but not $Cr^{3+}$ or the alkali metal cations, $Na^+$, $K^+$, $Cs^+$ and $Rb^+$, which were not taken up (Tobin *et al.*, 1984). The main site of actinide uptake is the cell wall. Both adsorption and precipitation of hydrolysis products occurs with uranium (Tsezos and Volesky, 1982*a*) while coordination with cell wall nitrogen was the main mechanism of thorium biosorption (Tsezos and Volesky, 1982*b*, Tsezos, 1983). In *Saccharomyces cerevisiae*, uranium was deposited as a layer of needle-like fibrils on cell walls, reaching up to 50% of the dry weight of individual cells (Strandberg *et al.*, 1981).

Fungal phenolic polymers and melanins contain phenolic units, peptides, carbohydrates, aliphatic hydrocarbons, and fatty acids and therefore possess many potential metal-binding sites. Oxygen-containing groups in these substances, including carboxyl, phenolic and alcoholic hydroxyl, carbonyl and methoxyl groups may be particularly important in metal binding (Gadd, 1988). Melanin from *Aureobasidium pullulans* can bind significant amounts of metals like $Cu^{2+}$ and $Fe^{3+}$ (Gadd, 1984; Senesi *et al.*, 1987; Gadd and De Rome, 1988) as well as organometallic compounds, e.g. tributyltin chloride (Gadd *et al.*, 1990).

## Transport of Toxic Metals

The plasma and vacuolar membranes are the main transport membranes of fungi. Most work on metal ion transport in fungi has concerned $K^+$ and $Ca^{2+}$ and the transport of toxic metal species is still poorly understood. However, it is known that integral to the transport and intracellular compartmentation of metal ions is the operation of $H^+$-pumping ATPases at the plasma and vacuolar membranes which generate electrochemical proton gradients across the membranes which indirectly energize uptake of a variety of charged solutes (Figure 3). Metabolism-dependent intracellular uptake or transport of metal ions into cells can be inhibited by low temperatures, the absence of an energy source (e.g. glucose), and by glucose analogues, metabolic inhibitors and uncouplers (Norris and Kelly, 1977; Borst-Pauwels, 1981; Gadd, 1986; Starling and Ross, 1990). In certain fungi, especially yeasts, greater amounts of metal may be accumulated by such a process, in conjunction with internal sequestration, than by metabolism-independent processes. However, at the relatively high concentrations of metals frequently used in uptake studies, energy-dependent intracellular uptake may not be as significant a component of total uptake as general biosorption. This is particularly true for filamentous fungi where high values of biosorption can mask low rates of intracellular influx (Gadd and White, 1985; Gadd *et al.*, 1987). In growing fungal cultures, phases of adsorption and intracellular uptake may be obscured by changes in the physiology and morphology of the fungus and the physical and chemical properties of the growth medium (Gadd, 1988).

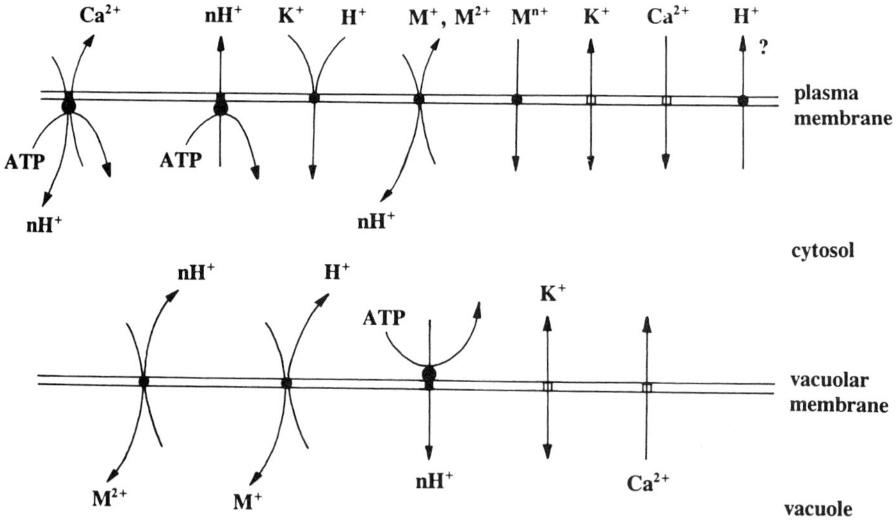


**Figure 3.** Simplified representation of metal ion and $H^+$ transport processes at the fungal plasma and vacuolar membranes. At the plasma membrane, systems shown are (left to right): $Ca^{2+}$-ATPase; $H^+$-ATPase; $K^+$-$H^+$ symport; $M^+$,$M^{2+}$-$H^+$ antiport for efflux of monovalent ($M^+$) and divalent ($M^{2+}$) metal cations; $M^{n+}$ uniport; $K^+$ channel; $Ca^{2+}$ channel; alternative $H^+$ efflux. At the vacuolar membrane, systems are (l to r): $M^{2+}$-$H^+$ antiport; $M^+$-$H^+$ antiport; vacuolar $H^+$-ATPase; $K^+$ channel; $Ca^{2+}$ channel. No attempt has been made to show stoichiometries, relationships with movements of anions or the different kinds of ion channels that may exist. Toxic metals may influence, or be influenced by, the transport systems shown for $H^+$, $K^+$ and $Ca^{2+}$; in some circumstances toxic metals may substitute in essential metal ion transport systems (see Borst-Pauwels, 1981; Okorokov, 1985; Sanders, 1988, 1990; Jones and Gadd, 1990; Klionsky *et al.*, 1990; Gadd, 1993).

For potentially-toxic metal cations, reduced transport into yeast cells is a frequently observed mechanism of resistance, e.g. in metal-resistant strains of *S. cerevisiae* (Gadd *et al.*, 1984; Gadd, 1986). However, a $Mn^{2+}$-resistant strain accumulated more $Mn^{2+}$ than the wild-type strain, possibly by means of a more efficient internal sequestration system (Bianchi *et al.*, 1981). Those external factors that can reduce uptake of toxic cations may also result in protection while protection of yeast cells by $Ca^{2+}$ from $Cd^{2+}$ toxicity was due to a reduction in $Cd^{2+}$ uptake (Kessels *et al.*, 1985). For $Li^+$, a low net entry into cells results from an electrogenic $H^+/Li^+$ antiport (Rodriguez-Navarro *et al.*, 1981). Extrusion pumps have also been demonstrated for divalent cations, e.g. $Ca^{2+}$, $Sr^{2+}$ (Theuvenet *et al.*, 1986). Earlier evidence indicated the existence of a $Ca^{2+}/H^+$ antiport on the plasma membrane although it is now proposed that $Ca^{2+}$ efflux is via a $H^+/Ca^{2+}$-ATPase with a stoichiometric ratio of at least $2H^+/Ca^{2+}$ in *Neurospora crassa* (Miller *et al.*, 1990).

**Intracellular Fate of Toxic Metals**

**Metallothioneins** Metallothioneins are small, cysteine-rich polypeptides that can bind essential metals, e.g. copper and zinc, as well as inessential metals like cadmium. Copper resistance in *S. cerevisiae* is mediated by the induction of a 6573-dalton cysteine-rich protein, copper-metallothionein (Cu-MT) (Hamer, 1986; Butt and Ecker, 1987; Fogel *et al.*, 1988). This protein normally functions to maintain low levels of intracellular copper ions and thus prevent futile transcription of the *CUP1* structural gene. Although yeast metallothionein can also bind cadmium and zinc *in vitro*, it is not transcriptionally induced by these ions and does not protect against them (Winge *et al.*, 1985).

The gene for *N. crassa* Cu-MT encodes a 26 amino acid protein which contains 7 different amino acids (28% cysteine) (Munger *et al.*, 1987). It appears that copper is present in cuprous (Cu(I)) form and all the cysteines are ligated to form Cu(I) complexes with 6 copper molecules per mole of protein (Lerch and Beltramini, 1983). The isolated protein from *S. cerevisiae* contains 8 mol of copper (Cu(I)) ligated to 12 cysteines per mole protein (Winge *et al.*, 1985).

**Metal γ-Glutamyl Peptides** Another group of fungal metal-binding molecules are short, cysteine-containing γ-glutamyl peptides, encompassed by the trivial name "phytochelatins" (Winge *et al.*, 1989; Rauser, 1990). They are also sometimes referred to as "cadystins" (Mutoh *et al.*, 1991). They have the general formula $(\gamma\text{Glu-Cys})_n\text{-Gly}$ $[(\gamma\text{-EC})_n\text{G}]$ (Grill *et al.*, 1986). Peptides of n = 2 through n = 5 appear to be the most common and metal ions are bound within a metal-thiolate cluster composed of an oligomer of peptides (Winge *et al.*, 1989). Phytochelatins are apparently not synthesised by *S. cerevisiae* but *Schizosaccharomyces pombe* is now known to synthesise at least seven different homologous phytochelatins in response to metal exposure (Grill *et al.*, 1986). The originally discovered two γ-glutamyl peptides in *S. pombe*, cadystin A and B (n = 2 and 3 respectively) were synthesised in response to cadmium though other metals including $Cu^{2+}$, $Pb^{2+}$, $Zn^{2+}$ and $Ag^+$ were also effective (Hayashi *et al.*, 1988; Reese *et al.*, 1988). However, the greatest amount of peptide synthesis occurs with cadmium (Hayashi *et al.*, 1988).

Cadmium-induced (but not copper-induced) γ-glutamyl peptides contain labile sulphur. Cadmium stimulates the production of $S^{2-}$ in *S. pombe* and *Candida glabrata* and a portion of this may be incorporated within the Cd-γ-glutamyl peptide complexes to levels ranging from 0.1 to 1.5 mol (mol peptide)$^{-1}$ (Winge *et al.*, 1989). Sulphide incorporation enhances the stability and metal content of the peptide complex which may increase the efficacy of the complex in maintaining a low intracellular concentration of $Cd^{2+}$. The extracellular and cell-associated CdS colloid formed by interaction of $Cd^{2+}$ with $S^{2-}$ may also contribute to $Cd^{2+}$ detoxification. *Candida (Torulopsis) glabrata* is unique in expressing both metallothionein and γ-glutamyl peptide synthesis for metal detoxification (Mehra and Winge, 1991).

It is now been shown that other non-metallic inducing agents may induce γ-glutamyl peptide synthesis in yeasts. The antifungal agents tetramethylthiuram disulphide (TMTD) or dimethyl dithiocarbamate (DMDTC) induce γ-glutamyl peptide synthesis to a lesser degree than $CdCl_2$ (but more than $ZnCl_2$ or $CuSO_4$) in *S. pombe* (Mutoh *et al.*, 1991). Such results may indicate biochemical functions other than that of metal detoxification including intracellular transport of essential metal ions, e.g. $Zn^{2+}$, $Cu^{2+}$, detoxification of xenobiotics, and scavengers of active oxygen (Mutoh *et al.*, 1991).

**Vacuolar compartmentation** The fungal vacuole is a highly important organelle with functions including macromolecular degradation, storage of metabolites, and cytosolic ion and pH homeostasis (Jones and Gadd, 1990; Klionsky *et al.*, 1990). A vacuolar $H^+$-pumping ATPase has been identified in several fungi, notably *S. cerevisiae*, *S. carlsbergensis* and *N. crassa* (Okorokov, 1985; Anraku *et al.*, 1989). The electrochemical proton gradient generated across the vacuolar membrane energizes transport of monovalent and divalent cations into the vacuole, as well as other substances including basic amino acids. The main mechanisms of transport appear to be *via* proton antiport and ion channels, although there is some evidence of a pyrophosphatase activity in *S. carlsbergensis* which may be responsible for a $PP_i$-dependent pH gradient (Klionsky *et al.*, 1990). The fungal vacuole has an important role in the regulation of cytosolic metal ion concentrations, both for essential metabolic functions and the detoxification of potentially-toxic metal ions (Gadd and White, 1989*a*; Avery and Tobin, 1992). Polyphosphates, the only macromolecular anion found in the vacuole, have an important role in maintaining ionic compartmentation, and their biosynthesis accompanies vacuolar $Mg^{2+}$ or $Mn^{2+}$ accumulation (Okorokov, 1985). It has been proposed that, in *S. pombe*, $S^{2-}$ is incorporated into a Cd-γ-glutamyl peptide in the vacuole (Ortiz *et al.*, 1992). It has also been shown that changes in the amino acid pools of yeast can occur in response to nickel exposure with the formation of nickel-histidine complexes in the vacuole being proposed as a survival mechanism (Joho *et al.*, 1990, 1992).

## Metal Transformations

Fungi, as well as other microorganisms, can effect chemical transformations of metals by, e.g. oxidation, reduction, methylation and dealkylation (Gadd, 1992*b*). Some enzymatic transformations may be involved in survival since certain transformed metal species are less toxic and/or more volatile than the original species. Reductions carried out by fungi and yeasts include $Ag^+$ to metallic $Ag^0$ (Kierans *et al.*, 1990) and the reduction of Cu(II) to Cu(I) by copper reductase in *D. hansenii* (Wakatsuki *et al.*, 1991). There is only a limited amount of evidence for reduction of $Hg^{2+}$ to $Hg^0$ in fungi and yeasts in contrast to bacteria. Fungal reductions of metalloids include reduction of tellurite to elemental tellurium, which can appear as a black deposit and located on the endoplasmic reticulum (Corfield and Smith, 1970) and the reduction of selenate or selenite to form amorphous elemental selenium (Konetzka, 1977).

$Hg^{2+}$ methylation, which involves methylcobalamin (vitamin B12) and arsenic biomethylation, by transfer of carbonium ions ($CH_3^+$) from S-adenosylmethionine (SAM) has been demonstrated in several fungi and yeasts (Thayer, 1988). Selenium biomethylation appears to be by an analogous mechanism to that of arsenic, and several fungi can produce, e.g. dimethylselenide from inorganic selenium compounds, a phenomenon of relevance to the bioremediation of Se-contaminated soils (Thayer, 1988; Thompson-Eagle and Frankenberger, 1992). Certain fungi may be capable of the production of volatile dimethyl telluride from tellurium salts (Konetzka, 1977). Methylated metal(loid) species are usually more volatile and may be lost from the environment (Gadd, 1992*c*).

Organometal degradation involves breaking of the metal-C bonds through direct and indirect fungal action: metal-C bonds can be disrupted by physico-chemical means, although fungal activity may enhance the conditions necessary for abiotic attack by, e.g. alteration of pH, metabolite excretion. Organotin degradation involves sequential removal of organic groups from tin atom, a process generally resulting in a decrease in toxicity (Cooney and Wuertz, 1989). Organomercurials may be detoxified by organomercury lyase, the resulting $Hg^{2+}$ being then reduced to $Hg^{o}$ by mercuric reductase (Tezuka and Takasaki, 1988).

## MYCORRHIZAS AND MACROFUNGI

The responses of mycorrhizal fungi to toxic metals are of relevance to the reclamation of polluted land and their influence on plant growth and productivity. Although there is a wide diversity in responses between different plant-fungal symbioses, amelioration of metal phytotoxicity by mycorrhizal fungi has been widely demonstrated (Bradley *et al.*, 1982; Gildon and Tinker, 1983; Dixon, 1988).

Elevated concentrations of heavy metals and radionuclides often occur in the fruiting bodies of higher fungi when growing in polluted environments. Higher levels of Pb, Cd, Zn and Hg are found in macrofungi from urban or industrial areas although there are wide differences in uptake abilities between different species and different metals (Lepsova and Mejstrik, 1989).

## BIOTECHNOLOGICAL ASPECTS OF FUNGAL METAL ACCUMULATION

The removal of radionuclides, metal or organometal(loid) species, compounds or particulates from solution by microbial biomass is now often referred to as "biosorption", particularly where predominant interactions are physico-chemical. Biosorption is an area of increasing biotechnological interest since the removal of potentially toxic and/or valuable metals and radionuclides from contaminated effluents can result in detoxification prior to environmental discharge (Gadd, 1988; McEldowney, 1990). Furthermore, appropriate treatment of loaded biomass can enable recovery of valuable metals for recycling or further containment of highly toxic and/or radioactive species. Although basic features of metal accumulation are common to most microbial groups, fungi possess a number of attributes which reflect their morphological and physiological diversity. The majority of fungi exhibit structural and biochemical differentiation with the filamentous growth form enabling efficient colonisation of substrates. Different cell forms may have different uptake capacities and a wide spectrum of sensitivities towards potentially toxic metals are encountered. In addition, many fungi can be extremely tolerant of toxic metals in comparison with other microbial groups (Gadd, 1986). Fungi and yeasts are, in general, easy to grow, produce high yields of biomass and several are amenable to genetical and morphological manipulation. Their use in a variety of industrial fermentations also means that waste biomass is available for which the biotechnology of metal removal may provide a good use. *Aspergillus niger* biomass arises in substantial quantities from citric acid production while *S. cerevisiae* is available from food and beverage industries. Other kinds of biomass arising from certain industrial fermentations include *Rhizopus arrhizus*, *Aspergillus terreus* and *Penicillium chrysogenum* (Volesky, 1990).

### Biosorption of Metals and Radionuclides by Fungi

As described previously, a variety of mechanisms occur in fungal cells for the removal of metals and radionuclides from solution. These mechanisms range from physico-chemical interactions ("biosorption") such as adsorption, to processes dependent on cell metabolism. In some circumstances, fungi can adsorb insoluble metal compounds, e.g. sulphides, zinc dust and $Fe(OH)_3$ ("ochre") and this may represent another area of potential biotechnological application. *A. niger* oxidized copper, lead and zinc sulphides to sulphates, the sulphide particles in the medium being adsorbed onto mycelial surfaces (Wainwright and Grayston, 1986). Magnetite-loaded fungal biomass was susceptible to a magnetic field which has implications for biomass removal from solution (Wainwright *et al.*, 1990).

**Industrial Considerations**

An important consideration, whether living or dead cells are used, is the form and kind of biomass to be used. For industrial applications biomass has some disadvantages. It is generally of low strength and density which can limit the choice of suitable reactors and make biomass or effluent separation difficult. For use in packed bed or fluidized bed reactors, immobilized or pelleted biomass is of greater potential. Immobilized biomass, whether within or on an inert matrix, has advantages in that high flow rates can be achieved, clogging is minimized, particle size can be controlled and high biomass loadings are possible (Tsezos, 1986). Fungal examples of laboratory-scale immobilized systems include *Aspergillus oryzae* and *R. arrhizus* on reticulated foam particles (Lewis and Kiff, 1988), *Trichoderma viride* packed in molochite and used for Cu removal from simulated effluents (Townsley *et al.*, 1986), *S. cerevisiae* immobilized in a sand column and used for removal of Cu(II) (Huang *et al.*, 1990), microfungal biomass incorporated into paper- and textile-fibre derived filters (Wales and Sagar, 1990), and alginate-encapsulated *S. cerevisiae* for U, Sr and Cs removal (De Rome and Gadd, 1991).

Filamentous fungi may grow in pellet form in culture, and such pellets may have similar advantages to immobilized particles. Pellets (4 mm diameter) of *A. niger* have been used in a fluidized bed reactor for uranium removal. A simple ion-exchange process predominated, where $UO_2^{2+}$ ions replaced protons within the cell wall. It was found that such a system was 14 times more efficient than the commercial ion-exchange resin IRA-400 (Yakubu and Dudeney, 1986). Certain industrial process streams containing actinide elements are extremely acidic (pH < 1). Despite the low pH, biomass from several fungal species removed thorium from solution in 1M $HNO_3$, pH 0-1 (Gadd & White, 1989*b*; White and Gadd, 1990). Air-lift reactors removed approximately 90-95% of the thorium supplied over extended time periods and exhibited well-defined breakthrough points. Of the species tested, *A. niger* and *R. arrhizus*, used as mycelial pellets, were the most effective with loading capacities of approximately 0.5 and 0.6 mmol $g^{-1}$ respectively (116 and 138 mg g dry $wt^{-1}$) at an inflow concentration of 3 mM thorium (as nitrate) (White and Gadd, 1990). Metal-tolerant fungi proved successful for the removal of cadmium when grown in a continuous flow air-lift fermenter, with >97% removed from the simulated effluent of 6 mg Cd $l^{-1}$ (Campbell and Martin, 1990).

Biotechnological exploitation of metal accumulation by microbial biomass, including fungi, may depend on the ease of metal recovery for subsequent reclamation, containment, and biosorbent regeneration (Tsezos, 1984; Volesky, 1990; Gadd and White, 1992). The means of metal recovery may depend on the ease of removal from the biomass which in turn can depend on the metal species involved and the mechanism of accumulation. Metabolism-independent biosorption is frequently reversible by non-destructive methods whereas energy-dependent intracellular accumulation and compartmentation is often irreversible requiring destructive recovery (Gadd, 1988). Most work has concentrated on non-destructive desorption which should ideally be highly efficient, economical and result in minimal damage to the biosorbent. Of several elution systems examined for uranium desorption from *R. arrhizus*, sodium bicarbonate ($NaHCO_3$) exhibited > 90% efficiency of removal and high uranium concentration factors (Tsezos, 1984).

It is clear that fungi and fungal products can be highly efficient removal agents and exhibit high uptake capacities though many phenomena remain unexplored in a biotechnological context, e.g. metal-binding proteins and peptides, extracellular precipitation, and transformations. Microbial biomass is generally highly efficient at dilute external concentrations and therefore may not necessarily replace existing technologies, but it may serve as a "polisher" after an existing treatment that is not completely efficient. It is also pertinent to point out the potential interaction of biosorption technology with microbial leaching of metal ions from ores and wastes. While the bulk of leaching information is derived from bacterial systems, it has been demonstrated that *Penicillium* sp. are capable of zinc leaching from industrial wastes and mixtures of metal oxides (Schinner and Burgstaller, 1989; Franz *et al.*, 1991). Further work is needed in several areas to establish or realize the full potential of fungal systems for environmental bioremediation.

## REFERENCES


Adjimani, J.P. and Emery, T. (1987) Iron uptake in *Mycelia Sterilia* EP-76. J. Bacteriol. 169, 3664-3668.

Arnebrant, K., Bååth, E. and Nordgren, A. (1987) Copper tolerance of microfungi isolated from polluted and unpolluted forest soil. Mycologia 79, 890-895.

Avery, S.V. and Tobin, J. (1992) Mechanisms of strontium uptake by laboratory and brewing strains of *Saccharomyces cerevisiae*. Appl. Environ. Microbiol. 58, 3883-3889.

Anraku, Y., Umemoto, N., Hirata, R. and Wada, Y. (1989) Structure and function of the yeast vacuolar membrane proton ATPase. J. Bioenergetics Biomembr. 21, 589-603.

Babich, H. and Stotzky, G. (1980) Environmental factors that influence the toxicity of heavy metal and gaseous pollutants to microorganisms. CRC Crit. Rev. Microbiol. 8, 99-145.

Babich, H. and Stotzky, G. (1985) Heavy metal toxicity to microbe-mediated ecologic processes: a review and potential application to regulatory policies. Environ. Res. 36, 111-137.

Bewley, R.J.F. (1980) Effects of heavy metal pollution on oak leaf microorganisms. Appl. Environ. Microbiol. 40, 1053-1059.

Bewley, R.J.F. and Campbell, R. (1980) Influence of zinc, lead and cadmium pollutants on microflora of hawthorn leaves. Microbial Ecol. 6, 227-240.

Bianchi, M.E., Carbone, M.L. and Lucchini, G. (1981) $Mn^{2+}$ and $Mg^{2+}$ uptake in Mn-sensitive and Mn-resistant yeast strains. Plant Sci. Lett. 22, 345-352.

Borst-Pauwels, G.W.F.H. (1981) Ion transport in yeast. Biochim. Biophys. Acta 650, 88-127.

Bradley, R., Burt, A.J. and Read, D.J. (1982) The biology of mycorrhiza in the Ericaceae. VIII. The role of mycorrhizal infection in heavy metal resistance. New Phytol. 91, 197-209.

Butt, T.R. and Ecker, D.J. (1987) Yeast metallothionein and applications in biotechnology. Microbiol. Rev. 51, 351-364.

Campbell, R. and Martin, M.H. (1990) Continuous flow fermentation to purify waste water by the removal of cadmium. Wat. Air Soil Poll. 50, 397-408.

Collins, Y.E. and Stotzky, G. (1989) Factors affecting the toxicity of heavy metals to microbes, in "Metal Ions and Bacteria" (Beveridge, T.J. and Doyle, R.J., Eds), pp. 31-90. Wiley, New York.

Collins, Y.E. and Stotzky, G. (1992) Heavy metals alter the electrokinetic properties of bacteria, yeasts and clay minerals. Appl. Environ. Microbiol. 58, 1592-1600.

Colpaert, J.V. and Van Assche, J.A. (1987) Heavy metal tolerance in some ectomycorrhizal fungi. Functional Ecol. 1, 415-421.

Cooney, J.J. and Wuertz, S. (1989) Toxic effects of tin compounds on microorganisms. J. Ind. Microbiol. 4, 375-402.

Corfield, P.S. and Smith, D.G. (1970) The endoplasmic reticulum as the site of potassium tellurite reduction in yeasts. J. Gen. Microbiol. 63, 311-316.

De Rome, L. and Gadd, G.M. (1987) Copper adsorption by *Rhizopus arrhizus, Cladosporium resinae* and *Penicillium italicum.* Appl. Microbiol. Biotechnol. 26, 84-90.

De Rome, L. and Gadd, G.M. (1991) Use of pelleted and immobilized yeast and fungal biomass for heavy metal and radionuclide recovery. J. Ind. Microbiol. 7, 97-104.

Dixon, R.K. (1988) Response of ectomycorrhizal *Quercus rubra* to soil cadmium, nickel and lead. Soil Biol. Biochem. 20, 555-559.

Duxbury, T. (1985) Ecological aspects of heavy metal responses in microorganisms, in "Advances in Microbial Ecology" (Marshall, K.C., Ed.), pp. 185-235. Plenum Press, New York.

Fogel, S., Welch, J.W. and Maloney, D.H. (1988) The molecular genetics of copper resistance in

*Saccharomyces cerevisiae* - a paradigm for non-conventional yeasts. J. Basic Microbiol. 28, 147-160.

Franz, A., Burgstaller, W. and Schinner, F. (1991) Leaching with *Penicillium simplicissimum*: influence of metals and buffers on proton extrusion and citric acid production. Appl. Environ. Microbiol. 57, 769-774.

Freedman, B. and Hutchison, T.C. (1980) Effects of smelter pollutants on forest leaf litter decomposition near a nickel-copper smelter at Sudbury, Ontario. Can. J. Bot. 58, 1722-1736.

Gadd, G.M. (1984) Effect of copper on *Aureobasidium pullulans* in solid medium: adaptation not necessary for tolerant behaviour. Trans. Brit. Mycol. Soc. 82, 546-549.

Gadd, G.M. (1986) The responses of fungi towards heavy metals, in "Microbes in Extreme Environments" (Herbert, R.A. and Codd, G.A., Eds.), pp. 83-110. Academic Press, London.

Gadd, G.M. (1988) Accumulation of metals by microorganisms and algae, in "Biotechnology - A Comprehensive Treatise, Volume 6b, Special Microbial Processes" (Rehm, H-J, Ed.), pp. 401-433. VCH Verlagsgesellschaft, Weinheim.

Gadd, G.M. (1990) Fungi and yeasts for metal binding, in:"Microbial Mineral Recovery" (Ehrlich, H. and Brierley, C.L., Eds.), pp. 249-275. McGraw-Hill, New York.

Gadd, G.M. (1992*a*) Microbial control of heavy metal pollution, in "Microbial Control of Environmental Pollution" (Fry, J.C., Gadd, G.M., Herbert, R.A., Jones, C.W. and Watson-Craik, I., Eds.), pp. 59-88. Cambridge University Press, Cambridge.

Gadd, G.M. (1992*b*) Metals and microorganisms: a problem of definition. FEMS Microbiol. Lett. 100, 197-204.

Gadd, G.M. (1992*c*) Molecular biology and biotechnology of microbial interactions with organic and inorganic heavy metal compounds, in:"Molecular Biology and Biotechnology of Extremophiles" (Herbert, R.A. and Sharp, R.J., Eds.), pp. 225-257. Blackie and Sons, Glasgow.

Gadd, G.M. (1993) Interactions of fungi with toxic metals. New Phytol. 124, 1-35.

Gadd, G.M. and Griffiths, A.J. (1978) Microorganisms and heavy metal toxicity. Microbial Ecol. 4, 303-317.

Gadd, G.M. and De Rome, L. (1988) Biosorption of copper by fungal melanin. Appl. Microbiol. Biotechnol. 29, 610-617.

Gadd, G.M. and White, C. (1985) Copper uptake by *Penicillium ochro-chloron*: influence of pH on toxicity and demonstration of energy-dependent copper influx using protoplasts. J. Gen. Microbiol. 131, 1875-1879.

Gadd, G.M. and White, C. (1989*a*) Heavy metal and radionuclide accumulation and toxicity in fungi and yeasts, in:"Metal-Microbe Interactions" (Poole, R.K. and Gadd, G.M., Eds.), pp. 19-38. IRL Press, Oxford.

Gadd, G.M. and White, C. (1989*b*) The removal of thorium from simulated acid process streams by fungal biomass. Biotechnol. Bioeng. 33, 592-597.

Gadd, G.M. and White, C. (1992) Removal of thorium from simulated acid process streams by fungal biomass: potential for thorium desorption and reuse of biomass and desorbent. J. Chem. Technol. Biotechnol. 55, 39-44.

Gadd, G.M., Stewart, A., White, C. and Mowll, J.L. (1984) Copper uptake by whole cells and protoplasts of a wild-type and copper-resistant strain of *Saccharomyces cerevisiae*. FEMS Microbiol. Lett. 24, 231-234.

Gadd, G.M., White, C. and Mowll, J.L. (1987) Heavy metal uptake by intact cells and protoplasts of *Aureobasidium pullulans*. FEMS Microbiol. Ecol. 45, 261-267.

Gadd, G.M., Gray, D.J. and Newby, P.J. (1990) Role of melanin in fungal biosorption of tributyltin chloride. Appl. Microbiol. Biotechnol. 34, 116-121.

Gildon, A. and Tinker, P.B. (1983) Interactions of vesicular-arbuscular mycorrhizal infection and heavy metals in plants. I. The effects of heavy metals on the development of vesicular-arbuscular mycorrhizas. New Phytol. 95, 247-261.

Grill, E., Winnacker, E-L. and Zenk, M.H. (1986) Synthesis of seven different homologous phytochelatins in metal-exposed *Schizosaccharomyces pombe* cells. FEBS Lett. 197, 115-120.

Hamer, D.H. (1986) Metallothionein. Ann. Rev. Biochem. 55, 913-951.

Hayashi, Y., Nakagawa, C.W., Uyakul, D., Imai, K., Isobe, M. and Goto, T. (1988) The change of cadystin components in Cd-binding peptides from the fission yeast during their induction by cadmium. Biochem. Cell Biol. 66, 288-295.

Huang, C-P., Huang, C-P. and Morehart, A.L. (1990) The removal of Cu(II) from dilute aqueous solutions by *Saccharomyces cerevisiae*. Wat. Res. 24, 433-439.

Hughes, M.N. and Poole, R.K. (1991) Metal speciation and microbial growth - the hard (and soft) facts. J. Gen. Microbiol. 137, 725-734.

Joho, M., Inouhe, M., Tohoyama, H. and Murayama, T. (1990) A possible role of histidine in a nickel resistant mechanism of *Saccharomyces cerevisiae*. FEMS Microbiol. Lett. 66, 333-338.

Joho, M., Ishikawa, Y., Kunikane, M., Inouhe, M., Tohoyama, H. and Murayama, T. (1992) The subcellular distribution of nickel in Ni-sensitive and Ni-resistant strains of *Saccharomyces cerevisiae*. Microbios 71, 149-159.

Jones, R.P. and Gadd, G.M. (1990) Ionic nutrition of yeast - the physiological mechanisms involved and applications for biotechnology. Enzyme Microb. Technol. 12, 402-418.

Kessels, B.G.F., Belde, P.J.M. and Borst-Pauwels, G.W.F.H. (1985) Protection of *Saccharomyces cerevisiae* against $Cd^{2+}$ toxicity by $Ca^{2+}$. J. Gen. Microbiol. 131, 2533-2537.

Kierans, M., Staines, A.M., Bennett, H. and Gadd, G.M. (1991) Silver tolerance and accumulation in yeasts. Biol. Metals 4, 100-106.

Klionsky, D.J., Herman, P.K. and Emr, S.D. (1990) The fungal vacuole: composition, function and biogenesis. Microbiol. Rev. 54, 266-292.

Konetzka, W.A. (1977) Microbiology of metal transformations, in "Microorganisms and Minerals" (Weinberg, E.D., Ed.), pp. 317-342. Marcel Dekker Inc., New York.

Lepšová, A. and Mejstřík, V. (1989) Trace elements in fruit bodies of fungi under different pollution stress. Agric. Ecosyst. Environ. 28, 305-312.

Lerch, K. and Beltramini, M. (1983) *Neurospora* copper metallothionein: molecular structure and biological significance. Chem. Scripta 21, 109-115.

Lewis, D. and Kiff, R.J. (1988) The removal of heavy metals from aqueous effluents by immobilised fungal biomass. Environ Technol. Lett. 9, 991-998.

McEldowney, S. (1990) Microbial biosorption of radionuclides in liquid effluent treatment. Appl. Biochem. Biotechnol. 26, 159-180.

Mehra, R.K. and Winge, D.R. (1991) Metal ion resistance in fungi: molecular mechanisms and their related expression. J. Cell. Biochem. 45, 30-40.

Miller, A.J., Vogg, G. and Sanders, D. (1990) Cytosolic calcium homeostasis in fungi: roles of plasma membrane transport and intracellular sequestration of calcium. Proc. Nat. Acad. Sci. U.S.A. 87, 9348-9352.

Minney, S.F. and Quirk, A.V. (1985) Growth and adaptation of *Saccharomyces cerevisiae* at different cadmium concentrations. Microbios 42, 37-44.

Mowll, J.L. and Gadd, G.M. (1985) The effect of vehicular lead pollution on phylloplane mycoflora. Trans. Brit. Mycol. Soc. 84, 685-689.

Munger, K., Germann, U.A. and Lerch, K. (1987) The *Neurospora crassa* metallothionein gene. Regulation of expression and chromosomal location. J. Biol. Chem. 262, 7363-7367.

Mutoh, N., Kawabata, M. and Hayashi, Y. (1991) Tetramethylthiuram disulfide or dimethyldithiocarbamate induces the synthesis of cadystins, heavy metal chelating peptides, in *Schizosaccharomyces pombe*. Biochem. Biophys. Res. Comm. 176, 1068-1073.

Nieboer, E. and Richardson, D.H.S. (1980) The replacement of the nondescript term "heavy metals" by a biologically and chemically significant classification of metal ions. Environ. Poll. 1, 3-26.

Norris, P.R. and Kelly, D.P. (1977) Accumulation of cadmium and cobalt by *Saccharomyces cerevisiae*. J. Gen. Microbiol. 99, 317-324.

Ochiai, E.I. (1987) "General Principles of Biochemistry of the Elements", Plenum Press, New York.

Okorokov, L.A. (1985) Main mechanisms of ion transport and regulation of ion concentrations in the yeast cytoplasm, in "Environmental Regulation of Microbial Metabolism" (Kulaev, I.S., Dawes, E.A. and Tempest, D.W., Eds.), pp. 339-349. Academic Press, London.

Ortiz, D.F., Kreppel, L., Speiser, D.M., Scheel, G., McDonald, G. and Ow, D.W. (1992) Heavy metal tolerance in the fission yeast requires an ATP-binding cassette-type vacuolar membrane transporter. Embo J. 11, 3491-3499.

Rauser, W.E. (1990) Phytochelatins. Ann. Rev. Biochem. 59, 61-86.

Reese, R.N., Mehra, R.K., Tarbet, E.B. and Winge, D.R. (1988) Studies on the γ-glutamyl Cu-binding peptide from *Schizosaccharomyces pombe*. J. Biol. Chem. 263, 4186-4192.

Rodriguez-Navarro, A., Sancho, E.D. and Perez-Lloveres, C. (1981) Energy source for lithium efflux in yeast. Biochim. Biophys. Acta 640, 352-358.

Ross, I.S. (1975) Some effects of heavy metals on fungal cells. Trans. Br. Mycol. Soc. 64, 175-193.

Sanders, D. (1988) Fungi, in "Solute Transport in Plant Cells and Tissues" (Baker, D.A. and Hall, J.L., Eds.), pp. 106-165. Longman, Harlow.

Sanders, D. (1990) Kinetic modelling of plant and fungal membrane transport systems. Ann. Rev. Plant Physiol. Plant Mol. Biol. 41, 77-107.

Schinner, F. and Burgstaller, W. (1989) Extraction of zinc from industrial waste by a *Penicillium* sp.. Appl. Environ. Microbiol. 55, 1153-1156.

Senesi, N., Sposito, G. and Martin, J.P. (1987) Copper (II) and iron (III) complexation by humic acid-like polymers (melanins) from soil fungi. Sci. Total Environ. 62, 241-252.

Starling, A.P. and Ross, I.S. (1990) Uptake of manganese by *Penicillium notatum*. Microbios 63, 93-100.

Strandberg, G.W., Shumate, S.E. and Parrott, J.R. (1981) Microbial cell as biosorbents for heavy metals: accumulation of uranium by *Saccharomyces cerevisiae* and *Pseudomonas aeruginosa*. Appl. Environ. Microbiol. 41, 237-245.

Sutter, H.P., Jones, E.B.G. and Walchli, O. (1983) The mechanism of copper tolerance in *Poria placenta* (Fr.) Cke and *Poria caillantii* (Pers.) Fr.. Material und Organismen 18, 243-263.

Tezuka, T. and Takasaki, Y. (1988) Biodegradation of phenylmercuric acetate by organomercury-resistant *Penicillium* sp. MR-2. Agric. Biol. Chem. 52, 3183-3185.

Thayer, J.S. (1988) "Organometallic Chemistry, An Overview", VCH Verlagsgesellschaft, Weinheim.

Theuvenet, A.P.R., Nieuwenhuis, B.J.W.M., Van de Mortel, J. and Borst-Pauwels, G.W.F.H. (1986) Effect of ethidium bromide and DEAE-dextran on divalent cation accumulation in yeast. Evidence for an ion-selective extrusion pump for divalent cations. Biochim. Biophys. Acta 855, 383-390.

Thompson-Eagle, E.T. and Frankenberger, W.T. (1992) Bioremediation of soils contaminated with selenium. Adv. Soil Sci. 17, 261-310.

Tobin, J.M., Cooper, D.G. and Neufeld, R.J. (1984) Uptake of metal ions by *Rhizopus arrhizus* biomass. Appl. Environ. Microbiol. 47, 821-824.

Tobin, J.M., Cooper, D.G. and Neufeld, R.J. (1990) Investigation of the mechanism of metal uptake by denatured *Rhizopus arrhizus* biomass. Enz. Microb. Technol. 12, 591-595.

Townsley, C.C., Ross, I.S. and Atkins, A.S. (1986) Copper removal from a simulated leach effluent using the filamentous fungus *Trichoderma viride*, in "Immobilisation of Ions by Bio-sorption" (Eccles, H. and Hunt, S., Eds.), pp. 159-170. Ellis Horwood, Chichester.

Tsezos, M. (1983) The role of chitin in uranium adsorption by *Rhizopus arrhizus*. Biotechnol. Bioeng. 25, 2025-2040.

Tsezos, M. (1984) Recovery of uranium from biological adsorbents - desorption equilibrium. Biotechnol. Bioeng. 26, 973-981.

Tsezos, M. (1986) Adsorption by microbial biomass as a process for removal of ions from process or waste solutions, in "Immobilisation of Ions by Biosorption" (Eccles, H. and Hunt, S., Eds.), pp. 201-218. Ellis Horwood, Chichester.

Tsezos, M. and Volesky, B. (1982*a*) The mechanism of uranium biosorption by *Rhizopus arrhizus*. Biotechnol. Bioeng. 24, 385-401.

Tsezos, M. and Volesky, B. (1982*b*) The mechanism of thorium biosorption by *Rhizopus arrhizus*. Biotechnol. Bioeng. 24, 955-969.

Volesky, B. (1990) "Biosorption of heavy metals", CRC Press, Boca Raton.

Wainwright, M. and Grayston, S.J. (1986) Oxidation of heavy metal sulphides by *Aspergillus niger* and *Trichoderma harzianum*. Trans. Br. Mycol. Soc. 86, 269-272.

Wainwright, M., Singleton, I. and Edyvean, R.G.J. (1990) Magnetite adsorption as a means of making fungal biomass susceptible to a magnetic field. Biorecovery 2, 37-53.

Wakatsuki, T., Hayakawa, S., Hatayama, T., Kitamura, T. and Imahara, H. (1991) Solubilization and properties of copper reducing enzyme systems from the yeast cell surface in *Debaryomyces hansenii*. J. Ferment. Bioeng. 72, 79-86.

Wales, D.S. and Sagar, B.F. (1990) Recovery of metal ions by microfungal filters. J. Chem. Technol. Biotechnol. 49, 345-355.

White, C. and Gadd, G.M. (1990) Biosorption of radionuclides by yeast and fungal biomass. J. Chem. Technol. Biotechnol. 49, 331-343.

Winge, D.R., Nielson, K.B., Gray, W.R. and Hamer, D.H. (1985) Yeast metallothionein - sequence and metal binding properties. J. Biol. Chem. 260, 14464-14470.

Winge, D.R., Reese, R.N., Mehra, R.K., Tarbet, E.B., Hughes, A.K. and Dameron, C.T. (1989) Structural aspects of metal-$\gamma$-glutamyl peptides, in "Metal Ion Homeostasis: Molecular Biology and Chemistry" (Hamer, D.H. and Winge, D.R., Eds.), pp. 301-311. Alan R. Liss Inc., New York.

Winkelmann, G. (1992) Structures and functions of fungal siderophores containing hydroxamate and complexone type iron binding ligands. Mycol. Res. 96, 529-534.

Yakubu, N.A. and Dudeney, A.W.L. (1986) Bisorption of uranium with *Aspergillus niger*, in "Immobilisation of Ions by Bio-sorption" (Eccles, H. and Hunt, S., Eds), pp. 183-200. Ellis Horwood, Chichester.

## INDEX